CW01313836

IN SEARCH OF ARCTIC BIRDS

IN SEARCH OF ARCTIC BIRDS

by Richard Vaughan

Illustrated by
Gunnar Brusewitz

T & A D POYSER
London

© T & A D Poyser 1992

ISBN 0–85661–071–2

First published in 1992 by T & A D Poyser Ltd
24–28 Oval Road, London NW1 7DX

United States Edition published by
ACADEMIC PRESS INC.
San Diego, CA 92101

All rights reserved. No part of this book may be reproduced, stored in a retrieval system, or transmitted in any form or by any means, electronic, mechanical, photocopying or otherwise, without the permission of the publisher

Text set in Garamond
Typeset by Phoenix Photosetting, Chatham, Kent
Printed and bound in Great Britain by
Mackays of Chatham PLC, Chatham, Kent

A catalogue record for this book
is available from the British Library

Contents

Preface and acknowledgements		xiii
Chapter 1	The Arctic and its bird life	1
	The Polar ice caps. What is the Arctic? Countries of the Arctic. The Arctic climate. The Arctic through the year. Arctic environments. Composition of the Arctic avifauna. Arctic bird populations. Residence and migration among Arctic birds. Adaptations to the Arctic environment. Breeding strategies of Arctic waders.	
Chapter 2	Birds and Arctic peoples	20
	Birds in Eskimo mythology and ritual. Eskimo knowledge of birds. Birds as food. Birds for clothing and bedding. Methods of bird hunting. Bird hunting activities of different Inuit groups. Bird hunting activities of the Nganasan. Birds eaten by European settlers in Greenland. Egging. The impact on populations. Conservation. Man-made nest-sites for Arctic birds.	
Chapter 3	Whalers and discovery ships	49
	Historical background. Early voyagers to Spitsbergen. Birds eaten by Dutch whalers at Smeerenburg. Hamburg's contribution to Arctic ornithology. New birds from west Greenland. Naval bird collectors in Canadian waters. Birds shot for the pot. Egg-collecting whalers. "I shot the albatross". The ornithology of Smith Sound and northwards. Newcomb of the Jeanette. *The cruise of the* Corwin *in 1881. Nordenskiöld and Nansen. Italians in the Far North. The voyage of the* Zarya, *1900–1902.*	
Chapter 4	Falcons from the Arctic	80
	The natural history of the Gyrfalcon. The Gyrfalcon in medieval history. Subspecies and migrations. Sources of supply. Catching falcons. Gyrfalcons by the hundred. An egg-collector in Lappland. Skins for the museums. A twentieth-century falconer in Iceland. Present-day falconry in North America. Gyrfalcons for the sheiks?	
Chapter 5	Bird collecting in North America	98
	The Hudson's Bay Company. Samuel Hearne. Franklin's Narrative *of his first land expedition, 1819–1822. Robert Hood's bird paintings. Franklin's second land expedition, 1825–1827. John Richardson. The Hudson's Bay company again: Roderick MacFarlane and the Eskimo Curlew. The*	

vi   In search of Arctic birds

U.S. Army and the birds of Alaska. Knud Rasmussen and the Danes. The Hooper Bay Expedition of 1924. George Miksch Sutton.

Chapter 6   Britons in the Russian and Siberian Arctic, 1850–1914   130
The spell of Lappland. Norway and Russia. The Great Land Tundra. Bird collecting on the Yenisey. Cut off on Kolguyev Island. The shores of the Barents Sea. A woman takes the field.

Chapter 7   Ornithological treasure trove   157
The eggs of the Knot. Arctic birds' eggs in Newton's collection. Ross's Gull.

Chapter 8   Goose puzzles in North America   172
Samuel Hearne's geese. Hutchins's Goose. Blue geese and snow geese. Ross's Goose.

Chapter 9   Arctic ornithologies and their authors   181
The Arctic as a whole. Regional ornithologies. Bailey on the birds of Arctic Alaska. Snyder on the birds of Arctic Canada. Todd on Labrador birds. Salomonsen and the birds of Greenland. Løvenskiold on Spitsbergen birds. Uspensky on the birds of the Soviet Arctic.

Chapter 10   Bird studies in the Russian and Siberian Arctic   197
Ornithology in the Soviet Union. The birds of Franz Josef Land. Sandpipers and stints. Pioneers of Russian Arctic ornithology. Seabirds of the Kola Peninsula. The Chukchi Peninsula and Wrangel Island. Aleksandr Aleksandrovich Kishchinsky. The Magadan team. Wrangel Island again. The geese and the fox. Birds of the Yamal Peninsula. The better-known north?

Chapter 11   Ornithological research from Alaska eastwards to Spitsbergen   239
Arctic research institutes. Huxley's Red-throated Divers. Tinbergen on Snow Buntings in Greenland. The Ptarmigan's diet. Bird studies in the Brooks Range, Alaska. Some Arctic research problems. Snowy Owls on Baffin Island. Little Auks of Horse Head Island. The Thick-Billed Murres of Prince Leopold Island. Birds of prey in West Greenland. Peregrines and Gyrfalcons in Arctic Canada. Breeding Long-tailed Skuas. Spitsbergen goose studies. Greenland white-fronts. Lapland Buntings and other birds at Barrow, Alaska.

Chapter 12   The birdwatcher's Arctic: where to go and how to get there   297
The allure of the Arctic. Arctic vexations. A taste of the Arctic: Iceland and North Norway. Svalbard or the

Spitsbergen Archipelago. Greenland. Canada. Alaska. Russia and Siberia.

Chapter 13	Conservation	334
	International action. Threats to the environment. Marine and air pollution. Oil and mineral exploration. Tourism. Resource management and mismanagement. Ecosystem conservation. National parks, nature reserves and bird sanctuaries in Spitsbergen. The world's greatest national park? Canada's migratory bird sanctuaries. Alaska: oil or wildlife? Zapovedniks and zakazniks. Endangered Arctic species. Danish bird protection in Greenland.	

References	370
Scientific names of plants and animals mentioned in the text	395
Index	405

List of Maps

1	Countries of the Arctic	3
2	Peoples of the Arctic	22
3	Places named by early voyagers to Spitsbergen	52
4	The Northwest Passage: discovery ships and whalers in Greenland and Canada	59
5	Explorers north of Russia and Siberia	69
6	Gyrfalcons in the west	84
7	Gyrfalcons in the east	88
8	Bird collecting in the Canadian Arctic	100
9	Bird collecting in Alaska	118
10	Places where birds were collected on the Fifth Thule Expedition	122
11	Lappland	132
12	Britons in Russia and Siberia	136
13	Some Ross's Gull localities	165
14	The Russian and Siberian Arctic	198
15	The Kola Peninsula	214
16	The Chukchi Peninsula and Wrangel Island	220
17	Natural regions of the Russian and Siberian Arctic	237
18	Research in Canada	240
19	Research in Spitsbergen	245
20	Research in Alaska	256
21	Research in Greenland	269
22	Research in the North Atlantic area	280
23	Iceland	303
24	The Varanger Peninsula	305
25	Travel to Spitsbergen	307
26	The geographical position of Spitsbergen	308
27	Spitsbergen	309
28	Greenland	311
29	The Canadian Arctic	317
30	Alaska	322
31	Barrow and surroundings	323
32	The birdwatcher's Soviet Arctic	330
33	Circumpolar conservation	337
34	Oil drilling installations at Prudhoe Bay	340
35	Seabird bazaars of Novaya Zemlya	345
36	Conservation in Spitsbergen	350
37	Conservation in Greenland	356
38	National parks (NP) and bird sanctuaries (BS) in Canada	358
39	Conservation in Arctic Alaska	361
40	Zapovedniks and zakazniks	364

List of Photographic Plates

1	Arctic scenery	7
2	Semipalmated and Ringed Plovers	13
3	Black Guillemots between ice-floes	16
4	Eggs and down of the Eider	28
5	Flock of moulting Greater Snow Geese	32
6	Food birds of the Eskimo: Eider and Ptarmigan	37
7	Arctic Redpolls' nests in man-made sites	46
8	Wintering Brent Geese in the Netherlands	51
9	Fulmar	55
10	Sabine's Gulls	60–61
11	Dalrymple Rock	65
12	Spoon-billed Sandpiper with downy young	71
13	Grey and Red-necked Phalaropes	72
14	Gyrfalcons	82
15	Arctic Tern hovering over the tundra	103
16	White-billed Diver	104
17	Stilt Sandpiper	110
18	Dark-phase Arctic Skua	112
19	Baird's Sandpiper	113
20	Pectoral Sandpiper and nest	119
21	Male Snow Bunting	124
22	Drake Long-tailed Duck	128
23	Little Stint	139
24	Nest and eggs of Dotterel	145
25	Nest and eggs of Grey Plover	147
26	King Eider ducks incubating	149
27	Rough-legged Buzzard's nest	151
28	Curlew Sandpiper on nest	153
29	Grey Plover pair	154
30	Breeding Knots	161
31	Ross's Gull	169
32	Hutchins's or Richardson's Goose	174
33	Breeding Greater Snow Geese	177
34	Ptarmigan pair	190
35	Kittiwake colony in Greenland	201
36	Ivory Gull	202
37	Incubating Semipalmated Sandpiper	204–205
38	Incubating Purple Sandpiper	206
39	Temminck's Stint	206
40	Drake Steller's Eiders	209
41	Female Lapland Bunting and nest	210
42	Black Guillemot	216
43	Eider group feeding	217
44	Grey Phalarope at nest and Grey Phalarope pair	224–225
45	Emperor Goose at nest	227

46	Snowy Owl with lemming	233
47	Arctic Fox	234
48	Incubating Red-throated Diver	246–247
49	Snow Bunting pair	250
50	Male Ptarmigan	252
51	Food plants of the Ptarmigan	254
52	Snowy Owl pair and female incubating	258
53	Snowy Owl at nest and nest site	260
54	Little Auks	263
55	Arctic Fox at Little Auk colony	265
56	Brünnich's Guillemot and breeding colony	267
57	Long-tailed Skua in flight	276
58	Breeding Long-tailed Skuas	277
59	Dwarf Arctic Willow	283
60	Lapland Buntings	289
61	Grey Plover's distraction behaviour	293
62	Distraction behaviour of Knot and Baird's Sandpiper	294
63	Incubating Turnstone and Stilt Sandpiper's nest and eggs	298–299
64	Tame Red-necked Phalaropes and Ptarmigan	300
65	King Eiders	313
66	Incubating Baird's Sandpiper	315
67	American Golden Plover	319
68	Pomarine Skua and Snowy Owl	324
69	Breeding Pomarine Skua	325
70	Male Red-spotted Bluethroat	327
71	Tundra Swans	343
72	Brünnich's Guillemots	346
73	Glaucous Gull	348

Preface and Acknowledgements

This is a book about the pursuit of birds in the Arctic, not a monograph on the birds of the Arctic. It has no list of Arctic birds. It merely seeks to describe the way birds or their eggs have been hunted, killed and eaten, collected or studied in the Arctic. It also discusses the conservation of Arctic birds and the ways and means of Arctic birdwatching.

I am indebted to Aleksandr Andreev for two photographs of Gyrfalcons and to Pavel Tomkovich for the plates of White-billed Diver, Emperor Goose, Curlew Sandpiper, Purple Sandpiper, Spoon-billed Sandpiper, Ross's Gull and Ivory Gull. The remaining photographs are my own.

My thanks go to the following people who have kindly read through parts of the book and made valuable comments: Aleksandr Andreev, Hugh Boyd, Michael Densley, Will Higgs, Carl Jones, Ray Newell, David Parmelee, Brian Pashby, Eugene Potapov, Henning Thing, Pavel Tomkovich, my son John Vaughan, and Michael Wilson.

For useful information and printed material I am especially indebted to Gregers Andersen and the Danish Polar Centre, Aleksandr Andreev, Sven Blomqvist, Hugh Boyd, Henry Bunce, Leif Eldring, Will Higgs, Carl Jones, Eugene Potapov, Delbert Rexford, Henning Thing and the Department of the Environment and Wildlife Management of the Greenland Home Rule Government, Pavel Tomkovich, and the Canadian Wildlife Service.

I thank the authorities and staff of the following institutions which have helped to make the writing of this book possible and in many of which I have been allowed to work: the Reference Library of the American Museum of Natural History in New York; the Arctic Centre, Groningen, Holland; the Arctic Coast Tourist Association, Cambridge Bay, Canada; the Balfour and Newton Libraries, Cambridge, England; the Departments of History and University Libraries of Groningen University and of Central Michigan University at Mount Pleasant, Michigan; the Edward Grey Institute and Alexander Library, Oxford; the Scott Polar Research Institute, Cambridge, England; the Ukpeagvik Inupiat Corporation's National Arctic Research Laboratory, Barrow, Alaska; and the Kirkbymoorside Branch of the North Yorkshire County Library. To these should be added the inter-library loan services in the Netherlands, the United States of America, and Britain.

On the publishing side I thank Trevor Poyser for accepting this book in the first place and Andrew Richford for his friendliness and editorial help. To my especial delight, he was able to persuade Swedish artist Gunnar Brusewitz to illustrate the book. My heartfelt thanks go to everyone at T & A D Poyser who have contributed to its editing and production: they did a splendid job.

This book is a family affair. My wife Margaret typed the manuscript into the computer, revised and advised in numerous particulars, and helped make the list of scientific names and the index. Our daughter Nancy drew the maps, Margaret, Nancy and our son John have accompanied me on birdwatching trips to the Arctic. Another son, David, and my brother John, have sent me newspaper cuttings, and daughters Clare, Mary and Jane would have helped had they had the opportunity. I thank them all.

CHAPTER 1

The Arctic and its bird life

THE POLAR ICE CAPS

Apart from relatively brief ice ages at irregular intervals, the planet Earth has normally been a good deal warmer than it is now. It seems during most of its past to have entirely lacked permanent ice at the Poles. Only 5 or 6 million years ago the Polar regions were temperate at sea level and had been for tens of millions of years, supporting extensive forests, and along the shores of an ice-free Arctic Ocean lived mastodons and other large land animals. But the slow cooling that heralded the onset of an ice age had already begun. The Arctic that we know today was created between 1 and 2 million years ago, when two of its most characteristic features, the massive Greenland ice cap and the sea ice cover of the Arctic Ocean, were probably formed. By this time the southern Polar ice cap had already expanded to its maximum extent. The Arctic and the Antarctic are of course very different in important respects. The Antarctic has penguins, which the Arctic lacks, but has neither polar bears nor Eskimos or any other native peoples. The Antarctic is a good deal colder than the Arctic and has much more ice, and, while the Arctic is made up of a frozen sea surrounded by continents, the Antarctic is a continent surrounded by frozen seas (Ley 1971; Stonehouse 1971; Sugden 1982; on the topic of this chapter in general, see Irving 1972 and Young 1989).

WHAT IS THE ARCTIC?

The Arctic was named from the Greek word *arktos* meaning bear, which was applied to the distinctive constellation in the northern sky now called the Great Bear. No one disputes that the North Pole is at the centre of this region, but there has been much discussion about its boundaries, or rather its one and only southern boundary, since Schalow in 1904 considered the problem at length. One possibility (Remmert 1980) is the Arctic Circle, at about 66° 31′ North, north of which the midnight sun can be seen in summer and continuous darkness occurs in midwinter. But most experts maintain that the Arctic Circle merely expresses an abstract astronomical concept and has no biological or climatic significance. Nor is the permafrost line helpful, for it extends much too far south in some areas. For example, in the Soviet Union it underlies much of the boreal coniferous forest, or taiga, as far south as Lake Baikal, which no one would place in the Arctic. More satisfactory and quite widely accepted is the line connecting places where the mean temperature in the warmest month of the year, namely July, is 10°C (50°F) (Stonehouse 1971). On this definition the Arctic is the area within which the warmest summer month is no warmer than London in November or February. But the 10°C July isotherm has the disadvantage, when used as the southern boundary of the Arctic, that it takes a wide sweep southwards in the Bering Sea so as to include in the Arctic some of the Aleutian Islands and part of the Kamchatka Peninsula. On the other hand, it does have the advantage of nearly coinciding with the northern limit of trees (Freuchen and Salomonsen 1958; Ogilvie 1976), though the tree line is often vague and ill defined and would bring Iceland into the Arctic, a suggestion firmly rejected by Schalow (1904). Clearly there is no single universally applicable boundary of the Arctic (Armstrong *et al.* 1978) and it should be added that it is not recognised as one of the world's zoogeographical regions, but is now considered a mere zone, the Arctic Zone, comprised of the northernmost parts of two zoogeographical regions, namely the Palearctic and Nearctic. The definition of the Arctic is further complicated by some authors' insistence on adding to the south of it a wide belt called the sub-Arctic (Stonehouse 1971) or on splitting it into two zones, the High and Low Arctic (Ogilvie 1976). These terms are useful, but more so for present purposes is a definition of the land areas that will be considered as lying within the Arctic in the pages that follow.

COUNTRIES OF THE ARCTIC

The Soviet Union, by applying the sector principle of drawing straight north–south lines between the North Pole and the extreme easterly and westerly

Map 1 Countries of the Arctic.

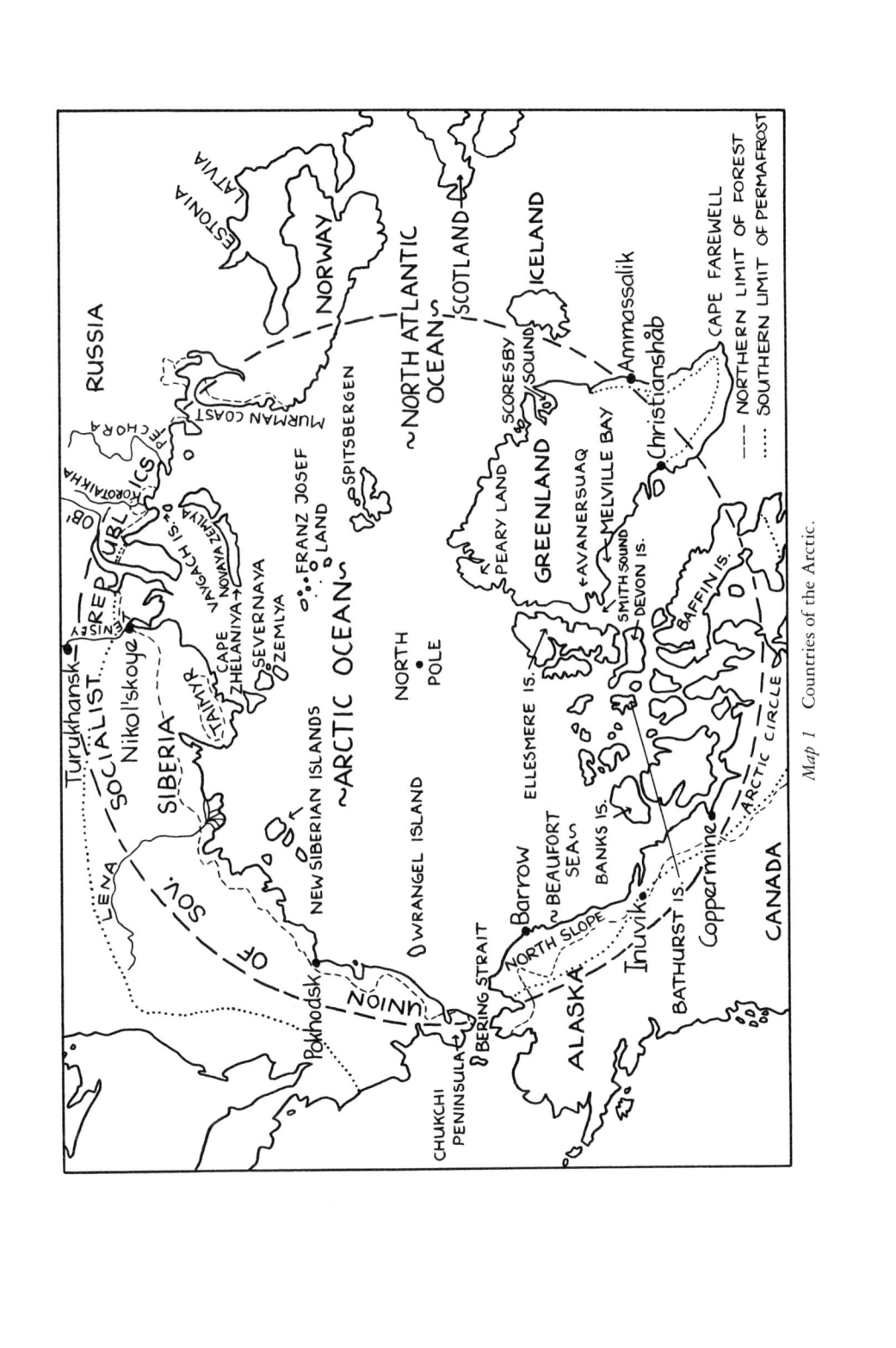

points of its territory, has laid claim to almost half the Arctic. Its Arctic regions, which are within the borders of the Russian republic, comprise a strip of land bordering the Arctic Ocean from Norway at one end of the country to the shores of Bering Strait at the other, and five substantial islands or island groups, namely Franz Josef Land, Novaya Zemlya, Severnaya Zemlya, the New Siberian Islands and Wrangel Island. The first three of these are extensively glaciated, as is the Spitsbergen Archipelago which, though the Norwegians claim it and call it Svalbard, probably currently has more Russian than Norwegian inhabitants. On the Soviet Arctic mainland great northward-flowing rivers, frozen in winter, the Yenisey, Ob', Pechora and Lena and others, intersect the low-lying plains which are interrupted by mountains only in the Taimyr and Chukchi Peninsulas. In geographical rather than political terms, the Ural Mountains divide Russia from Siberia.

Canada, along with which for present purposes we may consider Alaska, is the second great Arctic country. Indeed, some authorities credit Canada alone with more genuinely Arctic territory than the Soviet Union (Sage 1986). Alaska has only a narrow coastal strip north of the Brooks Range, called the North Slope, within the Arctic. This coastal plain continues eastwards into Canada to include large areas of the mainland on either side of Hudson Bay. But the greater part of the Canadian Arctic is formed by 14 large islands, and many small ones to the north making up the Canadian Arctic Archipelago, prominent among them the partly glaciated Baffin and Ellesmere Islands.

Greenland, four-fifths of it covered by an ice sheet rising to a height of 3300 m (10 800 feet) above sea level, is the third major Arctic country, though its southern tip, Cape Farewell, reaches far south of the Arctic Circle to the latitude of Oslo and Leningrad, and much of its west coast is reckoned Low rather than High Arctic. Greenland's rocky ice-free coastal land is roughly equally divided between the west coast, where most of the inhabitants live, the northeast, and the extreme north. Iceland, just south of the Arctic Circle, between Greenland and Norway, is, as has just been pointed out, doubtfully Arctic and will not figure prominently in these pages.

THE ARCTIC CLIMATE

The climate of the Arctic is characterised by low temperatures, so low that they remain continuously below the freezing point for more than half, and in many places three-quarters, of the year. High winds are relatively frequent and precipitation is mostly extremely meagre, making parts of the Arctic as arid as the Sahara Desert. In northwest Greenland the annual precipitation is around 76 mm (3 inches) as compared with 612 mm (24 inches) in London. A striking feature of the temperature in the Arctic is that it fluctuates seasonally rather than daily. This means that, in summer in the Arctic, the temperature remains more or less constant over any given 24-hour period regardless of the time: it is nearly as warm at midnight as it is at midday, for the sun remains above the horizon day and night. In the continuous darkness of the Arctic winter, too, diurnal temperature variations are non-existent: it is always dark and always cold. In sum, the Arctic climate is one of long cold winters and short cool summers with only around 50 days or fewer between the last frosts of spring in June and the first autumn frosts in August.

Although every expert agrees that the Arctic has been warming up, in a somewhat erratic way, since the end of the last ice age some 10 000 years ago, the precise fluctuations of the last 100 years or so are not always easy to particularise. However, it seems fairly clear that the first half of this century was a period of warming and that the second half has so far been one of cooling or perhaps stability.

THE ARCTIC THROUGH THE YEAR

Avanersuaq, the region of northwest Greenland formerly known as the Thule district, between 76° and 79° N, which is the most northerly land ever colonised by man before very recent times, is reasonably representative of the High Arctic. It was taken as such by the Danes Peter Freuchen and Finn Salomonsen in their book *The Arctic year* (1958). In Avanersuaq the year begins in complete darkness and it is only on about 20 January that, for the first time, colours begin to be discernible in the noon twilight, and not until 20 February or thereabouts is the sun first seen above the horizon. Meanwhile, among the local Inuhuit Eskimos the winter sledging season is in full swing, especially during periods of full moon, when the extraordinary beauty of the Arctic night has been expatiated upon by many an Arctic traveller. In spite of the warming effect of the sun and the rapidly lengthening days, February and March are the coldest months of the year. Unless one is lucky enough to be near one of the few places where there is always running water, in order to obtain a drink one has either to a dig a hole through a metre or more of ice on a lake or chop ice from an iceberg and then melt it. The worst blizzards often occur in February or March, when it may be virtually impossible to go out of doors for hours or even days on end. Only in April do temperatures rise noticeably and summer birds begin to arrive. The first Snow Buntings find the land still snow covered, and indeed in Avanersuaq the snowmelt does not get under way until June, by which time the breeding waders have eggs, the sun no longer dips below the horizon at night, and the Arctic flora is in full bloom. Only in July, however, does the fjord ice melt, break up, and disperse seawards, allowing walruses and narwhals to penetrate inland and the Inuhuit to use their motor boats in pursuit of them and other game. The brief Arctic summer, when sunbathing is sometimes possible, can be marred by heavy rain or blizzards, and on calm days by hordes of mosquitoes. It is soon over. By the end of August the summer birds have almost all left Avanersuaq, save for the young Fulmars which, abandoned by the adults, leave the breeding ledges early in September. Frost soon kills the plants and insects and the lakes freeze over, to be followed in October by the fjords. When the sun fails to appear above the horizon one day in October, the Arctic winter has set in once more.

ARCTIC ENVIRONMENTS

In the Polar regions there are three environmental systems, formed respectively by the ice, the sea and the land. The ice, or glacier system (John 1979), is of limited interest for the ornithologist because no birds live there, though some fly over parts of it. The only large area of ice remaining from the ice age is the ice sheet or ice dome that covers most of Greenland; there are smaller ice caps in the Spitsbergen Archipelago, northern Novaya Zemlya, the eastern Canadian Arctic islands and elsewhere, and there are some small ice shelves on the northern coast of Ellesmere Island.

Much more important for bird life is the sea. Although the Arctic Ocean (Rey and Stonehouse 1982; Baker and Angel 1987) is permanently ice covered, this ice is nowhere continuous, forming only a thin, often fragmented, layer of floating ice. Open leads and areas of water, or polynyas, are present in all parts of the ocean and at all times of the year, especially round its periphery. The sea or pack ice shrinks enormously in summer, when large quantities of melt-water form pools all over its surface. This Arctic marine environment is four times larger than the Mediterranean and makes up about 3% of the world's oceans. Most of it is comprised of very cold water, around the freezing point on the surface, and the ice which floats on it circulates slowly in a clockwise direction round the Pole, sending off numerous floes which drift down the east shore of Greenland or along the east coast of Baffin Island, carried south by currents of cold Polar water.

Most of the Arctic Ocean is relatively poor in plant and animal life, which, in the form of phytoplankton and zooplankton respectively, is concentrated in areas where the cold ocean currents mix with warmer water and in certain polynyas which, because of their repeated occurrence in one particular place, are known as recurring polynyas (Dunbar 1982, 1987). Here, as the long Polar winter darkness comes to a close, algae growing on the underneath of the ice proliferate under the sun's influence and are soon being browsed by shrimp-like crustaceans which, in their turn, are voraciously eaten by the diminutive Arctic cod and other fish. Then, as the sunlight increases, the free-floating phytoplankton in the open waters of the polynyas and along the ice edge surges into activity, the crustaceans and other animals forming the zooplankton begin to feed on it, and fishes, birds and mammals are soon attracted to these food-rich areas. No wonder nearly all the largest seabird colonies in the Canadian Arctic are situated near recurring polynyas or ice edges. What is probably the world's largest concentration of breeding Little Auks or Dovekies is found along the eastern boundary of the famous recurring polynya named by the nineteenth-century whalers the North Water, which lies off Greenland's northwest coast north of Melville Bay, in the entrance to Smith Sound.

Plate 1   (a)   Sea ice in Wolstenholme Fjord, Avanersuaq, northwest Greenland, in June, with well-developed shore lead between floating pack ice and shore-fast ice in foreground.

(b)   Coastal tundra in northwest Greenland: low marshy ground strewn with innumerable lakes. Two stranded icebergs on the horizon mark the shoreline.

(a)

(b)

The third Arctic environment is provided by the large areas of ice-free land, especially in Canada and the Soviet Union. Certain parts of this, in the far north, mainly on the more northerly Arctic islands, for example northern Ellesmere Island and northern Novaya Zemlya, have been designated Polar desert because most of the bare rocky ground here supports virtually no plants or animals larger than microscopic algae and tiny invertebrates that feed on them. Here and there in this Polar desert patches of lichens and mosses may occur. It is essentially birdless. Other much smaller land areas are provided by nunataks, rocky islands surrounded by ice that protrude here and there from ice sheets. Somehow or other a few species of flowering plants have colonised some of Greenland's nunataks, which have also recently been found to offer breeding places for the Ivory Gull (Salomonsen 1981). Otherwise they are largely devoid of life.

By far the largest part of land in the Arctic is taken up by the tundra (Bliss *et al.* 1981; Chernov 1985), named from the Lapp word *duoddâr* meaning a hill both by the Russians (*tundra*) and the Finns (*tunturi*). The tundra, or Barren Lands or Grounds in Canada, is an undulating plain to the north of the taiga, or coniferous forest, with which its southern boundary merges in an intermediate zone called the forest tundra. The word tundra is also used for any vegetated area in the Arctic. The tundra is the northern equivalent of steppe or heath and more than half of it is in the Soviet Union. The permafrost (Bruemmer 1987) which underlies the tundra and the snow which covers it for more than 6 months in the year is responsible for many of its characteristics, causing its surface in summer, in spite of the arid climate, to be largely waterlogged: every hollow becomes either a lake or a bog. Surface drainage on a large scale is non-existent, but there are numerous dry stone ridges between the moist areas. The tundra is thought to be the youngest of the world's great ecosystems, having evolved only since the ice age, namely in the last 3000 to 15 000 years, and this is said to explain in part the paucity of species living in it (Webby 1978). The High Arctic flora, for example, comprises in many parts fewer than 100 species: Iceland has 500 against Spitbergen's 150 (Sage 1986.) Food webs tend to be uncomplicated and bird and animal numbers can vary cyclically, especially those of the various lemming species. This small rodent, along with voles and larger animals like the hare, caribou and muskox, are the typical Arctic herbivores. The smaller species are preyed on by foxes as well as by Snowy Owls, Gyrfalcons and other birds: Russian workers (Uspenskii 1986) have found lemmings and other rodents in the stomachs of King Eiders. The larger herbivores fall victim to wolves. A short-lived surge of insect and other invertebrate life on the tundra, which includes the already-mentioned millions of mosquitoes, enables waders and passerines to breed. Besides supporting mammalian herbivores, the richer vegetation of tundra bogs is grazed by numerous geese, some species of which – the Barnacle and Red-breasted Goose, for example – have taken to cliff breeding perhaps because of predation by foxes (Stonehouse 1971; Sage 1986).

COMPOSITION OF THE ARCTIC AVIFAUNA

The most striking characteristic of the Arctic avifauna is the paucity of species. According to Sage (1986: 78), "The total number of species currently breeding in the Arctic is 183", but he concedes that 39 of these, including the Sedge Warbler, Chiffchaff and Willow Warbler, are only of "marginal occurrence"; that is, they have only bred occasionally in the Arctic. Removing these reduces his total to 144—not far off Salomonsen's (1972: 27) "total of 141 Arctic bird species, i.e. species regularly breeding within the Arctic zone". Of course it all depends on how one defines the Arctic and how one defines an Arctic bird. Johansen (1956, 1958) had reduced the total to 80 by rejecting birds like the White Wagtail, which breeds near the Arctic Circle around Ammassilik in southeast Greenland and far to the north of it, up to 72° N, in southern Novaya Zemlya, as not genuinely Arctic. According to him, no doubt rightly, this and several other species which indubitably breed in the Arctic are recent intruders from the south. Still, it seemed a trifle ungenerous not to give them numbers in his numbered list of 80 Arctic breeders. Even on the most generous count, of Sage, relatively few bird species breed in the Arctic—fewer, in the whole of that vast region, than the 200 or so regularly breeding in Britain. Less than 2% of the world's bird species breed in the Arctic and, in any one part of that zone one would be most unlikely to find more than 50 breeding bird species. Uspensky (1986) placed 91 species in the category of regularly breeding in the Soviet Arctic. Snyder (1957) counted 72 species as regularly occurring in the Canadian Arctic; they included the Little Auk and Ross's Gull, which were only later found breeding in Canada.

The Arctic avifauna can be regarded as composed of two groups, autochthonous or indigenous species, as it were the original inhabitants of the Arctic that had evolved before the ice age and have lived in the Arctic ever since, and allotochthonous species—recent intruders from the south. These, according to Snyder, make up about a quarter of Canada's Arctic birds. The most characteristic autochthonous Arctic species are also more or less endemic: truly Arctic in that they do not breed anywhere else, though some have close relatives or offshoots that have evolved further south. Among these most authentically Arctic birds are at least three species of eider (two belonging to the genus *Somateria*), the Brent and Barnacle Goose, most of the *Calidris* sandpipers, two *Stercorarius* skuas, the Ivory Gull and Ross's Gull, the Little Auk, the Snowy Owl and two buntings—the Snow Bunting and the Lapland Bunting or Longspur. It must be emphasised, however, that the ranges of most of these Arctic endemics do extend south here and there into the sub-Arctic. For example, the Glaucous Gull, Brünnich's Guillemot and the Little Auk all breed in Iceland. Other Arctic birds, also among the autochthonous species, are no longer endemic there. Apparently brought southwards in Europe and elsewhere by the ice age and contriving thereafter to adapt to the warmer post-glacial climate, they have remained there ever since. These relict Arctic species, which now breed far south of the Arctic in the temperate climatic zone in parts of their ranges, remain strictly Arctic species elsewhere. In this group are the Eider, Long-tailed Duck, Ringed Plover, Turnstone,

Dunlin, Kittiwake, Arctic Skua and Arctic Tern. Finally, there are the allotochthonous species, the recent or relatively recent intruders into the Arctic from further south. Besides the White Wagtail already mentioned, these include the Peregrine, Shore Lark or Horned Lark, and Wheatear.

Most of the birds which the experts (Voous 1960; Salomonsen 1972) have placed in the Arctic "faunal category" or "faunal type" are widely distributed in the Old as well as the New World Arctic. Some, like the Turnstone, Sabine's Gull and the Snow Bunting, breed more or less continuously all round the Pole. In others, like Bewick's Swan, the Grey Plover and the Pomarine Skua, the continuous circumpolar distribution is interrupted by a gap in the North Atlantic region. Other birds are confined to one part of the Arctic only: the Curlew Sandpiper, apart from occasional breeding near Point Barrow, Alaska, only breeds in the Soviet Union between the western Taimyr and the western Chukchi Peninsula, while the Stilt Sandpiper's range is limited to northern Alaska and parts of northern Canada west of Hudson Bay. In considering the varying distributions of different Arctic species one has to bear in mind that, during the last ice age or "glacial maximum", also known as "the Wisconsin" or "Würm glaciation" depending on whether one is referring to the New or the Old World, there was a land bridge over the Bering Strait between Asia and America. The implication of this is that Arctic land birds could spread, as the ice receded, across from Asia into America or, like Baird's Sandpiper, vice versa, and could then extend their distribution almost all the way round the Pole. The only insurmountable gap was caused by the icy seas and extensive ice sheets round the North Atlantic and it is this that explains the "North Atlantic gap" in the range of several species. Furthermore, the last great glaciation, though reaching as far south in Europe as northern England, left extensive areas in the far north uncovered by ice (Ploeger 1968). The most important of these so-called refugia in the High Arctic were Beringia, a vast region of tundra between Siberia and Alaska constituting the land bridge just mentioned; Banksia, the northwestern Canadian Arctic archipelago centred on most of Banks Island; and the Peary Land refugium in north Greenland. In these refugia Arctic species could survive through the ice age and then later extend their ranges outwards as the ice receded. Salomonsen (1972) sees the Snow Goose, Steller's Eider, Baird's Sandpiper and the Long-billed Dowitcher as typical birds of Beringia or Banksia.

One cannot help noticing that the Arctic favours large birds: according to Snyder more than half of Canada's Arctic species are pigeon-sized or larger. This is supposed to reflect the fact that in large birds and animals the outer surface is relatively smaller than in small species, so that heat loss is reduced because of the comparatively reduced surface area. There are, however, important exceptions to this so-called Bergmann's rule, especially among Arctic birds. Thus the Ptarmigan becomes smaller, not larger, in the northern parts of its range and the only truly Arctic breeding swan, Bewick's or the Tundra Swan, is smaller than the other three members of the genus *Cygnus*, namely the Mute, Trumpeter and Whooper Swan (132 cm or 52 inches as against 152 cm or 60 inches for the other three).

The Arctic likewise favours waterbirds; after all a great deal of it is ocean and the tundra is dotted everywhere with lakes. More than three-quarters of Arctic bird species live on or near water and it is no coincidence that almost half the Arctic avifauna belongs to a single one of the 29 orders of recent birds, namely the Charadriiformes, comprising shorebirds or waders, gulls and alcids. In his *Arctic birds of Canada* Snyder (1957) included 34 birds from this order. The next best represented order in Arctic Canada is the Anseriformes, providing 14 species of swan, goose and duck. Passeriformes comes next with 11 species. Besides these, to make up his total of 72 Canadian Arctic species Snyder included four divers, four birds of prey, two ptarmigans or grouse, a crane, a fulmar, and two owls. Such too, in broad lines, is the composition of the Arctic avifauna as a whole. One of its most remarkable features is the paucity of passerine species. They make up 60% of the world's bird species—over 5000 out of a total of about 8700 (Welty 1975)—but form a mere quarter (35) of the 144 birds listed by Sage as regularly breeding in the Arctic. Even then, more than half of these 35 Arctic breeding passerines are, like the Meadow Pipit, Yellow Wagtail and Dipper, extremely marginal there. In most of the High Arctic there are only half a dozen passerine species—the Shore Lark, Wheatear, Raven, Arctic Redpoll and the Lapland and Snow Buntings—but these, and especially the last two, often account for a high proportion of the total number of birds in any one High Arctic area.

This account of the composition of the Arctic avifauna has shown that it has been changed in the last 10 000 years or so since the end of the ice age by the intrusion of species from the south. There is evidence, too, of changes brought about in the last 100 years by the climatic fluctuations mentioned earlier in this chapter, namely a warming of the Arctic from about 1850 up to about 1950 (Salomonsen 1948) and a cooling since then. During the first half of the present century the number of Long-tailed Ducks and Little Auks breeding at the extreme south of their ranges, in sub-Arctic Iceland, declined dramatically; presumably the warmer climate caused them to withdraw northwards (Fisher 1954: 149). The same happened to Willow Grouse breeding in Estonia, Latvia and Lithuania; this bird's breeding range in Russia moved northwards from a northern limit in 1899 on the Korotaikha River to the south of the Yugorskiy Peninsula at 69° to Cape Lyamchina on Vaygach Island (ostrov Vaygach) at 70° north in 1957 (Slessers 1968). In this same period in west Greenland, some Low Arctic species may have moved north into the High Arctic, namely northwards into Avanersuaq from south of Melville Bay. These are the Great Northern Diver or Common Loon, the Peregrine, the Wheatear and the Lapland Bunting or Longspur (Salomonsen 1950, 1967). In Russia in this period even the House Sparrow, aided by the spread of human settlements and forest clearance, moved into the Arctic. In the Yenisey valley it was breeding as far north as Turukhansk in 1911 (66° N); by 1960 it had moved down river northwards to Nikol'skoye (70° N) (Slessers 1968).

Southward shifts among Arctic birds may have resulted from the cooling after 1950. In 1967–1982 Snowy Owls, Purple Sandpipers, Lapland Buntings and Shore Larks all bred in Scotland and pairs of Sanderlings were twice seen on Scottish mountains in summer (Cumming 1979; Dennis 1983).

Have any truly Arctic species dropped out of the Arctic avifauna in recent historic times by becoming extinct or nearly so? In the Nearctic, the Whooping Crane and the Trumpeter Swan, both of which seem now to have been narrowly rescued from extinction, do not breed in the Arctic as defined in this book, but the Eskimo Curlew certainly did—between Inuvik and Coppermine in the far north of Canada's Northwest Territories, and probably in northern Alaska too. The fate of this msyterious bird, which looks like a small Whimbrel, still hangs in the balance. Once locally very common, it was shot in large numbers on migration and had become very rare by about 1900. No nest has been found since 1866, in spite of recent searches, but Eskimo Curlews were seen in 25 of the 41 years between 1945 and 1985, mostly in their winter quarters in the United States, and a flock of 23 was identified in Texas in 1981 (Gollop *et al.* 1986; Iversen 1989). The species must still breed in very small numbers, almost certainly in the Canadian Arctic somewhere on the Barren Grounds. One indubitably extinct bird, which nested until the early nineteenth century on both sides of the North Atlantic, in Newfoundland, Iceland and Scotland, was the Greak Auk. But, in spite of the 1986 discovery of three Great Auk bones at a 4000-year-old site near Christianshåb (Qasigiannguit) in West Greenland (*Newsletter*, Commission for Scientific Research in Greenland, June 1987; Meldgaard 1988), this was a boreal or at best sub-Arctic species (Grieve 1885; Nettleship and Birkhead 1985). One other claimant for near extinction from the Arctic avifauna is the Siberian White Crane (Stewart 1987). While the western population on the lower reaches of the River Ob' may have been or may still be in danger of extinction, the 1984 count of a flock of 1350 birds wintering in China (Don H. Messersmith, pers. comm., 1985) may indicate that the population breeding in Yakutia is at reasonable strength. Thus no species seems to have entirely dropped out of the Arctic avifauna in the last few hundred years.

A curious and unexplained feature of the Arctic avifauna is the high proportion of polymorphic species, that is, species having two or more different varieties of plumage called morphs (Salomonsen 1972). Characteristic examples are the Fulmar, Snow Goose, Rough-legged Buzzard, Gyrfalcon and three species of skua. In all of these, some individuals have a predominantly white or light-coloured plumage, while others are dark. At least four Arctic birds occur in two closely similar forms which are currently regarded as separate species. These pairs of closely related species are the Ringed and Semipalmated Plover, the Pacific and American Golden Plovers formerly combined in one species as the Lesser Golden Plover (Connors 1983; Knox 1987), the Redpoll and Hoary or Arctic Redpoll (Knox 1988), and the Black-throated

---

Plate 2 (*a*) Semipalmated Plover photographed at Cambridge Bay in July. Examination of the bird in the hand is needed to see the partial webbing between the toes. Field characters that help to distinguish it from Ringed Plover are the larger white patch above bill, narrower black band across breast, and distinctive *chi-vee* call.

(*b*) Ringed Plover in northwest Greenland in July. Although usually regarded as two distinct species, some authorities treat the Semipalmated and Ringed Plovers as morphs of the same polymorphic species.

The Arctic and its bird life 13

(a)

(b)

and White-necked or Pacific Divers (Kishchinsky and Flint 1983), the second of which is known in North America as the Arctic or Pacific Loon. An example of two similar forms *not* regarded as constituting separate species are the yellow-billed Eurasian Bewick's Swan and the black-billed North American Tundra Swan, both *Cygnus columbianus*. In those species which belong to the Arctic avifauna but whose ranges also extend far south of the Arctic, the northern populations can often be morphologically distinguished. Thus Arctic subspecies of the Peregrine, Herring Gull and Shore Lark have been recognised (Uspenskii 1986; 197).

ARCTIC BIRD POPULATIONS

Just as the Arctic avifauna comprises relatively few species, so the numbers of breeding birds in the Arctic are likewise low. Once can trudge for hours over likely-looking tundra and scarcely see a single bird. In southern Baiffin Island in 1953, Watson (1963) reckoned on 61 adult birds per 40 hectares (or 100 acres), but this was a sort of Low Arctic oasis. The birds were mostly passerines: Lapland and Snow Buntings, Water Pipits and Shore Larks. Bird densities here were comparable to those found for rough grazing land in England (70 birds per 40 hectares) or dry grass veld in South Africa (65) (Welty 1975: 362), and much higher than those for more typical High Arctic tundra, where there were only 8–11 pairs per square kilometre on Devon Island, Canada, and as few as 1.3 pairs per square kilometre at Scoresby Sound, east Greenland (Remmert 1980: 128). In the Soviet Low Arctic, on the southern island of Novaya Zemlya and northeastern Vaygach Island, Uspensky (1986) recorded only seven species with a breeding density as high as 1–2 pairs per hectare. They were the Snow Bunting with a maximum density of 2 pairs per hectare, the Shore Lark, with 1.5 pairs per hectare, the Lapland Bunting with 1 pair, and three species of wader each with a maximum density of 1 pair per hectare, namely the Purple Sandpiper, Little Stint and Ringed Plover (see also Sage 1986: 95–100). For comparison, the density of breeding Blackbirds in the Oxford Botanic Garden in 1953–1955 was around 5.48 pairs per hectare (Simms 1978: 181), while a breeding density of 1–5 pairs per hectare has been given for breeding Lapwings in favoured localities in Britain (Spencer 1953: 32).

Arctic bird numbers tend to be unstable, and to vary from year to year and from place to place. Birds are exposed to occasional decimation by climatic disasters. The unusually late break-up of the ice in the Beaufort Sea in 1964 made it difficult for seabirds to feed and caused the death by starvation of an estimated 100 000 migrating King Eiders, perhaps wiping out 10% of the breeding population (Barry 1968). Breeding success among Arctic birds is also dependent on the vagaries of a notoriously unstable climate and on variations in the numbers of other birds and animals. Autumn counts of Brent Geese have shown that in some summers virtually no young birds are raised. In autumn 1972, when the geese arrived in their winter quarters hardly any young birds of the year could be seen, reflecting a very poor breeding season in the Arctic. However, 1973 was an extremely favourable summer in the Arctic and about 60% of the Brent Geese in the flocks that autumn were young birds (Owen 1980; 180–181).

Although little is known about fluctuations in the numbers of waders breeding in the Arctic (Meltofte 1985), these do occur quite widely, especially perhaps in certain species, like the Sanderling, Turnstone and Ringed Plover, and at certain places. In the neighbourhood of Thule Air Base in Avanersuaq, northwest Greenland, in the three successive Junes of 1983–1985 the numbers of breeding Ringed Plovers and Baird's Sandpipers varied by up to a factor of 3 (Vaughan 1988). On the other hand, waders seem less affected by the so-called "non-breeding" phenomenon in the Arctic than swans and divers, which may not be able to breed at all if lakes remain frozen too late in the summer, and skuas, Rough-legged Buzzards and Snowy Owls, which may not breed at all in years of very low lemming numbers.

RESIDENCE AND MIGRATION AMONG ARCTIC BIRDS

The Arctic in winter, not so much because of the cold itself but rather because the darkness and the ubiquitous ice makes it impossible to obtain food, is no place for birds. Thus in Peary Land, in the extreme north of Greenland at 83° N latitude, every single one of the 21 summering species is said to leave in winter, while in temperate Denmark 64% of bird species leave, and in tropical rain forests all are resident (Salomonsen 1972: 45).

Do any species regularly overwinter in the High Arctic? Opinions vary and the evidence is conflicting. In Spitsbergen at 78° N "total darkness" has been said to last from 27 October to 16 February (Greve 1975), but this is probably an exaggeration because the stars, the aurora borealis, the moon, and midday twilight all provide some illumination. To gain an exact knowledge of bird life in a Spitsbergen winter one can consult the extracts printed by Abel Chapman (1897: 343–350) from the diaries of his friend Arnold Pike, who overwintered in 1888–1889 near Amsterdam Island (Amsterdamøya, Map 3) at 79° north. In the wooden hut he occupied with seven Norwegians, Pike recorded, lamplight "was required *all day* for exactly one hundred days". The first day when this happened was 31 October. However, already on 25 January the twilight was "appreciable at noon", and on 29 January large type could be read in the middle of the day. Pike's bird observations that winter are as follows:

> 3 November 1888. Snowy Owl seen yesterday for the last time this year, and three Eiders shot to-day from a flight of thirteen.
>
> 2 and 3 January 1889. Birds heard both these days, and some seen, but too dark to distinguish species. The men thought they were Brünnich's Guillemots.
>
> 11 January. Saw Tystie [Black Guillemot] on water, and heard Eiders and Brünnich's Guillemots crying and diving close inshore.
>
> 10 February. Ryper [Ptarmigan] droppings observed. The Spitsbergen Grouse make long burrows beneath the snow; and since snow falls deep long before the severe frosts set in, the autumnal crop of berries, seeds, etc., is thus preserved

beneath it, and provides the grouse with food. This explains the problem of how game-birds can survive the Arctic winter.

20 February. Many Tystie observed in small flocks, plumage very white; also three Fulmars. A glint of sunshine was observed for the first time at noon, impinging on the mountain-tops.

Of the species mentioned by Pike, the Snowy Owl and Eider probably do not regularly overwinter in the High Arctic, but Brünnich's and Black Guillemots do so wherever there is open water, as do Ivory Gulls. Russian geographer V. T. Butev (1959) spent the winters of 1955–1956 and 1956–1957 near Cape Zhelaniya (mys Zhelaniya) at the northern tip of Novaya Zemlya and recorded Long-tailed Duck, Ivory Gull, Brünnich's Guillemot, Little Auk and Black Guillemot. The Ptarmigan, too, certainly overwinters in the High Arctic and so does the Raven. The American ornithologist S. A. Temple (1974) analysed 684 Raven pellets from a winter roost in Arctic Alaska at 69° 22' N and found that the birds' diet was obtained in about equal proportions by preying on small mammals like voles and lemmings and by scavenging, especially at caribou and Ptarmigan carcasses and among garbage. This represents the sum total of occasionally or regularly resident Arctic birds, apart from the Arctic Redpoll, which, astonishingly enough, at least in some areas can apparently find grass and other small seeds enough to survive the winter darkness even in very low temperatures and in spite of the snow cover.

Plate 3 Black Guillemots between the ice floes in northwest Greenland. This bird survives the harsh Arctic winter by feeding in the few patches and leads of open water. Here two are utilising the ever-present break caused by the tide between the ice-pack and the shore-fast ice.

One might think the High Arctic breeding birds would migrate southwards in autumn only so far as necessary to find food. This seems to be the case, for example, with the Eiders breeding in Avanersuaq. On 26 July 1928 the manager of the trading station at Thule settlement, Hans Nielsen, landed on Dalrymple Rock (Plate 11) at 76° 30′ N and ringed 30 well-grown Eider ducklings. Twelve were recovered in the next 4 years, 11 of them between October and May along the west Greenland coast between 65° and 70° N (Bertelsen 1932, 1948), showing that the birds only moved southsoutheast in winter some 1000 km or more down the Greenland coast presumably to the nearest ice-free waters where they could find sufficient food. In striking contrast is the behaviour of the Arctic Tern, which migrates right across the tropics from the Arctic to the Antarctic, and winters there. Some Arctic Terns probably circle the entire Antarctic continent before returning to their Arctic breeding grounds (Remmert 1980: 80–84; Mead 1983: 90–91). Welty (1975: 476) called the Arctic tern "the champion long-distance migrant" and claimed that it "may travel as far as 36000 km each year in migration flights".

Arctic migrants do not only use their wings to escape a deteriorating late summer environment. In their hurry to move from their Arctic breeding colonies to suitable feeding grounds along the edge of the pack ice, adult and still flightless young Brünnich's Guillemots undertake a swimming migration of up to 500 km across the open sea. To do this, birds from the Murman Coast (Murmanskiy bereg, Map 15) colonies travel northwards, those breeding in Novaya Zemlaya swim westwards, and Spitsbergen and Franz Josef Land breeders swim southwards (Salomonsen 1972). Birds breeding in northwest Greenland swim southwards down the west Greenland coast (Remmert 1980: 83). At least one Arctic species has been credited with a walking migration: Buturlin (Pleske 1928: 220) claimed that the young Ross's Gulls hatched around Pokhodsk on the Kolmya River at about 69° N made their way 50 km (30 miles) northwards on foot to the shores of the Arctic Ocean on 26–31 July, while still with down on the head.

Among the migration systems of Arctic birds one of the most remarkable is that of the Snow Buntings breeding in Greenland. Those breeding in west and southeast Greenland winter around the Great Lakes in central North America, while those breeding in northeast Greenland winter in the south Russian steppes north of the Caspian Sea (Salomonsen 1981).

ADAPTATIONS TO THE ARCTIC ENVIRONMENT

Arctic birds (and mammals) have adapted in many different ways to the conditions of life in high latitudes. In some cases the plumage is white or partly white, at least in winter. The Snowy Owl and Ptarmigan exemplify this, but the Raven is a conspicuous exception. Experts are not in agreement over the cause of this so called "Polar whiteness". Has it evolved because of its camouflage value against a background of snow or to minimise heat loss (Salomonsen 1972; Dyck 1979)? Specific insulation devices among Arctic birds are the feathered leg and foot of the Snowy Owl and the extra dense plumage of the Ptarmigan. Another notable adaptation is the shortened or accelerated breeding cycle. Middendorf noted long ago (1853) that in Arctic Siberia, on the tundras of the Taimyr River valley (Map 40: Nizhnyaya (lower) and Verkhnyaya (upper) Taymyra), Golden Plovers were only present from 29 May to 4 August (67 days) and Dotterels from 4 June to 15 August (72 days) (apart from a sick juvenile killed with a stone on 27 August). He found a King Eider's nest containing fresh eggs on 25 June—only 9 days after the bird's arrival. Special moult strategies have evolved in Arctic birds to enable them to shorten their stay in the north: in some, moulting begins extra early, as early as March or April in the case of the Ivory Gull; in others, for example the Grey Plover, the moult is suspended in the autumn and resumed and completed in a favourable area en route for the winter quarters or in the winter quarters themselves (Ginn and Melville 1983). Changes in the circadian rhythm have also helped to accelerate Arctic breeding cycles. Since the pioneer papers of Armstrong (1954) and Cullen (1954) on the behaviour of land and sea birds respectively in continuous daylight, it has been shown that some land birds breeding in the High Arctic can lengthen the period of feeding and other activities to 20 or even 23 hours, while some Arctic-breeding seabirds show virtually no intermission of activity at all during the "night" (Uspenskii 1986: 96–102).

BREEDING STRATEGIES OF ARCTIC WADERS

The social systems adopted by *Calidris* sandpipers and other waders in the brief Arctic summer are of particular interest (Myers 1989). In the majority, the breeding birds hold fairly large territories and are monogamous, the male and female sharing incubation more or less equally. This is true, for example, of the Knot and Dunlin. On the other hand, in some species, notably the Sanderling and Temminck's Stint, the female has been found on occasion to lay two clutches in quick succession, one being then incubated by the male and the other by the female. This was established for the Sanderling on Bathurst Island in 1968–1970 (Parmelee and Payne 1973), but in northeast Greenland in 1974 the Sanderlings were found to be producing only one clutch per pair and sharing incubation between the sexes (Pienkowski and Green 1976). In other *Calidris* species, the Curlew Sandpiper for example, polygyny appears to be the rule and only the female incubates (Sage 1986: 90–92). All this needs further elucidation. In spite of much research, often carried out in extremely difficult conditions, knowledge of the social systems of Arctic birds, like that of Arctic birds as a whole, remains patchy. One of the aims of this book is to try to show how that knowledge has been built up, by whom, and in what circumstances.

CHAPTER 2

Birds and Arctic peoples

BIRDS IN ESKIMO MYTHOLOGY AND RITUAL

Among all Arctic peoples a fund of stories has been transmitted by word of mouth from generation to generation. The Eskimos or Inuit often told such stories in the winter houses after banquets of raw, frozen meat. Birds and other animals figured largely in them because, for the Inuit, a special bond existed between men and animals. Some stories describe sexual relations between men and animals. A gull flies down and carries an unmarried girl to his cliff abode to be his wife, but she escapes (Kroeber 1899: 175). Though hunted and killed, animals were respected and they enjoyed a status in the spirit world alongside men and women. The diver, swan, goose, eider, falcon, skua, gull, Little Auk and Snow Bunting all appear in Eskimo legend, as well as others, but the most popular bird was undoubtedly the Raven. Often it is associated with other birds, as in the fable of the Raven who married two geese. When the autumn came, his wives told him that they had to go away, far away. The Raven insisted on accompanying them and slept at night on an iceberg while they floated on the sea. Eventually they arrived in an area where there were no icebergs, and the Raven tried to sleep lying across his two wives while they floated in the water. But they decided to swim apart, and the Raven fell into the sea, shouting in vain to his goose wives to come back (Holtved 1951a: 230–231; 1951b: 92–93: Rasmussen 1908: 162–164; ten Berge 1976: 263–264). The Raven, like other birds, was credited with the ability of turning itself into a human being and back again, and the Inuhuit of Avanersuaq in northwest Greenland described how the Ptarmigan and the Snow Bunting had once been people (Kroeber 1899: 172; Holtved 1951b: 98–99). In a Canadian story, a blind Eskimo boy became a loon and regained his sight (Brody 1987: 71). Many Eskimo fables feature this interchange between men and animals. For example, the story of the owl that was too greedy recounts how a man out walking came to a cave where an owl, a Raven, a gull, a falcon and a skua were living together in human shape. Anxious to play host, each bird did its best to obtain food for the man: the gull brought some fish and the falcon caught eiders and Little Auks. But the owl tried to chase two hares at once and was practically torn in two when they ran off in different directions (Rasmussen 1908: 164–165; Holtved 1951a: 232–238; 1951b: 93–95). Other stories figure a Ptarmigan and a sea urchin, an owl and a Ringed Plover and an owl and a Snow Bunting in conversation with one another (ten Berge 1976). Inuhuit animal lore also featured songs of the Little Auk and Snow Bunting.

Birds figured among the ceremonial masks used for animal dances by the Eskimos of western Alaska. Around 1880, Edward W. Nelson (1899: 402, 404) collected an elaborate carved wooden mask representing a Sandhill Crane. On the bird's breast, which fitted over the wearer's face, was carved a face with eye-slits. Above this extended the Crane's long neck and forward-pointing bill. The maker told Nelson that the mask represented the Crane's *inua* or spirit. Later, in the 1920s, Brandt (1943: 85) saw masks of loons, swans and eider, besides those of various animals. A plate in Bailey's *Birds of Arctic Alaska* (1948) shows an Eskimo man wearing a ceremonial head-dress in which some bird feathers and the beak of a White-billed Diver were used. In the Field Museum of Natural History in Chicago there is an amulet from Coronation Gulf in northern Canada made from the head and bill of a White-billed Diver. It was a favourite Eskimo bird. At Wainwright on the north coast of Alaska in December 1924 the school teacher watched "the dance of the loon" being performed in her schoolroom. The head-dress worn for this was made of sealskin, attached to which was "the great yellow beak" of a White-billed diver (Richards 1949: 125). In about 1898–1900 Charles Brower (1950: 185) saw an Inupiat whale dance at Barrow performed by men naked to the waist except that "each wore a loonskin on his head". During some of these Alaskan Eskimo dances the women held wands made of a Golden Eagle's wing feathers (Nelson 1899: 414, 415). According to Ray (1885: 108, 109), at Barrow Golden Eagle skins obtained somewhere east of the Colville River were "in great repute as talismans or charms for securing good luck in whaling", while the dried skins of Grey Plovers were used as "talismans to secure good luck in deer-hunting". At the same place, in 1853, Raven skins were being used for this purpose Bockstoce 1988a: 210).

The Alaskan Eskimos sometimes themselves imitated birds in their dances. Thus at Nome, Alaska, in 1913, anthropologist Diamond Jenness (1985: 26)

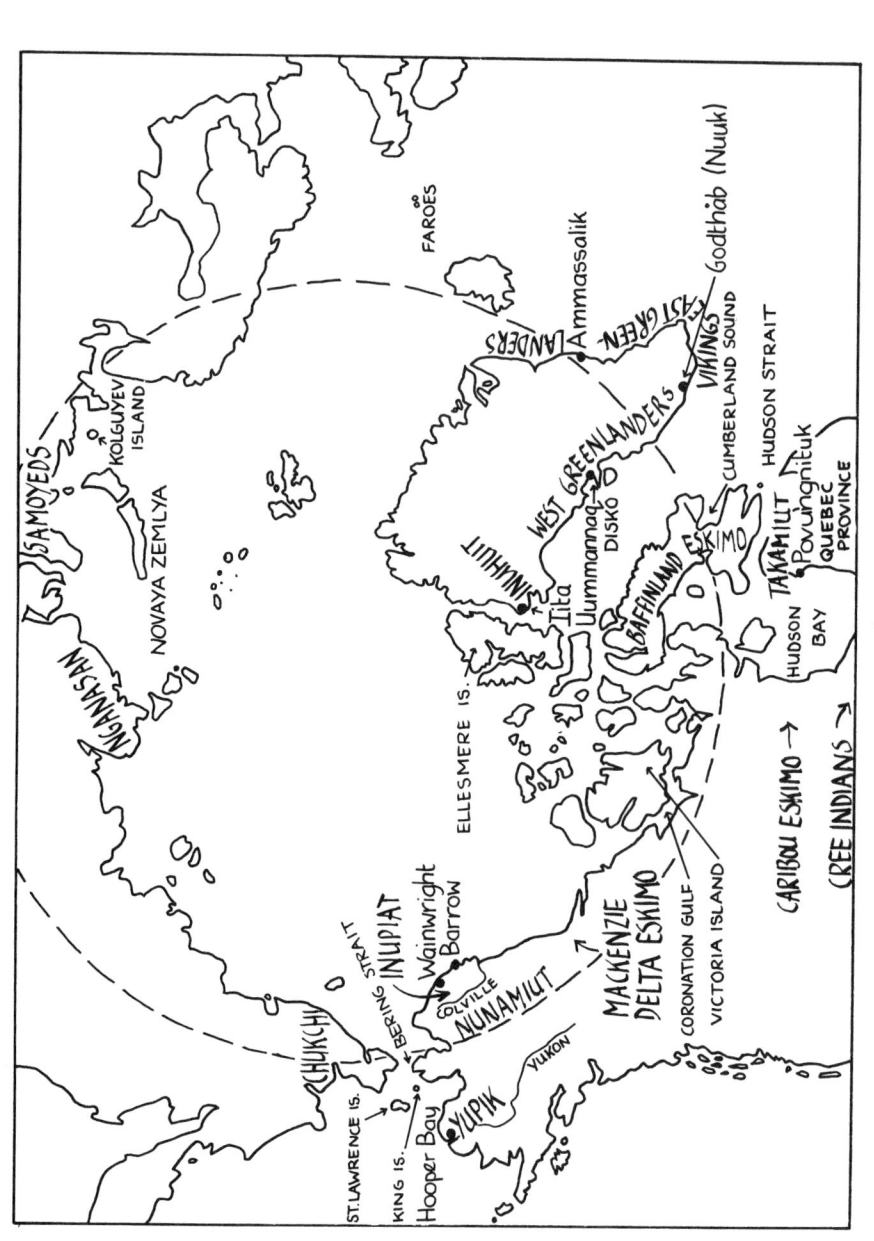

Map 2 Peoples of the Arctic.

watched an Eskimo from King Island staging "a regular ballet", in which he "portrayed with amazing fidelity the fluttering efforts of a mother Ptarmigan to lure an enemy away from the fledglings in her nest".

ESKIMO KNOWLEDGE OF BIRDS

Among the former subsistence-hunting Inuit communities knowledge of birds was often of importance for survival and almost every bird had its name. Eva A. Richards, the above-mentioned school teacher at Wainwright on the north coast of Alaska in 1924–1926, made a list of bird names and other ornithological terms given to her in Inupiat by her Eskimo friend Tooruk. The readily identified species are set out in Table 1. Among the terms recorded by Eva Richards were Inupiat names for a bird's nest, birds in a flock, a gull chick, a pair of loons, a covey of ptarmigan, a swan with young and Snow Buntings in flight and in a flock.

Also at Wainwright, while studying Inuit hunting techniques on the sea ice in 1964–1966, R. K. Nelson (1969) was able to list over 30 birds with their Inupiat names. To do this, he had to secure the help of the older men, for the younger generation of hunters no longer knew so many species. Not that the Eskimo had any concept of species; he merely gave a name to each recognizably different kind of bird. At Wainwright, both the Pacific and Red-throated Divers were called *kahraok* in Nelson's spelling, though the Eskimos knew the differences between them. It seems that in some cases the Inupiat Eskimos used a kind of generic name as well as a specific one. Thus all three skuas were called *isungaq*, though each had its own name too, and the four species of eider

Table 1 Some Inupiat bird names from Wainwright, Alaska (Richards 1949: 264–265).

English name	Inupiat name
White-billed Diver	toolik
Sandhill Crane	tah-tis'a-rook
Tundra Swan	kook-ru'it
White-fronted Goose	kanga-yuk
Eider	amau-lik
King Eider	king'a-lik
Steller's Eider	mit-tik
Pintail	kook-a-ruk
Long-tailed Duck	ah-hol'lik
Grey Plover	e-lak-tal'lik
American Golden Plover	too-lik
Sabine's Gull	kook-'ar-rook
Glaucous Gull	now-yuk
Arctic Tern	too-rit-ko'yuk
Ptarmigan	kah-wah
Snowy Owl	ook-pik
Snow Bunting	a-mau'look

were collectively called *kaugak* at Wainwright. On the other hand, at Point Barrow the drake King Eider was called *kingaling* or *kinalin*, meaning "nosey bird" from its prominent orange frontal shield, and the duck was called *annabia* (Ray 1885: 120–122), while, at the other end of the North American Arctic, the Baffinland Eskimos had three names for the Eider: *amaulik* for the male, *arnaviak* for the female, and *metik* or *mittek* for both sexes (Snyder 1957: 83).

Irving (1953) studied the naming of birds by a group of Nunamiut Eskimos living in tents at the summit of Anaktuvuk Pass (Map 20) in the Brooks Range of northern Alaska in the late 1940s and early 1950s. Fortunately for him, one man, Simon Paneak, spoke English, and he was able to give names to 89 out of the 103 species recorded from Anaktuvuk up to 1953. The Nunamiut had three different names for the Gyrfalcon, depending on the age and plumage of the bird, but only one name for the two kinds of redpoll, and one for the two kinds of scaup. D. B. MacMillan (1918: 403–411), during a 4 year stay at Iita in Avanersuaq, northwest Greenland, in 1913–1917, listed 34 species of bird seen there and recorded Eskimo, in this case Inuhuatut, names for 18 of them. But the Inuhuit Eskimos certainly had names for more birds than those noted by MacMillan—the Fulmar and Turnstone, for example.

Few of those who have recorded Inuit names for birds have been both ornithologists and linguists. While staying at Chesterfield Inlet and Baker Lake (Map 10) in summer 1967, ornithologist E. O. Hohn was able to enlist the aid of a missionary, E. Fafard, as interpreter. With his help he compiled a list of about 46 species with their Caribou Eskimo names. These he transcribed in a somewhat eccentric manner, to be pronounced as if German, which was Hohn's native tongue. A noteworthy feature of Hohn's list is that he explains the meaning of each name, many of which are onomatopoeic. Some examples are set out in Table 2. The Caribou Eskimo, just like present-day taxonomists, had problems with the two rather similar species of grouse—the Ptarmigan and Willow Grouse. They used the names *achigik* and *achigiak*, and variants. It

Table 2 Some Caribou Eskimo bird names with their meanings (Hohn 1969).

English name	Caribou Eskimo name	Meaning of Eskimo name
Great Northern and White-billed Diver	tudlik	From the spring call
Snow Goose	knaguk	From the call
Long-tailed Duck	agiadshuk, a'lier	From the male's spring call
Gyrfalcon	kigavik	The grasper
Sandhill Crane	tatigak	The long one
American Golden Plover	tull'iuk tullik	From the alarm call
Arctic and Long-tailed Skua	ishunga	Large-winged
Snowy Owl	ugpik	Wide open eyes
Raven	tulugak	Pecks very often
male Snow Bunting	amauligak	Seems to be wearing a black *amaut* or woman's cloak

seems that the commoner of the two species in a given area was called *achigik* while for the other a variant name was used.

Eskimo knowledge of birds was put to excellent use in 1924 by American ornithologist Herbert Brandt, who in that year led an expedition to Hooper Bay, southwest Alaska, an important object of which was to "collect specimens of birds, their eggs and nests" (1943: ix). This was achieved in large measure because of the nest-finding skills of the local Eskimos, who "are among the best bird observers and detectors of ground nests that I have ever met" (p. 143). Stimulating this natural aptitude with gifts of tea and tobacco, Brandt was able to enlist the aid of many of the younger men and the women in the task of finding nests. A characteristic entry in his journal, that for 16 June 1924, reads in part (pp. 250–251): "My Eskimo boys returned from the marshes with the announcement that they had found three nests of the Pacific Eider containing eggs, four of the Spectacled Eider, and three nests with six eggs each of the Cackling [Canada] Goose. They discovered also four nests containing complete complements of eggs of the Red-throated Loon, and like number belonging to the Pacific Loon".

BIRDS AS FOOD

The knowledge of birds and their habits almost everywhere evinced by Inuit and other native peoples in the Arctic is a direct result of their dependence on birds for subsistence and survival. First and foremost, of course, birds were and in many places still are, an important food source. Danish doctor Peder Helms (1981), who studied changes in the diet of the Ammassalik Eskimos between 1945 and 1978, found that in 1945 three-quarters of their calorie intake was provided by seal meat, fish, and other local or Greenlandic products, whereas by 1978 only one-fifth came from that source and the rest was imported from Denmark. But in 1978, each inhabitant was still consuming exactly the same weight of seabirds as in 1945, namely 4 kilograms per annum. The population of Ammassalik more than doubled in this period from 1000 to over 2000, so that more than twice as many birds were being killed in 1978 as in 1945. As to the number of birds killed, in 1978 this must have been well over 10 000. In terms of percentages, birds made up 1.8% of the food requirements by weight of the Ammassalik Eskimos in 1945 and 2.7% in 1978, the increase being due to the reduction in the total weight of food consumed because of the much increased use of calorie-rich shop foods. At other times and in other places birds have been a far more important element in the diet of native Arctic and sub-Arctic peoples. The Cree Indians of northern Quebec were thought in 1976 to obtain 18% by weight of the food they harvested annually from geese (Boyd 1977). In northwest Greenland, according to Ekblaw (1919), before they acquired kayaks in the 1860s the Inuhuit Eskimos of Avanersuaq were almost totally dependent on the Little Auk for food during the summer when the fjords and coast were ice free and they could not hunt sea mammals from the ice edge (see too Vaughan, in press).

Table 3 Average weights of some food birds of Arctic peoples (Uspenskii 1986: 328).

Species	Weight (grams)
White-fronted Goose	500–900
Brent Goose	1400–2200
King Eider	2000–3200
Long-tailed Duck	1200–2300

The average weights of some of the Arctic birds most frequently hunted and eaten by Arctic peoples are set out in Table 3 after the Russian ornithologist Savva Uspensky.

The same author (Uspenskii 1986: 334) carefully weighed 16 Brünnich's Guillemots collected on 28 July 1949 on Novaya Zemlya and found that the average weight was 981.5 grams. After removal of feathers, skin, bones and other inedible parts he found that just over half (50.7%) by weight of each bird remained to be eaten. The Norwegian worker Kåre Rodahl (1949) took the trouble to measure the amounts of vitamin C (ascorbic acid) in some of the Eskimos' food birds. He found that a freshly killed Long-tailed Duck contained 1 milligram of vitamin C in every 100 grams eaten, exactly the same proportion as in tinned pears and fresh muskox meat.

It would be difficult to draw up a list of bird species used as food by Eskimos because they normally hunted or trapped and ate whatever birds they could. A few mostly colonially-nesting species were a principal element in the diet in a few places. Thus in west Greenland the Little Auk and Brünnich's Guillemot have been systematically exploited by Eskimos settled, sometimes for that purpose, near their colonies. Elsewhere different species of ducks and geese have been a major item of Eskimo diet. A handful of birds, common throughout the Arctic, are or were a staple in the diet of all Arctic peoples: the Long-tailed Duck and the Ptarmigan leap to mind. The Ptarmigan, because it is present the year round, is perhaps the most eaten of all Arctic birds. The east Greenlanders call or called it "Nakatogak" or "Mitigak", meaning "he at whom stones are thrown" (Gelting 1937: 8). It was not only, as this name implies, easily obtainable, but it was highly prized for the vegetable matter usually found in its crop and stomach.

Archaeological investigation of the ancient dwelling sites and other structures of the prehistoric inhabitants of eastern Ellesmere Island, Canada, has provided information about the food birds of the predecessors of the present-day Inuit. At one site on the western shore of Kane Basin (Map 4), dated to 1900–1600 B.C., bones of Snow Goose, Brent Goose, King Eider, Eider, Gyrfalcon, Arctic Tern, Little Auk and Arctic Redpoll were found. At two other, much later, sites, dated to A.D. 800–1000, many of the same species were being eaten, namely Snow and Brent Geese, King and common Eiders and Arctic Terns, but now the Ivory Gull appears as an item of diet (Schledermann 1990: 63, 70, 216, 251).

In the Bering Strait area birds were normally eaten cooked, either roasted on an open fire or boiled, by the local Eskimos (Ray 1975: 117). Boiling seems to have been the most widespread method of cooking birds in the Arctic. The Inuhuit of northwest Greenland ate their Little Auks raw, either fresh or frozen; or boiled in a soapstone cooking pot or iron kettle (Vaughan, in press). They also liked them "high". For this purpose up to a hundred or more freshly-killed birds were stuffed tightly into a bag made of sealskin with the blubber still attached, and this was cached under a pile of rocks. The fermented birds were eaten raw, months later, after the feathers had been brushed off them (Holtved 1967: 102–113). An example of a bird eaten raw was the Snowy Owl which American explorer E. K. Kane (1856b: 202) watched two Inuhuit lads fighting over at Iita. "Before I could secure it", he wrote, "they had torn it limb from limb, and were eating its warm flesh and blood, their faces buried among its dishevelled feathers."

BIRDS FOR CLOTHING AND BEDDING

Although the native peoples of the Arctic have exploited birds mainly for food, they put them to other purposes too. Ptarmigan feathers were used for bartering with Russian traders by the Samoyeds (Nordenskiöld 1880: 74; Purchas 1906: 205–215). They were used for cleaning by the Bering Strait Eskimo, as well as for babies' nappies (Damas 1984: 289). Diamond Jenness (1985: 15) describes how, in 1913, his Eskimo hostess on the north coast of Alaska gave him and his companions a "serviette" of ptarmigan feathers to clean greasy fingers and used it afterwards to wipe out their cups. For cleaning blood and grease from their hands and faces, the Baffinland Eskimos of Cumberland Sound wiped them with Raven's feathers (Kumlien 1879: 80). On the Alaskan coast igloos were swept out with a bird's wing (Richards 1949: 265); on the south shore of Hudson Strait a Snowy Owl's wing served for this purpose (Graburn 1969: 24). The Nunamiut Eskimos of interior Arctic Alaska used Gyrfalcon feathers for making arrows and spears and these feathers were also an important commodity for trading with Eskimos living along the coast where Gyrfalcons are rare (Irving 1953). Snowy Owl primary feathers were favoured for arrows by the Cumberland Sound Eskimos (Kumlien 1879: 81). Among the "ethnological specimens" collected at Barrow in Arctic Alaska by the American International Polar Year Expedition in 1881–1883 (Ray 1885) were a bunch of eagle's feathers worn as an ornament on a fur jacket and an ivory rod tipped with feathers for marking a cache.

Whether or not the Inuit or other Arctic peoples ever used eiderdown as a filling before woven cloth was made available to them by European and American traders is doubtful (Reed 1986b). Probably they did not. Nor did they collect down for sale until the trade in feathers was organised in the eighteenth century in west Greenland and the Canadian Maritime Provinces. According to Salomonsen (1970), in the 1820s the quantity of eiderdown purchased annually from the West Greenlanders by the Royal Greenland Trading Company corresponded to over 100 000 nests.

In Iceland, on the southern fringe of the Arctic, eiderdown has been

Plate 4 Eggs and down of the Eider in a nest on Lille Ekkerøya, Varanger Peninsula, Norway. 6 July 1972.

collected and exported by the local people, who are of Viking descent, for some centuries (Doughty 1979). Exactly when this trade began seems not to be known with certainty, but feathers, and in particular eiderdown, were certainly used for stuffing cushions and bedding in Iceland before the end of the middle ages (Jones 1986; 270–271). Apparently most Icelanders did not themselves sleep in feather beds (Faber 1826: 316), and in early times the birds were probably mainly used as a source of food. Iceland's Eider colonies have been so intensively protected and farmed over the centuries that the birds have virtually been domesticated. A nineteenth-century account of a visit to the island of Vigur (Map 23) in northwest Iceland (Shepherd 1867) described the Eider colony there as "the most wonderful ornithological sight imaginable". Along the bottom of a 3-foot-high stone wall almost every one of the spaces formed by leaving out alternate stones was occupied by a sitting duck. On a grassy bank near the farmhouse ducks were sitting everywhere in square patches where the turf had been cut out. The farmhouse itself seemed to have been "converted into one large *duckery*. The earthen wall surrounding it, and the window embrasures, were filled with ducks; on the ground, encircling the house, was a ring of ducks; on the sloping roof were seated ducks; and a duck was perched on the door-scraper."

Bird skins have in the past been used throughout the Arctic by native peoples for clothing and rugs. Nelson (1899) recorded Eskimo coats in western Alaska made from skins of cormorants, auklets, scaups, guillemots, eiders and loons. The Inuhuit of northwest Greenland wore an inner coat or shirt of Little Auk skins, which was gradually replaced by cotton underwear and shirts in the second quarter of the present century (Holtved 1967: 39; Vaughan, in press). In west Greenland south of Melville Bay (Map 1) the inner coat was made from the skins of Brünnich's Guillemots (Helms 1929) or of the Eider. This eiderskin inner coat or shirt required 15–20 skins and was renewed annually (Salomonsen 1950: 129). One particular birdskin inner coat from west Greenland merits special mention. It was recovered from the mummified body of a young Eskimo woman who was interred in a boulder-covered grave with five other women and two children at Qilakitsoq, across the fjord from Uummannaq, possibly after their umiak had capsized and they had drowned. This was probably in about 1475. The coat in question was made from the feather-covered skins of five species of bird. The hood was made from two Red-throated Divers' skins; the two breast pieces were of White-fronted Goose skins; Cormorant skins were used for the shoulders and lower front; duck Eider (or possibly King Eider) skins formed the two side pieces; and the sleeves were of duck Mallard (Hansen et al. 1985: 153–154; and see Hansen and Gulløv 1989). In southwest Alaska the Eskimos of Hooper Bay favoured the "soft blue-gray colours" of the Emperor Goose or the King Eider for their birdskin parkas (Brandt 1943: 279, 106). Near Barrow on Alaska's north coast, Charles Brower (1950: 127) once came across in an ancient ice-house the seated corpse of an Eskimo who had been shot through his heart at close range. The man had been in this Arctic tomb for a very long time and was "dressed in curious loonskin clothing". Brower had evidently never seen the like on living Eskimos. In Canada the Inuit living on the east side of Hudson Bay used eider skins for blankets and clothing. The outer feathers were usually removed, leaving the inner down, and the skins were treated to make them pliable and soft before they were sewn together. These eider-skin blankets were sometimes trimmed round the border with head skins of Eider and King Eider drakes (Driver 1974: 72–73). In northern Quebec Province inner socks were made from eider skins (Reed 1986b). In the Chukchi Peninsula in the far east of Siberia, the local inhabitants used to trim the collars and cuffs of a fur coat with up to 40 or 50 skins of the heads of drake Spectacled Eiders, and Portenko (1981) also mentions a small rug decorated with 10 heads of King Eiders and Eiders.

Rugs or mats were commonly made of bird skins. The Inupiat of northern Alaska made feather mats from birds' breasts to sit on while sealing or fishing from the ice (Richards 1949: 265) and slept on mattresses of swan skins (Jenness 1985: 76). The production of eider-skin rugs in Greenland was transformed by the Danish Royal Greenland Trading Company into a regular industry. More than a hundred skins were used in making a good-sized rug, which was made with the skins of bodies of the birds with the feathers removed but the down still adhering. The skins of the heads of Eiders and King Eiders were used to decorate the borders. The trade reached a peak in the 1920s, when up to 1300 rugs were being purchased annually by the Trading Company from West Greenlanders and sold on the Danish market; it was stopped by the authorities in Greenland in 1939 (Salomonsen 1950: 129–130).

Bird skins were also used to make bags. Among the Eskimos of Bering Strait, swan and diver skins were used to make small decorated storage bags (Ray 1975: 117) and in 1924 Brandt (1943: 88) saw a Hooper Bay Eskimo woman who kept her sewing kit inside a Pacific Loon's skin. Driver (1974: 73) mentions a small bag made by an Eskimo in the Hudson Bay region that had four pockets, each made from the cured head skin of a drake Eider. On the south shore of Hudson Strait the Takamiut or Tahagmiut Eskimos made bags from the skin of the Snowy Owl (Graburn 1969: 23–24).

Even the beaks of birds had their uses. In southern Novaya Zemlya in 1897 the English ornithologists Pearson (1899: 119) and Feilden noticed in a Samoyed store house "some long strings of the upper mandibles of Bean and White-fronted Geese, made as playthings for the children". On St. Lawrence Island fur garments were sometimes ornamented with rows of the horny bill sheaths and crests of Crested Auklets sewn along the seams (Nelson 1899).

METHODS OF BIRD HUNTING

It would take a whole book to describe all the various and ingenious methods used by Arctic peoples to catch or kill birds. A few examples only will have to suffice here. Although some Inuit peoples used bows and arrows to bring down birds in flight, others lacked this weapon, and the bolas was perhaps in more general use for this purpose among the Inuit of the North American Arctic than the bow and arrow: it was used by the Eskimos of Quebec, by the Caribou Eskimo, and by the Eskimo of the north coast of Alaska, Bering Strait and St. Lawrence Island.

In the mid-1960s at Wainwright, on the Arctic coast of Alaska, the shotgun had long replaced the bolas; but R. K. Nelson (1969: 158–159) talked with a local Eskimo, then in his seventies, who had learnt to handle the bolas before he could use a shotgun. On the basis of his information, Nelson was able to describe this device and its use. It was known as a *killamittaun* and consisted of a bunch of about nine wing feathers, each of them doubled over, tied together. Attached to this "handle" and firmly held inside the angles of the folded-over feathers, were six or seven braided sinews each some 75 cm (30 inches) in length and having a weight made from bone or walrus ivory tied to its end. During the spring and autumn passage of geese and ducks every man, woman and child carried several of these bolases with them. When ducks or geese came in sight the weights were held in the left hand and pulled away from the handle to straighten the strings. The *killamittaun* was then whirled overhead a few times and released so that, as it flew through the air towards the oncoming waterfowl, the weights spread apart to form a circle up to 150 cm (5 feet) in diameter. If any part of the bolas hit a bird, the weights wrapped themselves round it and brought it down instantly. The *killamittaun* was said to have had a range of 25–35 m and to have been capable of bringing down more than one bird at a time. Chances of success were increased if the hunter lay in waiting concealed on a known flight line.

Further along the Alaskan coast, at Barrow, the local Eskimos were still hunting migrating eiders in this way when Charles Brower (1950: 72) arrived there in the 1880s. While those with guns "blazed away from far down the lagoon" and killed a good many, many of the remaining birds swerved "directly over the sandspit where other natives could use their ancient slings or *kalumiktouns*. These consisted of small ivory balls attached to 3-foot lengths of braided sinew fastened together at one end. Swung once around the head then thrown into a flock of ducks, the spreading weights curled around wings and necks and brought down many a fine bird that had escaped the shotguns". Much earlier, in May 1854, the surgeon on board H.M.S. *Plover*, John Simpson, had described the Inupiat eider shoot in the following terms (Bockstoce 1988b: 382):

> It was a novel sight to see 200 (counted 183) people all ranged in line with balls attached to string, throwing them up at the flocks of birds as they came on—and at each 3–4 to 15 might be seen dropping. Each flight as they received the unexpected assault would rise higher waver & fly along the line entire & there reform in line to proceed on their journey. The flocks became less numerous towards 5 P.M. and many of the Natives left their station on the beach to the West of the Village. I think not less than 600 that is twelve to each hut were taken by the Natives.

The Chukchi, who live in the peninsula named after them, also used—and may still use—a bolas for hunting waterfowl in flight. Called *eplicathet* in the Chuckchi language, it was described by Russian ornithologist L. A. Portenko (1981: 247–249), who saw it in use at Uelen (Map 16) in Bering Strait, the most easterly mainland point of the Soviet Union, in the early 1930s. At that time shotguns were also in use. Indeed Portenko was woken up by the noise made by an early-morning eider-shooting party on his second day at Uelen. The shouting and yelling, accompanied by the noise of shots, convinced him that the entire village was rioting. The shotgun had the disadvantage that it frightened the walruses away, while the *eplicathet* was silent. It thus continued in use. The Chukchi *eplicathet* as used at Uelen was very similar to the *killa-mittaun* of the Eskimos at Wainwright, a little less than 700 km (about 400 miles) to the northeast. The egg-shaped weights were made from walrus teeth and were attached by cotton strings up to 90 cm (3 feet) long to a handle made of wing feathers of Long-tailed Ducks bound together. The *eplicathet* when not in use was suspended inside the yaranga, the Chukchi dwelling; it was also carried on the head, the weights forming a sort of crown round the forehead.

Besides the bolas, bow and arrow, sling and shotgun, birds in flight were brought down with a net mounted on the end of a long stick. This method was used at breeding colonies of Little Auks, auklets and Brünnich's Guillemots in various parts of the Arctic, as it still is to catch Puffins on Mykines in the Faeroe Islands. The catcher takes up a position in a traditional spot on the cliffs or scree slopes and sweeps his net deftly upwards into the path of oncoming birds. An experienced catcher at a crowded auk colony can take up to 1000

Plate 5 Flock of moulting Greater Snow Geese in Avanersuaq, northwest Greenland, 11 July 1984. Easily driven into a pen by several people, geese in this condition can escape a solitary hunter by running.

birds in a single day in this way. As to birds on the ground or in the water, snares and three-pronged bird darts or spears were widely used to take these. The Baffinland Eskimo placed a line of nooses fixed to a thin strip of whalebone in the water of a lake to catch swimming and diving ducks (Kemp 1984). Towards the close of the nineteenth century Inupiat Eskimos trudged inland from Barrow in northern Alaska with their kayaks on their backs. They launched them on the lakes and rivers, and young and moulting eiders on the water in the late summer were killed with a light spear tipped with three iron barbs and thrown from the kayak (Ray 1885; Damas 1984: 291). A similar three-pronged bird dart was thrown at birds with a 2-foot-long throwing board with a handgrip at one end and a socket for the dart at the other, by the Takamiut Eskimo on the southern shore of Hudson Strait (Graburn 1969: 49).

Geese are especially vulnerable to the hunter when, in the late summer, they become flightless for several weeks while their wing feathers are moulted and re-grown. In Arctic Quebec the Eskimos around Povungnituk drove the moulting geese into stone enclosures, which were then closed round the birds (Saladin d'Anglure 1984). In the Yukon Delta the pound was made of salmon nets and the geese driven into it were killed with sticks (Nelson 1899: 135). A Samoyed goose hunt on Kolguyev Island was watched on 18 July 1894 by English ornithologist Aubyn Trevor-Battye (1895: 217–227), who described it in meticulous detail. A large proportion of the 59 inhabitants of the island first assembled together on a low-lying stretch of shore where thousands of geese could be made out in the mist crowded together on the offshore sandbanks. On a small flat island close inshore the goose trap was first set up. From two posts on the water's edge, 30 yards apart, strips of netting, forming a sort of fence about 4 feet in height, led inland, gradually converging until, 40 yards from the entrance, they were not more than 5 yards apart. From this point the netting fences curved outwards to form a circular cul de sac. Next, seven rowing boats set out along the tidal creeks to work their way behind the geese, and they were followed by reindeer teams dragging sleighs. While the boats drove the massed geese slowly inshore, the sleighs ensured that they did not escape on either flank. Many of the grey geese present were not yet in moult and flew off, but many other geese remained on the ground. Soon a dense mass of birds arrived close to the shore. Some ran with heads up and wings outstretched, some crouched down with their heads and necks stretched out on the ground until forced to move on or be killed by the Samoyeds, and all the while the main body of swimming Brent Geese was inexorably driven towards the entrance to the trap. These Brents made no effort to fly or dive, but some of the swimming grey geese reacted to the oncoming boats by stretching their heads out forwards to their full extent and then sinking themselves into the water. Eventually, when all the birds were inside the trap, the two posts were moved together to close the front entrance, and the entrance to the cul de sac was also closed. Birds were then allowed to move in batches into the cul de sac where they were killed by being picked up by the head and swung quickly around, which broke their necks. Once killed, the birds were thrown outside the netting. It took 5 hours from the start of the operation to get all the geese into the trap and another 5 hours elapsed before all the geese were despatched.

The day's bag was: 3300 Brent Geese, 13 Bean Geese and 12 White-fronted Geese. Total 3325 geese. On the following day the geese were loaded onto sleighs and transported to the Kolguyev mainland to be cached for the winter. In a place where the cloudberry grew thickest, turves were cut round with an axe, rolled up by hand, and removed. On the patch of bare earth thus exposed, the geese were stood on their tails with their heads tucked in and closely packed together. The turves were then unrolled and a double layer was placed over the closely stacked goose carcasses, forming a mound some 3 or 4 yards in diameter.

Decoys were widely used by Inuit subsistence hunters. At Hooper Bay in southwest Alaska a dead Willow Grouse would be set up on the tundra by means of a stick pushed through its neck. The hunter concealed himself nearby and imitated the bird's rattling call. Taking up the challenge, another male Willow Grouse would arrive on the scene and promptly fall victim to an Inuit arrow (Brandt 1943: 201). Two members of the genus *Lagopus*, namely the Ptarmigan and Willow Grouse, in North America known as the Rock Ptarmigan and Willow Ptarmigan respectively and often indeterminately described as "ptarmigan", were important food birds for the Inuit and other Arctic peoples. In 1883 in northwest Alaska, inland from the coast between Capes Lisburne and Krusenstern (Map 20), Brower (1950: 32–33) reported that the Eskimo women used two methods to catch ptarmigan. They "set snares of braided deer-sinew among the willows where the birds had favourite runways" and they used "long nets laid on top of the snow, often stretching several hundred feet. Women and children got behind and drove the birds slowly ahead. If the driving wasn't too fast, sometimes a whole flock got entangled in the meshes." At Anaktuvuk Pass in the Brooks Range, Alaska, the local Nunamiut Eskimos depended for survival in poor caribou winters on systematic Willow Grouse snaring (Gubser 1965: 247–8).

At Point Barrow in 1882–1883 the Inupiat Eskimos often shot Glaucous Gulls, which they were fond of eating, along the shore in autumn (Ray 1885: 123). But they also caught them with bait attached to a line.

> They are also occasionally caught with a baited line in the autumn when there is a light snow on the beach. A little stick of hard wood, about 4 inches long and sharpened at both ends, has attached to its middle a strong line of deer sinew. The stick is carefully wrapped in blubber and meat and exposed on the beach, while the short line is securely fastened to a stake driven into the sand and carefully concealed in the snow. The gull picks up the tempting morsel and swallows it and of course is caught by the stick, which turns sideways across the gullet, and his struggles to escape fix it more firmly.

BIRDS HUNTED BY DIFFERENT INUIT GROUPS

Different groups of Arctic peoples have of course exploited different species of bird, but geese, ducks and auks were the universal favourites. In his study of "The food birds of the Smith Sound Eskimos" W. Elmer Ekblaw (1919) mentioned 12 species more or less regularly eaten by the Inuhuit of Avanersuaq in northwest Greenland at the beginning of this century: the Fulmar, Snow Goose, Brent Goose, Eider, Glaucous Gull, Kittiwake, Ivory Gull, Brünnich's Guillemot, Little Auk, Ptarmigan, Raven and Snow Bunting. By far the most important of these was the Little Auk, and several of the Inuhuit settlements are sited near large Little Auk colonies (Vaughan, in press). The Eider and Brünnich's Guillemot also nest colonially in the district in huge numbers and formed an important element in the Inuhuit diet. Fulmars were eaten early in the spring, for they returned some weeks before the more palatable auks.

The Eskimos of St Lawrence Island, in Bering Strait some way south of the Arctic Circle, were able to hunt birds the whole year round (Hughes 1962: 117–131). In 1954–1955 the year's bird hunting began in January with the occasional shooting of Long-tailed Ducks along the edge of the shore ice. More birds were shot during the spring migration and also in June–July, during the breeding season, when geese, ducks and auks were shot and the eviscerated carcasses hung up to dry along with fish. Auklets were netted at the breeding colonies in July. Ducks, geese and seagulls again became the targets of shotguns during the autumn migration. In November an effort was made to shoot as many gulls as possible to freeze them for subsequent mid-winter use.

On the mainland coast of Alaska opposite St. Lawrence Island, Klein studied the role of birds, and specifically waterfowl, in the economy of the Yupik Eskimos living in the Yukon-Kuskokwim Delta in 1964–1965, in the sub-Arctic rather than the Arctic proper. He thought that the 9521 persons living in this area in 1963 represented the largest concentration of Inuit anywhere in the world, and he reckoned that they consumed annually some 83 000 geese, 38 000 ducks, 5500 swans and 1000 cranes. Communal drives of moulting flightless geese in the late summer had almost ceased by the time of Klein's study and serious bird hunting was confined to the spring and autumn migration periods, spring being the more important. Five species of geese, three of ducks, one swan and one crane species were reported shot by these Eskimos, as the summary of Klein's data set out in Table 4 shows.

Table 4 Approximate numbers of birds reported in 1964–1965 as killed by Yukon-Kuskokwim Delta Eskimos (Klein 1966).

Species	No. shot in spring	No. shot in autumn	No. shot in the year
Tundra Swan			5 585
White-fronted Goose	13 500	9 100	22 600
Snow Goose	5 400	400	5 800
Emperor Goose	6 500	1 700	8 200
Canada Goose	20 000	18 200	38 200
Brent Goose	2 500	5 500	8 000
Mallard	4 700	4 800	9 500
Pintail	12 000	10 500	22 500
Eider	3 300		3 300
Sandhill Crane			1 033

One may well question the accuracy of these figures (see Usher and Wenzel 1987). The survey was conducted with the aid of an interpreter by interviewing specially-convened meetings of male inhabitants in each of 23 different villages. The author claims to have obtained better co-operation and more information, and thus higher figures of kills, than the personnel of the United States Fish and Wildlife Service, which would not be surprising.

At Wainwright in Arctic Alaska nearly 1000 km (600 miles) north of the Yukon Delta, Nelson (1969: 150–170) studied bird hunting by the local Eskimos at the same time as Klein but without trying to emulate his numerical precision. Although in the past Black Guillemots were netted in winter in the leads between the shore and sea ice, and Glaucous Gulls were still being shot "as they ride the winds southward along the ocean cliffs" in September and October, serious bird hunting at Wainwright was confined, as in the Yukon Delta, to the spring and autumn migration of waterfowl along the coast. Using shotguns, the hunter would conceal himself on the shore or among the jumbled slabs of shore-fast ice and await the often tightly-grouped low-flying flocks of geese and ducks. Eiders and King Eiders were the main bird species of economic importance to these Eskimos, the carcasses being stored against the winter in specially made underground ice cellars. Long-tailed Ducks were sometimes shot, but did not taste particularly good. In the fall, Brent Geese were shot in numbers along the coast southwest of Wainwright, especially at Icy Cape (Map 20). It was here that, in September 1921, Alfred M. Bailey (1948) came across 200 Brent Geese hanging up to dry by an Eskimo tent and saw thousands more either flying over or assembling in dense flocks on the shore to feed on heaps of seaweed. The Eskimos were killing the birds with shotguns, often concealing themselves in blinds made of turf.

The Canadian anthropologist Derek G. Smith (1984: 354) has compiled a list of the bird species hunted and utilized by the Eskimo groups living in the Mackenzie Delta. There are 16 of them: ten species of duck, one swan (the Tundra Swan); two grouse (the Willow Grouse and Ptarmigan), and three geese (Snow, Canada and Brent). The carcasses were stored in permafrost pits along with fish and the meat of mammals.

BIRDS HUNTED BY THE NGANASAN

The 700 nomadic Nganasan of the Taimyr Peninsula were probably "the most unspoiled native tribe in Siberia" when Russian ethnologist A. A. Popov studied them in the late 1930s (Chard 1963: and see Storå 1968). They were thought to represent a remnant of the ancient tundra propulation of this area.

Plate 6   (*a*)   Drake Eider on the shore of the Varanger Fjord, Norway.

        (*b*)   Ptarmigan pair near Cambridge Bay. The Ptarmigan and Eider are among the commonest and most widely used food birds of the Inuit; both are abundant, circumpolar in distribution, and exellent eating.

Birds and Arctic peoples 37

(a)

(b)

Their economy was still based on the subsistence hunt, and especially on the wild reindeer hunt. The annual round of bird-hunting activities began early in the spring when other food was scarce. Ptarmigan were driven into nets 10–15 metres long made of white thread and set up with stakes, or lured into them by a female decoy bird. At the start of summer a similar net set up in shallow water was used to catch ducks, which were either driven into it or lured into it with decoys. Sometimes a decoy duck was tethered in front of the net, or pieces of turf were placed upside-down in the water to resemble floating ducks. Birds were also shot with special two-headed arrows using a bow made of separate pieces of larch lashed together, and sitting geese were trapped on the nest. The most important bird hunt of the year was the co-operative goose drive in the moulting season, starting late in July. A long V-shaped net with a square pen at the apex was stretched between poles on the shore of a lake used by the moulting geese. The birds were then driven ashore by men in dugout canoes and guided into the net by hunters hidden on either side. Several hundred could be caught at a time. The goose's necks were wrung and they were left in a heap with the net over them to protect them from seagulls. Later the women went out and fetched in the bag, which they then processed: the fat was carefully rendered down and stored in reindeer stomachs and the meat was laid out on sleighs to dry in the sun.

BIRDS EATEN BY EUROPEAN SETTLERS IN GREENLAND

Although the Inuit, the Samoyeds, the Nganasan and the Chukchi, or their ancestors, moved into the Arctic and have remained there since, other people have attempted to settle there in the last thousand years, usually without permanent success. There have been two successive European colonisations in Greenland, both initiated by Norwegians. The first was the Viking settlement of parts of southwest and west Greenland inaugurated by Erik the Red shortly before the year A.D. 1000; the second was the colony established in west Greenland by Hans Egede in 1721. The Viking settlers disappeared in about 1500; Egede's Norwegians, soon replaced by Danes, never really settled in the true sense of the word of living uninterruptedly in one spot generation after generation. Both groups of settlers may, at any rate for the purposes of this chapter, be considered among Arctic peoples. Recent archaeological research has enabled us to answer the question—what birds did they eat?

At the head of Ameralik Fjord, 80 km (50 miles) inland from the coast of west Greenland at Godthåb (Nuuk), the rubbish tips or middens of three Viking homesteads or farms were excavated by Danish researchers. These yielded nearly 30 000 bone fragments dating from the beginning of Viking occupation around A.D. 1000 until about 1400 or 1500. Since these farms were within a circle of about 20 km (12 miles) in diameter, it makes sense to consider the material as a single unit. The birds specifically identified are listed in Table 5.

Table 5 Bird bones or bone fragments found in Norse middens at the head of Ameralik Fjord (Eqaluit, Niaqussat and Nipaitsoq) (Møhl 1982).

Species	No. of bone fragments found
Red-throated Diver	1
Great Northern Diver	1
Whooper Swan	4
White-fronted Goose	6
White-tailed Eagle*	9
Ptarmigan*	201
Iceland Gull	3
Kittiwake	3
Brünnich's Guillemot*	124
Razorbill	1
Black Guillemot	3
Raven	1

*Found in all three middens.

These birds all occur nowadays in the Nuuk area, but the Whooper Swan is a rare visitor only and has never been recorded breeding in Greenland. Perhaps it did so in the middle ages? It is hardly surprising that quantities of Ptarmigan were eaten, but the large number of Brünnich's Guillemots is strange, for this seabird seldom if ever penetrates inland far up the fjords. Moreover, there are no large colonies today around Nuuk. Could these birds have been traded from the Eskimos? Or did these Norse farmers make long summer excursions along the outer coast?

Long after the abandonment of the Viking homesteads at the head of Ameralik Fjord, the missionary Hans Egede settled in 1721 on what he called Hope Island, at the mouth of the fjord, on the outer west Greenland coast. After only 7 years, Egede moved his colony to Godthåb, now Nuuk, some 15 km (9 miles) inland, and Hope Colony has ever since remained deserted, named by the Greenlanders *Igdlueruumerit* or "the place where there used to be houses" (Gulløv and Kapel 1979: 22). Jeppe Møhl, of Copenhagen University's Zoological Museum, author of the article cited above, has also given us a list of the bird bone fragments recently unearthed by Danish archaeologists from the floor of the dwelling house at Hope Colony, dated historically to 1721–1728. The birds identified generically, specifically, or by family in a total sample of 1211 bone fragments of all sorts of animal, are listed in Table 6.

The preponderance of seabirds in this list is not surprising considering the colony's situation on the outer coast. One of them, the Great Auk, became extinct in 1844. The Great Shearwater breeds in the Tristan da Cunha island group in the south Atlantic and, after its breeding season, migrates northwards across the equator. It is common off Greenland's west coast as far north as Disko, especially in July and August. The fact that only a single piece of Ptarmigan bone was found shows that these bones are in no way representative

Table 6 Bird bones or bone fragments found in the floors of the dwelling house at Hope Colony, near Nuuk (Møhl 1979).

Species	No. of fragments
Great Shearwater	9
Cormorant	4
eider, *Somateria* sp.	30
Long-tailed Duck	4
Red-breasted Merganser	1
Gyrfalcon	2
Ptarmigan	1
Great Black-backed Gull	1
Iceland Gull	9
Glaucous Gull	13
Kittiwake	1
gull, *Larus* sp.	3
Great Auk	4
alcids, *Alcidae*	114
Black Guillemot	28
Little Auk	3
Raven	2

of the birds killed and eaten by the Norwegian colonists: historical sources mention a total of 68 Ptarmigan as shot on five different occasions, and on at least 15 other occasions "some" Ptarmigan were shot. Whatever did happen to their bones, they were not dropped, like so many others, inside the house, to form part of the detritus on or under the floor boards.

EGGING

So far we have discussed bird hunting without mentioning egging. It goes without saying that birds' eggs have almost everywhere been collected and eaten by Arctic peoples, though among the Mackenzie Delta Eskimo there was a widespread taboo against them (Smith 1984: 341). Uspensky (1986: 332–333) reports that Brünnich's Guillemot's eggs are twice as big and just as good to eat as hen's eggs, and yield 1988 calories per kilogram as opposed to "medium quality beef" which yields only 1358 calories per kilogram. Since H. B. Cott in 1953–1954 reviewed the exploitation of wild birds for their eggs throughout the world, new information has become available, some of it very detailed, and new insights into how this should be interpreted have been gained. Besides auks, many common Arctic breeding birds have relatively large and palatable eggs, notably eiders, various goose species, and waders, and all of these have been, and in places still are, an important food resource for native peoples. Some of them, like the eggs of the Snow Goose, can be collected in quantity from easily accessible colonies. The eggs of waders, whose nests are dispersed on the ground in the open tundra and are extremely difficult to find, are sought out because of their palatability. In Canada in 1986 clutches of American Golden and Grey Plovers' eggs disappeared mysteriously and my wife and I thought they had been taken by Eskimos camping nearby. Brandt (1943: 372) reported that in southwest Alaska the eggs of the Grey Plover were considered superior as food to those of any other species, though the local Eskimos also relished eggs of the Tundra Swan, Eider, Sandhill Crane and Sabine's Gull.

The eggs of colonially nesting birds were often stored by the Inuit for use during the winter. In northwest Greenland the only eggs collected in quantity were those of the Eider, which nests in thousands on a few low-lying offshore islands. They are still being taken nowadays: in 1975 over 4000 eggs were taken at one colony, and in the early 1980s the same colony was thought to be providing the local Eskimos with 2000–2500 Eider eggs annually. In the old days the Inuhuit used to cache these eggs under heaps of rocks. Months later the frozen eggs were eaten raw with much relish after the shell had been removed. Alternatively they were stored by being made into sausages by "sucking them, masticating the whites and yolks together, and expectorating them into casings prepared from seal intestines" (Ekblaw 1927–1928: 189; Vaughan, in press).

While in northwest Greenland the Eider was the only important egg bird for the local people, further south on the west Greenland coast Brünnich's Guillemot took over this role: Salomonsen (1950, 1970) claimed that 10 000 Brünnich's Guillemot eggs were taken annually in the Upernavik district. In many or perhaps most other parts of the Arctic different species of goose have been the targets of native egg gathering. In the coastal tundra of the Yukon Delta it was in the first place the Emperor Goose, and secondly the Canada Goose, that provided the bulk of the local people's substantial harvest of some 40 000 eggs per annum (Klein 1966). At Eskimo Point (Map 10), on the western shore of Hudson Bay, it was the Snow Goose that was mainly exploited, and here, too, the Canada Goose took second place. In 1977 the Polar Gas Project based in Toronto sponsored a study of the resource harvest of the 1000 or so mainly Eskimo people then living at Eskimo Point. This showed that 61% of the 7529 eggs gathered were those of the Snow Goose and 26% belonged to Canada Geese. Besides these goose's eggs, 867 Ptarmigan or Willow Grouse eggs were taken and only 137 eggs of other species, probably including divers and eiders (McEachern 1978).

THE IMPACT ON POPULATIONS

It is easy enough to establish that birds and their eggs have formed a relatively small but none the less important natural resource for the native peoples of the Arctic. It is much harder to assess the impact of Inuit and other native people's resource harvesting on bird numbers. In Greenland, birds have been ringed more or less systematically since the 1920s and a high proportion of the recoveries have been due to shooting by Greenlanders. It could therefore be assumed that the percentage of ringed birds recovered reflected human hunting pressure. Thus, according to Freuchen and Salomonsen (1958: 41–42), ringing showed that in the 1950s more than one-third of the annual hatch of young Cormorants was being shot, and this was considered a serious threat to the species. On the other hand, the same authors concluded that Brünnich's Guillemot was not in danger. Even though the Greenlanders were shooting more than 200 000 annually "banding has shown that this figure constitutes less than 4% of the population"; which was thus some 5 million birds. However, recent studies (Kampp 1988) point to a decline in numbers at least in part due to over-exploitation by Greenlanders. The proportion of the population of some other typical "target" species shot by Greenlandic Eskimos, "as established by ringing", namely the percentage of ringed birds recovered shot, is shown in Table 7.

Meticulous recent research in the North American Arctic, much of it carried out by government agencies like the Canadian Wildlife Service and the United States Fish and Wildlife Service, has yielded quantitative information about the harvest of waterfowl by native peoples and its likely impact on the populations of the different species. The general consensus, as expressed at international symposia like the North American Wildlife and Natural Resources Conference, is that the bird kill by native hunters has so far been of negligible importance over the continent as a whole except in one or two more or less localised instances. As long ago as 1963 Thompson and Person counted eiders passing in autumn over or near Pigniq, the already mentioned "Duck Camp" at the base of the narrow peninsula that juts northwards at Barrow, Alaska, into the Arctic Ocean (Map 31). The average number of birds per hour over 46 days of observation was 987. The average number of Inupiat Eskimo shooters was 4.1 per hour. On average, these hunters fired 11.5 shots which killed 3.5 eiders and failed to recover a further 1.5 cripples per hour. The author calculated that only 0.5% of all southbound eiders were killed by Point Barrow shooters. More recently (1968), using figures based on censuses of breeding eiders in the western Canadian Arctic, T. W. Barry of the Canadian Wildlife Service calculated or surmised that the annual native kill of eiders

Table 7 Proportion of the population of certain species shot by Greenlanders in the period 1945–1965 (Salomonsen 1967, 1970).

Species	Percentage of population shot
Red-throated Diver	8
White-fronted Goose	7
Eider	22
Long-tailed Duck	17
White-tailed Eagle	35
Purple Sandpiper	14

along the Beaufort Sea (Map 1) migration route "is at most only one per cent" of the total Beaufort Sea eider population. As far as the sub-Arctic is concerned, the Cree Indians living round the shores of James Bay in the Canadian provinces of Ontario and Quebec are among the native peoples for whom subsistence hunting remains of vital importance. It has been calculated that in 1974–1976 they killed 2.9% of the total number of Lesser Snow Geese migrating south along the shores of James Bay in autumn. The 22 700 ducks killed annually by these natives were thought to be of no importance when related to the total number migrating through the area. On the west side of James Bay, in Ontario, the kill of geese by Cree Indians at least doubled between the mid-1950s and the mid-1970s, because of a doubling of the number of hunters. However, during this same 20-year period the Canada and Snow Goose populations exploited by these native hunters more than doubled (Prevett et al. 1983). George Finney of Environment Canada has recently reviewed the native waterfowl harvest there and concluded that no species of goose or duck is endangered by this harvest in Canada (Finney 1990).

CONSERVATION

The native peoples of the Arctic, interested in birds and partly dependent on them, are increasingly becoming involved in their study and conservation. When Finn Salomonsen (1956; Mattox 1970) reorganised bird marking in Greenland after the Second World War the Greenlanders were enlisted to carry out the ringing. They were paid for this work according to a sliding scale varying with the species: 10 crowns for a White-tailed Eagle or a Snowy Owl; 10 cents (øre) for a Snow Bunting. At the same time in the North American Arctic, Eskimo collaboration, especially as hunters, observers and guides, had become a crucial element in research into bird populations. Simon Paneak's contribution to Irving's study was mentioned at the beginning of this chapter. Another example, taken from the "Acknowledgements" section of the important study by Parmelee and others (1967) of the birds of southeastern Victoria Island, is the reference to "Eskimo David Koomyuk" who "with a superb knowledge of the birds of the region . . . guided us to many excellent areas".

Closely linked to the study of birds is their conservation. In North America determined attempts are now being made to put right the wrongs of the past by involving the native peoples in the management of what is now regarded as their own resource. They were to a large extent deprived of this resource, in the eyes of the law, as a result of the Convention for the Protection of Migratory Birds, which was signed in 1916 by the United States and Great Britain. This Convention included a general prohibition of the hunting of all migratory game birds between 10 March and 1 September both in Canada and in the U.S.A., with two relevant exceptions: "Indians may take at any time scoters for food but not for sale" and "Eskimos and Indians may take at any season auks, auklets, guillemots, murres and puffins." Alcidae are not regarded elsewhere as game birds.

Native subsistence hunting in Alaska had, up to the implementation of this Convention in the U.S.A. in 1918, been fully exempted from the wildlife conservation legislation: the first Alaska Game Act of 1902 had clearly stated that "The Indians and Eskimos may at all times kill game animals or birds for their food or clothing." Now, after 1918, summer hunting of waterfowl became illegal, though in 1925 the prohibition was relaxed to the extent that "any Indian or Eskimo" may take "birds during the close season when he is in absolute need of food and other food is not available". Even this concession was removed in a 1944 regulation of the U.S. Bureau of Biological Survey, which added the phrase "except migratory birds" in brackets after the word "birds". Since nearly all the species utilized in the native peoples' subsistence hunt were migratory, this effectively again made the subsistence hunt totally illegal in Alaska. However, no attempt was made in the ensuing years by the U.S. Bureau of Biological Survey, nor its successor the U.S. Fish and Wildlife Service, to enforce the ban against native subsistence hunters. Their (criminal) activities quietly continued until, in May 1961, Fish and Wildlife Service law enforcement agents charged some Inupiat Eskimo eider shooters at Point Barrow with taking migratory wildfowl in the closed season. A protest meeting of some hundred Eskimos followed, after which each presented to the enforcement officers an eider he had shot and a signed statement admitting his action, demanding to be arrested. Subsequently a petition signed by 300 Barrow Eskimos was sent to President Kennedy seeking amendments to the convention and new legislation to authorise their spring and summer subsistence hunt. Although no action followed, enforcement agents remained thenceforth inactive. At last, in 1978, the Convention Concerning the Conservation of Migratory Birds and their Environment between the U.S.A. and the Soviet Union authorised spring and summer subsistence hunting in Alaska and the amendment of the convention with Canada to permit this. Such an amendment was negotiated by the U.S.A. and Canada in 1979 but has not been presented to the U.S. Senate for ratification because of objections by sporting organisations like the National Wildlife Legislative Fund of America, Ducks Unlimited and the Waterfowl Habitat Owners Alliance.

Alaska's native hunters might still now be in a limbo of criminality had it not been for a "biological emergency" which arose in the Yukon-Kuskokwim Delta in 1983, when available data indicated a dramatic decline in the number of breeding Canada, White-fronted, Brent and Emperor Geese. Negotiations between the U.S. Fish and Wildlife Service and the Association of Village Council Presidents, representing the 56 Yupik Eskimo villages in the delta and along its shores, led early in 1984 to the so-called Hooper Bay Agreement, subsequently renamed the Yukon-Kuskokwim Delta Goose Management Plan, in which the local subsistence hunters voluntarily agreed to curtail their shooting and egging activities. The legality of this goose management plan was challenged by Alaskan sportsmen in the U.S. District Court in Juneau, but the judge ruled against them. At last the native hunters of Alaska, or some of them at least, had been given a measure of control over and responsibility for their own resource (Bartonek 1986, Mitchell 1986).

Because of the American failure to ratify the 1979 protocol, native subsistence hunting of geese and other waterfowl (except scoters) has remained illegal throughout Canada between 10 March and 31 August, in accordance with the clauses of the convention of 1916. The Canadian Wildlife Service in its Waterfowl Management Plan identifies the need to transform the situation by amending the convention, acknowledging "the importance of harvest by subsistence users" and the need for their increased involvement in waterfowl management (Murray 1986), but some provincial governments, which share the responsibility for wildlife management, are unwilling to accept any changes in the convention and would prefer to see it abandoned. The principle that native hunters have a right to harvest migratory birds and a role in waterfowl management is fully established under the programme of land claim settlements now being negotiated by the Canadian government with all northern native groups. The first of these settlements was the James Bay and Northern Quebec Agreement of 1975, which defined the rights and privileges of 14 000 native people, some 10% of whom were active goose hunters, in an area almost as large as Norway, Sweden and Finland combined (Boyd 1977). Not only were these people in principle granted the "right to harvest" but their representatives were to sit on the "Co-ordinating Committee" alongside government specialists to manage hunting, fishing and trapping in the area (Drolet 1986). Ten years later, in a 1985 land claim settlement, the Mackenzie Delta Eskimos were given the exclusive right to hunt migratory gamebirds on their own territory. In future we can be sure that, throughout the North American Arctic, native peoples will become increasingly responsible for collaborating in the conservation of their own resources in wildlife (Cournoyea and Bromley 1986).

MAN-MADE NEST-SITES FOR ARCTIC BIRDS

Deserted Inuit winter houses are sometimes utilised by birds as nest sites. Brandt (1943: 292) described how the Hooper Bay Eskimos of southwest Alaska unintentionally provided homes for Red-breasted Mergansers when they deserted their villages along the coast at the end of May to move inland for fishing and egging. The bird enters the deserted dwelling through the opening in the roof "and makes its down-lined nest of grass and sticks in a dark corner". Snow Buntings have also used deserted Inuit houses as nest sites. In 1984 my son and I found a nest containing young birds in the remains of a stone shelter built by a group of American explorers on the coast of northwest Greenland in 1854.

Ludwig Kumlien (1879: 76–77) often found Snow Bunting's nests around Cumberland Sound in the cairns erected by Eskimos over their dead and had "even seen a nest built *in an Eskimo cranium*". British naturalist David Haig-Thomas (1939: 49) was told of a nest in northwest Greenland placed inside the open mouth of a dead Eskimo. On Kolguyev Island, Aubyn Trevor-Battye (1895: 420) found that a pair had nested in 1893 "under the wooden covering of a Samoyed grave". A Snow Bunting's nest that became famous was found on

46 *In search of Arctic birds*

(a)

(b)

29 August 1824 on an island in the north of Hudsons Bay and described in his journal by Captain George Francis Lyon. It was placed on the neck of the coiled-up skeleton of an Eskimo child which had been buried under a pile of stones. Sir John Barrow subsequently persuaded "an accomplished lady" friend of his, who signed herself Georgina, to compose a poem on this nest that he thought worthy of printing in a footnote to his second volume of British Arctic explorations (Sutton 1932: 7).

Snow Buntings have also made use of nest sites provided by the white man in modern Arctic settlements, as well as man-made song-posts like house roofs and telegraph poles. At Barrow they nest in nestboxes. In 1989, at the former United States Naval Arctic Research Laboratory (NARL) three miles northeast of Barrow, 15 nestboxes had been put up by the personnel of the North Slope Borough's Department of Wildlife. But only one or two had been taken over for nesting purposes by the numerous local Snow Buntings, perhaps because so many other excellent sites had been unwittingly provided for them. They were nesting in trash bins, on ledges or in holes in buildings, in rolled-up sections of fencing and in holes in the ground under discarded pieces of hardboard. Hereabouts they had been enabled by man to colonise the miles of hummocky, marshy tundra where they would normally be absent, for rocks, with holes in or under them, are usually required by the Snow Bunting to nest in. But here they were nesting in artefacts left abandoned on the tundra by NARL personnel, such as a deserted building, a piece of rusting machinery, a box-like structure on a wooden platform, and a ledge on an iron ammunition truck mounted on a giant sledge. A pair were even nesting inside the roller of a large bright yellow road roller parked across a closed tundra road.

Plate 7   (*a*)   Nest of Arctic Redpoll behind an outside staircase.

        (*b*)   Arctic Redpoll nesting in a metal cylinder jammed on top of a post. The buildings and other installations at the former United States Naval Arctic Research Laboratory near Barrow, Alaska, have enabled Arctic Redpolls to colonise what was once an area of bare marshy tundra.

The NARL facilities for nesting passerine birds in 1989 also included nest sites for the local Arctic Redpolls. A metal cylinder placed on top of a fence post could be regarded as a sort of artificial bush, and a nest had been built in it. Another Arctic Redpoll's nest was under the top step of a steel outside staircase and others were sited elsewhere on NARL's buildings.

CHAPTER 3

Whalers and discovery ships

HISTORICAL BACKGROUND

The Arctic was first penetrated by sea and the first voyages there were made by English and Dutch mariners in the second half of the sixteenth century. Their discovery ships were not exploring the Arctic itself but seeking a passage through it to the Far East or the Pacific. From the sixteenth to the start of the twentieth century repeated efforts were made to penetrate or circumvent this barrier of ice. Three routes were tried: the Northwest Passage round the north of North America; the Northeast Passage north of Eurasia; and, surprisingly, a Polar Passage over the North Pole. This last, suggested in 1527, was first attempted in 1607 when "Henry Hudson was set forth, at the charge of certain worshipful merchants of London, to discover a passage by the North Pole to Japan and China" (Phipps 1774: 4). The whalers followed in the wake of the explorers. From 1611 on they hunted the big, relatively slow-moving Bowhead or Greenland right whale, first in Spitsbergen waters, then off the west Greenland coast and, finally, in the eastern Canadian Arctic and in the Beaufort Sea north of Alaska (Kugler 1983; Vaughan 1983). At the end of the nineteenth century, when the Northwest and Northeast Passages had been explored and the Polar Passage abandoned, new goals for Arctic maritime expeditions were proposed. One was to take a ship as far north as possible in order to reach the North Pole on foot. Another was to search for hitherto undiscovered lands. Nearly everyone involved in these voyages, both whalers and explorers, shared some interest in birds; but the birds they encountered were fortunate if they were only *recorded* by the naturalist who sometimes accompanied these Arctic expeditions: usually they were shot for the pot or for a museum collection. But in all this maritime history of the Arctic, ornithology remained a by-product, or sideline, only.

EARLY VOYAGERS TO SPITSBERGEN

Pride of place in any account of ornithology in the Arctic must surely go to the Dutchman Gerrit de Veer whose classic account of Dutch Arctic exploring voyages published in Dutch in 1598 was translated into English in 1609 with the rather long-winded but colourful title:

> The True and Perfect Description of Three Voyages so strange and woonderfull, that the like hath neuer been heard of before: Done and performed three yeares, one after the other, by Ships of Holland and Zeeland, on the North sides of Norway, Muscouia, and Tartaria, towards the Kingdoms of Cathaia & China; shewing the discoueries of the Straightes of Weigates, Noua Zembla, and the countrie lying under 80 degrees; which is thought to be Greenland: where neuer any man had bin before: with the cruell Beares, and other Monsters of the Sea, and the unsupportable and extreame cold that is found in those places. And how that in the last Voyage, the Shippe was so inclosed by the Ice, that it was left there, whereby the men were forced to build a house in the cold and desart Countrie of Noua Zembla, wherin they continued 10 monthes togeather, and neuer saw nor heard of any man, in most great cold and extreme miserie: and how after that, to saue their liues, they were constrained to sayle aboue 350 Duch miles, which is about 1000 miles English in litle open boats, along and ouer the maine Seas, in most great daunger, and with extreame labour, unspeakable troubles and great hunger.

On 21 June 1596, on the third of these expeditions, the first of a long series of discoveries of familiar European winter birds on their Arctic breeding grounds was made. After chasing a white bear a mile out to sea in the ship's boats near the northwest point of Spitsbergen in what the English called Fair Haven, the Dutch seamen made for a couple of small islands where they could see geese sitting on their nests. One of these birds, afterwards cooked and eaten, they managed to kill with a stone, and a good 60 eggs were collected and taken back to the ship. Disturbed from their nests, the birds flew off calling *rot, rot, rot*. These were Brent Geese, which the Dutch call *Rotgans* from this call. Gerrit de Veer was perfectly familiar with them because he thought they were the identical birds that came every winter in large numbers to the neighbourhood of Wieringen in North Holland, where they were caught. Until that moment, he tells us excitedly, nobody knew where they nested. It was even claimed that in Scotland the fruits of trees whose branches overhung water "hatched" into goslings after they had fallen. The English translator of de Veer's account mistakenly had these geese cry *red, red, red*, and made them into non-existent "red geese", although the Brent was a perfectly well-known winter bird in England, too (Løvenskiold 1964: 136, Naber 1917a: 52, 53).

In 1607 the English navigator Henry Hudson was in Spitsbergen waters in the ship *Hopewell* of 80 tons with a crew of ten men and a boy. One of the ten, the seaman John Pleyce, kept a journal of this voyage, but was no ornithologist: he has the merit at least of mentioning birds, but cannot identify them specifically. We can assume that the "wild Geese" seen ashore in King's Bay (now Kongsfjorden) were Brent Geese, and that the "red-billed Bird" killed in or near Bell Sound (Bellsund) was a Puffin. Moreover, there seems little doubt that the "small flocks of Birds, with blacke Backes and white Bellies, and long speare Tayles" seen on 25 June at 75° N shortly before Spitsbergen was sighted, were Long-tailed Ducks – the first published record of this species. As to the other birds seen at sea "with blacke backes and white bellies in form much like a Ducke", one may hazard the guess that these were Brünnich's Guillemots (Purchas 1906; Løvenskiold 1964: 379; Powys 1928).

Plate 8 Wintering Brent Geese feeding on the Dutch coastal marshes, Schiermonnikoog. De Veer was wrong in thinking these were the birds he found nesting in Spitsbergen. In fact the Brent Geese that winter in Holland, figured here, belong to the Dark-bellied race that breeds in Arctic Russia and Siberia. Spitsbergen breeding Brents are Light-bellied and winter mainly in Denmark.

52 *In search of Arctic birds*

The next Spitsbergen voyager was the pioneering whaling captain Jonas Poole who, after noting some of the birds of Bear Island (Map 25) in 1604–1606, was able to add to the Spitsbergen list on his voyages there in 1610 and 1611 (Purchas 1906), though in most cases, again, not specifically. Either his "Willockes" or his "Noddies" or both were certainly Brünnich's Guillemots; his "white Partridges" were Ptarmigans; and his "small land bird, like a Sparrow, partly white and partly browne" was surely the Snow Bunting.

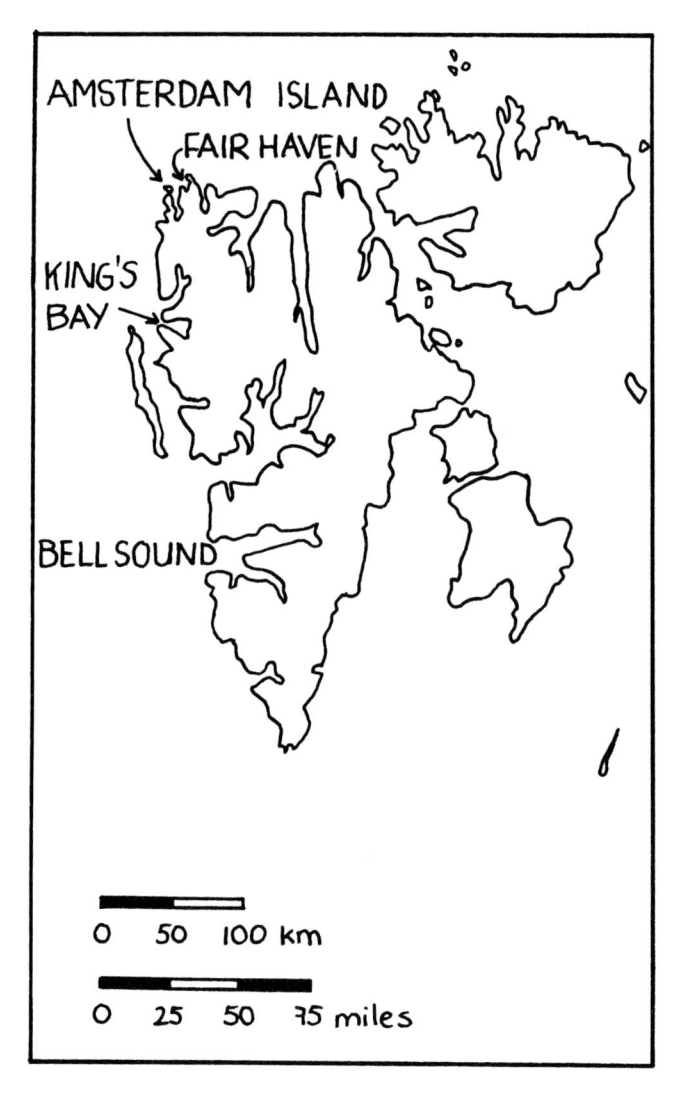

Map 3 Places named by early voyagers to Spitsbergen.

Equally surely the "white Fowle with a greene bill, the top of the bill of it and the eyes were redde, with black feet" must have been an Ivory Gull. His "Ice Birds" were perhaps Little Auks, his "Allen" the Arctic Skua, and his "seapidgeons" Black Guillemots (Løvenskiold 1964: 380; and names listed in Swann 1913 and Hare 1952). But what is one to make of the "Fowle with a combe and a tayle like a Cock" (another Ptarmigan?) or the "redde fowle of the bignesse of a pidgeon", for which the Grey Phalarope has been plausibly suggested?

One cannot claim that much was added to the knowledge of Spitsbergen's avifauna in the account of William Baffin's voyage there in 1613, usually attributed to Robert Fotherby (Markham 1881, Barthelmess 1987). This author makes some rather off-hand remarks about the birds in his chapter entitled "A Briefe Description of the Country of Greenland, otherwise called King James his New Land", which is his rather roundabout way of referring to the country Dutch explorer Willem Barents had already dubbed Spitsbergen. He merely records that "Upon this land ther be manie white beares, graie foxes, and great plentie of deare; and also white partridges, and great store of white fowle, as cueluerduns, wilde geese, sea pigeons, sea parots, willocks, stint, guls, and diuers others, werof some are unworthy of nameing as tasting" (Markham 1881: 71). The birds referred to here could be, in the same order, Eiders, Brent Geese, Black Guillemots, Puffins, Brünnich's Guillemots, Purple Sandpipers (Løvenskiold 1964: 380; and see Swann 1913), and Glaucous Gulls.

BIRDS EATEN BY DUTCH WHALERS AT SMEERENBURG

In the seventeenth century Spitsbergen was visited annually by fleets of whalers from English and Dutch ports and from elsewhere. No permanent settlements were established, but the different Dutch whaling ports built tryworks and put up buildings on Amsterdam Island in the northwest corner of Spitsbergen where they tried out the blubber from the killed whales. The place became known as Smeerenburg, which has been loosely translated Blubbertown, and the six tryworks and 15 or so buildings were annually occupied in the summer by some 150 sailors and workers (Hacquebord 1983, 1984). In 1633–1634 seven men over-wintered in a hut near the Middelburg tryworks and their commander Jacob Segersz van der Brugge recorded their experiences in a journal (Naber 1930; Conway 1904). No birds were seen in the winter months, though they did occasionally hear a noise like the croak of a Raven. They spent a great deal of their time shooting at the numerous Polar bears which scavenged around their hut, though bear meat was never on their menu: they killed the animals for their fat and skins. However, in the spring, after months on a diet of salted meat, stockfish and other provisions brought from Holland, a few birds did not come amiss. On 6 April 1634, the day they ran out of shot and had to melt their tin drinking-can and the lead gutter of the Middelburg tryworks shed to make some more, they launched their rowing boat and shot four or five Black Guillemots (*duyfkens* or *duykertjes*). On the next day they got some more. On 30 April they enjoyed a meal of Glaucous Gulls (*groote grauwe meeuwen* or *borgemeesters*) which they often shot while they were feeding on the carcasses of the dead Polar bears which now littered the ground around their hut. From mid-May onwards, too, they began to shoot Eiders, which they called *eyndvogels* or *bergeynden*. These the English translator rendered "mountain ducks", which is simply a too literal translation of *Bergeend*, the Shelduck, *berg* meaning mountain and *eend*, duck.

That these "shelducks" were in fact Eiders is scarcely open to doubt. The drake Eider and the Shelduck are both large white ducks frequently seen on the sea. The seven over-winterers at Smeerenburg would have been unfamiliar with the Eider, which only began to breed in the southern North Sea around 1800, and in the Netherlands in 1906 (Teixeira 1979). Before then it must have been a (probably scarce) winter visitor only, when most of the birds are in brown plumage. But they probably had some knowledge of the Shelduck and would perhaps naturally have called the large white sea ducks they saw and shot at Smeerenburg "Shelducks". Their own observation, that these birds "dive in an incredible manner", points to Eiders rather than Shelducks. To clinch the matter, while the Eider is common on the Spitsbergen coast, the Shelduck has never been recorded there (Løvenskiold 1964).

During excavations at Smeerenburg in 1979–1981 by the Dutch archaeologist-historian Louwrens Hacquebord (1983), the bird and animal bones were carefully collected and identified in an attempt to throw light on the whalemen's summer diet. Here, too, the evidence points to a distinct preference for salted meat and other home products, but local birds were also killed and eaten. Some of these were not specifically identifiable: the bones of the Brent and Barnacle Goose could not be separated. But the following bird species have been identified among these remnants of the seventeenth-century Dutch whalers' meals: Fulmar, Kittiwake, Ivory Gull, Glaucous Gull, Brünnich's Guillemot, Black Guillemot, Little Auk, Eider, Arctic Tern and domestic chicken (Wijngaarden-Bakker 1987, revising and completing Wijngaarden-Bakker and Pals 1981; Wijngaarden-Bakker 1984).

HAMBURG'S CONTRIBUTION TO ARCTIC ORNITHOLOGY

On either side of 1700 two distinguished inhabitants of Hamburg (Map 25), then an imperial free city, profited from her citizens' frequent whaling voyages to the Arctic to add substantially to existing knowledge of Spitsbergen and Greenland. The first of these, Friedrich Martens, sailed to Spitsbergen in person in 1671 as ship's surgeon on the Hamburg whaler *Jonas im Walfisch* or *Jonas in the Whale*. His book was published in Hamburg in 1675. Its unusual merits caused its translation into Italian (two editions in 1680), Dutch (1685 and three eighteenth-century editions), English (1694 and 1695) and French (1715). Its modest title, *Description of a journey to Spitsbergen*, belied its contents. The entire account of the voyage occupies only the first few pages. The rest of the book is taken up with a description of Spitsbergen, of the sea there, of the ice, of the air, and of the island's plants and animals.

Martens' account of the birds of Spitsbergen as he knew them or got to know about them in 1671 is quite astonishing. He accurately identifies and describes 15 different species and makes interesting and original observations about many of them. The Fulmar (Martens 1855: 75–78) he reckons "the first and commonest bird of all you see", and he describes how these birds descend in dense flocks onto the whale carcasses and peck out large pieces of blubber. They were such a nuisance while the whales were being flensed, that is while the blubber was being stripped off them, that "we were forced to kill them with sticks and with broad nets in frames, such as they use in the Tenis Court, to be rid of them". He mentions the Fulmar's curious bill and, noting that many of the Fulmars about Spitsbergen were grey all over, whereas those around the North Cape and England were grey on their backs and wings but had white heads and bellies, he suggests that these plumage differences represent two different forms of the bird and not, as was then thought, two different age groups, the grey ones being supposed to be the older birds. This is the first published reference to the polymorphism of the Fulmar.

Martens (1855: 78) does not tell us how far north he saw his Gannet, which he calls John of Ghent (*Jan van Gent* is the modern Dutch name), but if it was in Spitsbergen waters, it would be the only record from so far north. His description clinches the identity of the bird. It was "very handsom", "as big as

Plate 9 The Fulmar, an abundant breeder in the Spitsbergen Archipelago, probably ranges as far north as any other bird and is in many ways typically Arctic. This bird was photographed in northwest Scotland.

a stork and of the same shape, with white and black feathers" and it "shoots down from a great height into the water". Martens was shown a flock of Brent Geese in flight, and his very detailed account of the *Berg Enten*, here too translated "mountain ducks", confirms that these were Eiders. He says that when they see men "they hold up their heads and make a very long neck" and that "they make their nests upon the low islands . . . of the feathers of their bellies". But he mistakenly thought that these Spitsbergen birds were not the same as the birds the Icelanders call *Edder*, the down of which was exported from Iceland to Hamburg (Martens 1855: 61, 72–73).

As in his account of the Gannet, Martens often skilfully characterises a bird in a few sentences, enabling us to identify it beyond doubt. Of the Purple Sandpiper (*Schnepfe*) he says that "It is of the colour of a lark; but when the sun shines upon it, it shews blueish, very like those two colours observed on our ducks' necks". Of the Arctic Skua (*Strunt-jager*): "his tayl, which is like unto a fan, hath this mark, that one feather thereof stands out before all the rest: he is black on the top of his head: his eyes are black, about his neck he hath a dark yellowish ring or circle, his wings, as well as his back, are brown; underneath his belly he is white". Martens goes on to describe the Glaucous Gull, Ivory Gull, and Kittiwake in sufficient detail to put their identity beyond doubt, and the same goes for the Arctic Tern (the *Kirmewe*, which has "a thin sharp-pointed bill, as red as blood"), Brünnich's Guillemot, the Black Guillemot, the Little Auk, the Puffin, and the Snow Bunting (Martens 1855: 57–79).

Besides his remarkable descriptions of the birds' plumage, habits and nests, Martens has the knack of conveying their vocalisations. Thus "The calling or crying of the rotges {Little Auks] amongst one another sounds almost, at a distance, as if you hear a great many women scolding together" (p. 69) and the Fulmars "cry all together, and it sounds afar off as if they were frogs" (p 77). He always endeavoured to obtain specimens of the birds he saw and made recognisable drawings of at least four of them—the Fulmar, Puffin, Little Auk and Arctic Tern. Remarkably, there is only one bird described by him (p. 59) that is not readily identifiable, and that is his Ice-bird (*Eiss-vogel*), which Løvenskiold improbably makes into a Ptarmigan; perhaps it was a Grey Phalarope?

> I saw also in English Haven [probably Fair Haven, in the northwest] a very beautiful *ice-bird*, which was so tame that we might have taken him up almost with our hands; but we would not go too near him with our gun, for fear that we should shoot him all in pieces, and so spoil his curious feathers; so we missed him, and he flew away.
>
> The sun shined at that time upon him, which made him look like gold, so as it dazled our eyes almost. He was as big as a small pigeon. I would willingly have delineated him, if we could have catched him. I saw but this one of the kind.

After the sparkling first-hand species accounts of Friedrich Martens, the contribution to knowledge of our second Hamburg Arctic ornithologist may seem pedestrian. For Johann Anderson, who was born in 1674, three years after Martens' voyage, compiled his *Reports from Iceland, Greenland and Davis Strait* published in German at Hamburg in 1746 and in Dutch at Amsterdam in 1756, from information supplied by others. During a busy and learned life, much of it in the service of his home town, where he became mayor in 1723, he travelled to Holland, Switzerland and France, but declined King George I's invitation to go to England in 1715 as a royal councillor. In off-duty periods at Hamburg he took a special delight in inviting skippers sailing to Iceland and Greenland to his home and quizzing them about the natural history and other curiosities of those countries. Many of them were Hamburg whaling captains. Some misinformed him. One told him he had noticed, when looking at Ptarmigans's nests, that the birds collected some of the small round leaves of their food plant together and kept them as a food supply against the winter. Another described how the young guillemots were escorted down to the rocks or sea below the breeding cliffs by both parents, one flying below the young bird to give it a soft landing on its back, the other above the young bird to protect it from birds of prey. But he went on to give an excellent description of the old bird accompanying the young out to sea. Anderson was also correctly told that Black Guillemots and Little Auks, called *Rottjens* or *Rattjens* because they resembled little rats, nested among jumbled boulders.

This information, or misinformation, may have originated from either Greenland or Spitsbergen, because Spitsbergen was also called Greenland at that time and was indeed supposed by many to be a peninsula of Greenland (Map 25). But in 1733 the mayor of Hamburg was brought a living Fulmar, or *Mallemuk*, as the Dutch called this bird, perhaps meaning "stupid gull", by a ship returning from "Davis Strait", namely from west Greenland, and of this bird he gives a detailed first-hand description (Anderson 1756: 149–154). He kept it alive for some time to observe its behaviour, which included catching hold of a cat by the tail with its beak and making it shriek, then strangled it so as to describe and measure it in detail before dissecting it. He describes many of the internal parts, including the cornea, which he found acted as a magnifying glass, and concludes that the bird was a kind of gull. "Furthermore", he writes, "I avail myself of the privilege usually accorded to anyone who is the very first to describe a bird or animal and I give this bird the following name: *Larus marinus maximus ex albo, nigro et fusco varius, Groenlandicus*—a variegated Greenland gull of the largest kind with white, black and brownish-yellow feathers" (p. 54; author's translation). Mayor Anderson's cumbersome name for the Fulmar was never recognised. Carl Linnaeus used *Larus marinus* for the Great Black-backed Gull in 1758, and the Fulmar, which Linnaeus correctly established was not a gull, became *Fulmarus glacialis*.

NEW BIRDS FROM WEST GREENLAND

Although whalers from Dutch and German ports were soon followed through Davis Strait and up the west coast of Greenland by those from London, Hull and other places on the east coast of Britain, it was not until 1818 that reports on the local natural history comparable to those of Hamburg's learned mayor were made known in English. These reports, which included mention of hitherto unknown bird species, were published in 1818 as a result of voyages made in 1817 and 1818.

The first of these voyages was made by a certain Bernard O'Reilly, who took ship on the whaler *Thomas* of Hull on 8 March 1817 for the sake of science and the study of natural history. "This was his purpose", he tells us himself, "in undertaking a voyage hazardous in the extreme, cooped up with uninformed, unsociable beings. Nature was the grand object of his choice, and his sole consolation" (O'Reilly 1818: 5). In his journal he took the trouble daily, besides describing the weather, to mention the birds seen, and he also devoted a section of his chapter on "Arctic zoology" to birds. All the typical seabirds of the west Greenland coast are well described: the Eider, Cormorant, several gull species including the Ivory Gull, the Arctic Skua, the Black Guillemot, Brünnich's Guillemot and the Little Auk. Of the "fulmar petrel, or mallemuk" O'Reilly describes (pp. 118–121) four colour phases in exactly the same terms as used much later by James Fisher (1984: 268). His claims to be describing new species were not quite so dubious as those of Johann Anderson. But he had no right whatsoever to claim the "roch" or Little Auk as newly described and to give it a specific name: Linnaeus had called it *Alca alle* long before and Thomas Bewick had figured it with his usual accuracy in 1804 from a specimen "caught alive on the Durham coast". O'Reilly had not realised that the "Little Auk" of these authors, also called "Little Black and White Diver, Greenland Dove, or Sea-Turtle" according to Bewick, was the same bird as the *Rottjens* of the Dutch, corrupted into rotge, roche or roach by the English whalers. He had better reason to claim that his description (pp. 121–122) of "*Larus maximus* (burgomaster, or the white-winged gull)", namely the Glaucous Gull, was the first, because he does seem to have been the first English writer to mention it. However, it had already been noted and described from west Greenland by Otto Fabricius in 1780 in his *Fauna groenlandica* (Helms 1929: 142–143), which O'Reilly mentions. O'Reilly's real claim to fame is apparent when one looks up the Great Shearwater in a modern bird book. This bird's Latin name is: *Puffinus gravis* (O'Reilly). Actually O'Reilly's name was "*Procellaria gravis*" and he goes on "(the cape hen). This familiar name is given by the sailors to a *new species* of petrel, seen only in the latitude of Cape Farewell and Staten Hook, and somewhat farther eastward in the summer months" (O'Reilly 1818: 121). This one was indeed a new species. The Great Shearwater is still common in summer off the west Greenland coast. It was not until years after O'Reilly wrote that its south Atlantic breeding places were found in the Tristan da Cunha group of islands. Staten Hook (Statenhoek) was the Dutch name for Cape Farewell (Kap Farvel), the southern tip of Greenland.

Following in the wake of the whalers, the explorers, in the shape of the Royal Navy, sailed northwards up Greenland's west coast. The voyage of H.M. Ships *Isabella* and *Alexander* "for the purpose of exploring Baffin's Bay" in 1818 was commanded by John Ross (1819). The *Alexander*'s surgeon, Alexander Fisher (1819), compiled a list of 13 bird species encountered on this expedition in the general area of Greenland, and his captain described 20. Many of these, following the official instructions from the admiralty "to collect and preserve such specimens of the animal, vegetable and mineral kingdoms, as you can conveniently stow on board the ships" (Ross 1819: 11–12), were

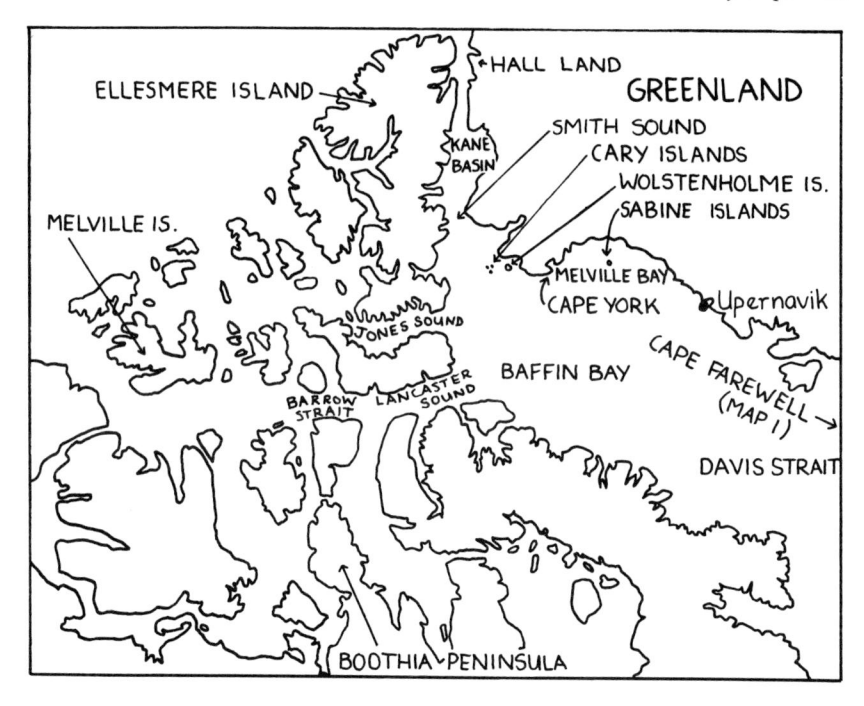

Map 4 The Northwest Passage: discovery ships and whalers in Greenland and Canada.

shot. But it was Captain Edward Sabine, of the Royal Artillery, who had been sent with the expedition on the recommendation of the Royal Society as naturalist and scientist, who made its one really important ornithological discovery. Landing on some islands in Melville Bay on 25 July 1818, which Ross named after Sabine, he discovered a mixed breeding colony of Arctic Terns and elegant fork-tailed gulls with dark grey heads. Specimens of these were sent home forthwith on a returning whaler and before the end of the year Sabine's Gull was made known to science at a meeting of the Linnaean Society of London by the captain's brother Joseph Sabine (Sabine, J. 1818). The bird was for a time placed in a special genus of its own *Xema*, but the latest authorities have returned it to the genus *Larus*. Its full scientific name, *Larus sabini* (Sabine), commemorates both brothers. After his return, Edward Sabine (Sabine, E. 1818) published "A memoir on the birds of Greenland, with descriptions and notes on the species observed on the late voyage of discovery in Davis Strait and Baffin's Bay" which appeared in the same volume of *Transactions* of the Linnaean Society as his brother's description of the new species. In this he described how a Glaucous Gull when shot disgorged a Little Auk and proved on dissection to have a second one in its stomach.

60 *In search of Arctic birds*

NAVAL BIRD COLLECTORS IN CANADIAN WATERS

John Ross, "Captain, Senior Officer, and Commander of the Expedition" as he emphatically described himself in 1818, took with him as commander of H.M.S. *Alexander*, Lieutenant William Edward Parry. These two, professional rivals and both of them subsequently knighted admirals, led expeditions into the Canadian Arctic from 1818 onwards to search for the Northwest Passage. Both of them collected birds and both habitually published an ornithological appendix to go with their published journals. Thus Edward Sabine (1824) listed the birds recorded on the 1819–1820 Parry (1821) voyage in the supplement to the appendix to Parry's journal of his 1819–1820 voyage; John Richardson (1825) contributed an appendix on zoology including birds to Parry's (1824) journal of the 1821–1823 voyage; and Ross's nephew James Clark Ross (1826) included 29 bird species in his appendix describing the zoology of Parry's (1826) third expedition of 1824–1825 (pp. 96–108) and 41 species in the appendix to John Ross's (1835) second expedition of 1829–1833 (J. C. Ross 1835). In this way a knowledge of the avifauna of the Canadian Arctic archipelago was slowly built up by what Alfred Newton subsequently described as "the far too meagre natural history supplements to the several 'voyages' of Parry and of Ross—works which excite regret at the glorious opportunities so ingloriously missed through the absence of special naturalists, and only redeemed from utter opprobrium by the zeal of volunteers" (Newton 1875: 94).

(*a*)

(b)

Plate 10 (*a*) Incubating Sabine's Gull, photographed on 5 July 1986 near Cambridge Bay. It belonged to a small colony of five pairs breeding on an islet in a shallow tundra lake. (*b*) The author's wife mobbed by Sabine's Gulls disturbed from their nests near Cambridge Bay. Referring to the discovery of this species on the west Greenland coast in 1818, John Ross (1819: lvii) recorded how, "uttering the same clamorous notes" as the Arctic Terns, it "flew, without fear, above its nest close to the head[s] of the party".

Besides these more or less official publications of Parry and Ross, we sometimes have accounts by other members of the expeditions. For example, the Alexander Fisher (1821) already mentioned was with Parry in 1819–1820 and frequently mentions birds in his journal. Passing near North Rona (Map 25) on 22 May 1819 they saw Fulmars and Kittiwakes and on 1 June, when they were 376 miles from Greenland, several Snow Buntings flew about the ship and Arctic Skuas were seen. On 2 June, in boisterous weather, Storm Petrels and shearwaters appeared in flocks and on 18 June, when the voyagers reached the ice, they began to see Little Auks, "the little divers, called rotges", and Black Guillemots, "called by the seamen, dovekey", and on 24 June an Ivory Gull. On land at or near Lancaster Sound, Ravens, Ringed Plovers, Snow Buntings and "ptarmigans" were recorded. After the winter spent on the south coast of Melville Island the first ptarmigan was seen on 12 May, and on 15 May a Raven and the first Snow Bunting were recorded. The author was convinced that the ptarmigans had not wintered in Melville Island, where, as a member of the inland expedition in June 1820, he saw ptarmigans, Arctic Skua, a Raven, Brent Geese, Sanderlings, American Golden Plovers, King Eiders, Long-tailed Ducks and Glaucous Gulls. Of particular interest is his mention of "some bank swallows" seen on 12 June. Could these have been Sand Martins? There are no other records of this species from so far north (Snyder 1957: 291).

Another unofficial journal, though by a less good naturalist than Alexander Fisher, was kept by the steward William Light on Ross's voyage of 1829–1833, when four successive winters were spent in the Canadian Arctic on or near Boothia Peninsula. Light evidently bore a grudge against Ross, but its expression was surely not his sole purpose in handing over what was in fact a detailed and interesting journal of events to the hack writer Robert Huish, who published it in 1835, together with other relevant material, in a 716-page book. Somewhere off Cape Farewell at the end of June and early in July 1829, they shot shearwaters and a Solan Goose or Gannet from the ship (p. 120). The average weight of the shearwaters was 32 ounces (907 g) and that of the Gannet, 6lb. 2 ounces (2783 g). The weight of the Gannet is about right; those for the shearwaters are correct for Great Shearwater, which is the only shearwater regularly frequenting Greenlandic waters. On 1 September Captain Ross shot an Ivory Gull, of which the steward gives an accurate description (p. 133). On 9 March 1830 a Black Guillemot was shot and, again, an accurate description of the bird's plumage is given, as well as of its breeding habits (p. 301). "It generally lays two eggs, about the size of a pullet's, of a dirty white with black spots. It makes its nest in the holes, which are found in the fragments of the rocks on the shore". In June 1830 Light gives an enthusiastic account of the beautiful King Eider and adds that "the sportsmen seldom returned without bringing either plovers, grouse, snipe, buntings, gulls or ducks". He complains that Captain Ross insisted that "every bird which was killed by any of the crew, should be brought into the cabin" so that he could obtain a specimen of every species encountered and "complete the ornithological history of the country". But he describes how this rule was easily circumvented, so that two crew members had better specimens of the King Eider than Ross himself and a gull and a plover "were cooked for the evening's repast, without having been subject to an examination in the [captain's] cabin, respecting their fitness to be received into the cabinet of curiosities" (Huish 1835: 434–436).

The culminating achievement of the naval bird collectors came later, in the middle years of the nineteenth century, when the ornithological zeal of individual officers was effectively coordinated by Mr John Barrow, son of that grand old man of the Royal Navy, Sir John Barrow, Bart., F.R.S., second secretary of the Admiralty from 1803 to 1845, author of a two-volume history of Arctic voyages and of a best-seller, *The eventful history of the mutiny and piratical seizure of H.M.S. Bounty*, and principal founder of the Royal Geographical Society. His son persuaded a group of naval officers engaged in Arctic exploration to procure specimens of birds for him and these, collected in 1848–1855, were later presented to the University Museum at Oxford. This "Arctic collection of birds" comprised 83 specimens of 48 species in 43 glass cases. Among these Case 9 was pre-eminent. It contained specimens of 11 birds of eight species collected in 1849 by Commander Moore of H.M.S. *Plover* around the shores of Providence Bay (bukhta Provideniya) (Map 5) on the south coast of the Chukchi Peninsula. One of these was a Spoon-billed Sandpiper in summer plumage, a bird at that time scarcely known to science and quite unknown in its breeding dress (Harting 1871, giving erroneous locality; see Portenko 1981: 4, 393).

BIRDS SHOT FOR THE POT

Not every commander of a British naval expedition to the Canadian Arctic in the nineteenth century followed Captain Ross's example in collecting specimens for ornithological purposes. On board H.M.S. *Assistance*, Captain H. W. Austin, in 1850–1851, the emphasis was on shooting for the pot. The handwritten newspaper *Aurora Borealis* published on board the *Assistance* on the fifteenth of every month, carried a report on "Articles of food found in the Arctic regions" (Anon. 1852: 340–343), the first part of which ran as follows:

> The Esquimaux, from dire necessity, have been taught the value of a well-filled larder; and these seal-clad residents of the Arctic circle employ the constant light of summer in laying in a store for the long dreary night. But without entering into the question as to the possibility of Europeans maintaining life upon the productions of the country, it will not be uninteresting or uninstructive to consider what are the varieties that have appeared upon the Arctic refreshment bill. To the feathered tribe we are chiefly indebted, and foremost in the list for flavour and delicacy of fibre stand the ptarmigan (*Ptarmichan tetrao lagopus*), and the willow grouse (*Tetrao saleceti*). These may be used in pie, stewed, boiled, or roast, at pleasure, and are easily shot. Next in gustatory joys the small birds rank, a kind of snipe and a curlew sandpiper, both are however rarely met with, and do not repay the trouble of procuring them.
>
> The brent goose (*Anser torquatus*) is excellent eating, and its flesh is free from fishy taste; then follow the little auk, or rotge (*Mergulus melanoleucus*), the dovekie, or black guillemot (*Uria grylle*), the loom, or thick-billed guillemot (*Uria brunnichii*). The first two are better baked with a crust, and the last makes, with spices and wine, a soup but little inferior to that of English hare. All these are found together in flocks, but the easiest method of obtaining them is either to shoot them at the cliffs, where they breed, or as they fly to and fro from their feeding-ground. The ducks now come upon the table, and are placed in the following order by most Polar epicures. The long-tailed duck (*Fuligula glacialis*), the king-duck (*Somateria spectabilis*), and the eider duck (*Somateria mollissima*). They require to be skinned before roasting or boiling, and are then eatable, but are always more or less fishy.
>
> The divers are by some thought superior for the table to the ducks but the difference is very slight. The red-throated diver (*Colymbus septentrionalis*) was most frequently seen, but few were shot; and of the great northern diver (*Colymbus glacialis*) none were brought to table, two only having been seen. Some of the gulls were eaten, and pronounced equal to the other sea-birds; they were the kittiwake (*Larus tridactylus*), the tern (*Sterna arctica*), and the herring or silver gull (*Larus argentatus*).

Over 20 bird species figure in the game lists also published in *Aurora Borealis* (Anon. 1852: 186, 347). Hardest hit were the auks: 1438 Little Auks fell victim to the ship's sportsmen, mostly off the coast of northwest Greenland, and 1080 Brünnich's Guillemots were killed, many of them at the Cary Islands. The lists also include 19 Snow Buntings, 64 Arctic Terns, 74 Ivory Gulls and a Snowy Owl. A Gyrfalcon was wounded but not obtained.

In September 1852 H.M.S. *Resolute* penetated westwards through Lancaster Sound and Barrow Strait as far as the south coast of Melville Island, where she wintered. Between 3 September 1852 and 9 September 1853 the ship's officers and men shot 711 ptarmigan (Ptarmigan or Willow Grouse?), 128 geese, 229 ducks and 16 plover. These birds surely made a significant and welcome contribution to the ship's company's fare, though its fresh meat came mostly from 114 muskoxen, 95 caribou and 146 hares (M'Dougall 1857).

The whalers, too, made depredations on the Arctic avifauna for culinary purposes. On the Dundee whaleship *Arctic*, in 1873, Commander Albert H. Markham, R.N., Fellow of the Royal Geographical Society, who had signed on for the voyage as second mate, noted in his journal (1875: 116) for Tuesday, 10 June, that:

> At 2 a.m. we came up to where the water was swarming with innumerable rotges [Little Auks]: a couple of boats were lowered, and four guns sent away to shoot for the "pot". They returned in less than twenty minutes with between three and four hundred. The little birds were swimming and flying in such thick clusters that *forty-five* were killed at one discharge from a gun, and *thirty three* at another! It sounds very much like murder.
>
> The flesh of these birds is excessively sweet, and they afford a very pleasing contrast at our meals to the continual beef and potatoes.

This was somewhere between the Cary Islands off northwest Greenland and Coburg Island in the entrance to Jones Sound.

EGG-COLLECTING WHALERS

Nineteenth-century English and Scottish whalers raided the seabird colonies along the west and northwest coasts of Greenland for fresh eggs. The most sought-after eggs were those of the Eider, which nested in great numbers on small low-lying offshore islands. In 1854 off Upernavik the famous Hull whaleship *Truelove* sent off a boat to collect duck's eggs which returned after 20 hours "with 30 dozen eggs and a quantity of eider ducks" (Barron 1890: 56). Egg-collecting raids on the Eider colony at Dalrymple Rock near Wolstenholme Island off the northwest Greenland coast by Dundee steam whalers must have been almost annual in the last quarter of the nineteenth century and early in the twentieth, to judge from the rather fragmentary information set out in Table 8.

In some years the whalers arrived too early. Thus in 1885 there were no eggs to be found on Dalrymple Rock on 14 June (*Polynia*); in 1895 the *Esquimaux* found none there on 2 June; and in 1900 on 10 June Captain William Adams

Whalers and discovery ships 65

Plate 11 Dalrymple Rock off the Avanersuaq coast in northwest Greenland, now a bird sanctuary, formerly a favourite target for egg-collecting whalers because of the large Eider colony there.

Table 8 Exploitation of Eiders and their eggs at Dalrymple Rock northwest Greenland, 1876–1904, by Dundee steam whalers. From logbooks and journals at the Broughty Castle Museum, Dundee, Town Docks Museum, Hull and Scott Polar Research Institute, Cambridge.

Date	Ship	Source of information	Takings
1876	*Erik*	T. F. Miller	200 dozen eggs
1884	*Nova Zembla*	Matt., Camp, surgeon	Eggs
1885	*Esquimaux*	log	Eggs
1886	*Esquimaux*	log	Eggs
3–4 July 1887	*Esquimaux*	Captain Milne	200 dozen eggs
1894	*Eclipse*	private journal, Savours 1960	Eggs
12 June 1900	*Esquimaux*	Captain McKay, log	2 dozen eggs
27 June 1903	*Diana*	Alexander Lamb	600 dozen eggs
			48 ducks
30 June 1904	*Diana*	Alexander Lamb	135 dozen eggs
			33 ducks
3 July 1904	*Diana*	Captain Adams, log	140 dozen eggs

in the *Diana* was too early. Competition between the whalers' egg-collecting and native exploitation of this source of food sometimes occurred. Thus in 1903 local Inuhuit Eskimos complained to the members of a Danish expedition that the Scottish whalers were rifling their caches of Eider eggs as well as taking eggs from nests.

"I SHOT THE ALBATROSS"

It was on a Peterhead steam whaler that the most remarkable ornithological exploit of the British whaling industry occurred, namely the shooting of the most northerly albatross ever encountered. The entry in the logbook of the *Eclipse*, Captain David Gray, for 15 June 1878, reads as follow (Lubbock 1937: 408):

> Lat. 80° 11'N; Long. 4° E. Shot an albatross, the only one I suppose ever seen here, a black-browed; spread of wing 6 feet 10 inches, length 2 feet, weight 10 lbs.

J. A. Harvie Brown visited Peterhead Museum in 1882 and found the bird there, correctly identified and labelled; Captain Gray had presented it to the museum (Løvenskiold 1964: 73 and 404). The Black-browed Albatross is a bird of the southern oceans; this example was shot nearly 100 km (60 miles) west of the northwest tip of Spitsbergen, further to the north than the species has ever been seen to the south. Apart from a second bird seen by Captain Gray in 1885 at 74° N, the Black-browed Albatross was not seen again in or even near the Arctic until 1935, when one was shot at sea off west Greenland at about 66° N. Since then there have been at least three more west Greenland records and in 1960 one was seen off the Lofoten Islands, Norway, at about 68° N (Bourne 1967). Thus, after more than a century, Captain Gray's record of the most northerly albatross still stands.

THE ORNITHOLOGY OF SMITH SOUND AND NORTHWARDS

In spite of repeated attempts to reach the North Pole by sea via Spitsbergen or Franz Josef Land, it was the so-called "American route", northwards through Smith Sound, which proved the most rewarding in terms both of exploration and ornithology. The first expedition to this area to publish an avifauna of any value was that organised by the United States government and led by C. F. Hall in the U.S. Ship *Polaris* in 1871–1873, which wintered north of 81° N. This contribution to Arctic ornithology was made by the German medical doctor Emil Bessels, who after studying medicine at Heidelberg University and zoology at those of Jena and Stuttgart, was appointed Chief Scientist of the expedition (Loomis 1972: 251–252). His annotated list of the birds of Hall Land, northwest Greenland, comprised 24 species, of which some 15 were thought to be breeding. He seems to have found nearly all the species that could be expected in that remote High Arctic area except possibly the Ringed

Plover. Perhaps his list (Bessels 1879: 311–312; incomplete in Bessels 1875) should have included one more species, the American Golden Plover, for the official account of Hall's expedition (Davis 1876: 391) describes how, on 12 July 1872, "The cook killed three dovekies [Black Guillemots], and Dr. Bessels bagged a golden plover". Further south, not far from Cape York in northwest Greenland on 19 June 1873, Bessels (1879: 478) identified a Ross's Gull at sea. It was with some Ivory Gulls, was smaller than them, and its movements reminded Bessels of a tern. Through the telescope he could "clearly distinguish the wedge-shaped tail and black ring on the neck". The difficulties encountered by Hall's expedition, which led to the loss of nearly all the eggs and bird skins Bessels had collected, also caused the explorers to live off the land. A typical entry in the journal (Davis 1876: 394) reads: "During low water, the men were allowed to go hunting on the plain. They brought in seven brent-geese, three goslings, six dovekies, the nest of a gull with five young ones, and two lemmings." On another occasion, on 4 June 1873, seven hungry men breakfasted on a ragout made from 40 Little Auks (p. 463).

Soon after the failure of this American expedition to reach the North Pole via Smith Sound, the Royal Navy sent a two-ship expedition with the same goal and along the same route, at the behest of the British government. The attainment of their principal aim, of reaching the North Pole, was not to prevent the pursuit of science, and a veritable team of naturalists, four men strong, led by British Ornithologists' Union member Henry W. Feilden, was divided equally between the two ships, and between naval (the two surgeon-naturalists E. L. Moss and R. W. Coppinger) and civilian personnel. The civilians, Feilden and H. Chichester Hart, one on each ship, were described as "two gentlemen" appointed "to serve as naturalists, with the same pay, clothing, and emoluments as the lieutenants of the expedition" (Feilden 1878: 314; see too Nares 1878a and b; Levere 1988). This team was issued with a closely-printed 86-page volume containing *Instructions for the use of the Scientific Expedition to the Arctic Regions, 1875, suggested by the Arctic Committee of the Royal Society*, and a 783-page *Manual of the natural history, geology, and physics of Greenland and the neighbouring regions* (Jones 1875). In this, ornithology was represented by two of the best-known experts of the day, P. L. Sclater and Alfred Newton. In the *Instructions* (p. 45) Sclater urged the collection and preservation of specimens of every species of bird met with. "Every specimen collected should be carefully labelled with a small paper or parchment ticket attached to the foot, on which the exact locality, date, and collector's name, as likewise the sex as ascertained by dissection, should be stated." The expedition's naturalists were to pay special attention to possible changes "in the general character of the avifauna as the expedition proceeds north", to recording all species seen and the extreme northern limit of each, and to the arrival and departure dates of migrants. Moreover, they were to keep a special lookout for certain little-known Arctic breeding birds whose eggs were rare or about whose nidification little or nothing was known, namely: the Grey Plover, Sanderling, Grey Phalarope and Knot among waders, and Sabine's and Ross's Gulls. Newton's contribution to the *Manual, Notes on birds which have been found in Greenland* (pp. 94–115), included the Eskimo names of many of them and a description of "The principal features by which each may be distinguished". His prediction that the expedition naturalists would probably find fewer than 30 species – the number hitherto recorded in Spitsbergen—turned out to be accurate: Feilden's list (Feilden 1877, Nares 1878b: 206–217), mainly from Ellesmere Island, contained the same number of species as Bessels's from Hall Land, namely 24. The two lists were largely identical, but Feilden had Ringed Plover, Grey Phalarope and an unidentified diver from the Ellesmere Island side of Nares Strait, not met with by Bessels in northwest Greenland, while Bessels had Purple Sandpiper, Sabine's Gull and Arctic Skua in Hall Land, missed by Feilden.

The British Arctic Expedition of 1875–1876 had failed to reach the Pole but had found breeding Knots and Sanderlings and brought back eggs of the Sanderling. It had also found the most northerly nests ever discovered of better-known species like the Brent Goose, Snowy Owl, and Snow Bunting, all at 81° 33' N, the Arctic Tern at 81° 50' N, and the Gyrfalcon at 79° 41'. It also had the merit of bringing back its ornithological collections and publishing its observations in full. Besides these ornithological successes, it too had eaten birds. The game book of one of the ships, H.M.S. *Alert*, mentioned 200 guillemots, 18 Black Guillemots, 14 Little Auks, 99 Eiders, 16 King Eiders, 10 Long-tailed Ducks, 207 Brent Geese and 27 Ptarmigans (Markham 1878: 424–425). One of the "gentlemen naturalists", H. C. Hart (1880), reported that "The eggs of the Brent Goose are perfectly delicious eating", and that the bird itself was "excellent eating", being "preferred on board H.M.S. *Discovery* to all other Arctic game".

NEWCOMB OF THE *JEANNETTE*

After so many expeditions had failed to reach the North Pole via Smith Sound and the strait between Canada and Greenland it was hardly surprising that James Gordon Bennett, proprietor of the *New York Herald*, who had sent two successful exploring expeditions to Africa under Henry M. Stanley, should agree to something new: an attempt on the North Pole via Bering Strait. Having bought the English steam yacht *Pandora*, which had twice voyaged in the Arctic, he re-named her *Jeannette*, refitted her and provisioned her for 3 years, and sent her off on 8 July 1879 from San Francisco crewed by officers and men of the United States navy under the command of Lieutenant George W. De Long (Guttridge 1988). Besides these professionals were two "civil scientists", one of whom was the expedition's official naturalist Raymond L. Newcomb (Newcomb 1888). He had been personally recommended by the honoured chief of American ornithology of the day, Professor Spencer Fullerton Baird and, though he turned out a little disappointing as an ornithologist, was by no means lacking in enthusiasm. In a letter to his wife on 13 July the commander of the *Jeannette* was able to report that as soon as Newcomb found his sea legs "he had his lines over the side fishing for albatross, and no sooner had he caught a good one, measuring seven feet across the wings, than he skin-

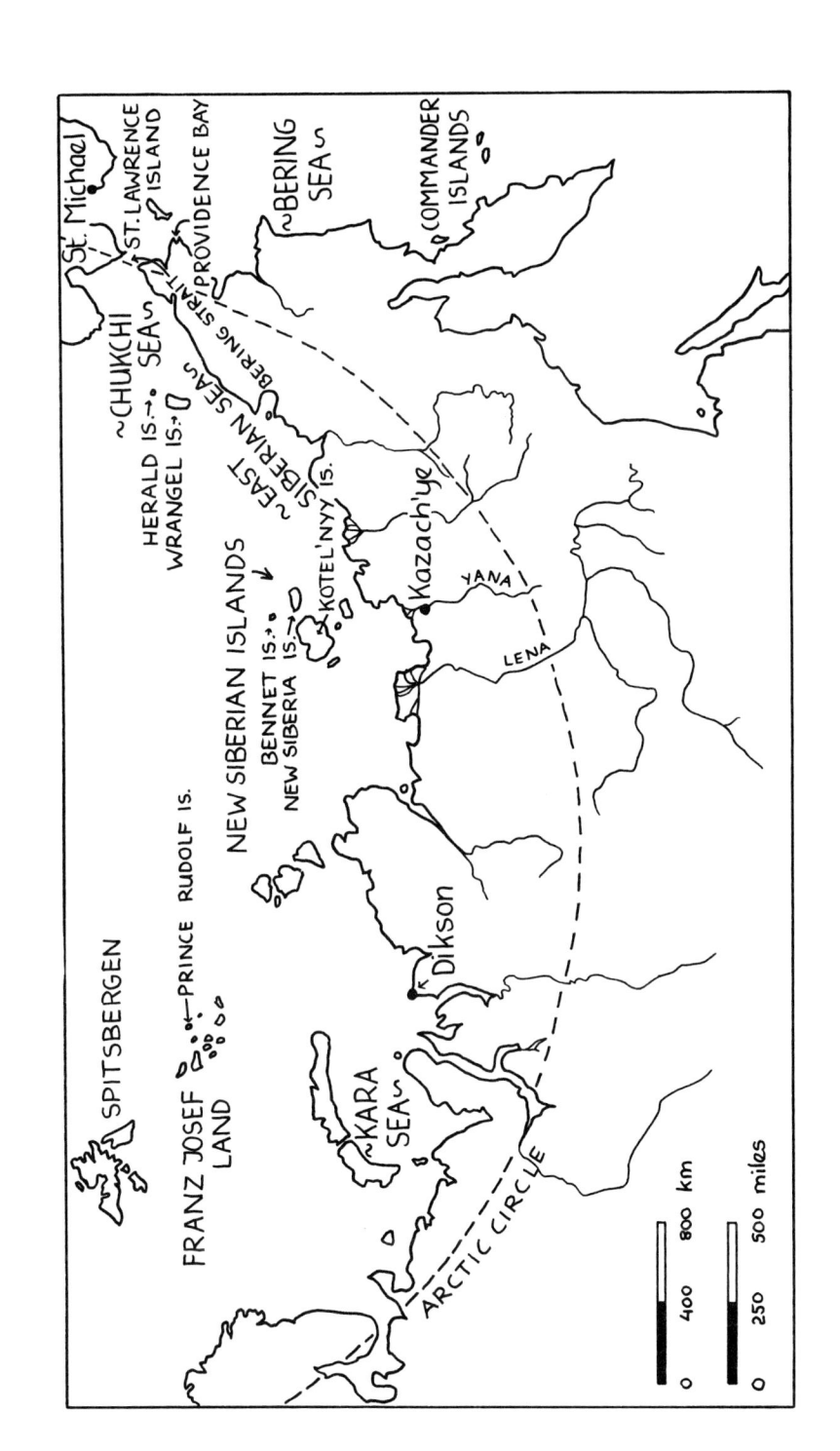

Map 5 Explorers north of Russia and Siberia.

ned it and got it ready for mounting" (De Long 1884: 80–81). Near Herald Island (ostrov Geral'd) at the beginning of September Newcomb watched his first Arctic birds: phalaropes in flocks of 6, 10 and 12, swimming buoyantly near the ship were "so interesting that I could watch them for hours"; but they were not specifically identified (Newcomb 1888: 279). He also saw "the lovely ivory gull in both adult and immature plumage" and, later that autumn, still not far from Herald Island, he obtained his first Ross's Gulls (p. 282). During the summer of 1880, while the ship zig-zagged to and fro some 200 miles north of the nearest land, which was Wrangel Island (ostrov Vrangelya), returning in November almost to the same position she had reached in April, Newcomb recorded Ross's and Ivory Gulls as well as Kittiwakes, Black and Brünnich's Guillemots, Grey Phalaropes, a sandpiper of uncertain species and two "stragglers from shore". One of these last was described as "a small sparrow"; the other was a Yellow Wagtail identified by Newcomb as *Budytes flava* (pp. 288–291).

By the following summer, 1881, the *Jeannette* had drifted in a northwesterly direction right across the East Siberian Sea and had reached a hitherto undiscovered group of small islands scattered far out in the Arctic Ocean to the northeast of the New Siberian Islands (Novosibirskiye ostrova). On one of these on 31 May Newcomb shot a male Snow Bunting. But his summer's ornithological work was brought to a sudden halt on 12 June when the *Jeannette* was crushed by ice and sank. Thereafter her crew were struggling over the ice and water for their very lives and birds were only good for eating. Nevertheless, Newcomb continued his ornithological observations as he and his colleagues made their way past the New Siberian Islands en route for the Lena Delta. At Bennett Island he saw Black Guillemots perched on patches of green vegetation on top of rock pinnacles cropping out from the mountainside and noisy Brünnich's Guillemots sitting in long rows on the cliff ledges. He shot 41 of these for the pot and another 125 were killed with stones thrown by the men. Later, he saw flocks of Long-tailed Ducks and shot 12 Sanderlings. In the first week of September, while they passed along the south coast of Kotel'nyy Island "a large white owl (*Nyctea nivea*), sitting alone and silent" could be seen at intervals (Newcomb 1888: 317). Besides this Snowy Owl, Newcomb recorded eiders, gulls and phalaropes along the island's coast. He was one of the lucky ones who survived the nightmare journey to the Lena River delta and was subsequently rescued. Though his bird notes were incomplete and fragmentary, none the less they were of value as the first published bird records from the East Siberian Sea and the New Siberian Islands.

THE CRUISE OF THE *CORWIN* IN 1881

In the spring of 1880 American attention was focussed on the Arctic seas north of Bering Strait by the failure of two whalers to return in the fall of 1879. As a result, the United States revenue cutter *Thomas L. Corwin* was ordered to search for these vessels and keep a look out for traces of the *Jeannette* during her routine cruise in Alaskan waters. She found nothing and was prevented by ice from reaching Herald and Wrangel Islands. A year later three relief expeditions were mounted by the United States government, but still there was no serious public anxiety about the *Jeannette*, and the aim of one of the expeditions, another cruise by the *Corwin*, was limited to landing on these two islands, which the *Jeannette* had planned to visit and where her commander might have left messages. This time, fortunately for ornithology, the *Corwin* stopped at St Michael on the west coast of Alaska on 30 June and took aboard Edward W. Nelson of the U.S. Army Signal Service who had been stationed there for 4 years. Then in his twenties, Nelson was later to put together important bird collections from Alaska and Mexico and serve a term as president of the American Ornithologists' Union. Though bound for home in summer 1881, he can have had no objection to enjoying, first, an extended cruise around the shores and islands of the Bering and Chukchi seas. The *Corwin* visited Provideniya in Providence Bay on the Chukchi Peninsula, St. Lawrence Island, Herald and Wrangel Islands, and then steamed along the Alaskan coast to Point Barrow (Map 9). While at St Michael, Nelson had gathered material on the local birds and this he wrote up in 1883 with the *Corwin* cruise observations in his *Birds of Bering Sea and the Arctic Ocean* (see too Newcomb 1888: 39–53 and Muir 1917 on the *Corwin* cruise).

For Nelson, the ornithological highlight of the *Corwin's* cruise in 1881 occurred on 26 June in Providence Bay on the south coast of the Chukchi Pen-

Plate 12 Adult Spoon-billed Sandpiper with recently-hatched chick photographed at the Belyaka Spit in the Chukchi Peninsula, Siberia, on 14 July 1987 by Pavel Tomkovich.

72 *In search of Arctic birds*

(a)

(b)

Plate 13 (a) Female Grey Phalarope in breeding plumage near Cambridge Bay. (b) Female Red-necked Phalarope in breeding plumage in northwest Greenland. In both species the female's plumage is more colourful than the male's. These elegant little birds are circumpolar in distribution, the Grey ranging further north than the Red-necked.

insula. On that day he secured "a fine adult female in breeding plumage" of the rare and unusual Spoon-billed Sandpiper, a bird to be mentioned again below when Nordenskiöld's explorations in the *Vega* come under notice. Nordenskiöld's and Nelson's specimens, collected in 1879 and 1881 respectively, were apparently the first in summer plumage since the already-mentioned bird in John Barrow's collection was shot at Providence Bay in 1849. Nelson points out that 23 of the 24 specimens of this species known to James E. Harting in 1869 were from southern India and were "doubtless all in winter plumage". Harting learned at that time that not all known specimens were authentic: he was told that the Spoon-billed Sandpiper supposed to be in the Paris Museum "is nothing less than a *Tringa* with the hind toes cut off and bill remodeled with the aid of some warm water" (Nelson 1887: 112).

Another noteworthy observation of Nelson's while on the *Corwin* was the presence of late summer feeding flocks of phalaropes far out to sea north of the shores of Siberia and Alaska, which confirmed Newcomb's record from the *Jeannette*. They were met with wherever there were leads of open water among the fields of ice, even in the ice-choked seas around Wrangel and Herald Islands. Both Grey and Red-necked Phalaropes occurred together, but the Grey became more numerous eastwards along the Siberian shore and northwards: "The few vessels which break the monotony of these northern waters in summer find dotting the waves on every hand these buoyant and graceful birds, their quick, agile, and elegant movements attracting attention, while their numbers render them conspicuous as they wheel and circle in flocks about the vessel, their wings flashing in the sunlight". Nelson goes on to explain that the American whalers in this region call both species "bowhead birds" from their habit of "feeding upon minute animalculae which afford the right whale or bowhead its food. Hence a community of interests attracts these pigmies and the largest cetacean of the North to prey on the same fare. A logical deduction follows, based upon experience, by which the whalers predict the presence of whales wherever this elegant bird is to be found in great numbers".

On 12 August 1881 Nelson and others made the first recorded landing on Wrangel Island, which lies some 100 miles (160 km) north of the Siberian coast. They also landed on Herald Island, 50 miles (80 km) east of Wrangel. On Wrangel Island a sailor handed Nelson the dried remains of a bird he had found on the hillside well above the tideline, which turned out to be a juvenile Brown Shrike—far north of its usual Siberian range. The live birds seen on or near these two islands in July–August 1881 can be inventoried by going through Nelson's above-mentioned work. The resulting first bird lists from these islands are set out in Table 9.

NORDENSKIÖLD AND NANSEN

In 1879, at the very moment when the *Jeannette* steamed northward through Bering Strait on her ill-starred cruise, the Swedish explorer A. E. Nordenskiöld brought his 300-ton ship *Vega* southwards through the same strait having successfully negotiated the Northeast Passage or Northern Sea Route

Table 9 Birds recorded on and near Herald and Wrangel Islands by E. W. Nelson (1883). Breeding species are italicised.

Herald Island, 30 July 1881	Wrangel Island, 12 August 1881
	Snow Bunting
	Snowy Owl
Turnstone	Turnstone
Grey Phalarope	Grey Phalarope
	Pacific Golden Plover
	King Eider
Red-faced Cormorant	Red-faced Cormorant
Kittiwake	*Kittiwake*
Glaucous Gull	*Glaucous Gull*
Glaucous-winged Gull	
Pomarine Skua (numerous)	Pomarine Skua
Horned Puffin (1)	
Crested Auklet (a few)	Crested Auklet (2 or 3)
Black Guillemot	*Black Guillemot*
Pigeon Guillemot (abundant)	*Pigeon Guillemot* (abundant)
Brünnich's Guillemot	*Brünnich's Guillemot*
(breeding in thousands)	(breeding in thousands)

from Tromsø to Yokohama. This dedicated natural scientist, veteran of over 20 years of Arctic explorations, naturally paid attention to ornithology. The Swedes had learnt to appreciate the culinary as well as scientific importance of birds on earlier Arctic expeditions. Medical officer Dr A. Envall, reporting to the Swedish government after the 1872–1873 Swedish Arctic Expedition had wintered in Spitsbergen, noted that the 150 Ptarmigan they shot in the autumn "formed a welcome delicacy". Rating the island's seabirds in order of palatability, Brünnich's Guillemot was given first place and the Little Auk second (Leslie 1879: 396).

In his somewhat discursive classic *The voyage of the* Vega *round Europe and Asia*, Nordenskiöld (1880: 100–125, 1881a: 107–132) gives a 25-page digression on the bird life of the Arctic, mostly based on his Spitsbergen experiences. Here too, attention is given to edibility: the egg of the Glaucous Gull is said to be delicious when boiled, and the white meat of the young is compared to that of a chicken. Elsewhere in the book he describes the low-lying lake-dotted landscape of Gusinaya Zemlya or Goose Land (Map 35) in southern Novaya Zemlya, breeding haunt of numerous swans, geese and waders. The landscape here was dotted with the very conspicuous nests of Bewick's Swans, which could be seen a long way off. They were built up to a height of more than half a metre with moss which the bird had pulled from the ground all round to a distance of 2 metres from the nest.

The *Vega* all but passed through the Northeast Passage in summer 1878. She was forced to winter only a few kilometres short of Bering Strait, near Kolyuchin Bay (Kolyuchinskaya guba) in the Chukchi Peninsula (Map 16).

Here in 1879 Nordenskiöld noted the return of bird life in the summer. The first Snow Bunting was seen on 23 April. Early in May the *Vega*'s rigging and decks thronged with newly-arrived Arctic Warblers, attracted no doubt by the fact that the ship was the only ice-free spot within sight. As spring advanced Spoon-billed Sandpipers were seen ashore, and at times this bird was so common around the ship "that it was twice served at the gunroom table, for which after our return we had to endure severe reproaches . . ." (Nordenskiöld 1881b: 44). Nordenskiöld attributed the disappearance of these birds at the end of June to the fact that they had migrated further north, but the author of the *Birds of the Chukchi Peninsula and Wrangel Island*, Russian ornithologist L. A. Portenko (1981: 394), draws the more obvious conclusion in implying that the Swedes, who had collected 20 of these birds as well as eating many, had wiped out the small local population.

In 1893 the Norwegian explorer Fridtjof Nansen took his specially constructed ship *Fram* along the northern coasts of Russia and Siberia more or less in the footsteps of Nordenskiöld as far as the New Siberian Islands but then deliberately allowed his ship to become beset in the ice. After the *Fram* had remained locked in the ice, drifting slowly northwestwards, during two successive winters and the intervening summer, Nansen himself left the ship with Hjalmar Johansen on a bid to sledge to the North Pole. This failed, and while the two explorers returned to civilisation via Franz Josef Land, the *Fram* with her crew drifted on, reaching 85° 55' N before turning southwards until she emerged from the ice off the north coast of Spitsbergen on 13 August 1896, on the very same day that Nansen and his companion arrived in Norway from Franz Josef Land on the English yacht *Windward*.

Originally a biologist working on the nervous system of a group of worms and subsequently on the embryology of whales (Reynolds 1949, Shackleton 1959), Nansen saw to it that the scientific results of the Norwegian North Polar Expedition of 1893–1896 were published in full in English in a beautifully-produced series of quarto volumes which vied with the five volumes of scientific observations made by the *Vega* expedition published in Swedish by Nordenskiöld. As far as birds are concerned, however, the slim volume which Nansen co-authored with Norwegian ornithologist Robert Collett, simply recording observations, cannot compare with the 270-page masterpiece on the avifauna of the Eurasian shore of the Arctic Ocean by Swedish ornithologist J. A. Palmén, which made up Vol. 5 of Nordenskiöld's results. The most notable contribution to ornithology of Collett and Nansen's 55-page booklet (Nansen 1899) was its announcement of the surprisingly varied and abundant bird life encountered in the frozen wastes of the Arctic Ocean 300 km (almost 200 miles) north of Franz Josef Land and north of 84° N, further north than any birds hitherto recorded. This was between 14 May and 14 September 1895, after Nansen had left the *Fram*. The ten species seen that summer were the Fulmar, a skua and a black-backed gull of uncertain species, the Kittiwake, Ross's Gull, the Ivory Gull, Arctic Tern, Black Guillemot, Little Auk and surprisingly, Snow Bunting. Most northerly of all these was a Fulmar at 85° 5' N. Snow Buntings were seen on four dates between 22 May and 19 June, single birds on three occasions, two together on one. In the following summer, when the *Fram* was drifting south between 83° and 81° N directly towards Spitsbergen and some 300 km from it, a pair of Snow Buntings took up residence on the ship for a time and were seen carrying nest material. They fed on the ship's refuse piled on the ice and when it drifted away from this were not seen again. Other birds noted this summer were Eider, Ringed Plover, Grey Phalarope, Pomarine Skua, Sabine's Gull and Puffin.

ITALIANS IN THE FAR NORTH

In 1899 an Italian expedition arrived in Franz Josef Land in a Norwegian sealer re-named *Stella Polare* or *Pole Star* (Abruzzi 1903). It was intent on voyaging as far north from Franz Josef Land as possible, when parties would then sledge over the ice to the North Pole. Its leader was a renowned mountaineer and Himalayan explorer, Prince Luigi Amedeo of Savoy, duke of the Abruzzi. Lieutenant Umberto Cagni, of the Italian navy, succeeded in travelling further north than any other man before him, beating Nansen's farthest north by 22 or 23 miles, but the Pole eluded him. In 1903 the expedition's scientific observations were published in a quarto volume (Cagni and Cavalli-Molinelli 1903). Among the extensive meteorological and other data was an eleven-page appendix on birds by Tommaso Salvadori (pp. 597–607). This was an annotated list by a well-known Italian ornithologist of the 38 specimens of ten species collected during the expedition by Dr Cavalli, chiefly from the most northerly island of the archipelago, namely Prince Rudolf Island (ostrov Rudol'fa). Although they failed to find the coveted Ross's Gull, the Italians added six species to the two already known from this very seldom-visited island. Besides the Little Auk and Black Guillemot which had been noted there by an Austrian expedition, they were able to add Fulmar, Ivory Gull, Kittiwake, Glaucous Gull, Arctic Skua and Snow Bunting. They also collected the first Pomarine Skua seen in Franz Josef Island. All this amounted to but little, but at least it was published in full.

THE VOYAGE OF THE *ZARYA*, 1900–1903

Not to be outdone by the achievements of the Scandinavian explorers along the northern shores of their country, the Russians mounted a Polar Expedition in 1899–1900 under the leadership of veteran Arctic explorer Eduard Vasil'evich Toll, alias Baron Eduard von Toll, who was born in Tallinn, Estonia, and whose mother-tongue was German though he referred to himself and his colleagues as "we Russians". The declared aim of the expedition (Barr 1981), which was organised by the Imperial Academy of Sciences at St Petersburg, later Leningrad, was to search for Sannikov Land, a supposedly large and mountainous land or archipelago first seen early in the nineteenth century by Russian trader Yakov Sannikov to the north of Kotel'nyy Island in the New Siberian Islands. The expedition travelled in a Norwegian sealer, a schooner with auxiliary steam power, which was ice-strengthened by Colin Archer, designer of the *Fram*, and re-named the *Zarya (Dawn)*. Toll, who was trained as a geologist, at first proposed taking only three scientific colleagues—an astronomer, a meteorologist and a surveyor. He ended up with all of these, but took an accomplished professional biologist from the Academy's Zoological Museum along too, in the person of A. Birulya.

At first, all went well. At Kotel'nyy Island the expedition came across a scientific Robinson Crusoe in the shape of a left-wing student Constantin Volosovich. Exiled on political grounds to Yakutsk in 1898, Volosovich had obtained permission in 1899 to withdraw alone to Kotel'nyy Island and set up a biological observatory there. When Toll arrived in autumn 1901 Volosovich had seen no human beings at all for 2 years, apart from a couple of Yakuts who had brought him supplies each summer. During his lonely vigil he had kept notes of all the birds he had seen. The baron and the student struck up a friendship and sledged inland together in January 1902. But then misfortune struck the expedition. Its physician, Hermann Walter, who had been responsible for ornithological observations, died of a heart attack apparently caused by an inadvertent overdose of digitalis, administered by himself. Then, the plans for field work in the New Siberian Islands in summer 1902 went awry. The *Zarya*, which should have collected Toll and four companions from Bennett Island and Birulya with his companions from the island of New Siberia (ostrov Novaya Sibir), at the end of the summer, was prevented by ice from reaching them. Toll's division of the expedition was lost without trace, but Birulya, after lingering on the west coast of New Siberia Island until December for the ice to thicken, and having in the early autumn shot sufficient reindeer to feed himself, his companions and the dogs, successfully led his group to safety in the town of Kazach'ye on the River Yana, covering in 25 days a distance of 600 km (nearly 400 miles) as the crow flies.

In spite of these disasters and the failure to find Sannikov Land, which proved to be yet another Arctic mirage, the Russian Polar Expedition of 1900–1903 was ornithologically quite outstanding. Naturally birds were eaten as well as collected. Already while crossing the Kara Sea, on 8 August, its members were savouring Long-tailed Duck soup. At Dikson a week later Toll brought in "a White-fronted Goose for the table and a Purple Sandpiper for the zoologists". At the end of August, on the Taimyr coast, Dr Walter returned with his "game bag full of waders, of which only a few rarer ones found their way into the collection: most went to the kitchen and were served for our midday meal today" (Toll 1909: 72). The expedition's ornithologist seems to have preferred deer-stalking to other field activities but, on one of what Toll calls his zoological hunting expeditions, he shot 12 geese on a single lake.

The ornithological star of the expedition, turned out to be not its official ornithologist, Dr Walter, but the zoologist of the Imperial Academy's Zoological Museum, the A. Birulya just referred to, whose official responsibilities were "mammals (his specialty at the museum), all invertebrate animals and botany" (Pleske 1928: 113), namely, everything except birds. While still busy with expedition papers on the biology of the reindeer and of decapods, Birulya brought out his remarkable *Sketches of the bird life of the Siberian Polar shores* in 1907, as the second in the series of scientific publications of the expedition. This 157-page large quarto book, illustrated with excellent photographs, contained the first detailed accounts of the breeding behaviour of the Curlew Sandpiper and Knot, but it remained tucked away in Russian in the publications of the Zoological Museum until a large part of it appeared in English in 1928 in Theodore (alias Fedor) Pleske's magnificent volume *Birds of the Eurasian tundra*. That work represents the official and definitive publication of the scientific results of the Russian Polar Expedition of 1900–1903. In it full details were given of the 16 eggs of the Curlew Sandpiper collected by the expedition and safely conveyed to the St Petersburg Zoological Museum. Up to that time, only two clutches had ever been taken. Even more highly prized were the seven sets of Knot's eggs and 26 young in down brought back by the expedition.

A year after the disappearance of Toll and his companions, who most probably lost their lives after falling through thin ice while crossing the 150 km (100 miles) of frozen sea between Bennett Island and Kotel'nyy Island, a list of the birds of Bennett Island, seen between 3 August and 8 November, was recovered at his last camp. This document shows that the explorer still believed in Sannikov Land, and, now that we know Sannikov Land does not exist, points to the possibility of migration across the Arctic Ocean which we now know *does* exist (Toll 1909: 591; see Uspenskii 1986: 222–223):

The following birds lived on the island: five species of gull, among them Ross's Gull, but only juveniles; two species of guillemot, a phalarope, the Snow Bunting. As migrants the following appeared: a White-tailed Eagle, which flew from south to north, a Peregrine which came from the north and flew southwards, and flocks of geese which likewise flew from north to south. Because of the obscured horizon, the land from which these birds came could no more be seen than could Sannikov Land during the *Zarya*'s cruise in the previous year.

CHAPTER 4

Falcons from the Arctic

Highly prized for falconry and favourite subject of bird artists, the fast-flying hunting falcon has always enjoyed a secure place in the literature and imagination of mankind. One species, the Gyrfalcon, is a true Arctic product which has long been valued by falconers, collectors, and museum curators and admired by birdwatchers and ornithologists. The rare, large and legendarily beautiful white form of the Gyrfalcon has lent the species an extra lustre. There can scarcely be any more impressive sight for an ornithologist than one of these spectacular birds winging its way over the desolate north Greenland landscape in the teeth of an Arctic gale.

THE NATURAL HISTORY OF THE GYRFALCON

Through much of its range the Gyrfalcon breeds further north than any of the falcons. It is found all round the Pole along the shore of the Arctic Ocean and inland from it, reaching as far north as almost any other bird in North America and Greenland. It does not, however, nest on any of the Eurasian Arctic archipelagos except perhaps Novaya Zemlya. Its breeding range extends southwards well into the sub-Arctic, namely south to about 60° N in North America and in Norway, and to 55° N in the Kamchatka Peninsula (Glutz von Blotzheim *et al*. 1971). Breeding Gyrfalcons are scattered over a vast area of Arctic seaboard, tundra and mountain, and in parts of the Soviet Union and elsewhere they penetrate the northern fringe of the taiga, breeding in trees. In Mount McKinley National Park in southern Alaska, now called Denali, there are five known eyries, but no more than three pairs of Gyrfalcons ever breed in the area, which has over 5000 km^2 (2000 square miles) of suitable habitat. Greater densities of breeding Gyrfalcons have been found in Iceland, but even there, in a record year, eyries are so spread out that there is likely to be only one in every 150 to 300 km^2 (58–115 square miles) (Cade 1982: 78–79).

With a length of 50–60 cm (20–24 inches) and a weight of 1–15 kg (Cramp and Simmons 1980) the Gyrfalcon is comparable in size to a Mallard. Its daily food requirement has been put at about 200 or 250 g (Dement'ev *et al*. 1966; Dementiew 1960). Its habitual prey on the tundra is the Ptarmigan or Willow Grouse, but it is quite able to rear a healthy brood of young on a diet of lemmings or, in the absence of lemmings, Snow Buntings and Arctic hares (Summers and Green 1974). Where it breeds near seabird colonies, it naturally feeds on seabirds. At Kharlov Island, one of the Seven Islands (Sem' ostrovov), on the east Murman Coast, the diet of the local Gyrfalcons included Razorbills, Puffins, Black Guillemots, Kittiwakes and Herring Gulls (Dementiew 1960; Bannerman 1956). At a plucking place near a Gyrfalcon's nest on a bird cliff in west Greenland hundreds of carcasses of Kittiwakes and Glaucous Gulls were found (Salomonsen 1950). Exceptionally, swans, Snowy Owls and even young Arctic Foxes have fallen prey to the Gyrfalcon (Glutz von Blozheim *et al*. 1971) and in wooded areas they hunt the Hazel Grouse and capercaillies, as well as smaller species.

The Gyrfalcon habitually flies fast and low over the ground when hunting birds, rapidly covering long distances so that the chance of taking its prey by surprise is enhanced. Birds and mammals may be swooped onto directly from the air while they are on the ground, but the aerial stoop is likewise often employed, the prey being seized or struck down with the talons. It is said that, before feeding, the Gyrfalcon always breaks the neck of its victim with its beak. It may tear the head and legs off a bird it has struck down and, after eating the meat off the breast and smashing the breast bone in several places with its powerful beak, may leave the wings lying on the ground, often still joined together, with the breast bone upward.

Although some authors, among them John Wolley to be quoted later in this chapter, have maintained that the Gyrfalcon occasionally builds its own nest, this is unlikely. In common with most other falcons, it probably always lays its eggs in a scrape on the bare ground or in the old nest of another bird. In the case of the Gyrfalcon the bare ground is usually that of a broad cliff ledge protected from spring snowfalls by a substantial overhang. Salomonsen (1950) suggests that in northern Greenland nest sites may face north "to take advantage of the

82 *In search of Arctic birds*

(a)

(b)

Plate 14 Young Gyrfalcons ready to fly. Photographed at Pokhodskaya Edoma, a rocky outcrop, in the Kolyma Delta, by Aleksandr Andreev.

heat of the sun at midnight". Five nests found by Danish ornithologist A. L. V. Manniche in Germania Land all faced north, but the only nest I saw in Avanersuaq (Map 1) faced south. Further south down the west coast, around Uummannaq, 26 nests faced in all different directions, but at Disko, Salomonsen's nests faced south or west (see Map 2). The position of an occupied cliff eyrie is marked from afar by a white patch immediately below the nest caused by the birds' droppings or mutes. Gyrfalcons often return to the same site but not necessarily year after year: on Kharlov Island falconers knew of the eyrie there as early as the seventeenth century (Dementiew 1960). At an eyrie in west Greenland the eggs were on a layer of guano over 2 m (6 feet) thick, apparently all of it produced by the falcons over a period of many years or even centuries (Burnham and Mattox 1984). Where the cliff-breeding Gyrfalcon lays in an old nest this is nearly always that of the Raven. In areas where nests are in trees, old nests of the Rough-legged Buzzard are often used (Glutz von Blotzheim *et al*. 1971).

The Gyrfalcon lays three to five rounded eggs of a mainly reddish-brown colour from the first half of April in the south of its range; in May further north. The provisioning of the female or falcon by the male or tiercel begins during courtship, before the first egg is laid, and continues until the young fledge. The incubation period is about 5 weeks, the young fledge around 7 weeks after hatching, and the falcon is totally dependent on her mate for food during most of this period (Cramp and Simmons 1980).

THE GYRFALCON IN MEDIEVAL HISTORY

Because of its role in falconry, the Gyrfalcon figures predominantly in the historical record. Not that one can say that it was always the favourite bird of this sport; after all, the Merlin was preferred for hunting Skylarks in England; the Golden Eagle was and perhaps still is flown at wolves from horseback in central Asia (Campbell and Lack 1985: 204); and the bird on King Harold's fist in the Bayeux Tapestry has been identified as a Goshawk (Yapp 1987: 30–31). Still, the Gyrfalcon was invariably given precedence in the manuals of falconry in western Europe. Thus the emperor Frederick II of Hohenstaufen, in his famous thirteenth-century treatise *On the art of hunting with birds (De arte venandi cum avibus)* explains that "Out of respect for their size, strength, audacity, and swiftness, the gerfalcons shall be given first place in our treatise" (Wood and Fyfe 1943: 111). He also says of the Gyrfalcon that it was imported from "a certain island lying between Norway and Gallandia [Greenland], called in Teutonic speech Yslandia" and that "in our experience the rare white varieties from remote regions are the best" (p. 121) and this is confirmed in numerous medieval references to Gyrfalcons both in literary works and in accounts and other documents.

In medieval literature, Gyrfalcons, both the white and the grey, *Des gerfaus, de blans et de bis* as a famous fourteenth-century French poem on falconry has it

84　*In search of Arctic birds*

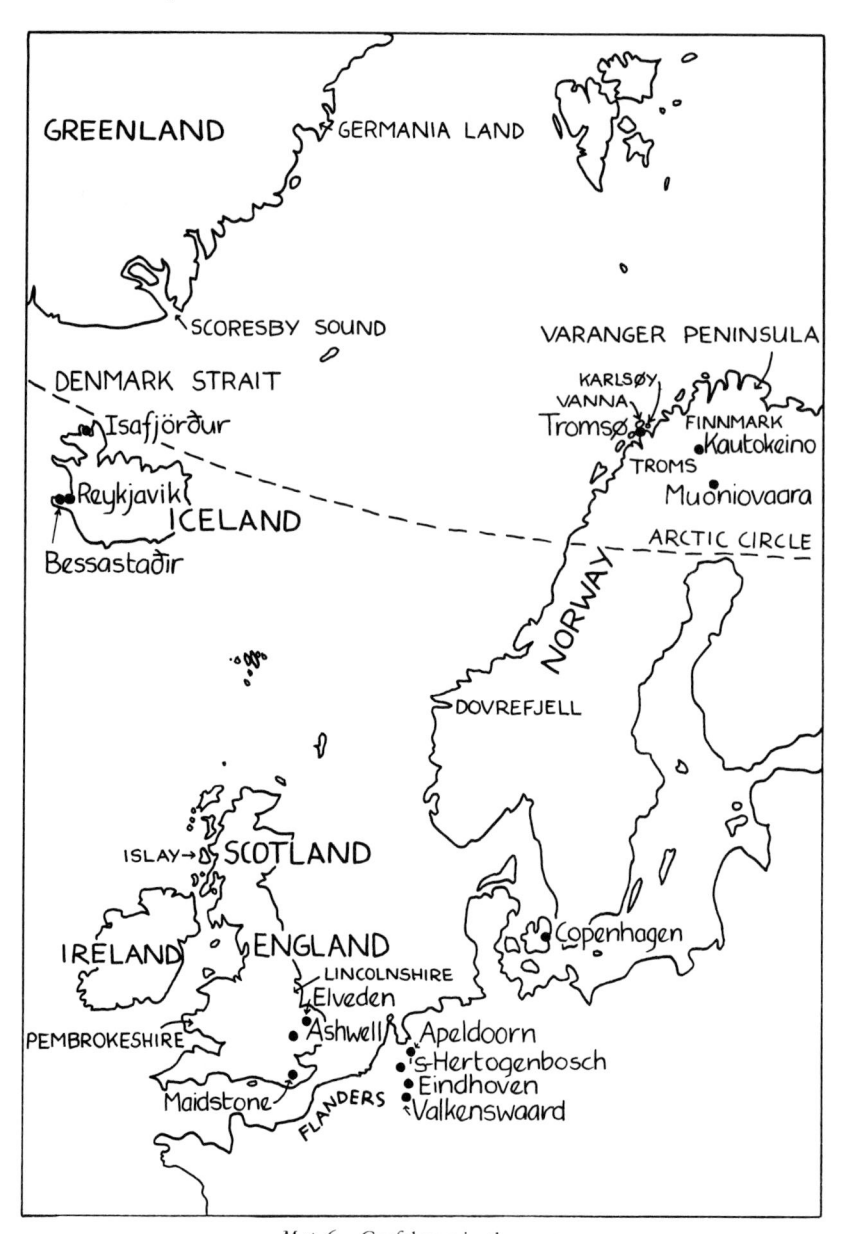

Map 6 Gyrfalcons in the west.

(Blomqvist 1951; and see Hensel 1909: 634), are often mentioned. In Britain around A.D. 1200, Gerald of Wales knew of the import of Gyrfalcons from Iceland (Dimock 1867: 95–96) while Alexander Neckham thought they came from across the sea with the wild geese (Wright 1863). A contemporary of

Frederick II, the German friar Albertus Magnus or Albert the Great, in his work on animals, mentions the Gyrfalcon and the white falcon which "comes from the north and the ocean sea, from the regions of Norway and Sweden and Estonia and the neighbouring woods and mountains" (Oggins 1980: 458), and the Venetian traveller Marco Polo refers to Gyrfalcons in his famous description of Asia as being produced in large numbers in the far north and used for falconry by the Great Khan, Kublai Khan, who was supposed to have had 200 of them (Vaughan 1982: 329).

Besides these literary references to Gyrfalcons in the middle ages there are plenty of documents to show that they were being exported from both Iceland and Norway and that the king of Norway made a habit of presenting these precious birds to his fellow rulers (Vaughan 1982: 331). Again white falcons were the most esteemed. In the fourteenth century, popes were wooed with gifts of white and grey Gyrfalcons by Scandinavian rulers (Hofmann 1957; Tillisch 1949: 81). There are references to Gyrfalcons in the voluminous archives of English medieval kings. In 1139 a certain Duti of Lincolnshire had to pay what seems to have been an extraordinarily heavy fine to the exchequer of 100 Norwegian hawks (probably Goshawks) and 100 Gyrfalcons. Six of the Gyrfalcons and four of the hawks were to be white ones, and if he could not obtain the white hawks, he was to substitute white Gyrfalcons (Gurney 1921: 42). King Henry II used Gyrfalcons for falconry in 1172 in Pembrokeshire (now Dyfed). In 1212 King John killed seven cranes near Ashwell in Cambridgeshire with Gyrfalcons, and in that or the following year, he took nine more in Lincolnshire (Dementiew 1960: 51; Gurney 1921: 49).

The highly-prized white Gyrfalcon appeared in legend, too, in the middle ages. According to tradition, a certain Baaf or Bavo, who lived in Flanders and is supposed to have died on 1 October 655, was accused of stealing a white falcon. He was sentenced to death by hanging but saved in the nick of time by the re-appearance of the lost bird, which swooped down and perched on the gallows prepared for his execution (Swaen 1937: 30–31; Dementiew 1960: 46). Similar in many ways is the much better documented story of Ivan the Terrible's youthful falconer Trifon Patrikeyev. While he was hawking at the village of Naprudny, now incorporated into Moscow, the tsar's favourite white falcon flew off and disappeared. Ivan gave him three days to recover it. At the very last moment Trifon's patron saint and namesake Trifon (who had been martyred at Nicaea under Decius in 251 A.D.) appeared to him in a vision with the lost falcon on his fist, and told him where to find it. Hence the many later representations of Saint Trifon with a white Gyrfalcon on his gloved hand (Dementiew 1960: 46–47).

SUBSPECIES AND MIGRATIONS

How did the medieval falconers obtain their birds and in particular, where did the valued white ones which, on average, are larger than the grey ones, come from? It should be explained at the outset that there has been much discussion among experts and others about the different varieties of Gyrfalcon, for this bird varies in size, and in colour from white to dark grey-brown. In spite of assertions to the contrary (for example, Wood and Fife 1943: 509), the whiteness has nothing to do with age. Nor is it at all acceptable nowadays to classify white birds as a separate subspecies or race. Earlier authors, up to and including Russian ornithologist Georgy Petrovich Dement'ev, whose monograph on the Gyrfalcon, already referred to several times, was published in German in 1960 under the name Dementiew, split the species into distinct geographical races. His list of Gyrfalcon subspecies is shown in Table 10.

Since the publication of Dement'ev's monograph ornithological opinion has swung towards regarding the Gyrfalcon as monotypic, that is, not divisible into subspecies or races (Cramp and Simmons 1980). Nevertheless, within this single species it is now accepted that three more or less clearly defined colour phases or morphs occur: the Gyrfalcon is trimorphic. In general, dark grey-brown birds predominate in the Boreal and sub-Arctic or more southerly parts of the range; grey birds predominate in the middle part of the range; and white birds predominate in the far north, namely in the High Arctic. This is true of the Gyrfalcons breeding between Labrador and Avanersuaq in northwest Greenland, where there is a south–north gradation or cline from dark grey-brown to pure white. But in Eurasia the gradation from dark to light birds also moves from west to east, white birds becoming increasingly numerous as one travels eastwards along the north Siberian coast. Refinements and standardisation of measuring techniques may eventually make it possible to identify the geographical origin of Gyrfalcons wandering outside their breeding range but this cannot with certainty be done at the moment (Glutz von Blotzheim *et al.* 1971).

Table 10 Races of the Gyrfalcon *Falco rusticolus* according to Dement'ev (Dementiew 1960) with added detail from Glutz von Blotzheim *et al.* 1971.

Subspecies name	Plumage colour	Average wing length of females (mm)	Breeding range
rusticolus	Grey birds only	396	Scandinavia to Kanin Peninsula
intermedius	Pale grey, some white birds	397.5	Pechora basin to Yenisey
grebnitzkii	Very pale grey 50% white birds	410	Lena Delta to Kamchatka and Commander Islands
obsoletus	Dark grey; very few white birds	393.6 405.8	Arctic North America Labrador and south Greenland
candicans	Almost exclusively white birds	400.3	Middle and north Greenland
islandicus	Grey birds only	402.8	Iceland

In the case of the Gyrfalcon, as with many other birds in the northern hemisphere, the more northerly breeding populations are migratory, while the more southerly ones are not; and the intensity and distance of the migration varies from year to year. It is the predominantly white Gyrfalcons from the far north that travel southwards from their breeding range in winter. Thus, out of 86 Gyrfalcons recorded in Ireland up to 1965, 81 were white birds assumed to have come from the northern half of Greenland (Glutz von Blotzheim *et al.* 1971). The same goes for the Eurasian white Gyrfalcons. The "large white falcons" reported to the authors of *Falconry in the British Isles* (Salvin and Brodrick 1855) by travellers to have been annually caught while on passage over the Caspian Sea, and believed by them to have been Greenland falcons, were in fact probably birds from the far northeast of Siberia.

The migratory flight of white Gyrfalcons is best documented from Greenland and Iceland. Between 30 August and 18 September 1937 some two or three hundred white falcons were seen flying south from a ship beset in the ice off Scoresby Sound (Glutz von Blotzheim *et al.* 1971). These were evidently birds which had bred in northeast Greenland, making their way down the coast and crossing Denmark Strait to their winter quarters in Iceland. A similar phenomenon occurs on the other side of Greenland. In autumn 1967 two American ornithologists trapped 14 Gyrfalcons at Disko on the west Greenland coast. Most of them were white birds, evidently migrating southwards from northwest Greenland, where all the Gyrfalcons are white (Harris 1979: 119; Burnham and Mattox 1984: 23).

SOURCES OF SUPPLY

Falconers have through the centuries made use both of young birds taken from the nest shortly before fledging and of juvenile, immature or fully adult falcons caught with decoy and net while on passage. It seems virtually certain that most of the white Gyrfalcons obtained by western European falconers were trapped in the autumn or winter in Iceland and that they originated from east or northwest Greenland. Danish royal official Niels Horrebov described in 1752 how the falcons in Iceland are of three colours: white, half-white, and grey, and he explains that, besides the birds nesting in Iceland, some falcons come over in the winter from Greenland which are mostly white (Horrebov 1752: 147–154).

The export of Gyrfalcons from Iceland to Denmark and elsewhere in western Europe is well attested from the sixteenth century on. Still, it must not be assumed, when we read of Gyrfalcons being acquired "from Norway" by the duke of Burgundy in the fifteenth century (Le Glay *et al.* 1873: 175) or of an official in Brabant sending "to Denmark" in 1502 to obtain falcons, including Gyrfalcons (Oorschot 1974: 15), that these invariably originated in Iceland. Norway itself was also an important source of supply for falconers who needed Gyrfalcons. By the middle of the sixteenth century falcon-trapping stations in Finnmark and Troms, north Norway, were being rented out by the king.

88   *In search of Arctic birds*

Traces of them have been found in the Varanger Peninsula and on the islands of Vanna and Karlsøy north of Tromsø. There are even references to white falcons from this area (Bratrein 1986), though only three records of white Gyrfalcons in Norway were known to ornithology when Svein Haftorn's work on Norwegian birds was published in 1971. Further south, on the Dovrefjell in about 1840, when E. C. Newcome was digging foundations for huts for falcon-catching, he came across the remains of some much older huts. The Dutch falconers he employed here caught 3 Gyrfalcons in one year and 10 or 12 in the following year (Lascelles 1892: 86). In 1876 the Scottish falconer John Barr caught 10 Gyrfalcons in Norway for the Falconry Club but almost all died within a year (Lascelles 1892: 88).

Another source of Gyrfalcons, including white ones, was northern Russia and Siberia. Already in the fourteenth century the king of Hungary was using Russian falconers and Russian Gyrfalcons. In 1475 the duke of Milan sent to Russia for white Gyrfalcons. At this time the entrepot for these and other products of the Eurasian Arctic was Novgorod: Ivan III was presented with seven Gyrfalcons when he visited the city in 1476–1477. Later records reveal exactly where Russian Gyrfalcons were obtained. In May 1670 an imperial official issued instructions for the falconer Nesterka Yevdokimov to go with nine companions to the places along the Murman Coast where Gyrfalcons were taken (probably as eyasses), including the Seven Islands and the River Voron'ya. The birds were to be fed on reindeer meat and transported carefully to Kholmogory near Archangel (Arkhangel'sk) on the first leg of their journey by sleigh to Moscow. By 1675, too, Gyrfalcons were being taken further east,

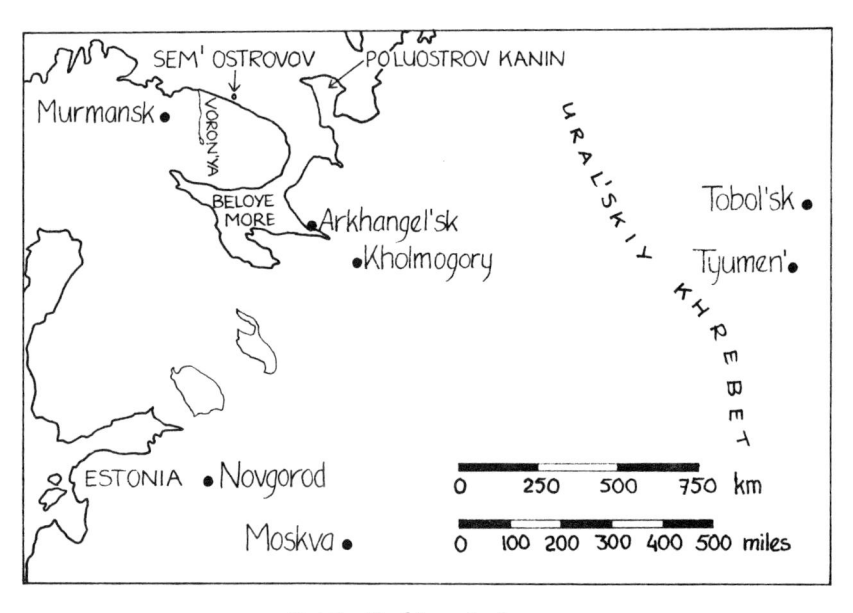

Map 7   Gyrfalcons in the east.

beyond the Urals (Ural'skiy khrebet), in the country round Tobol'sk and Tyumen' where the species may have ranged further south then than it does now (Dementiew 1960).

CATCHING FALCONS

In western Europe, from the seventeenth to the nineteenth century, the trapping of migrant falcons was virtually in the hands of the Dutch. They first developed this skill on the open heathlands of Noord-Brabant, especially those south of Eindhoven, where migrating Peregrines were the principal objective, though Gyrfalcons were occasionally caught. For example, in the nineteenth century, three Gyrfalcons were caught among the Peregrines at Valkenswaard, the third in 1878 (Lascelles 1892: 88), and in 1890 one was taken at Cromvoirt near 's-Hertogenbosch (Oorschot 1974: 252). In the course of the seventeenth century Dutch falconers became active as trappers of migrating falcons in Iceland, where the Gyrfalcon was the sole objective, Peregrines being virtually unknown there, as well as in Norway, where both Gyrfalcons and Peregrines were caught. In view of this it is scarcely surprising that falcon trapping methods were very similar in Holland, Iceland and Norway. The most important differences were in the bird used as a decoy, usually a pigeon in Holland but a Ptarmigan in Iceland, and in the material used for the falconer's hut: peat in Holland, stone in Norway.

The old-time Dutch falcon-catcher established himself and his equipment out in the middle of an extensive area of open heath where passage falcons were frequently seen in autumn. He concealed himself in a round hut partly dug out of the soft peaty soil. Its walls were of turves and an old cartwheel often served to support the turf roof. The net, decoy and other necessaries were arranged so as to be visible from within these cramped but well-camouflaged quarters. Three live birds formed part of the falconer's outfit: two pigeons as decoys and a Great Grey Shrike as sentinel. The shrike was tethered to a small mound of peat heaped over an open-sided box. On the approach of a falcon, this sentinel bird gave the alarm and dived for cover into its box. This was the signal for the falconer to pull a string which was fastened to a pigeon concealed in a box in the ground and which continued onto the top of a pole beyond the box. When this string was pulled, the pigeon was yanked up skywards out of its box. Its fluttering attracted the falcon's attention, but before it could seize this potential prey the flaconer released the string, precipitating the decoy back into its box. While the frustrated falcon circled overhead, the falconer pulled a second string which drew the second pigeon out of a box and along a string fastened to the ground into the centre of the area to be covered by the net, which was a clap or bow net already set up lying on the ground. As soon as the falcon stooped on this second pigeon the falcon-catcher pulled hard on the third string, which operated the net, pulling it over both the falcon and its prey. There were many variations on this scheme and improvements were made, such as the use of more than one net. In the nineteen-thirties the remains of these huts were still to be seen on the heath at Valkenhorst near Valkenswaard which is now mostly planted up with tress (Slavin and Brodrick 1955; Oorschot 1974: 55–58). These time-honoured techniques are still used today in Israel and the United States to catch migrating raptors for ringing and other studies.

GYRFALCONS BY THE HUNDRED

These or similar trapping techniques were put to excellent effect in Iceland where, during the eighteenth century, a centralised trapping organisation set up as a monopoly by the Danish crown was able to procure up to a hundred Gyrfalcons for despatch to Copenhagen every year. The local Icelandic trappers had to bring their falcons to a special falcon house at Bessastaðir near Reykjavik at midsummer. From 1663 onwards 15 thalers were paid for a white Gyrfalcon, 10 for a half-white bird, and 5 for a grey one. The annual royal falcon ship arrived in July and set off early in August with the hooded Gyrfalcons perched below decks in two rows along each side of the ship. The record year was 1764, when 211 Gyrfalcons were taken on board and conveyed to Copenhagen. The food specially supplied on this occasion for the consumption of the falcons and their keepers comprised 72 oxen, 39 sheep and 65 lambs. These Gyrfalcons did not remain long in the Danish king's falcon house or hawking establishment in Copenhagen, nor were they sold commercially. They were distributed as diplomatic gifts or blandishments to the courts of Europe. In 1764, 50 went to the empress Maria Theresa at Vienna, 60 to the king of Portugal, 20 to the landgrave of Hesse and 2 to the French ambassador in Copenhagen. The king of Denmark kept 3 for himself and the remaining 45 or so were deliberately killed, apparently to avoid any reduction in the value of the product through flooding the market (Tillisch 1949)!

There is little doubt that it was the French Revolution and the wars which followed that brought the heyday of European falconry to a close at the end of the eighteenth century. But the number of Gyrfalcons imported from Iceland to Denmark had begun to decline in about 1765, long before the storming of the Bastille in 1789. The cause of this decline was apparently a decrease in the number of available birds rather than the above-mentioned policy of maintaining the scarcity value of the product. It is true that, after the superfluity of 1764 the quota of falcons purchased from Iceland was fixed at the reduced number of 100 grey birds "but as many white and half-white as they could get" (Tillisch 1949: 95, 134). Subsequently, the annual quota was fixed at 60–70 Gyrfalcons in all. But these quotas were seldom reached. Regular transport of Gyrfalcons from Iceland to Copenhagen stopped in 1793 and the last shipment was sent in 1806. The detailed Danish source material relating to the import of Gyrfalcons from Iceland between 1731 and 1793 was summarised by Dutch historian J. M. P. van Oorschot. He printed a table giving the annual number of falcons of each of the three colour phases recognised by the Danish royal officials, namely white, half-white and grey, and he reckoned that, of the around 4600 Gyrfalcons distributed as gifts in this period, about 4050 were grey, about 150 were half-white and some 400 were white (Table 11).

Table 11 Annual number of Gyrfalcons of three colour-phrases imported from Iceland to Denmark in 1731–1793 inclusive (Oorschot 1974: 296–297).

Year	Grey birds	Half-white birds	White birds	Total	Year	Grey birds	Half-white birds	White birds	Total
1731	75	3	8	86	1763	171	5	6	182
1732	82	–	9	91	1764	205	2	4	211
1733	91	2	10	103	1765	152	1	3	156
1734	91	3	11	105	1766	60	10	12	82
1735	105	–	5	110	1767	38	–	3	41
1736	73	3	2	78	1768	53	3	3	59
1737	74	1	12	87	1769	66	3	2	71
1738	61	1	2	64	1770	64	4	15	83
1739	41	–	2	43	1771	76	4	5	85
1740	69	1	3	73	1772	56	7	8	71
1741	89	–	–	89	1773	64	1	5	70
1742	96	–	3	99	1774	52	–	–	52
1743	148	3	4	155	1775	32	1	2	35
1744	167	2	6	173	1776	38	–	1	39
1745	129	11	21	161	1777	47	1	–	48
1746	96	6	4	106	1778	46	2	–	48
1747	40	2	2	44	1779	43	–	2	45
1748	48	4	1	53	1780	33	–	2	35
1749	36	–	2	38	1781	36	–	–	36
1750	44	1	5	50	1782	45	2	4	51
1751	50	11	42	103	1783	50	4	3	57
1752	85	12	18	115	1784	32	2	14	48
1753	121	4	22	147	1785	15	–	–	15
1754	144	4	10	158	1786				
1755	83	3	3	89	1787	39	3	4	46
1756	49	3	16	68	1788	39	–	5	44
1757	40	1	6	47	1789	37	–	–	37
1758	35	–	4	39	1790	45	1	2	48
1759	59	2	9	70	1791	30	2	5	37
1760	61	5	2	68	1792	15	–	–	15
1761	83	7	16	106	1793	28	–	2	30
1762	146	3	2	151					

The mass production of Gyrfalcons in order to oil the wheels of diplomacy was also practised in Russia, by the tsars, though here the peak was reached earlier, in the seventeenth century, and the purpose was more evidently to keep on good terms with aggressive or potentially dangerous neighbours. Thus Gyrfalcons, among them white birds, were sent to the Turkish sultan in Constantinople, the shah of Persia and the khans of the Crimea, as well as to Christian monarchs like the kings of Poland and Denmark (Dementiew 1960).

In 1556 a "large and faire white Jerfawcon" was sent by the tsar to Philip and Mary, king and queen of England, but was lost after the ship bringing it and other gifts was wrecked on the Scottish coast (Hakluyt 1907: 356–367).

The seventeenth-century tsar Alexei Mikhailovich was credited with receiving 200 Gyrfalcons per annum in Moscow and is said to have kept the largest hawking establishment in Europe. Throughout the seventeenth century the falconer responsible for the White Sea (Beloye more) area was supposed to supply the tsar with 100–105 Gyrfalcons annually. In 1669 he sent a hundred or more to Moscow—always in specially fitted-out sleighs lined with felt. In addition to these birds, by 1675 his colleague further east beyond the Urals was charged with the annual despatch of 50 Gyrfalcons to Moscow. So the legendary total of 200 Gyrfalcons annually may not have been a wild exaggeration. In the eighteenth century these numbers were reduced, the quota in 1731 for the White Sea and Murmansk area being fixed at 50 birds, while it seems birds were no longer being obtained from beyond the Urals (Dementiew 1960).

AN EGG-COLLECTOR IN LAPPLAND

After the French Revolution and the Napoleonic Wars the demand for Gyrfalcons for diplomacy and falconry remained limited. We hear of 15 Gyrfalcons being brought from Iceland in autumn 1845 by the falconer John Pells, son of John Pells of Valkenswaard, who had pursued his career of falconer in England, on behalf of the duke of Leeds, and in Holland with the Anglo-Dutch Loo Hawking Club, so-called from a place called Het Loo, meaning "the clearing in the woods", near Apeldoorn (Salvin and Brodrick 1855: 5, 89; Oorschot 1974: 222). Again, in 1869, an English expedition to Iceland, consisting of John Barr and his nephew James Barr brought back Gyrfalcons for falconry. This time 33 birds were obtained for the maharajah Dhuleep Singh and kept at Elveden, Suffolk, but most died shortly afterwards (Lascelles 1892: 86–88). But a new demand arose, from museums and private collectors, for the eggs and skins of Gyrfalcons. The British zoologist John Wolley wrote a graphic account of the taking of his "first clutch of four Gyrfalcon's eggs not far from Kautokeino in north Norway on 7 May 1854" (Newton 1864: 88–89):

> We had not long left the track on the river, when a Falcon flew up from the rock where the nest was supposed to be, and soon afterwards, turning back, settled on the trunk of a dead tree, once or twice uttering a cry. I now knew there was a nest, and in a few minutes more I saw it, looking very large, and with a black space about it, as though it were in the mouth of a little cave in the face of the rock. This was a joyful moment, but not so much so as when the hen bird flew off with somewhat cramped wings, and settled on a little stump, some thirty yards from the nest. I would not let Ludwig shoot. We were ascending the hill, and might be fifty yards off when she left the nest. I took off my shoes, though there was a deep snow everywhere except just on the face of the rock, and first tried it from above, but it seemed

scarcely practicable. Then I went below, and, with the Lapp to support my feet, and Ludwig to give me additional help with a pole, I managed to climb up. Just at the last bit I had to rest some time. Then I drew myself [up], and saw the four eggs to my right hand, looking small in the middle of a large nest. Again I waited to get steady for the final reach. I had only a bit of stone to stand upon, not bigger than a walnut, and frozen to the surface of the ledge, which sloped outwards. I put two of the eggs into my cap, and two into my pocket, and cautiously withdrew. The nest appeared to have been quite freshly made, and therefore by the bird herself. The sticks were thick, certainly more so than those used by Ravens or Buzzards, and, unlike the nests of the latter, which I saw the next day, they were barkless and bleached. The only lining was a bundle or two of coarsish dry grass. As I returned I touched the eggs on a point of rock above me, luckily without injuring them. I handed them down in a glove at the end of a pole, which the Lapp improvised after the fashion of a church collecting-bag; and when they were placed in a safe corner, my feet were put in the right places and I descended in safety. Luckily I had brought a box with hay, and on 12th May had the eggs safe at Muoniovaara. There were young inside, perhaps an inch and a half long, with heads as big as horse-beans.

During the years 1854–1861 Wolley and the local egg-collectors working for him took at least 27 clutches of Gyrfalcons' eggs in north Norway. In seven cases one of the adult birds was snared or shot at the nest; in one case both birds were taken. Most of the skins of these adults were given away. Possibly other clutches, not retained for the collection, were sold.

SKINS FOR MUSEUMS

In the nineteenth and twentieth centuries countless expeditions were sent by museums to the lesser-known parts of the world with the express object of obtaining specimens. Other explorers collected specimens for museums as a side-line. At Iita, Avanersuaq, in northwest Greenland (Map 21) a pair of pure white Gyrfalcons bred, and may still breed, almost every year. This remote spot is seldom visited nowadays, even by the Eskimos, but in the first half of the twentieth century almost every time Iita was visited by non-Eskimos the Gyrfalcons suffered. In 1908, 1915 and 1925 the eyasses were taken by American expeditions; in 1934 the tiercel was shot and found its way into the British Museum (Natural History) (MacMillan 1918: 409, Koelz 1929). In 1929 the bird collection of the University of Michigan included 81 Gyrfalcons from Greenland, nearly all juveniles (Koelz 1929). In his classic *Birds of Greenland*, published in 1950, Danish ornithologist Finn Salomonsen complains that American ornithologists had lumped all the Nearctic forms of the Gyrfalcon into one subspecies on the basis of only 190 specimens, 70 of them from Greenland (p. 442). He points out that Danish workers had examined some 800 Gyrfalcon skins, "of which more than 450 came from Greenland". These figures may be compared with the already-mentioned 4600 Gyrfalcons removed from Iceland (at least 10% of them Greenlandic birds) in 62 years in the eighteenth century.

A TWENTIETH-CENTURY FALCONER IN ICELAND

In 1913, because eggers and bird collectors were thought to have reduced the number of Icelandic Gyrfalcons, a law was passed in that country giving the species full protection (Dementiew 1960: 54–55). When the English hawking enthusiast Ernest Vesey decided to visit Iceland in 1936 to try to obtain some eyass Gyrfalcons, he had first to seek special permission. This old Etonian, who was blind in one eye and had only one arm, and who was said by a master at Eton to have been "inattentive and idle" at school, was astonishingly successful, not only in obtaining the eyasses he wanted but in observing and recording the breeding behaviour of the falcons and the bird life of northwest Iceland. He wrote a most readable book about his Icelandic experiences entitled *In search of the gyrfalcon: An account of a trip to northwest Iceland* under the pen-name Ernest Lewis, which he had also used for his book *The hill fox*.

Vesey arrived at Isafjörður by steamer from Reykjavik early in May with 300 feet (100 m) of rope and tramped the hills and dales of the northwest peninsula during the next 6 weeks searching for Gyrfalcons' eyries. He found at least seven, and identified seven species of bird among the falcons' prey: Eider, Mallard, Whimbrel, Ptarmigan, Kittiwake, Puffin and Black Guillemot. During this time Vesey lived to a large extent off the land, namely on birds' eggs. He never took more than half the eggs in a clutch and found those of the Great Black-backed Gull and Red-breasted Merganser to be among the tastiest; Eiders' eggs were "rather strong". He spent some time camping but often slept and ate at the local farms. Eventually, in the third week of June, helped by some Icelanders to climb to the eyries with rope and basket (and single arm), he collected six eyass Gyrfalcons. These were transported by ship in a large laundry basket to Scotland. On the way Vesey fed them on birds he had bought from an Icelandic friend: 30 freshly-killed Puffins and 20 live pigeons. All six eyasses were turned out to hack in Islay when a few days over 5 weeks old. One died of some illness, another drowned while trying to bathe in a concrete water tank, and a third disappeared. Of the skins of the first two, both tiercels, one went to the British Museum (Natural History) and the other to a private collector. That autumn, of 1936, the three surviving Gyrfalcons were successfully flown at pheasants, partridges, Red Grouse and Black Grouse. Ernest Vesey died in January 1937 while still only a young man; his book was published posthumously. His revival of the noble art of hawking with the noblest hawk of all, the Gyrfalcon, was lasting. Other British falconers since his day have handled Gyrfalcons (Stevens 1953; Wayre 1965) and there are Gyrfalcons in Britain today.

PRESENT-DAY FALCONRY IN NORTH AMERICA

Among American falconers after the Second World War, H. Webster, who coauthored *North American falconry and hunting hawks* (1964) with F. Beebe, was one of the few experienced in obtaining, training and keeping the Gyrfalcon, in his own words "the proudest possession a falconer can ever have" (p. 184). His description of the acquisition of his first nestling Gyrfalcon is reminiscent of Wolley's just-quoted account of his first clutch of eggs (pp. 178–179):

> The rolling upland tundras seemed to stretch out to infinity, mist-shrouded and utterly silent. Just the vast empty loneliness of the land leaves a profound impression that is more than a little frightening. On a tall, red-brown sandstone cliff rising from the rolling, damp and soggy tundra I was to see my first big grey Gyr. The eyrie was there plainly marked with "hawk chalk" for all to see. It was the jerkin or tiercel that appeared first, calling with the harsh low-pitched cry of the Gyr, so alike, yet so different from the voice of other falcons. He put in a few half-hearted stoops at me as I ascended the grassy slopes toward the base of the cliff, but nothing that could be considered alarming or dangerous. He was not much larger than a big Peale's Falcon and even when I began to work myself up the side of the tall, crumbly sandstone escarpment he did little more than fly back and forth with no real purpose. He eventually withdrew to the north where he took perch on some low rocks overlooking a vast, grassy headland.
>
> I found his performance disappointing to say the least, and I resumed my climb up the cliff. Then, quite suddenly, I became aware of a much larger bird bearing

down on me at terrific speed and with what appeared to be every intention of knocking me off the cliff-face. I flattened myself against the rock as she shot past only inches from my head with her big yellow feet menacing me. This was the falcon and she looked almost as big as an eagle and twice as dangerous. I had visited several hundred Prairie Falcon eyries as well as a series of Peale's Peregrine sites during my first twenty years as a falconer, but never had a falcon set at me with such determination. The first series of stoops came close enough to be really frightening. Eventually she left to perch briefly on top of the high bluff where she watched my progress toward her half-hidden eyrie. This was my cue to start moving again, and I inched up toward the nest that was only a few feet away. The movement did not go unchallenged. Down she came at me again, and this time she kept me pinned on the cliff face for twenty minutes before she left to join the jerkin some distance away on a pile of low rocks. Only then did I dare to move directly onto the nesting ledge where I hastily put the larger of the two eyasses inside my buckskin jacket. I now had the job of retreating as quickly and carefully as possible to the bottom of the cliff where I enjoyed a brief rest.

After this grey Gyrfalcon, Webster was determined to secure a white bird, and he describes how he eventually caught two fine white Gyrfalcons while they were on passage, using a live pigeon on a pole as a decoy. One of these birds later became the official mascot of the United States Air Force Academy; Webster kept the other himself for falconry and describes it as "one of the finest birds it has ever been my privilege to fly" (p. 184).

GYRFALCONS FOR THE SHEIKS?

The Gyrfalcon in the wild nowadays has the full protection of the law in Canada, Greenland and elsewhere. But it is no more difficult to breed in captivity than the other large falcons. As a result, falconers can obtain perfectly legal captive-bred Gyrfalcons for their sport. Nevertheless, some wealthy enthusiasts in the Middle East hold the Gyrfalcon in such high esteem for its size and speed that they are said to have been prepared to pay very large sums of money to obtain these birds either legally or illegally.

In the 1970s the United States Fish and Wildlife Service became increasingly concerned about the international smuggling of rare birds of prey. In 1972, American falconer John Jeffrey McPartlin was convicted of conspiring to smuggle two grey Gyrfalcons out of the United States. In 1981, when federal wildlife agents initiated a 3-year undercover investigation called Operation Falcon, McPartlin, now turned federal informant, played a key role by posing as an underground dealer in birds of prey. He was credited with trapping falcons with the full knowledge of the enforcement officers. These birds included at least 12 Gyrfalcons caught on the Montana prairies. He then persuaded various falconers in and outside the United States to purchase one or more Gyrfalcons at $4000 to $7000 apiece. When the first phase of Operation Falcon was wound up in June 1984, 150 law enforcement officers arrested about 30 people and searched many more, but few convictions followed. Among McPartlin's victims were two young West German brothers, Marcus and Lothar Ciesielski, accused of attempting to export these Gyrfalcons and three prairie falcons to Riyadh, the capital of Saudi Arabia, from Kennedy International Airport, New York, in December 1982. They and their father Konrad were said to control a network of falcon smugglers who procured falcons in Scandinavia, Canada, Iceland and elsewhere and sold them in the Middle East. Lothar Ciesielski escaped the federal authorities, but his brother Marcus was fined $10 000 before being allowed to leave the United States.

Operation Falcon, besides bringing this West German falcon-smuggling ring to book, also exposed a Canadian concern called Birds of Prey International, which was said to have sold $750 000 worth of raptors over a period of 2 years—to the Saudis among others. During 1983, McPartlin was in contact with a member of this Canadian concern, Glen Luckman, who claimed that he could sell Gyrfalcons in Saudi Arabia for $25 000 to $30 000 each and that a white female might fetch $100 000. His partner subsequently offered McPartlin $10 000 each and a Peregrine for three Gyrfalcons. Later in the same year Luckman is supposed to have told McPartlin that Birds of Prey International had obtained 24 nestling Gyrfalcons from eyries, apparently in the Canadian Arctic. In 1984 he was fined $23 000 by the U.S. District Court in Great Falls, Montana. Naturally the North American Falconers' Association felt sore about Operation Falcon, for it had brought about the arrest of some twenty of its members. As to the Saudis, they employed a Washington lawyer to deny every accusation against them and were reported not to be answering phone calls (Robbins 1985; Shor 1988).

Operation Falcon seems to have been badly bungled. Its victims were for the most part exonerated; those convicted were found guilty of trying to evade customs duty rather than of obtaining falcons from the wild. The supposed prices of the birds seem ridiculously high: a perfectly legal captive-bred Peregrine is worth around $2000 only. It may be true that wild Gyrfalcons are not now being obtained for falconry in significant numbers, but the existence of an international trade in them is not open to doubt. In August 1990 a German, Rudolf Sperr, was sentenced to 18 months' imprisonment at Maidstone Crown Court for attempting to smuggle four Gyrfalcon chicks, apparently of North American origin, into Britain (Prytherch and Everett 1990).

CHAPTER 5

Bird collecting in North America

For a long time, the collection of specimens was of the essence in the study of natural history. Knowledge of the avifauna of North America and especially of its extensive Arctic and sub-Arctic regions was laboriously built up by amateur naturalists whose bird-collecting activities took second place to their main occupation as explorers, officials or whatever. Their bird collecting, and the subsequent preparation of skins, was often only one facet of their work in the field; they collected plants, fishes, mammals and insects as well. They were all-round naturalists. About 1900 came the beginnings of specialisation and professionalisation: the scientific ornithologist was born. He concentrated entirely on the study of birds but often continued the time-honoured collecting tradition. It was as recently as 1982 that one of the last of these North American collector-ornithologists, George Miksch Sutton, died at the age of 84, and this chapter closes with some account of him and of his expeditions to the Canadian Arctic.

THE HUDSON'S BAY COMPANY

The London-based fur trading company which for centuries enjoyed a monopoly round the shores of Hudson Bay was established by royal charter in 1670

(Rich 1958, 1959). Its formal legal title was "The Governor and Company of Adventurers of England trading into Hudson's Bay". "Trading" was a euphemism; the company's profits were based on primitive bartering. By the middle years of the eighteenth century its 120 officials and servants and three or four ships were bringing up to £30000 worth of furs to England annually, with costs averaging under £20,000 per annum. Based at first on a chain of posts round the southern shores of Hudson Bay, the company slowly and hesitantly began explorations westwards and northwards from there into the vast forests and pairies of the hinterland and, eventually, into the Barren Grounds or North American tundra. During the course of the eighteenth century it developed an enlightened policy of encouraging its officers in the pursuit of knowledge, and of natural history in particular. This to such an extent that, in 1831, it could be said that almost everything known of the "Ornithology of the American Fur-Countries", namely the area north of 48°N, was due to the efforts of the Hudson's Bay Company (Swainson and Richardson 1831: ix).

As early as 1740 a certain Alexander Light was sent out by the company "on account of his knowledge of Natural History". In the 1740s, too, James Isham, who served as governor of various trading posts, collected birds. Both men brought or sent their specimens back to England and entrusted them to George Edwards who figured some of them in his celebrated *Natural history of birds*. Ten of Light's birds appeared in Vol. 1 of this work and 32 of Isham's in Vol. 3, published in 1750. Another eighteenth-century collector-official of the company was Andrew Graham, who was in charge at York Factory, Fort Severn and elsewhere. In 1768–1769 he took advantage of William Wales's visit to Hudson Bay to observe the transit of Venus across the sun, to send back with Wales to the Royal Society in London a collection of "quadrupeds, birds and fishes" he had made. He must have been gratified when, in 1772, J. R. Forster enumerated 57 or 58 species from this collection in his "An account of the birds sent from Hudson's Bay". It was apparently the interest aroused by Forster's publication, which appeared in the *Philosophical Transactions of the Royal Society*, that prompted the governor and committee of the Hudson's Bay Company to order that "objects of Natural History should be annually sent to England". This directive bore fruit in a collection of several hundred specimens of plants, animals and birds sent home by Humphrey Martin from the environs of Fort Albany in James Bay, where he was governor. It also led to the annual despatch of bird skins and other specimens to London by officers of the company, either to private individuals like Joseph Sabine or to the governor and committee of the company. Many of those sent to the governor and committee were placed by them in the "museum of the Hudson Bay productions, which is liberally open to the public"; others were presented to the British Museum or the Zoological Society (Swainson and Richardson 1831: ix–xi).

Fortunately there were some officers of the Hudson's Bay Company who did not rest content with shooting birds, skinning them, and sending the skins to England. Thomas Hutchins, who succeeded Humphrey Martin as Chief Factor of Fort Albany in 1774, kept a journal in which he entered descriptions of animals and birds, adding their native names and notes on their food,

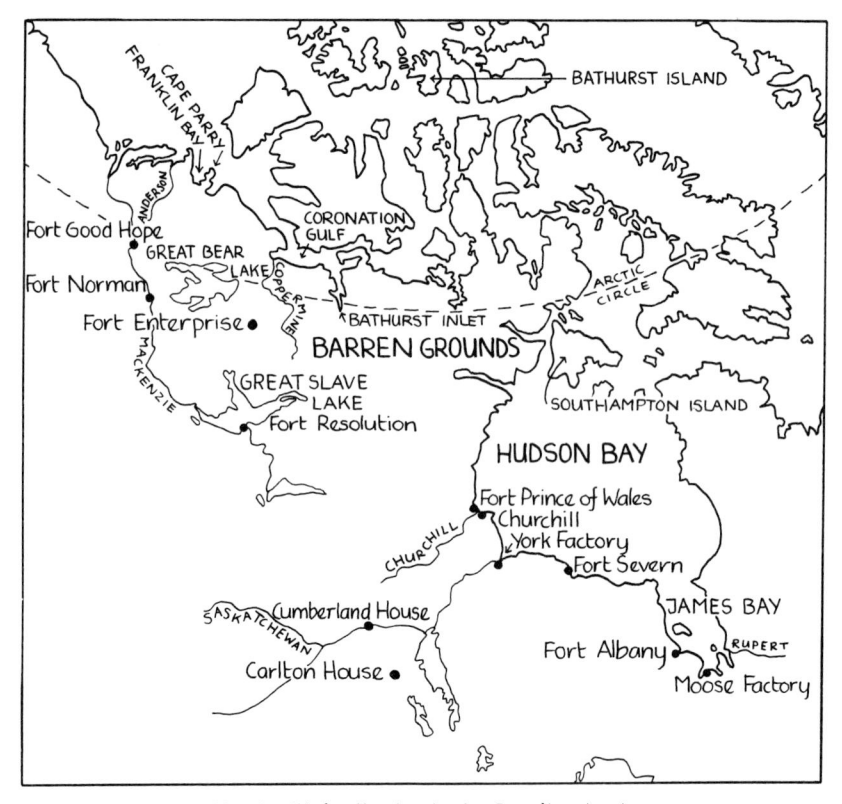

Map 8 Bird collecting in the Canadian Arctic.

behaviour and breeding. These observations "in one volume folio" were subsequently deposited "in the Library of the Hudson's Bay Company" in London. It was claimed in 1831 that they embraced "almost all that has been recorded of the habits of the Hudson's Bay birds up to the present time" (Swainson and Richardson 1831: x–xi). Apart from Swainson and Richardson, both John Latham and Thomas Pennant derived much of their information about the birds of the Hudson Bay area from Hutchins. He found and described the nest and eggs of Arctic-breeding waders like the Turnstone, Semipalmated Plover and Grey Plover. And it was Hutchins who made the first accurate notes on the Semipalmated Sandpiper, a bird which was subsequently made known in print by Alexander Wilson.

SAMUEL HEARNE

The only one of the early Hudson's Bay Company naturalists to find his way into print was Samuel Hearne, who was also a celebrated explorer. In 1766, after serving as a midshipman in the Royal Navy, he signed on as mate in the company's whaling sloop *Churchill*. Three years later, at the age of twenty-four, he was sent to explore northwestwards from Hudson Bay. He was to trace the course of the Coppermine River, referred to in his instructions as the river the Indians call Far Off Metal River, to its mouth and, if the environs of the mouth seemed of value, to take possession of the area for the company by cutting his name and the date on the rocks there. After two false starts, Hearne succeeded with Indian help in reaching the shores of the Arctic Ocean at the mouth of the Coppermine and even brought back a 4 pound (1.8 kg) lump of copper which eventually made its way to the British Museum. After a distinguished but brief career with the Hudson's Bay Company, during which he served as governor of Fort Prince of Wales at the mouth of the Churchill River and had to surrender it to a greatly superior French force in 1782, Hearne retired to England in 1787, aged 41, and died in 1792. His masterpiece of travel literature, based on the journals of his 1769–1772 journeys, only appeared in print posthumously, in 1795 (MacKay 1937: 105–114; Glover 1958; Speck 1963; Stone 1986a).

There are only a few mentions of birds in Hearne's narrative of his explorations and these show that, at the time he undertook his famous journeys, when in his mid-twenties, his knowledge of birds was patchy if not downright inadequate. For example, the ornithological part of his account of the geography and natural history of the area at the mouth of the Coppermine River is limited to the following laconic statement, the detailed interpretation of which is open to doubt (Glover 1958: 110):

I also observed several flocks of sea-fowl flying about the shores; such as, gulls, black-heads, loons, old wives, ha-ha-wie's, dunter geese, arctic gulls, and willicks. In the adjacent ponds also were some swans and geese in a moulting state, and in the marshes some curlews and plover; plenty of hawks-eyes (i.e. the green plover) and some yellow legs; also several other small birds that visit those Northern parts in the Spring to breed and moult, and which doubtless return Southward as the fall advances.

By the time he was preparing his journals for publication near the close of his short life Hearne had made substantial progress as an ornithologist. He included at the end of his book "An account of some of the principal birds found in the Northern Part of Hudson's Bay". This was the first published list of American Arctic birds by someone who had himself travelled in the Arctic and actually seen many of the birds. Moreover, Hearne was able to profit from the publication of Vol. 2 of Thomas Pennant's *Arctic zoology*, on birds, in 1785, which, he acknowledged in his preface, "enabled me to give several of the birds their proper names". His is no mere list of specimens; he includes first-hand observations on the habits, distribution and breeding biology of some typical Arctic birds. Thus of the Snowy Owl, he reports that it is common "as far north as the Coppermine River", that it lays 3–4 eggs but seldom hatches more than two, that it never migrates but even winters on the the Barren Grounds far north of the forest, and that it annoys sportsmen by making off with shot birds before they can be retrieved. For this last reason, the governors of the company offer a reward of a quart of brandy for every Snowy Owl killed. Of the Raven, which is "of a most beautiful glossy black, richly tinged with purple and violet colour", Hearne records that in winter they feed on a black moss that grows on pines, on the dung of deer and other animals, and on game caught in snares. He wonders how they can survive and surmises that they have a good sense of smell. "Their quills make most excellent pens for drawing, or for ladies to write with." Naturally Hearne was familiar with the Snow Bunting. "On their first arrival they generally feed on grass-seeds, and are fond of frequenting dunghills." He had kept them in the same room as Canaries and found that they sang well and lived long in confinement. He adds that they fly so far north that "their breeding-places are not known to the inhabitants of Hudson's Bay".

Hearne distinguished carefully between what are now known in North America as the Rock Ptarmigan and Willow Ptarmigan, our Ptarmigan and Willow Grouse (of which the Red Grouse is a subspecies) respectively, though he calls them partridges. He points out that the "Willow Partridge" is larger, that it feeds in woods and willow scrub in winter "where they hord together in a state of society" and that they burrow under the snow at night. He notes that "They are by no means equally plentiful every year"; in the winter of 1785 they were very common near Churchill. The "Rock Partridge" is smaller, never feeds on willow but prefers the buds of the dwarf birch. The sitting bird can sometimes be lifted off the eggs. The change from white winter plumage to brown in summer is mentioned as occurring in both species. In Hearne's time the Whooping Crane and the Trumpeter Swan were both widespread in the Canadian Arctic and he carefully distinguishes these larger, more southerly species from their commoner relatives the "Brown Crane" now called the Sandhill Crane, and the Tundra Swan. On gulls, Hearne is somewhat vague. He has "White Gulls" and "Grey Gulls", and "Black Gulls" which are evidently skuas. Nor did he succeed in attributing the correct name to the Arctic Tern, which he calls the "Black-head" and describes as the smallest gull he knows. Its eggs are good to eat but it dives at people taking them. "This bird may be ranked with the elegant part of the feathered creation, though it is by no means gay." It is found "as far North as has hitherto been visited". His description of it is excellent—scarlet bill, legs and feet, black crown, rest of plumage a light ash colour, tail much forked—except that he describes the quill feathers as prettily barred and tipped with black.

On geese Hearne is particularly interesting. Like other authors of the day, he considers the "White or Snow Goose" and the "Blue Goose" as distinct species. The "Snow Goose", also called the "Common Wavey" is the most numerous of all the birds frequenting the northern parts of Hudson Bay. In good springs 5000–6000 are killed at or near Churchill, but in the fall 7000–8000 is considered good. The breeding haunts of both these birds were unknown to Hearne, nor could he find any Indian or even Eskimo who knew where either the Snow or Blue Goose bred. Under the name "Horned Wavey", Hearne described Ross's Goose, and indeed was the first naturalist to do so. It is white all over save for its black wing-feathers and not much larger than a

Plate 15 Arctic Tern, widespread throughout the Arctic, here hovering over the tundra on the look-out for prey, just as it often hovers over water. When it dived down this bird sometimes picked the prey off the ground without settling, sometimes settled briefly. Cambridge Bay, July 1986.

Mallard. The bill is an inch long and studded round the base with pea-sized knobs. Rare at Churchill, Hearne had seen large flocks 200–300 miles (400–500 km) north of there. He continues (Glover 1958: 284–285):

> The flesh of this bird is exceedingly delicate; but they are so small, that when I was on my journey to the North I eat two of them one night for supper. I do not find this bird described by my worthy friend Mr. Pennant in his Arctic Zoology. Probably a specimen of it was not sent home, for the person that commanded at Prince of Wales Fort [Moses Norton] at the time the collection was making, did not pay any attention to it.

Besides his annotated list of the birds of the northern Hudson Bay area, Hearne included in the last chapter of his book notes on the area's "vegetable productions", on frogs and "a great variety of Grubbs, and other insects", on salt-water fish, on the sea-mammals of Hudson Bay and, of course, on the land animals. His powers as an observer are also reflected in the admirable descriptions of the natural history of the muskox, beaver and other animals which intersperse his narrative, and in the special chapter he devoted to the way of life and beliefs of the Chipewyan Indians with whom he travelled.

FRANKLIN'S *NARRATIVE* OF HIS FIRST LAND EXPEDITION, 1819–1822

One of the aims of Samuel Hearne's explorations had been to search for the Northwest Passage; and his search, interrupted by the Revolutionary and Napoleonic Wars and other involvements, was resumed by Britain, and by the British admiralty in particular, after the peace settlement of 1815. Besides the many expeditions sent by sea from 1818 onwards, the admiralty mounted a naval expedition by land in 1819 with the help of the Hudson's Bay Company. It was to repeat Samuel Hearne's journey from York Factory on the shore of Hudson Bay to the mouth of the Coppermine River, but to continue on eastwards from there to explore the unknown shores of the North American mainland. Although the pursuit of natural history did not figure prominently in the instructions received from the admiralty by the expedition's leader, Lieutenant John Franklin, R.N., nevertheless its other officers were appointed with this in mind. For midshipmen George Back and Robert Hood were "to make drawings of the land, of the natives, and of the various objects of natural history" (Franklin 1824a: xii) and surgeon John Richardson was to add to medicine the role of naturalist and "to collect specimens of minerals, plants and animals" (Houston 1984: xxiii). The expedition was in the field from 9 September 1819, when it struck off westwards inland from York Factory,

Plate 16 White-billed Diver in breeding plumage photographed by Pavel Tomkovich at the Belyaka Spit on the north coast of the Chukchi Peninsula, Siberia. 3 August 1987.

until 14 July 1822, when it returned there having reached its goal and explored some 500 miles of the Arctic coast of North America, namely the southern shore of Coronation Gulf and both shores of Bathurst Inlet. But it had experienced the horrors of starvation, had lost nearly half its members, and had narrowly escaped total disaster (Franklin 1823 and 1824; Nanton 1970).

In the circumstances it is perhaps surprising that Franklin and his men made any natural history observations at all. When the leader's *Narrative of a journey to the shores of the Polar Sea in the years 1819–1822* was published in 1823 it contained only passing references to birds. Thus the explorers were startled, when still in the forest, by the call of the Gray Jay or whiskey-johneesh which Franklin calls the "cinereous crow". A month later they came across a "flock of pelicans and two or three brown fishing eagles", and "several golden plovers, Canadian grosbeaks, cross-bills, wood-peckers, and pin-tailed grouse" were shot. In the autumn of 1820 at Fort Enterprise between the Great Slave and Great Bear Lakes, Franklin mentions what must have been a White-billed Diver: "The last of the water-fowl that quitted us was a species of diver, of the same size with the *colymbus arcticus*, but differing from it in the arrangement of the white spots on its plumage, and in having a yellowish white bill. This bird was occasionally caught in our fishing nets" (Franklin 1824b: 15). At the same place, in May 1821, Franklin noted the arrival of the first American Robin, the Red-breasted Merganser, the teal, "and lastly the goose". Bird records from the Arctic coast are equally terse and disappointing: "white geese" (Snow Geese), many ducks of a species locally called, from their cry "caccawees" (Long-tailed Ducks), Tundra Swans, and Sandhill Cranes. Most of them were killed and eaten by the hungry explorers.

The scattered references to birds in the text of Franklin's *Narrative* were supplemented by an appendix on birds (Franklin 1823: 669–703) by Joseph Sabine which was equally disappointing. Apparently based in part on Hearne's annotated list, it only mentions birds actually obtained by the expedition and by no means includes all of these. Moreover, Sabine evidently had no access to Richardson's field notes. Thus there is no mention of the White-billed Diver, Whooping Crane or Eskimo Curlew. But it did include such characteristic Arctic species as the Grey and Red-necked Phalarope, the Arctic Tern and the Arctic Skua, which had either been misidentified or altogether missed by Hearne.

The mentions of birds in Franklin's *Narrative* of his expedition, which was widely read and went into a second edition within a year, were in fact taken from Hood's or Richardson's journal, both of which remained unpublished while their leader's work was acclaimed. Because Franklin's own journal was lost on the expedition when his canoe overturned in some rapids, he had perforce to compile his own narrative from the other three officers' daily journals, namely those of John Richardson, George Back and Robert Hood, none of which was published in the nineteenth century. It was not until the last two decades of the present century, when the Canadian medical specialist and amateur ornithologist C. Stuart Houston, of the University of Saskatchewan, annotated and published the journals of Hood (Houston 1974) and Richardson (Houston 1984), that the true contribution to ornithology of Franklin's expedition became apparent, as well as the full extent of Franklin's often word-for-word copying of his colleagues' journals.

ROBERT HOOD'S BIRD PAINTINGS

Robert Hood, who lost his life towards the end of the expedition, included in his journal sensitive descriptions of the landscapes through which it passed. In some cases he gives useful details about birds: the "brown fishing eagles" of Franklin's narrative, more narrowly described by Hood, turn out to have been Ospreys; and migrating geese seen early in September near the Coppermine River are described well enough to be identified as Snow Geese (Houston 1974: 150):

> The gale brought with it, from their northern haunts, long flights of geese, stretching like white clouds from the northern to the southern horizon, and mingling their ceaseless screams with the uproar of the wind among the hills. They are of that description, called Wavy's in this country, being white, with yellow feet and some black feathers on each wing.

Along with Hood's journal Houston (pp. 167–188) also reproduced and commented on some of Hood's paintings, eight of which had been used as models by the engraver who illustrated the first edition of Franklin's *Narrative*. Four of these show groups of birds painted in exquisite and accurate detail. "Winter birds at Cumberland House" on the Saskatchewan River portrays, against a delicately sketched-in background of forest trees, a Yellow-headed Blackbird (actually a summer visitor), an Arctic Redpoll, a Black-backed Woodpecker, a Black-capped Chickadee, a Snow Bunting, a Sharp-tailed Grouse and two Two-barred Crossbills—all of them readily identifiable. Remarkably, the first three of these were at that time undescribed species, unknown to science. Specimens were, in one case certainly, in the other two probably, obtained by the expedition but never reached Sabine in London to be described in his appendix to Franklin's *Narrative*. In another particularly interesting painting of waterfowl in flight Hood included several typically Arctic species: the Long-tailed Duck, Snow Goose, Eskimo Curlew and Arctic Loon or Pacific Diver. Neither of the last two of these appeared in Sabine's appendix. The first was certainly and the second probably painted from a specimen that never reached him. Hood's bird paintings, which were done from field sketches and skins in January and May 1820 at Cumberland House, equal or surpass the best work of other bird artists of his day. Moreover, the species is in every case easily identifiable. The tragic death of this talented young artist was a sad loss to ornithology.

FRANKLIN'S SECOND LAND EXPEDITION, 1825–1827

Nothing daunted by their narrow escape from disaster and the loss of their companion Hood, Franklin, Back and Richardson gladly set out again in 1825 on behalf of the British admiralty. They descended the Mackenzie River, instead of the Coppermine, and then split into two sections to explore the Arctic coast in either direction. While Franklin journeyed westward towards Point Barrow, Richardson led a detachment eastward following the coast from the Mackenzie Delta to the mouth of the Coppermine River. This time Richardson contributed an account of his own Arctic coast explorations, which made up one-third of Franklin's book (Franklin 1828: 187–285), and his narrative contains some interesting notes on the birds encountered in July and August. At the outset, on 4 July, while still in the Mackenzie Delta, Richardson noted that "Our voyage amongst these uninteresting flats was greatly enlivened by the busy flight and cheerful twittering of the sand-martins, which had scooped out thousands of nests in the banks of the river, and we witnessed with pleasure their activity in thinning the ranks of our most tormenting foes the musquitoes". As they rowed along the coast in two open boats several "Swans, Canada and White geese, and Arctic ducks" were "killed by our sportsmen". Later Richardson mentions, still rather vaguely, seeing swans, Canada Geese, Eiders and King Eiders, "Arctic and surf ducks", Glaucous, silvery, black-headed and Ivory Gulls, terns and northern divers. On 22 July they came across what Richardson describes as a Kittiwake colony: the birds were breeding "in great numbers on the rocky ledges" but "their young were already fledged" (Franklin 1828: 237). This must be a mistake, the birds were surely Glaucous Gulls. It was on this same day, "whilst the party were seated at breakfast", that they watched a parent Snow Bunting feeding its four young "with the larvae of insects" in a nest in a large pile of driftwood. On 3 August, nearing the end of his journey to the Coppermine River, Richardson walked a few miles inland (Franklin 1828: 252).

> In my way I had occasion to wade through a small lake, when two birds, about the size of the *northern diver*, and apparently of that genus, swam, with bold and angry gestures, to within a few yards of me, apparently very impatient of any intruder on their domain. Their necks were of a beautiful pale yellow colour, their bodies black with white specks. I consider them to belong to a species not yet described, and regretted that, having left my gun at the tents it was not in my power, to procure one of them for a specimen.

The word "necks" must surely be a mistake for "beaks" and the birds in question must have been White-billed Divers. Curious, because Richardson himself had already written the first description ever of this species, from a specimen he had collected on the first Franklin expedition. This remained unpublished until 1984 (Houston 1984: 225) and Richardson's surmise about the birds he saw on 3 August 1826 was correct: the White-billed Diver was not described in print until 1859.

JOHN RICHARDSON

"The principal object of Dr Richardson's accompanying you", the admiralty explained in 1825 to Franklin in its instructions for his second expedition, is "that of completing, as far as can be done, our knowledge of the Natural History of North America" (Franklin 1828: xxiv). Though he had attended Robert Jameson's lectures at Edinburgh University on natural history while writing his M.D. thesis on yellow fever, Richardson knew little or nothing of that subject when he joined the first Franklin expedition in 1819 (Houston 1984: xxii–xxiii; Johnson 1976). He protests his "previous ignorance of Ornithology" as his reason for recording little of the habits of birds. He recalls that, while serving in the Mediterranean Fleet in the Napoleonic Wars, a tiny sparrow-sized owl came aboard his ship. But "having no acquaintance whatever with ornithology at the time" he had no idea what species of owl it was (Swainson and Richardson 1831: xvi–xvii, xx; presumably a Pygmy Owl). But this "ignorance" was no excuse for the unfair treatment meted out to Richardson during and after Franklin's first expedition. Most of the mammal and bird specimens collected in 1819–1820 by Richardson himself and Hood were sent back by the Hudson's Bay Company to London. There they were handed over to a civil servant and amateur ornithologist of repute, Joseph Sabine, who proceeded to prepare descriptions of the new species for publication. To Franklin's annoyance when he returned to London in October 1822, an account of three new species of ground squirrel sent home by the expedition had been communicated by Sabine to the Linnaean Society of London and was already in proof. Franklin was angry because he had intended that "an account of the subjects of Natural History collected by him during his expedition should accompany the narrative which he is preparing for the press" and not be published in advance separately. Richardson must have been angry too, and even more so when Franklin allowed Sabine to write the "Zoological appendix" to his *Narrative*. Probably Franklin's and Richardson's ire was only slightly palliated by the fact that Sabine had diplomatically named the three species of ground squirrel, wrongly identified as marmots, *Arctomys Franklinii*, *Richardsonii*, and *Hoodii* respectively (Houston 1984: 198, 246–247).

John Richardson bided his time. He greatly improved his knowledge of natural history and extended his personal experience of the northern parts of the North American continent on the second Franklin expedition. Thereafter, he became creator and author in large part of the monumental four-volume work entitled *Fauna Boreali-Americana or the zoology of the northern parts of British America* which really did succeed in "completing, as far as can be done, our knowledge of the Natural History of North America" (see Richardson and Swainson 1829–1837). It has remained famous ever since as a classic of regional zoology. The volume on mammals by Richardson assisted by William Swainson and William Kirby came out in 1829; the fishes, entirely by Richardson, came out in 1836; the birds, published in 1831, was proclaimed on its title-page as by William Swainson and John Richardson; and the insects, 1837, was written by William Kirby. As to the birds, both alphabetically and in terms of their relative contributions, Richardson's name should have been placed first on the title-page. After all, he wrote the main part of the text, comprising a 30-page introduction and accounts of 238 species, leaving to Swainson the 50 hand-coloured lithographs which illustrated the work, a short introduction and preface, the provision or revision of the specific names and some remarks on the "natural arrangement" of birds and other comments.

The real basis of the *Zoology of the northern parts of British America* is explained in its sub-title, where it is said to contain "descriptions of the objects of natural history collected on the late northern land expeditions under command of Captain Sir John Franklin, R.N.". In essence, *The birds* is a description of the specimens sent or brought back to England by the two Franklin expeditions and a few more besides. Some of Richardson's own field observations are included in it, though these are few and far between. On owls, for instance, he mentions having seen the Snowy Owl, which "frequents in summer the most remote Arctic lands that have been visited". He describes how the only known example of the pale, northern, form of the Great Horned Owl, which he calls the "Arctic or White Horned Owl", was collected by the second Franklin expedition. "It was observed flying at mid-day in the immediate vicinity of Carlton House, and was brought down with an arrow by an Indian boy." And he has this to say of an encounter with breeding Great Grey Owls:

> On the 23rd May, I discovered a nest of this Owl, built on the top of a lofty balsam-poplar, of sticks, and lined with feathers. It contained three young, which were covered with a whitish down. We got them by felling the tree, which was remarkably thick; and whilst this operation was going on, the two parent birds flew in circles round the object of their cares, keeping, however, so high in the air as to be out of gunshot. . . . The young ones were kept alive for two months, when they made their escape.

On the Franklin expeditions, Richardson admits, ornithology could be given limited attention only. Because of the difficulty of carrying bulky packages on overland marches and of preserving delicate bird specimens from injury, they tended to concentrate on mineral and botanical specimens. Birds could normally only be collected in the spring, when the expeditions were still in their winter quarters and before they set out on their northern travels. However, in 1822 a large collection of autumn passage waders was made at York Factory, though many of these specimens were "destroyed by moths in London" or lost there. Richardson has an interesting account in *The birds* of how the waders feed in spring on the moist prairies of the Saskatchewan. Then, when these begin to dry, they fly north to breed where the melting permafrost keeps the soil moist, returning in autumn to feed on the Hudson Bay mud flats. But of the breeding of these waders in the Arctic he has little first-hand information, having apparently himself never seen nests and eggs of such widespread Arctic species as Sanderling, Grey Plover, Semipalmated Plover or Semipalmated Sandpiper. A notable exception is the Eskimo Curlew, about which he writes (p. 378):

> This Curlew frequents the barren lands within the Arctic circle in summer, where it feeds on grubs, fresh-water insects, and the fruit of *Empetrum nigrum*. Its eggs, three or four in number, have a pyriform shape and a siskin-green colour, clouded with a few large irregular spots of bright umber-brown. The Copper Indians believe that this bird and some others betray the approach of strangers to the Esquimaux; and it is very probable that that persecuted people, always in dread of the treacherous attacks of their enemies, and accustomed to observe the few animals that visit their

(a)

(b)

Plate 17 (*a*) Stilt Sandpiper near its nest on a dry tundra ridge near Cambridge Bay. (*b*) Stilt Sandpiper feeding in shallow water. Although its body is the same length as the Dunlin's, the Stilt Sandpiper has much longer legs than that species. It is endemic to the North American Arctic and its range there is restricted to mainland coastal tundras between the North Slope of Alaska and the west side of Hudson Bay.

country with great attention, will be on the alert when they perceive a bird flying backwards and forwards over a particular spot. On the 13th of June, 1822 [*sic*, an error for 1821], I discovered one of these Curlews hatching on three eggs on the shore of Point Lake. When I approached the nest, she ran a short distance, crouching close to the ground, and then stopped to observe the fate of the object of her cares.

One Arctic-breeding wader which was the subject of a muddle in Swainson and Richardson's book was the Stilt Sandpiper. This bird, because of plumage differences between an adult collected on the Saskatchewan River in June 1827 and several first-winter birds collected near York Factory in autumn 1822, was wrongly made into two species, a supposedly new one named by Swainson "Douglas's Sandpiper", and the "Slender-shanks Sandpiper" already named *Tringa himantopus* by C. L. Bonaparte. A group of Arctic breeding birds which was nearly but not quite sorted out by Swainson and Richardson was that of the smaller skuas. They described three supposed species. First the "Pomarine Jager, *Lestris pomarina*", which is our Pomarine Skua. Second the "Arctic Jager, *Lestris parasitica*", which is our Long-tailed Skua. And third, what they took to be a new and separate species, which Swainson, in honour of his co-author, named "Richardson's Jager, *Lestris Richardsonii*"; but Swainson's plate (Plate LXXII) of this bird clearly portrays a dark-phase specimen of the Arctic Skua. The name "Richardson's" for the Arctic Skua stuck, and it is only in the last ten years or so that it has been entirely replaced in English ornithological literature by the name Arctic Skua. David Bannerman, in Volume 12 of his *The Birds of the British Isles*, published in 1963, headed his chapter on this species "Richardson's or Arctic skua" and refers to it as "Richardson's skua" throughout the text. In spite of its obvious inadequacies, Swainson and Richardson's book still remains, as regards both text and illustrations, one of the great nineteenth-century bird books.

John Richardson's already secure place in the history of Arctic ornithology was reinforced twenty years after the publication of *The birds* when his *Arctic searching expedition: a journal of a boat-voyage through Rupert's Land and the Arctic Sea* was published in 1852. This account of his and Dr John Rae's expedition in 1848 down the Mackenzie River and eastwards along the coast to the mouth of the Coppermine River in search of the missing John Franklin and his men has interesting references to birds. It was a virtual repeat of Richardson's journey on the second Franklin expedition, already described. Long before they reached the Arctic Circle, from 1 June onwards, the explorers heard the incessant song of the White-crowned Sparrow. Day and night, and loudest at midnight, it whistled loud and clear the first bar of "Oh dear, what can the matter be?". The Indians accompanying the expedition asserted that no one ever saw this invisible songster, but Richardson and Rae stopped to identify the bird and found its well-hidden grass-built nest in a hollow on the ground containing five eggs, "grayish, or purplish-white, thickly spotted with brown". As they moved north Ospreys and Bald Eagles were seen and Richardson reckoned that Bald Eagles' nests were spread about 20 or 30 miles apart along their route. He noticed this bird's territorial habits: "Each pair of birds seems to

112   *In search of Arctic birds*

Plate 18 This dark-phase Arctic Skua is settling onto its eggs in the Varanger Peninsula, north Norway. Long known in England as Richardson's Skua, the Arctic Skua's breeding range extends north–south from well beyond 80° N in Franz Josef Land to 52° N in the Aleutians and Kamchatka.

appropriate a certain range of country on which they suffer no intruders of their own species to encroach" (p. 59). Pelicans also claimed attention. They were met with as far north as the Great Slave Lake. Instead of diving for fish, they chose to float downstream on some rapid from the upper end and, while doing so, dipped their pouched beaks into the water to catch fish. This was the American White Pelican. When satiated with fish these birds were seen to stand on a boulder in mid-stream and to "air themselves, keeping their half-bent wings raised from their sides" (p. 60).

Brown or Sandhill Cranes, supposed by Audubon and others to be the young of the large white or Whooping Crane, were seen by Richardson and Rae along the banks of the Mackenzie from Fort Norman in latitude 65° N downstream to the Arctic coast. "They are in the habit of dancing round each other very gracefully on the sand-banks of the river" (p. 61). The fact that the white bird was unknown so far north helped to convince Richardson that the two cranes were distinct species. In this account Richardson again mentions the Sand Martins of the Mackenzie Delta, commenting on the fact that they cannot begin burrowing until the thaw is well under way at the end of June and must leave early in September when the insects they feed on are killed by frost. But he now entertains some doubt about the identification of these birds. Admitting that he failed to procure a specimen, he suggests that they were perhaps Northern Rough-winged Swallows, a bird which however does not range further north than central British Columbia. Later in his narrative, Richardson mentions Red-throated Divers, Lapland Buntings and Snow Buntings breeding on the Arctic Coast and describes a breeding colony of Sabine's Gulls, with the young already on the wing. "Mr. Rae shot some fine male specimens, whose plumage and dimensions agreed exactly with the description in the *Fauna Boreali-Americana*. The eggs are deposited in hollows of the short and scanty mossy turf which clothes the ground" (p. 159). But no such positive identification was made of the pair of supposed Ivory Gulls found breeding on 16 August near Cape Parry, the young of which "had ash-gray backs, and were nearly fledged". They were perhaps Glaucous Gulls. Once again Richardson shows limitations as a field ornithologist that were shared by almost all his contemporaries.

THE HUDSON'S BAY COMPANY AGAIN:
RODERICK MACFARLANE AND THE ESKIMO CURLEW

The birds collected by John Richardson and others on the two overland Franklin expeditions had been presented to the Zoological Society of London or the Museum of the University of Edinburgh. The arrival of an enthusiastic young American naturalist on the Mackenzie River in 1859 changed all that. For Robert Kennicott, with the full support of the Hudson's Bay Company, was working for the Smithsonian Institution in Washington and, during a 3-year stay in the Canadian Arctic "he managed to infuse into one and all with whom he had any intercourse more or less of his own ardent, zealous and indefatigable spirit as a collector" (MacFarlane 1891: 413). One of the Hudson's Bay Company officials who came under Kennicott's spell was the chief factor at Fort Resolution on the Great Slave Lake, Bernard R. Ross; another was Roderick MacFarlane. Both were soon collecting birds for the Smithsonian Institution, where the then doyen of American ornithology, Spencer F. Baird, encouraged them and others with a friendly and detailed correspondence. Knowledge of the Arctic birds of North America was soon considerably extended by these enthusiasts backed in the early 1860s by at least 15 other Hudson's Bay Company officers. Two new Arctic species were discovered and named. While working through some skins sent to the Smithsonian Institution from the Great Slave Lake by Kennicott and Ross, the 18-year-old ornithologist Elliott

114   *In search of Arctic birds*

Coues came across a hitherto unknown sandpiper. He named it, after his professor, "Baird's Sandpiper" (Mearns and Mearns 1988: 39). It is one of the commonest and most characteristic Arctic breeding waders of the New World, its range extending to northwest Greenland in the east and Wrangel Island in the west. In 1861 Bernard or, as he was familiarly known, "Barney" Ross scored another ornithological success when he sent a skin of an apparently new goose to the Smithsonian. John Cassin named it after its discoverer, "Ross's Goose" (Eifert 1962: 234–235). This was perhaps a little hard on Samuel Hearne, whose description of this goose had been published in 1795. He had even sent a skin to London but, as indicated earlier in this chapter, the skin never reached its destination, apparently because of the negligence of Chief Factor Moses Norton at Churchill.

The most assiduous and determined bird collector among the officers of the Hudson's Bay Company was surely Roderick Ross MacFarlane (Gollop *et al.* 1986; McFarlane 1891). He began his career with the company in 1852 as assistant clerk, aged 19. In 1854 he was placed in charge of Fort Good Hope on the Mackenzie River and in 1861 he built and established Fort Anderson on the Anderson River in the centre of what was then, and indeed still is, a

Plate 19   Baird's Sandpiper near its nest in dry lichen-covered tundra. Cambridge Bay, July 1986.

virtually unknown area east of the lower Mackenzie River. Based there from 1861 to 1866, newly-trained and inspired by Kennicott, MacFarlane, aided in every possible way by his employers, devoted as much time and energy as he could muster to collecting birds and their eggs and sending them to the Smithsonian Institution. During the long, cold, winter evenings he packed his specimens in boxes, labelled them and wrote out his notes on them. In the following spring they would be sent up the Mackenzie with the company's fur shipments and so, eventually, to Washington D.C. In June each year he set out on a protracted field trip of upwards of 100 miles (160 km) at the height of the mosquito season to collect breeding birds and their eggs, aided by a group of Indians. The mind boggles at the size and scope of MacFarlane's ornithological collection. He sent a total of 5000 specimens to the Smithsonian in the 5 years he was working the Anderson River area, often in extremely difficult conditions.

MacFarlane's zeal as an egg-collector is shown by the number of clutches taken of some of his "target" species as set out in Table 12. It is only fair to add

Table 12 Numbers of clutches of selected species taken by R. R. MacFarlane in the Anderson River area, 1862–1866 (from MacFarlane 1891, and, for Baird's Sandpiper, from Bent 1927: 194).

Species	No of clutches taken
Pacific Loon	165
Red-throated Diver	40
Long-tailed Skua	over 30
Bonaparte's Gull	37
Long-tailed Duck	over 100
Eider	over 1000 eggs
King Eider	200 eggs
Brent Goose	650 eggs
Tundra Swan	20
Red-necked Phalarope	about 70
Baird's Sandpiper	7
Least Sandpiper	about 20
Buff-breasted Sandpiper	about 20
American Golden Plover	170
Semipalmated Plover	20
Willow Grouse	nearly 500
Rough-legged Buzzard	70
Golden Eagle	12
Gyrfalcon	over 20
Lapland Bunting	83
Smith's Longspur	150
American Tree Sparrow	216

that he kept detailed written records of all his specimens and many of his observations and submitted these to the Smithsonian too. Years later he summarised this mass of valuable material in a paper published in 1891 entitled "Notes on and list of birds and eggs collected in Arctic America, 1861–1866".

It was in the north of his area, some 10 miles inland from the shore of Franklin Bay, that MacFarlane found the first authenticated Sanderling's nest on 29 June 1863. Composed of dry grass and leaves, placed in a depression in the ground, it contained four fresh eggs. The female was subsequently snared on her eggs, but MacFarlane "never afterwards succeeded in finding another nest". While the distinction of finding the first ever Sanderling's nest containing eggs goes to MacFarlane, he also takes credit, up to the time of writing, for finding the last Eskimo Curlew's nest with eggs, in June 1866. To MacFarlane, the Eskimo Curlew (Gollop *et al*. 1986) was one of the more abundant breeding waders of the Barren Grounds north of the forest in his area. "Among the many joyous bird notes which greet one while crossing these grounds, especially on a fine sunshiny morning, none seemed more familiar or pleasanter than the prolonged mellow whistle of the Esquimaux Curlew." Thirty nests were found and clutches taken of a bird which at that time rivalled the Passenger Pigeon in its abundance in North America. It almost rivalled the Passenger Pigeon in its rapid disappearance, too, but, miraculously, a few Eskimo Curlews survive to this day.

After the closing down of Fort Anderson in summer 1866 no white people and very few Indians visited the surrounding forest and tundra during the rest of the nineteenth century. At first, in the 1860s and 1870s, the Eskimo Curlew remained abundant as an autumn migrant on the Labrador coast and a spring migrant on the prairies of Kansas and Nebraska. A customs official in Labrador used to see them flying to and fro between the hills and the shore in flocks of 50 to 200 or 300 birds. His notes show that Hudson's Bay Company officials joined in the large-scale slaughter of these Eskimo Curlews, which many regarded as a delicacy (Bent 1929: 126):

> The Hudson's Bay Co.'s people at Cartwright annually put up large numbers of hermetically sealed tins for the use of the company's officials in London and Montreal. I have seen as many as 2,000 birds hung up in their store as the result of one day's shooting by some 25 or 30 guns. A fairly accurate idea of the plentifulness of these birds will be obtained from an account of my own experience. During the season I used to leave the cruiser at 6 a.m. and return at 9 for breakfast. I do not remember ever getting less than 30 or 40 brace during the two hours or so that I was shooting.

The greatest killings of Eskimo Curlews were reputed to have been those that took place in the spring on the western plains. One observer described how the hunters would drive out from Omaha, Nebraska, and fire away until they had filled a wagon with the sideboards on with Eskimo Curlew carcasses. Even after the wagons were full, more birds would be shot and dumped in huge piles on the prairie (Bent 1929: 133). Since these birds were also shot in their winter quarters in South America it is not surprising that their number declined rapidly. Between 1870 and 1890 the hordes of migrating Eskimo Curlews disappeared from the Atlantic coast and from the mid-west. After 1900 sightings became so irregular and sparse that A. C. Bent, writing in 1929, thought the species was "nearly or quite extinct". But Eskimo Curlew records continued. In the three decades since 1950 there were 4–6 sightings per decade and in the 1980s about 8 or more per annum. Nor were these later records all of migrating birds. From 1972 on, Tom Barry, of the Canadian Wildlife Service, searched the Fort Anderson area repeatedly both on foot and with a helicopter. On 24 May 1987 his Inuit guide and assistant Billy Jacobson saw an Eskimo Curlew on the tundra where MacFarlane had found them over a century before. He stopped his motorised sledge and watched the bird for some time. A second bird flew off when he re-started his skidoo (Anon. 1988; Iversen 1989). It looks as though we may not have to rely for very much longer on Roderick MacFarlane's meticulous notes for our knowledge of the breeding biology of the Eskimo Curlew.

THE U.S. ARMY AND THE BIRDS OF ALASKA

In describing the ornithological highlights of the 1881 cruise of the *Corwin* in Chapter 3 of this book, mention has already been made of the naturalist E. W. Nelson (Goldman 1935). He began collecting birds while still at school in Chicago and soon had a paper on the birds of Illinois to his credit. His lucky break came when he got to know Spencer Fullerton Baird, of the Smithsonian Institution, for Baird recommended Nelson, who had found employment in Chicago as a teacher, to the Signal Service of the U.S. Army for a post in Alaska. So, in 1877, 22-year-old Private Edward Nelson voyaged from San Francisco to St Michael, a remote outpost on the western coast of Alaska not far from the Yukon Delta. There he stayed 4 years, having for companions the three or four more-or-less permanent residents of the place. He had been ordered by the Chief Signal Officer to "take charge of the signal station" there and secure a continuous series of meteorological observations. He was also instructed to "obtain all the information possible concerning the geography, ethnology and zoology of the surrounding region" (Nelson 1887). He certainly made the most of these compulsory opportunities, for he published an important ethnological work on the Bering Strait Eskimos as well as a monumental *Report on the natural history collections made in Alaska between the years 1877 and 1881*, a 337-page quarto volume which covers mammals, fishes and butterflies as well as birds. It was published in 1887.

At St Michael, Nelson was by no means starting from scratch. Between them, the U.S. Army Signals Corps and Spencer Fullerton Baird had already in 1874 installed a highly competent ornithologist there in the shape of Lucien McShan Turner (Gabrielson and Lincoln 1959: 8–12). After devoting his first year at St Michael to studying and collecting in the immediate vicinity, Nelson made two prolonged trips to the Yukon Delta, on the second of which, in summer 1879, travelling by dog sledge and kayak, he obtained skins and

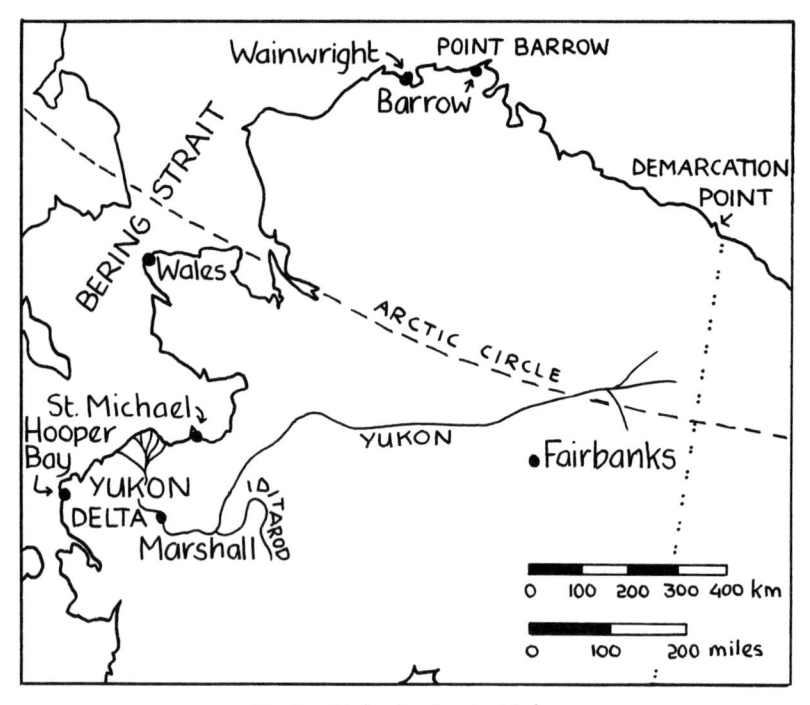

Map 9   Bird collecting in Alaska.

eggs of many breeding birds, including those of his special quarry, the Emperor Goose.

Although credited with sending over 2000 bird skins and 1500 eggs to the Smithsonian (Sherwood 1965: 94), Nelson was not only a collector but an excellent field observer too. He made the first observations ever of the courtship and breeding habits of that typical wader of marshy tundra, the Pectoral Sandpiper. The very first issue of the *Auk* (Nelson 1884: 218–221), journal of the American Ornithologists' Union, carried his account of the Pectoral Sandpipers he had encountered on an island in the Yukon Delta in May 1879:

> The night of May 24 I lay wrapped in my blanket, and from under the raised flap of the tent looked out over as dreary a cloud-covered landscape as can be imagined. The silence was unbroken save by the tinkle and clinking of the disintegrating ice in the rivers, and at intervals by the wild notes of some restless Loon, which arose in a hoarse, reverberating cry and died away in a strange gurgling sound. As my eyelids began to droop and the scene to become indistinct, suddenly a low, hollow, booming note fell upon my ear and sent my thoughts back to a spring morning in Northern Illinois, and to the loud vibrating tones of the Prairie Chicken. Again the sound arose nearer and more distinct, and with an effort I brought myself back to the reality of my surroundings and, resting upon one elbow, listened. A few seconds passed and again arose the note. A moment later and, gun in hand, I stood outside

Bird collecting in North America 119

(a)

(b)

Plate 20 (a) Incubating Pectoral Sandpiper in typical wet marshy tundra. (b) The same Pectoral Sandpiper's nest without a bird, showing the four eggs. Abundant in the mainland coastal tundras of the North American Arctic, the Pectoral Sandpiper also breeds in Siberia between the Yamal and the Chukchi peninsulas.

the tent. The open flat extended away on all sides with apparently not a living creature near. Once again the note was repeated close by and a glance revealed its author. Standing in the thin grasses, ten or fifteen yards away from me, was a male Pectoral Sandpiper. The succeeding days gave me opportunities to observe the bird, as it uttered its singular notes under a variety of situations and at various hours of the day or during the light Arctic night. The note is deep, hollow, and resonant, but at the same time liquid and musical, and may be represented by a repetition of the syllables *tōō-û, tōō-û-tōō-û, tōō-û*. . . .

Before the bird utters these notes it fills the aesophagus with air to such an extent that the breast and throat are inflated to twice or more the natural size, and the great air-sac thus formed gives the peculiar resonant quality to the note. . . .

. . . The male may frequently be seen running along the ground close to the female, its enormous sac inflated and its head drawn back and the bill pointing directly forwards; or, filled with spring-time vigor, the bird flits with slow but energetic wing-strokes close along the ground, its head raised high over the shoulders, and the tail hanging almost directly down. As it thus flies, it utters a succession of the booming notes adverted to above, which have a strange ventriloquial quality. At times the male rises twenty or thirty yards in the air and, inflating its throat glides down to the ground with its sac hanging below; again he crosses back and forth in front of the female, puffing out his breast and bowing from side to side, running here and there as if intoxicated with passion. Whenever he pursues his love-making his rather low but far-reaching note swells and dies in musical cadence, and forms a striking part of the great bird rising at that season in the North.

In his *Report* already mentioned Nelson reprinted this account with some minor changes. The *Report* included equally vivid accounts of other Arctic breeding waders, including the Grey and Red-necked Phalarope, the Western Sandpiper and the Dunlin. Years later he contributed a more popular description of his ornithological experiences in Alaska to an Audubon Society publication on the birds of Alaska (Ingersoll 1914). The climax of his distinguished career came in 1916–1927, when he was chief of the Bureau or Division of Biological Survey of the Department of Agriculture of the United States Government, now the Fish and Wildlife Service of the Department of the Interior. He died in 1934 (Goldman 1935).

The Signal Corps of the U.S. Army did Alaskan Arctic ornithology another important service when it attached Sergeant John Murdoch to its International Polar Year Expedition to Point Barrow in 1881–1883. He contributed a section on "Birds" to the official report of the expedition by its commander Lieutenant Ray (1885: 104–128), listing 54 species and subspecies recorded within 15 miles of the expedition's station between 8 September 1881 and 25 August 1883 (Gabrielson and Lincoln 1959: 12). This important paper on the birds of Barrow would have contained the first published account of the "eggs and breeding habits" of the Pectoral Sandpiper. But, Murdoch observes rather sourly, since he had written his description that of Nelson had been published in the *Auk*. He describes the "curious antics" of the male Buff-breasted Sandpiper, which walks along with one wing "stretched to the fullest extent and held high in the air". A bird would stretch itself to its full height with wings spread forward and, puffing out its throat, make a clucking noise, while one or two other birds "stand by and apparently admire him". Murdoch gives the first full description of the shooting of enormous numbers of migrating King Eiders by the local Eskimos as they fly over Point Barrow (see further Thompson and Person 1963), and the first record of the Spectacled Eider north of Bering Strait. Although he found no nest, he shot a duck Spectacled Eider on 19 June 1883 "not far from the station" which had a fully-formed egg in its oviduct. This bird was at that time only known from the neighbourhood of St Michael, where the very first specimen was collected by I. G. Vosnesensky. Sent to the Academy of Sciences at St Petersburg, it was named by the director of that institute's Zoological Museum, J. F. Brandt, in 1847, *fischeri* after the German zoologist and entomologist J. G. Fischer (Mearns and Mearns 1988: 151–152). Murdoch has two other claims to ornithological fame. His discovery and detailed description of the autumn migration of Ross's Gull at Point Barrow added significantly to knowledge of that little-known bird. Second, his report of a male Curlew Sandpiper shot near Barrow on 6 June 1883 long stood as the only western North American record of this species, which normally breeds on the coastal tundras of Arctic Siberia between the Taimyr and Chukchi peninsulas. However, to the surprise of ornithologists, between 1962 and 1974 Curlew Sandpipers bred near Barrow on several occasions, seven nests being found in the single year 1972 (Glutz von Blozheim *et al.* 1975: 676). Since then there has been no further trace of these birds there.

KNUD RASMUSSEN AND THE DANES

Bird collecting was by no means the main activity of the sledging expedition led by explorer–anthropologist Knud Rasmussen in 1921–1924 from Hudson Bay to Alaska, so vividly described in his *Across Arctic America* (1933). The expedition was primarily anthropological; its official title was The Fifth Thule Expedition, Danish Ethnographical Expedition to Arctic North America, 1921–24. It was the fifth of a series of expeditions led from, based on and financed by the Thule Trading Station, founded by Rasmussen in northwest Greenland in 1910. Its personnel did not include a professional ornithologist but its naturalist and cartographer, Peter Freuchen, did his best to supply this want, while Rasmussen and another expedition member, Helge Bangsted, shot and skinned birds whenever they could. The most important places where they did this are shown on Map 10. It was a fine achievement considering the privations and difficulties of transport, that upwards of 300 skins and an assortment of bits and pieces of birds—heads, skulls, wings, skeletons and the like—eventually arrived at the University of Copenhagen's Zoological Museum. There, curator Richard Hørring (1937) was faced with the task of sorting out, cleaning and identifying the material and deciphering Peter Freuchen's notebooks. "Many of the skins were received with the inside out and the feathers inside, greasy and injured, and damaged by salt and many of the labels were difficult or impossible to decipher." As to Freuchen's field notes, these were "arranged into a very complicated system and written in a hardly legible handwriting". Hørring made no attempt to go through Rasmussen's 30

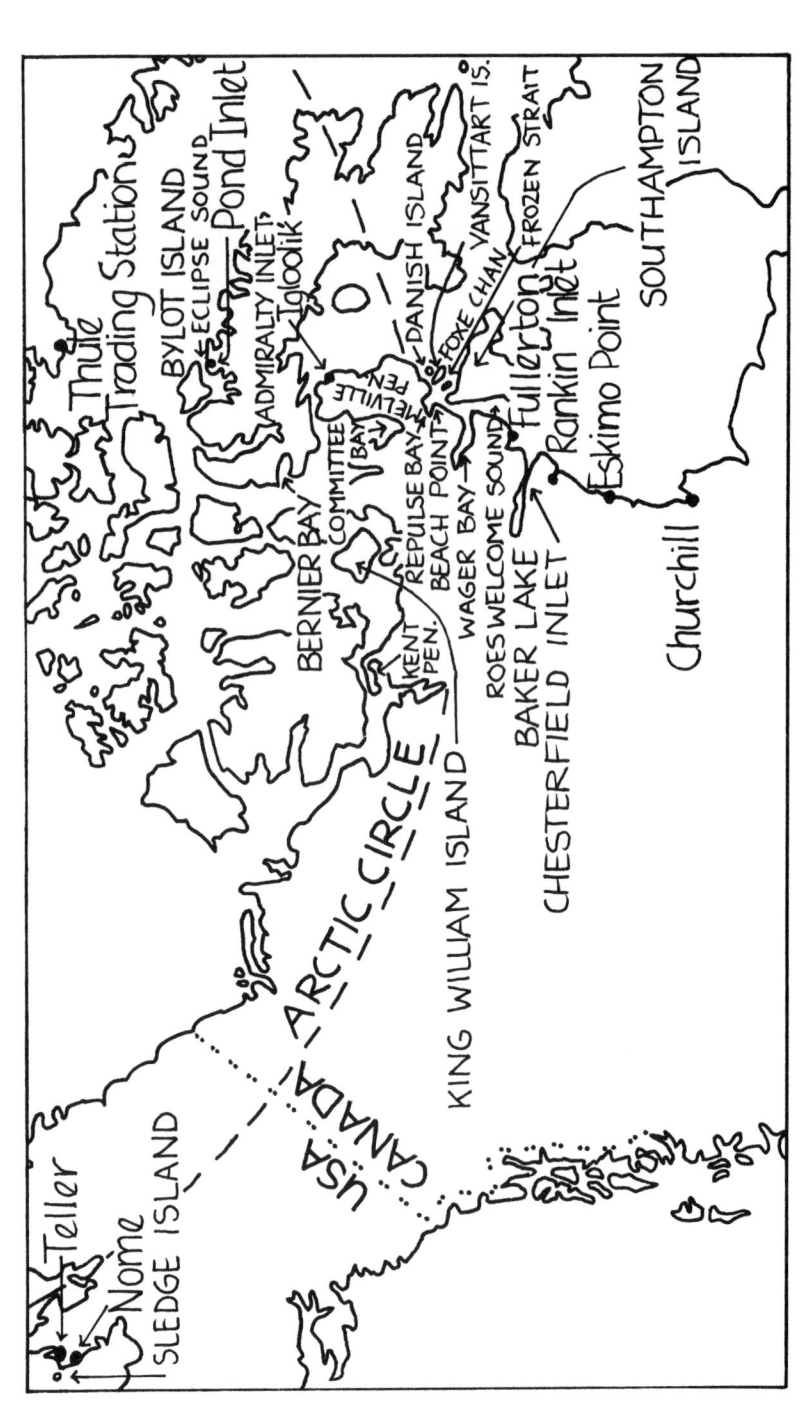

Map 10 Places where birds were collected on the Fifth Thule Expedition.

notebooks about the expedition; his handwriting would have been harder to decipher than Freuchen's. Naturally it was difficult to pinpoint the subspecies of every bird collected, but Hørring thought it was his duty to attempt this. He even managed to do it in the cases of a Redpoll, though all he had was a nest with three broken eggs and a locality, and a Canada goose, of which he had the skull only. His efforts bore fruit 13 years after the Expedition's return, in one of the ten published volumes of its reports.

The Fifth Thule Expedition's bird skins and other specimens represented 62 species, not all of them strictly Arctic: they included an American Robin from Churchill and a Slaty-backed Gull from Sledge Island in Norton Sound. All four North American divers were in the collection, which also included 11 of Canada's 17 or so Arctic breeding waders, two out of three skuas, and most of the seabirds. Absent were Snow Goose and Gyrfalcon, though these had been seen by expedition members, as well as Pectoral and Stilt Sandpiper, Pomarine Skua and Wheatear. Eggs were collected as well as skins, those of Long-tailed Duck, King Eider, Sandhill Crane, Semipalmated Plover, Thayer's Gull, Arctic Tern, Peregrine, Shore Lark, Redpoll, Snow Bunting and Lapland Bunting all reaching Copenhagen safely. For some reason the last-mentioned species received special treatment. In the case of most birds only one to six eggs were collected, but six complete clutches of Lapland Buntings eggs were collected by Bangsted and Freuchen during the expedition's last summer, of 1924.

Were it not for the inclusion of Peter Freuchen's field notes, Hørring's book would be little more than a dry inventory. Freuchen, either writing in quaint English or poorly translated, adds touches of colour and interest. For example he describes how he visited a Peregrine's nest on 18 June 1922 at the mouth of Wager Bay. The four eggs had been taken a few days before, but the adult bird "stood in the air just above the nest and screamed incessantly; it made looping the loop down towards us". Interestingly, "round the hole there were many remains of lemmings, but no remains of birds" (Hørring 1937: 101). In July 1922 on an islet off the Melville Peninsula coast he has this to say of Snow Buntings (Hørring 1937: 118):

> 12.7.1922. I saw a male Snow Bunting at a lake; it caught exclusively gnats on stones, moss and in the water; it hopped about in shallow water which went up over its feet; it gathered several grass leaves at a time with its bill and flew away with them.

> 13.7.1922. The Snow Buntings sleep sheltered by projecting stones. I roused several by surprising them when I walked silently over the rocks.

Watching Snowy Owls on Vansittart Island in October 1922, Freuchen saw one pursue an Eider. On another occasion a Snowy Owl and a Raven were watched in aerial combat. "The Snowy Owl was the attacker, and the Raven was whirled round in the air, screaming at every stroke it received". Later, however, a Snowy Owl perched on a boulder merely turned its head to face a Raven flying round it calling (Hørring 1937: 103).

124 *In search of Arctic birds*

Plate 21 A male Snow Bunting brings a beakful of tundra insects to feed its well-feathered young in a nest under a boulder which once formed part of the wall of an old Eskimo winter house. The place is the former Inuhuit settlement of Uummannaq, Avanersuaq, northwest Greenland, and the date, 9 July 1984.

THE HOOPER BAY EXPEDITION OF 1924

Although his time was increasingly taken up with administrative duties, especially after 1916 when he became head of the U.S. Bureau of Biological Survey, E. W. Nelson still maintained a lively interest in the birds of Alaska and it was he who took the initiative in organising a collecting expedition to the Yukon Delta coastal tundra based on the modern village of Hooper Bay, then an Eskimo settlement of under 200 persons called Napakiakamut. To effect this, Nelson enlisted the help of a noted amateur ornithologist, Herbert William Brandt (Schorger 1955), who was busy at the time in the field collecting material for his book *Florida bird life*. Brandt was a successful businessman with ample means. He became president of the family firm of wholesale grocers, The Brandt Coy., based at Cleveland, Ohio, and his profits and spare-time interests were reflected in his private Bird Research Foundation in Cleveland and in his nearly annual field expeditions to different parts of the American continent. So the Hooper Bay Expedition was a joint one of the U.S. Biological Survey and Brandt's Bird Research Foundation. Travelling 850 miles by dogsled from Fairbanks to reach Hooper Bay, the expedition was in the field from 30 April onwards throughout the summer of 1924, though Brandt himself set out on his return journey up the Yukon River on 26 June.

The Hooper Bay Expedition was notable in several ways. It was undertaken by a group of expert field ornithologists and bird collectors. Besides Herbert Brandt himself, there was Olaus J. Murie, of the Biological Survey, who had 3 years' field experience in Arctic Alaska and 3 years' in northern Labrador, and Henry Boardman Conover, of Chicago's famed Field Museum of Natural History. Its adventures and ornithological results were published in a solid de luxe quarto volume, beautifully illustrated, financed by the Bird Research Foundation and written by Brandt, entitled *Alaska bird trails. Adventures of an expedition by dog sled to the delta of the Yukon River at Hooper Bay*. Though originally written in 1926, it was not published until 1943. The expedition was brilliantly successful. Counts were made of offshore spring migrants: Brandt logged a total of 125 000 King Eiders flying north but missed many more. Data were obtained on some 1500 occupied nests of 60 species, and waterfowl and other birds were ringed in the late summer after Brandt's departure. A splendid collection of bird skins and eggs was distributed between the Washington D.C. collection of the Biological Survey, the Field Museum at Chicago and the Brandt Collection at Cleveland—all open to public inspection. Some idea of this material is given in Table 13.

Table 13 Numbers of bird skins (adults, juveniles and downy young) and eggs of selected species collected by the Hooper Bay Expedition (Brandt 1943)

Species	No. of bird skins	No. of eggs
Pacific Loon	7	37
Emperor Goose	30	34
Long-tailed Duck	31	50
Steller's Eider	37	40
Spectacled Eider	27	37
Willow Ptarmigan	22	100
Grey Plover	22	120
Black Turnstone	19	120
Dunlin	16	120
Western Sandpiper	20	120
Grey Phalarope	22	100
Pomarine Skua	17	14
Sabine's gull	6	54
Snowy Owl	7	90

The members of the Hooper Bay Expedition had begun their collecting before they reached their Hooper Bay destination. They saw their first nest on 9 April while still on the trail near the Iditarod River before they reached the Yukon. It was that of a Gray Jay. The bird was sitting on a clutch of four eggs and had been doing so since first discovered on 2 April, during which time the temperature had gone down to 35 degrees below zero Fahrenheit ($-37°$ C). The wall of this well-insulated nest, which was 3 inches thick, was "composed of dog fur and the feathers of ptarmigan and Spruce Grouse, all felted together with the consummate skill that the whole jay tribe exhibits in fashioning its abode" (Brandt 1943: 47). Later, on an island in the Yukon River near Marshall, five of the expedition's members were deployed with four shotguns and an egg basket between them to collect a pair of breeding Great Horned Owls and their eggs. "Since, owing to lack of specimens, the subspecific status of the horned owl in this lower Yukon region was unsettled, we were particularly anxious to obtain these birds", writes Brandt (p. 57). The female left the nest when they were still 75 yards away. Brandt climbed to it and collected four well-incubated eggs while his companions dispersed in the hopes of a shot at the adult birds. With some difficulty they shot the male. "To obtain the female, we set two fox traps in the nest and put in three precious hen's eggs". The male proved to belong to the pale "Arctic" form of the Great Horned Owl, while the female was a darker brown typical bird.

GEORGE MIKSCH SUTTON

When he died in 1982 aged 64, G. M. Sutton, affectionately known as "Doc" in his later years, was probably one of the best known and most liked of American ornithologists. His autobiography, published in 1980, described his experiences up to 1935 in a lively and most entertaining way. He was a distinguished bird painter, having been encouraged to paint birds and given an old paint box by the renowned artist Louis Agassiz Fuertes. He illustrated at least 15 other bird books besides his own. He was a museum man and a collector as well as a successful university teacher, beginning his professional life with W. E. Clyde Todd at the Carnegie Museum, Pittsburgh. In 1929 he registered as a graduate student at Cornell University, Ithaca, New York, but disappeared at once to the Arctic to do field work for his thesis on the birds of Southampton Island. For a decade after 1932 he was Curator of Birds at Cornell and towards the end of his career was for many years Professor of Zoology and then Research Professor of Ornithology at the University of Oklahoma (Pettingill 1984).

Sutton's collecting skills were proverbial. A superb shot, he was also a brilliant skinner and was credited with being able to skin, fill out and sew up a small bird in 15–20 minutes (Pettingill 1984). An admirer who watched him handle a bird he had just shot maintained that he could "preen" a bird better than the bird itself. He always carried a supply of paper cones for slipping dead birds into, including roadside kills. The same writer thought Sutton "must have painted a picture of every feather of every bird in the world". He could "walk through the woods and pick up a feather and identify not only its owner but the part of the body it came from" (Janovy 1978: 139–154).

Although his publications included books on the birds of Oklahoma, Mexico and Pennsylvania—for Sutton was a prolific author—the Arctic repeatedly claimed his attention and for years he planned to write and illustrate a book on Arctic birds as a whole. In 1936, in *Birds in the wilderness*, he described his trip to Churchill in the summer of 1931 to find the hitherto undiscovered eggs of Harris's Sparrow. Sutton and three other Cornell men scoured the countryside for a week in early June in competition with four Canadian ornithologists. The prize went to Sutton—he became the first white man ever to see the boldly-blotched reddish-brown eggs of Harris's Sparrow (Eifert 1962). In 1961, *Iceland summer. Adventures of a bird painter* was published, illustrated with a series of beautiful sketches of birds in down or just fledged, and a marvellous colour plate—perhaps Sutton's finest ever bird painting—of an Icelandic Gyrfalcon. In a later book *High Arctic. An expedition to the unspoilt north*, published in 1971, Sutton described a trip to Bathurst Island and other places in the Canadian Arctic archipelago with his former pupil David Parmelee and other ornithologist collectors. On this expedition, as well as blowing eggs and skinning birds, Sutton painted them, and some of his beautiful and evocative water colours of birds in an Arctic landscape are reproduced in the book, including a fine Purple Sandpiper and a Ptarmigan pair.

Sutton's most important contribution to Arctic ornithology was undoubtedly his doctoral thesis on the birds of Southampton Island. Published in 1932 in the Memoirs of the Carnegie Institute at Pittsburgh, this 275-page large quarto volume represented the first systematic modern treatment of the avifauna of a single Arctic locality, in this case a 39 000-square-mile

128 *In search of Arctic birds*

(101010 km^2) island at the northern end of Hudson Bay. The book was superbly illustrated with four fine watercolour plates of birds, exquisite watercolour paintings of the downy young of 13 species, and 48 black-and-white photographs. It was a tour de force, and the model for subsequent area studies.

Accounts of the 65 birds recorded from the island, 47 of them breeding but only 7 of them resident, take up most of the book. Under each species Sutton describes its status on Southampton Island, giving individual records if it is scarce, its annual routine, and its Eskimo name. He seems to describe almost every nest found and every bird collected. He portrays the birds and landscape just as vividly in writing as in painting. Witness the following excerpt from his account of the Long-tailed Duck (Sutton 1932: 61):

> The species was not recorded inland until June 13, on which date several pairs and a flock of eight were noted and a male and female collected. Once the species made its way in to the thawing lakes, the tundra began to ring and echo with the incessant *Ah, ah, ahng-owi* of the males. By the time the snow had disappeared, the birds became even more abundant, until flocks or pairs were to be seen in every pool.
>
> The courtship antics of the Old-squaw were very amusing. The males were violently attentive to the females, swimming about close to them all the time and

Plate 22 This handsome drake Long-tailed Duck allowed close approach one summer day in northwest Greenland. The bird breeds in marshy tundras all round the Arctic and as far south as Iceland and north Norway.

chasing them everywhere. They bowed, pointed their bills into the air, and laid their heads backward over their shoulders and backs, as they gave their amorous call-notes. Sometimes the females too bowed a little. Noticeable wiggling of the long tail-feathers accompanied many of these antics, and sometimes the tail was lifted so high in air that it pointed almost straight up. . . .

On June 18 at Itiuachuk I watched the birds mating. The males flew after the females everywhere, sometimes swooping with terrific speed at them, then shooting into the air to a height of a hundred feet to fall backward giddily, almost "looping the loop", and coasting to one side before resuming the pursuit. Usually these long chases, in which as many as three or four males joined in pursuit of one female, ended in the water. The females often seemed to be genuinely terrified, crying out hoarsely in objection to the violent attacks, and evidently doing their best to find a refuge somewhere. Sometimes all the birds flew higher and higher into the air until they were almost out of sight; then with a rush they returned on set wings, swinging from side to side with breath-taking speed, to plunge straight into the water, their wings open and even beating as they disappeared under the surface.

CHAPTER 6

Britons in the Russian and Siberian Arctic, 1850–1914

Towards the close of the nineteenth century the attention of British ornithologists, almost all of whom in those days were collectors of skins and eggs, was attracted northward and eastward from the British Isles towards the Arctic, the tundra and, ultimately, northern Siberia. Here, it was thought, new species were to be found in regions so far virtually unexplored. Moreover, the nests and eggs of birds familiar in Britain in winter or on spring and autumn passage might be discovered. Many of these, which evidently bred somewhere in the north, were waders, and the breeding places of some were entirely unknown. The enthusiasts who form the subject of this chapter combined an appreciation of the excitement of travel, often in harsh conditions; a certain mania for collecting which they shared with many of their contemporaries; a genuine interest in scientific ornithology, especially in the systematics and geographical distribution of birds; and a passion for publicity, for they vied with each other in publishing accounts of their travels. These came out in book form, being written for the general public, while their more strictly ornithological reports came out in the *Ibis*. This official organ of the British Ornithologists' Union was published from 1859 onwards, and the heroes of this chapter were all of them members of that organisation, often referring familiar to each other as "Ibises".

THE SPELL OF LAPPLAND

It was the Eton and Cambridge educated John Wolley, one of the co-founders in 1858 of the British Ornithologists' Union, who really began the craze for bird collecting in the far northeast. Born in 1823, he died in 1859 aged 36, in the very year in which Volume 1 of the *Ibis* was published. He took eggs himself, touring different parts of Europe for this purpose, bought eggs at auctions or from his fellow collectors, and employed others to collect eggs for him. Following a common practice of the time, more eggs were taken than were needed for the collection and those surplus to requirements were sold. In Wolley's case a "small proportion" of his eggs was sold annually at various auction rooms or at J. C. Stevens's in London (Newton 1864: xxx). Wolley laid the foundations for the magnificent collection which passed after his death to his friend Alfred Newton, most famous of all British nineteenth-century ornithologists. Newton not only added greatly to it but he also made himself responsible for the publication of Wolley's journals and records, as well as many of his own, in a magnificent multi-part work called *Ootheca Wolleyana*, issued in five parts between 1864 and 1905. Its subtitle was *An illustrated catalogue of the collection of birds' eggs, begun by the late John Wolley Jnr. and continued with additions by Alfred Newton*. Wolley's first egg-collecting excursion in Scandinavia was made in 1853 to Lappland. Based at Muoniovaara, a mile or two from Muonio, 100 miles (160 km) north of the Arctic Circle, he explored the forests and marshes on both sides of the Muonio River, which here forms the boundary between Sweden and Finland, with the help of an army of local volunteers. Wolley remained in Scandinavia for the winter and returned to Muonio in spring 1854. Late that summer, on a brief return visit to England, he brought back eggs of the Lesser White-fronted Goose, Bar-tailed Godwit and Spotted Redshank, most of them taken by Lapps working for him. Newton claimed that those of the Spotted Redshank "were certainly obtained for the first time": the two eggs "from Norway" figured in 1851 by German oologist F. A. L. Thienemann purporting to be of this species were not good enough for Newton (1864: 142), not being sufficiently authenticated. Wolley usually wrote on his eggs at the time of taking them to ensure their authenticity.

In 1855 Wolley travelled north to the far northeastern tip of Norway and spent the summer with Alfred Newton and a friend collecting in the Varanger Peninsula, especially around Vadsø and along the lower reaches of the Tana River. This area had been discovered ornithologically in 1840 by the German–Russian naturalist A. F. Middendorf (1843). Here they saw flocks of Steller's Eiders and found the nest and eggs of the Red-throated Pipit, long considered a dubious species (Newton 1864: xxxi-xxxii). But, wrote Newton exuberantly (1864: 212), "Of all Mr. Wolley's discoveries the one with which his name

132 *In search of Arctic birds*

Map 11 Lappland.

will be especially perpetuated is his unveiling the mystery that had hitherto surrounded the breeding-habits of the Waxwing". Wolley, his health already failing, made the exciting announcement at a meeting of the Zoological Society of London on 24 March 1857. A Lapp boy he employed had discovered "The first nest of Waxwing ever found for scientific purposes" (p. 213) on 11 June 1856 and snared the male bird on the nest. It contained five eggs and was some way northeast of Rovaniemi on the upper Kemi River in northern Finland, probably not far from the frontier with Russia. From the same locality four more clutches of Waxwing's eggs were taken in 1856 and five in 1857, one by Wolley himself. But it was the 1858 season which provided the most material for the illustrations of Waxwing's eggs at the end of the first volume of *Ootheca Wolleyana*: in that year the Wolley organisation in Lappland took a total of 84 clutches, mostly of five or six eggs.

The Jack Snipe, like the Waxwing, was well known in Western Europe as a winter visitor but its northern breeding haunts were undiscovered. Wolley filled in this gap in knowledge, too. On 17 June 1853 he flushed a bird in the "Great Marsh" at Muonioniska on the Swedish side of the Muonio River not far from Muoniovaara. Marking the place, he soon saw the nest containing four eggs and, flushing the bird again, shot it; ". . . in a minute I had in my hand a true Jack Snipe, the undoubted parent of the nest of eggs!". Next day Wolley organised "a large party of men and boys" to search the big marsh systematically. "I kept them as well as I could in line, myself in the middle. . . . Whenever a bird was put off its nest the man who saw it was to pass on the word, and the whole line was to stand whilst I went to examine the eggs. . . ." "As usual I took measures to let the whole party share in my gratification before I again gave the word to advance". These distributions of coin were evidently effective, because three more Jack Snipe's nests were found.

This time Wolley was surely right in regarding the Jack Snipe's as "a previously unknown egg". Although a certain Mr Hoy had taken supposed Jack Snipe's eggs on the heath of Valkenswaard in the Dutch province of Noord-Brabant in the 1840s, and sold some to Yarrell and other British collectors, these had rightly been treated with scepticism. It was only much later that the exclusively northerly breeding distribution of the Jack Snipe was appreciated; even as late as 1885 Howard Saunders was only "a little doubtful" whether it bred in Denmark or southern Sweden (Yarrell 1882–1884: 354). Of the clutch of four Jack Snipe's eggs taken on 17 June 1853 at Muonioniska, Wolley gave one egg to Alfred and his brother Edward Newton, sold two at Stevens's in London on 17 February 1854, and kept one for himself.

NORWAY AND RUSSIA

Although the first volume of *Ootheca Wolleyana* was not published until 5 years after Wolley's death, his oological finds were made known quickly and widely in the third edition of William Hewitson's *Coloured illustrations of the eggs of British Birds*, which appeared in 38 monthly parts between 1853 and 1856 (Mullens and Swann 1917: 292–294). Not only were the Jack Snipe's and other eggs collected by Wolley figured there in colour, but Hewitson also published long excerpts of Wolley's diaries describing his discovery of breeding Spotted Redshanks and Temminck's Stints, as well as the abovementioned Jack Snipes and other exciting birds. It was doubtless through the pages of Hewitson that Wolley inspired the young Scot John A. Harvie-Brown, who was 10 years old when Wolley brought his first batch of eggs from Lappland back to England. When he was 21, Harvie-Brown and a friend were able to buy eggs collected by Wolley from a Mr Baker at Cambridge. "For many years", Harvie-Brown (1905a) writes in the preface to his *Travels of a naturalist in northern Europe*, he "had studied the collecting experiences of the late Mr. John Wolley" and "treasured up in memory and in my notebooks . . . many sentences of Wolley's writings, regarding the discoveries he had made. . . ". Eventually, in 1871, Harvie-Brown was able to begin the realisation of his ambition to follow in Wolley's footsteps and even further. He undertook a fairly modest collecting trip in southern Norway with his friend Edward A. Alston. After several successful weeks they completed the packing of a wooden box containing 233 Fieldfare's eggs, "in addition to the six nests and eggs [of this species] in our big box". This represented only a small part of the eggs and skins brought home.

This "preliminary canter", as Harvie-Brown termed it, only whetted his appetite for more. "Dreams were dreamed" and plans were sketched out with the help of "many kindly Brethren of the B.O.U.—Professor Newton, H. E. Dresser and others". Two possibilities eventually held the field: northeast Finland or the eastern shore of the White Sea (Beloye more). In the outcome Russia won the day and in 1872 Alston and Harvie-Brown travelled to Archangel (Arkhangel'sk) and spent several weeks collecting among the islands of the Dvina Delta. The Moscow–Archangel railway had not yet been built, so after travelling to St Petersburg (now Leningrad) by train, the two naturalists covered most of the 750 miles (1200 km) from St Petersburg to Archangel by country cart. The nineteenth-century bird collector by no means travelled light. Harvie-Brown (1905a: 135) describes the "personal luggage" he took with him as follows:

> A large portmanteau, containing a suit, spare knickerbockers, light trousers, velvet coat and waistcoat, 3 pairs of long stockings, and 4 of socks, 3 flannel shirts with collars, 12 collars, ties, 18 pocket-handkerchiefs, 2 pairs of shooting boots, slippers, leggings, mosquito-veils, gloves, spare boot-laces and boot-nails, toilet requisites (brush, comb, tooth-brush and tablets, soap-box and soap, sponge), vols. ii. and iii. of Bree's book, Blasius' list, writing-case, gummed paper, indiarubber rings, botanical paper and boards, needle-book and thread-bag, box of pins, triangular needles,* Keating's insect powder, labels ready cut, wax candles, vestas and pipe-lights, spare pipes, tobacco (1½-lb. uncut, ½-lb. cut), climbing irons,* small fly-book and reel, flexible hat, cap, small powder and shot flasks for stickgun.
>
> A hand-bag contained my journal and three or four small note-books with straps, map and guide-book, novel, telescope, hunting-knife, ½-lb. cut tobacco, large flask, collecting-box, ink-bottle, compass, egg-instruments, etc.
>
> A bundle consisted of a plaid, macintosh, and rubber ground-sheet.
>
> The gun-case contained gun, cleaner, oil, and gun-sling and cartridges.
>
> Another package contained fishing-rods,* stick-guns, and ramrods.
>
> *These articles were afterwards found to be unnecessary.

The "vols. ii. and iii. of Bree's book" covered warblers, buntings, larks, finches and others. The book was *A history of the birds of Europe not observed in the British Isles* by C. R. Bree, published in 1864. J. H. Blasius was one of the leading German ornithologists of the middle years of the nineteenth century. Blasius's *Reise im europäischen Russland* was published in 1841; in 1862 Alfred Newton published his *List of the birds of Europe* in a handy format.

Harvie-Brown and Alston were the first two Britons to penetrate into Russia in quest of birds. German ornithologists had already been active there but the Dvina Delta was ornithologically virgin territory. It is true that the Swedish ornithologist W. Meves had visited some of the islands in summer 1869, but only briefly, and only in July–August, after the breeding season. Moreover, his observations had only been published in part and in Swedish; the full version of his report did not appear in German until privately printed in Vienna in 1886 (Mewes and von Homeyer 1886). Thanks in the main to the English ornithologist H. E. Dresser, author of the classic *Birds of Europe*, and his continental contacts, the two travellers were rendered every necessary assistance. At Archangel they found that a Mr Shergold had booked excellent lodgings for them at one-and-a-half roubles per diem each with Madame Nathalie Leitzoff, had employed a collector for them who handed them about 150 eggs that very day, 51 of them Terek Sandpiper's, and had arranged for a boat for them to explore the delta. With Mr Shergold's help they hired several more men, including a Polish exile named Ignati Nartzisovitch Piottuch who was retained to shoot and skin birds for 30 roubles a month. Then followed a month's work in the field from 17 June, mostly spent in the delta and along the coast on either side of it.

Alston and Harvie-Brown may have been mildly disappointed in the results of their expedition. Among the 331 bird skins and over 200 eggs they brought back, representing about 70 species, there was nothing spectacularly new and exciting. They had shot three fine Grey Plovers, but saw no sign of breeding. The prized Little Stint's nest they had also failed to find, although they had the skin of an adult female shot in the delta. Harvie-Brown did find a Yellow-breasted Bunting's nest containing five eggs which he collected, having first shot both birds. In all, they brought back 42 Yellow-breasted Bunting skins and 33 Little Buntings. A selection of the best of the skins and eggs were retained in the private collection which Harvie-Brown shared with fellow "Ibis" H. W. Feilden.

THE GREAT LAND TUNDRA

Convinced that he had merely touched "the fringe of the nesting distribution of the Little Stint" in 1872, and that an expedition further eastwards really would deliver the goods, Harvie-Brown (1905a: vi) continued planning—and dreaming. At last, in 1875, teaming up this time with Henry Seebohm, whose interest in northern birds had been stimulated by a visit to the Varanger Peninsula with Norwegian ornithologist Robert Collett in 1874, he set out to explore and collect specimens of the bird life along the lower reaches of the next great river to the east of the Dvina—the Pechora. Above all, he and Seebohm hoped to visit the western fringe of the Bol'shezemel'skaya tundra or Great Land Tundra, which stretched for 150 miles from the River Pechora to the northern tip of the Urals. This tundra was, for Seebohm (1901: 131) "the unexplored land, the land of promise". Explorer F. G. Jackson travelled across

136 *In search of Arctic birds*

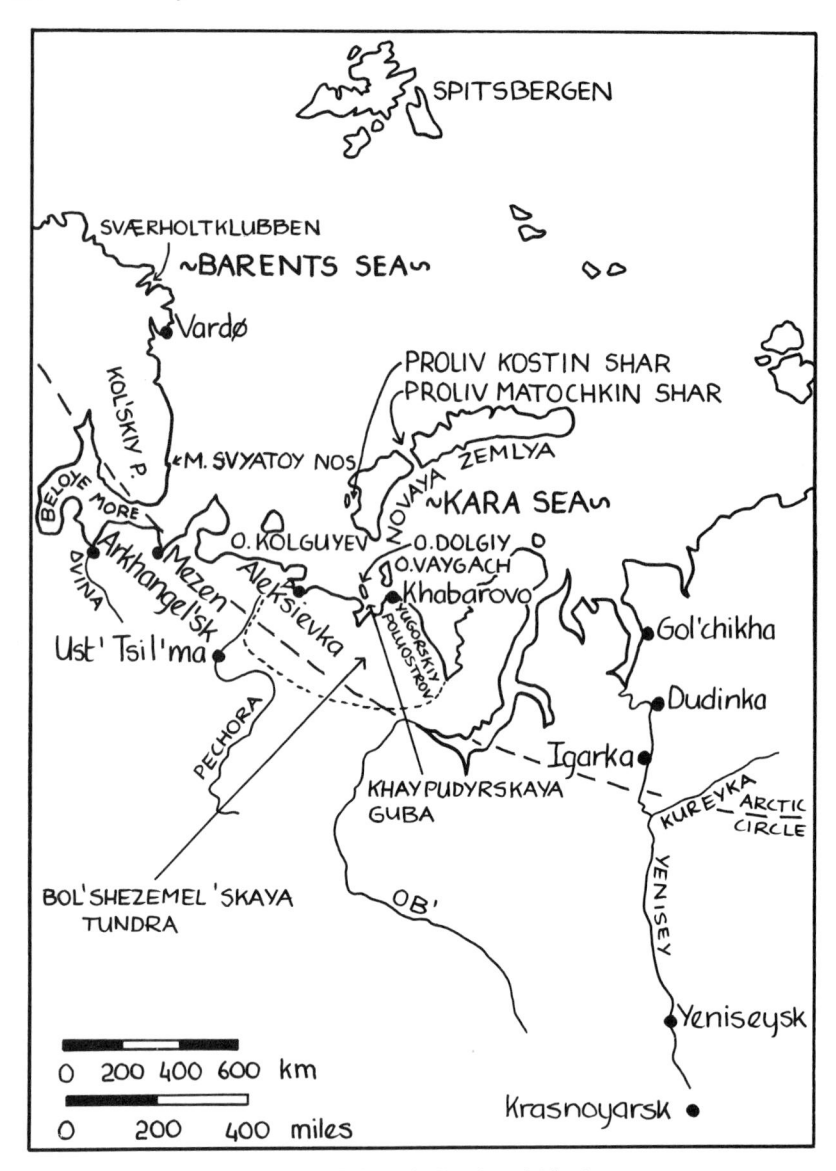

Map 12   Britons in Russia and Siberia.

it in winter 1893–1894 and called it *The Great Frozen Land* in the title of his 1895 book.

    Travelling via Archangel and Mezen by horse-drawn sledge, Harvie-Brown and Seebohm reached the River Pechora at Ust' Tsil'ma on 15 April 1875. Here, 200 miles upstream from its mouth as the crow flies, the great river was exactly one mile wide. It was still frozen hard and the two travellers established themselves in comfortable lodgings to await the spring break-up of the ice. Harvie-Brown (1905b: 264) describes their daily routine at Ust' Tsil'ma.

> At present we spend our days much as follows:—We breakfast about seven a.m. Rig up our hammocks for our morning pipe. Write journals, letters, or go out shooting, or receive visitors and show our guns and curios—or visit the Ispravnik or others.
> We lunch at 12 or 1 p.m.
> Piottuch skins birds in the outer room, where we also take our meals.
> Dinner at irregular times.
> At ten or eleven we sling our hammocks or turn in on the floor in our respective corners.
> Today we wrote letters, journals, etc., and strolled about the town, shooting Tree Sparrows. There was high wind and snow, but it was not very cold.

The travellers each had with them "a double-barrelled breechloader and a walking-stick gun" with 500 cartridges for each weapon. Harvie-Brown made full use of the contacts made and experience gained on his earlier expedition to the Dvina Delta. He again retained the services of the Pole Ignati Piottuch, who was to be paid on a sliding scale to encourage his exertions: 400 roubles for 700 birds skinned, 450 roubles for 900, and 500 roubles, which Harvie-Brown reckoned at £70, for 1000 skins. In the event Piottuch got his 500 roubles—for skinning 1019 birds!

As the spring thaw progressed more and more birds arrived, many of which fell to the guns of the collectors. Harvie-Brown describes how, on more than one occasion, he fired into a dense flock of Snow Buntings, bringing down 20 or 30 birds. Some of them were carefully skinned by Piottuch, and sexed. One morning they breakfasted on a dish of Snow Buntings. One evening, after dining particularly well on joints of bear and beef, reindeer tongues, and rice pudding, Harvie-Brown "loafed out into the yard with my stick-gun, and potted a female House Sparrow" (1905b: 308). Their first nest was brought to them on 27 April: a Siberian Jay's with three eggs. But it was only on 21 May that the ice on the Pechora broke up. Eventually on 10 June they were able to set out downstream on their northward journey to the river's mouth. Then at last, basing themselves at the then timber port of Aleksievka, they were able to set foot on the tundra. The elated Harvie-Brown (1905b: 377–378) recorded the results of their first full day there in detail:

> On Tuesday, the 22nd of June, the eggs obtained were 164 in number, as detailed *infra*:
>
> | | | | |
> |---|---|---|---|
> | Grey Plover | 16 | Bean Geese | 11 |
> | Dunlins | 7 | Wigeons | 17 |
> | Great Snipe | 4 | Redpolls | 16 |
> | Lapland Buntings | 25 | Redwings | 3 |
> | Yellow-headed [Citrine] Wagtail | 10 | Tringa Temmincki [Temminck's Stint] | 4 |
>
> The birds shot today were—
> 4 Grey Plover at nests
> 3 Dunlins
> 5 Lapland Buntings
> 1 Bean Goose

2 Red-throated Pipits
1 Yellow-headed [Citrine] Wagtail
1 Long-tailed Duck
1 Willow Grouse
2 Buffon's [Long-tailed] Skuas

By 7 July they had over 400 duck's eggs of 10 species, many brought to them against financial remuneration by local people. Among these last were a clutch of eight Smew's eggs and two Bewick's Swan's eggs. Their final success, right at the end of their trip, when they were camping in a wrecked sloop on the east coast of the estuary, was the discovery of breeding Little Stints. This was in the last week of July and the eggs were almost ready to hatch—which explains their very dilapidated condition after the naturalists' return home. They left the Pechora mouth in an English timber ship, the *Triad*, which was carrying larch to Kronshtadt near Leningrad. Disembarking at Helsingør in Denmark, they proceeded by train to Hamburg. There, Harvie-Brown, who lived at Dunipace in Stirlingshire, took ship for Leith, and Seebohm, who was a successful Sheffield steel manufacturer when not collecting birds, embarked in another for Hull, in his native Yorkshire.

(a)

(b)

Plate 23 (a) Little Stint facing the observer as it stands near its chicks among horsetails and other plants. (b) Breeding Little Stint in the Varanger Peninsula, north Norway. The Little Stint's breeding range is one of the more restricted among Arctic birds. Apart from a few pairs in north Norway, it is confined to the European Russian and Asiatic Siberian Arctic as far east as the River Lena and the New Siberian Islands; almost always north of the Arctic Circle.

Perhaps because Seebohm's book describing the expedition came out in 1880, 25 years before Harvie-Brown's, or because of his dominating personality and prominent position in British Ornithologists' Union circles, or because he was 10 years older than Harvie-Brown, Seebohm's name is traditionally placed first in mentions of their Pechora expedition: it is usually referred to as "the expedition of Seebohm and Harvie-Brown". But it was Harvie-Brown who was the initiator and inspirer of this British ornithological invasion of Arctic Russia and Siberia. Although their books recount, for the most part, little else than the wholesale shooting of birds and the triumphant plundering of their nests, both men had other interests and wrote other sorts of book. Best known of Seebohm's numerous books and papers on palearctic ornithology were his *History of British birds* (1883 on), his work on the *Geographical distribution of the family* Charadriidae, *or the plovers, sandpipers and snipes and their allies* (1887), and his posthumously published *Coloured figures of the eggs of British birds* (1896). As to Harvie-Brown, few modern ornithologists are unaware of his magnificent contribution to the knowledge of the avifauna of his native Scotland as author or co-author of the volumes on the vertebrate faunas of the different Scottish regions. His ornithological interests were in many ways ahead of his time. By 1875, as a result of correspondence with landlords, he had completed a census of rookeries in Caithness (Fisher 1954: 126). A few years later, in 1879, he and Lincolnshire ornithologist John Cordeaux organised the first national migration enquiry in Britain, sending printed questionnaires to over 100 lighthouses and lightships around the British Isles (Pashby 1985).

BIRD COLLECTING ON THE YENISEY

In 1876 the first consignment of English goods was brought by sea from the west to the River Yenisey. Captain Joseph Wiggins even took his 120-ton steamer the *Thames* almost a thousand miles up the river and delivered the first cargo of English goods ever seen there. Indeed the *Thames* was the first ocean-going vessel to ascend the Yenisey (Kinloch 1898; Johnson 1907). But Wiggins was too late in the season to take his ship home in the autumn of 1876. Reaching a tributary of the Yenisey called the Kureyka on 18 October, the *Thames* was soon solidly frozen in at its entrance hard by a settlement with the same name as the river. The captain secured lodgings for his crew there, and travelled by sledge and train to St Petersburg in the hopes of winning commercial support for his venture in the short time before he would have to set out on the tedious return journey back to his ship. Then he would hopefully extricate it when the ice broke up, pick up a cargo, and return to England. However, he was disappointed of commercial backing from Russian merchants, in spite of papers read for him in Russian to the Imperial Society for the Encouragement of Commerce and the Imperial Society for Naval Communications, while he was in St Petersburg, though both societies elected him to life membership. In England Captain Wiggins fared no better, except that a friend paid his travelling expenses from St Petersburg to London and back. But, though he had failed to arouse commercial interest in his venture, on 23 February "Mr. H. Seebohm, the well-known ornithologist, made his acquaintance" (Johnson 1907: 135). Seebohm (1901: 250) thought "that an opportunity of travelling with a gentleman who had already made the journey, and consequently 'knew the ropes' might never occur again", and, although the captain gave him only a week to get ready, decided to travel to the Yenisey with him.

On 1 March 1877 Wiggins and Seebohm were off. The travellers reached the Yenisey river at Krasnoyarsk on 2 April and were forced to exchange their sledges for wheeled carriages for the journey northwards to Yeniseysk. There Seebohm transacted two important items of business. First he hired a young Jewish exile named Glinski who spoke "bad German and bad Russian, and had an inconvenient habit of mixing up Hebrew with both these languages" to skin birds for him. He was less well paid than Piottuch had been, receiving 20 roubles a month and "an additional bonus of ten kopeks per skin" but he became equally skilful, and coped with a similar number of "more than a thousand" birds. After watching Seebohm skin a couple of redpolls, he made "a tolerable skin" of a Bullfinch and, after a week, "could skin better and quicker" than Seebohm himself. Also at Yeniseysk, Seebohm purchased a half share with Captain Wiggins in a small schooner which the local ship builder, a Heligolander called Boiling, had on the stocks. To be named the *Ibis*, she was to be brought down the Yenisey to Kureyka by Boiling as soon as the ice broke up. There, Wiggins promised to rig her and, while he took the *Thames* back to England, Boiling and Seebohm would sail down the Yenisey to the river mouth at Dudinka in the *Ibis* "ornithologising as we went along". There Seebohm might hope to have the choice of returning home by sea either in the *Ibis* or in the *Thames* (Seebohm 1901: 273–275).

Kureyka, lying just on the Arctic Circle, was in the taiga, or coniferous forest. Though Seebohm was anxious to travel northwards to the mouth of the river and the tundra as soon as possible, a series of unfortunate accidents upset his plans. Although the Yenisey ice broke up on 1 June, the *Thames* was not free until a week later. Then her rudder, which had been smashed by ice during the break-up, had to be replaced, and it took Wiggins and the ship's carpenter over a fortnight to make a new one, the Captain himself felling the necessary trees. The delay increased the difficulties likely to be faced by the *Thames*, with the *Ibis* now in company, because of the rapidly falling water levels as the great river's spring flood subsided. On 3 July at Igarka she ran hopelessly aground and had to be abandoned. At last, on 9 July, Seebohm was able to proceed down the Yenisey in the *Ibis*, which, drawing only 1 m (3 feet), passed easily over the shoals. The tundra was reached on 12 July not far below Dudinka but it was too late for eggs! A Dusky Thrush's nest contained "five young birds about a week old. This was very disappointing as the eggs of this bird were unknown" (p. 395). Later on, however, at another place, Seebohm was lucky enough to find a nest of the Asiatic, now called the Pacific, Golden Plover, containing four eggs, the identity of which he placed out of doubt by shooting the male bird. "These are the only authenticated eggs of this species known in collections" he wrote triumphantly.

In the event, Seebohm had only 6 days to work the tundra around Gol-'chikha, on the mouth of the river, before returning upstream in a passenger steamer to Yeniseysk, rather than risk crossing the Kara Sea in the *Ibis*. He found Ruffs, Red-necked Phalaropes, Dunlins and Little Stints, as well as Snow and Lapland Buntings, but the commonest bird on the tundra at Gol-'chikha were the Pacific Golden Plovers, most of which had young. "Had I been a fortnight earlier I should no doubt have obtained many of their eggs";

and he went on to state rather petulantly that he "had had to pay dearly for Captain Wiggins's blunders" (p. 409). This cut Captain Wiggins's biographer to the quick: he justifiably complains that Seebohm uses the word "blunder" for accidents and for "any arrangement or proceedings which failed to secure his approval" (Johnson 1907: 152). Seebohm left the tundra "with a feeling somewhat akin to disappointment and regret. My trip might be considered almost a failure, since I had not succeeded in obtaining eggs either of the knot, sanderling or curlew sandpiper" (p. 413). The rare Arctic waders of the tundra had eluded him.

Neither Captain Wiggins's nor Seebohm's reputation was tarnished by their Siberian adventures of 1877. In the following year the unstoppable Wiggins took the *Warkworth* up the River Ob' with a cargo of "salt, Sheffield goods, porcelain, glass, etc." and triumphantly returned to England with the first cargo ever brought through the Kara Sea from Siberia (Johnson 1907: 191–192). As to Seebohm, the scientific results of his Yenisey expedition were published in the *Ibis* in 1878–1880 and enhanced his ornithological reputation. Alfred Newton, who was professor of zoology at Cambridge University, wrote scathingly of his publications, but only after his death. According to Newton, referring to Seebohm's volumes on British birds, "The greater part of their text, when it is correct, will be found on examination to be a paraphrase of what others had already written. . . . Of downright errors and wild conjectures there are enough, and they are confidently asserted with the misuse of language and absence of reasoning power that mark all the author's writings . . ." (Newton 1896: 44). But his criticism was not directed at Seebohm's collecting activities, nor at his ornithological ability in the field. However, at least one famous ornithologist of the day, Heinrich Gätke, of Heligoland fame, who was himself a collector, expressed adverse opinions on these aspects of Seebohm's Yenisey expedition in April 1877, before he had even returned from it! In a letter to his English friend and fellow ornithologist John Cordeaux, Gätke wrote as follows (Pashby 1985: 62–63):

> It is a thousand pities that the good luck of the trip down the Yenisei . . . has not fallen to somebody more acquainted with birds in free life than Seebohm is: at 30–40 paces he can scarcely tell any living bird when at rest, when on the wing none; everything stirring is popped at, bagged, and the spoils sifted when quietly at home. . . .

CUT OFF ON KOLGUYEV ISLAND

In spite of the publication of Seebohm's books *Siberia in Europe* and *Siberia in Asia* in 1880 and 1882 respectively, there was no sudden rush of "Ibises" to the Russian Arctic. But, 10 years later, one of them with some experience of roughing it in the North American wilderness was casting about for an unexplored piece of country with ornithological promise. His name was Aubyn Trevor-Battye (1895, see too Stone 1986b). Colonel H. W. Feilden, who figured in Chapter 2 as a naturalist on the British Arctic Expedition of 1875, suggested Kolguyev Island in the southern part of the Barents Sea as a suitable place for his expedition. Seebohm, Harvie-Brown, Newton, Bowdler Sharpe, Howard Saunders and other luminaries of Britain's ornithological establishment encouraged him with their advice and experience. Trevor-Battye had the good sense to write to Lord Lilford, otherwise Thomas Littleton Powys, the then president of the British Ornithologists' Union, about his project, and received a letter from Lord Lilford suggesting that his nephew, Mervyn Powys, should join the expedition and offering to pay half the cost of chartering a steamer—an offer which Lilford more than made good (Trevor-Battye 1903: 270–271). Kolguyev is a tract of tundra stretching 60 miles (100 km) from north to south and nearly as broad, lying 40 miles off the coast of Arctic Russia. Its natural history was virtually unknown; it seemed just the place for some of those prized wildfowl and waders and other Arctic species that had captured the imaginations of Harvie-Brown and Seebohm and many others.

Trevor-Battye was undeterred by the negative results of the reconnaissance expedition he made to Archangel in 1893, during which he at least learned some Russian and obtained three blue kittens to take home. He had hoped to find a way of reaching Kolguyev from there, but discovered that the White Sea would be choked with ice until long after the start of the breeding season of the birds whose eggs he hoped to collect. The only information he could obtain there about Kolguyev was that it was a nasty place. Nevertheless, on 22 June 1894 he landed on the northern tip of the island with two companions, his bird skinner Thomas Hyland and his spaniel Sailor. He had disembarked from the yacht *Saxon*, chartered by Mervyn Powys, who promised to return in a month's time to take Trevor Battye off after he had done some walrus-shooting in Novaya Zemlya.

On board the *Saxon* Powys and the crew thought Trevor-Battye was crazy and tried to dissuade him from landing, but to no avail. Naturally, he was subsequently accused of "folly and rashness". When the *Saxon* returned to Vardø in north Norway on 8 August having failed to make contact with him on Kolguyev, two women, Mrs Leyborne-Popham and Mrs Ponsonby, set out from there in the steam yacht *Blencathra* to rescue him, but were driven back by storms. The Royal Geographical Society opened a subscription list. But on Kolguyev Island Trevor-Battye and his companions had made contact with the local Samoyeds, had enjoyed their hospitality for some weeks, and had been taken back to the mainland by the Russian trader who called at Kolguyev in August–September every year to collect a cargo of reindeer meat and skins.

After landing on the northern tip of the island and camping there a couple of days, Trevor-Battye and Hyland had cached their tent and a large barrel of stores, then set out southwards to find the Samoyeds. They were without tent, sleeping bags, or rucksacks. Their provisions and what they regarded as necessaries were carried in receptacles designed for other things. Thus a tin of 39 digestive biscuits, 6 pots of Bovril, 4 tins of potted meat, other edibles, a tin opener and a kettle lid were packed in a 300-cartridge magazine; the sextant case, instead of the sextant, held dried vegetables, raisins and spoons and forks; the camera case contained a pint bottle of methylated spirits, a "boiling machine", lint, plaster, quinine pills and cough tablets, and the precious egg-drill and blowers; and a pound of bacon was stowed in the tin case designed for botanical specimens. Their clothing was hardly what would now be termed "Arctic". Hyland was wearing rubber boots, cord breeches, a kettle tied round his waist with string, a long-tailed shooting coat and, over it, a canvas Norfolk jacket "covered with pockets of many kinds all bulged out to the utmost limit", and a flapped velvet stalking-cap. Besides his share of the food and other things, he also carried a pair of shooting-boots, a pair of Norwegian fur boots and a bundle of spare clothing (Trevor-Battye 1895: 87, 90).

In spite of this somewhat unspecialised equipment, and of the hardships of life on Kolguyev, Trevor-Battye collected and recorded much about the natural history of the island, and he also devoted himself to a perceptive study of the way of life and material culture of its Samoyed inhabitants. His description of the Samoyed goose hunt was mentioned in Chapter 2. His entertaining book *Ice-bound on Kolguev* has excellent appendices on the Samoyed language, on the 95 species of flowering plant he found, and, above all, on the birds. He managed to bring skins of 34 species of bird back home with him, as well as some eggs, including those of the Grey Plover and Little Stint. His list of 47 birds recorded on Kolguyev is a very creditable one: he seems to have missed nothing, though he found no rarities. Thirty years later a Russian scientific expedition was on Kolguyev from 9 July to 23 August 1925, but its ornithologist, the botanist Tolmachev (1928), identified a mere 30 species. He saw Black-throated Diver and Scaup with young, whereas Trevor-Battye only saw adult birds; but Trevor-Battye found Peregrine and Turnstone's nests while Tolmachev only saw the adults. Trevor-Battye recorded several species missed by Tolmachev, including Snowy Owl, and Bewick's Swan breeding. Each saw a single Wheatear. In 1896, Trevor-Battye joined Martin Conway's expedition to explore the interior of Spitsbergen and contributed a valuable paper on the birds of Spitsbergen to the *Ibis* for 1897, but seems not to have visited the Arctic thereafter (Brown 1920: 174–176).

THE SHORES OF THE BARENTS SEA

Henry J. Pearson and his brother Charles had made three successive summer bird-collecting expeditions to the north of Norway and one to Iceland when, in 1895, they decided to go further afield and make for Novaya Zemlya. In spite of her poor sailing power and small coal capacity, the steam yacht *Saxon*, which had taken Trevor-Battye to Kolguyev and his friend Powys to Novaya Zemlya in the previous year, was again pressed into service. With the Pearson brothers on this expedition was Colonel H. W. Feilden, already referred to more than once in these pages.

Starting from Bergen on 4 June 1895 the *Saxon* steamed up the Norwegian coast to Tromsø (Map 25), where an ice-master was engaged and the naturalists went ashore to collect eggs, securing those of Fieldfare, Brambling, Redpoll, and Reed Bunting. After coaling at Vardø in the Varanger Peninsula they laid course for Novaya Zemlya, but ice blocked their way and they decided to camp for a few days on the coast of the Kola Peninsula near mys

Svyatoy Nos or Holy Cape while the *Saxon* returned to Vardø for more coal. Here they found Rough-legged Buzzards' nests with eggs, a Gyrfalcon's nest with young, Arctic and Long-tailed Skua's nests, and nests with eggs of Great Snipe, Redshank, Dotterel and Temminck's Stint. On the *Saxon's* return on 27 June another attempt was made to reach Novaya Zemlya but, again, impenetrable ice blocked their way. After steaming along the edge of the ice at half speed for a day in dense fog, shooting a Black Guillemot, two Pomarine Skuas and several Brünnich's Guillemots, they again had to send the *Saxon* back to Vardø for more coal.

The question of where to fill in time while this was being done once again was easily solved because of the proximity of Kolguyev Island. Originally they had decided not to visit Kolguyev because Trevor-Battye had worked it thoroughly the year before (1894), but the Little Stint's and Grey Plover's eggs he had collected there encouraged them to change their minds. With a non-

Plate 24 Nest and eggs of the Dotterel on a north Norwegian fell. Only in the Taimyr Peninsula perhaps can the Dotterel be regarded as a genuine Arctic bird. A pair about to breed at Point Barrow, Alaska, were shot by Charles D. Brower in June 1930 (Gabrielson and Lincoln 1959: 327). The author found the most southerly breeding Dotterels ever on the summit of Monte Amaro in the Montagna della Maiella, Italy, in 1952 (Vaughan 1952).

146 *In search of Arctic birds*

chalance bordering on foolhardiness only comparable to Trevor-Battye's the year before, they disembarked on the western coast of the island, a part not much visited by Trevor-Battye, trusting that the *Saxon* would be able to return within a few days to take them off, and set up camp on 5 July.

During the next 10 days a systematic assault was made on the local bird life, while Feilden studied and collected flowers. Nothing was seen of the inhabitants, but the heaps of driftwood the local Samoyeds had stacked up ready for collecting along the beach were appropriated for the camp fire by Pearson and his friends, who left the Samoyeds in return "a collection of empty tins, a few needles, and other trifles, doubtless of more value to them than the wood we burnt" (Pearson 1899: 23). Though Pearson described the island as "one great heap of glacial mud with a few low sandhills" (p. 20), he found plenty of birds. On his very first day in the field, he was overjoyed to find a Little Stint's nest "containing four pretty eggs. I shot the bird, that the eggs might be identified without doubt. There are a few joyful moments that stand out clear and sharp in the memory and will never be forgotten while life lasts—the finding of this first Little Stint's nest is one of them" (pp. 24–25).

Another Kolguyev coup was the collection of a family of Bewick's Swans: an adult female and four cygnets. Pearson believed that "these cygnets were the first of their species brought to England in down; two were presented to the British (Natural History) Museum, and are now in the Swan Case of the Bird Gallery" (p. 33). The finding of Grey Plover's nests with eggs also caused elation. Seven clutches were taken. Eight of these eggs were almost hatching;

(a)

(b)

Plate 25   (a) Grey Plover's nest and eggs on a dry ridge in the Canadian tundra near Cambridge Bay. The nest is indeed hard to find, but the bird breeds commonly in suitable tundra all round the Arctic except in the so-called "Atlantic gap" between Canada and Russia.

(b) This three-egg Grey Plover clutch was laid on the sort of low stony ridge which is a favourite nesting place of the species. Not a mile from Cambridge Bay, 8 July 1986.

it took the brothers Pearson 3 hours' hard and very cold work in camp to dissect and extract the chicks from the shells. They were cheeping when found and already had their beaks through the shells. Another triumph was the shooting, after several attempts, of "a beautiful old male" Snowy Owl.

The *Saxon* returned on 16 July, fortunately unhindered by ice, and on 17 July a landing was at last made in Novaya Zemlya, or, more accurately, on a small island in Kostin Star Strait (proliv Kostin Shar), off the coast of southern Novaya Zemlya. Here, a colony of breeding Brünnich's Guillemots was raided for eggs. Able-bodied seaman Morrison "climbed over a large part of the loonery with the aid of a rope" and "196 eggs were taken fit for specimens" (p. 39). "The labour of blowing, washing, drying and marking these 196 eggs was considerable. My brother and I worked at them steadily for nine hours" (p. 40). Six days later, as they were leaving Novaya Zemlya, 244 more eggs were taken "the whole forming a most interesting series" (Pearson 1899: 40):

> Here were 440 eggs taken from one small cliff, laid by birds of the same species, feeding, as far as we knew, on the same food, certainly over the same area; yet the eggs varied from pure white without a spot, through all the shades of ochre and maroon-brown to the rich colours of a Razorbill's egg, through many shades of blue-green with all varieties of pencillings and blotching, and through an equal number of shades of yellow-green, to specimens as richly blotched with black as some of the finest of eggs of the Common Guillemot taken on the Yorkshire coast. This subject of the coloration of eggs is truly one of which we may be said to know nothing yet.

Steaming further up Kostin Shar Strait, they saw male King Eiders on the water and landed on islands in the hope of finding a nest. But all the female eiders they shot off nests on these islands proved to be common Eiders. However, "the Common Eider is a useful bird on the table, and little inferior to the Mallard" (p. 42). The whereabouts of the King Eider nests remained an unsolved problem; probably they were well inland, on marshy tundra. On one island three young Glaucous Gulls in down were secured but, for once, not killed and skinned. Two were safely taken home alive. One was kept for a year by Charles Pearson before it died after eating deadly nightshade berries. The other was still alive in full adult plumage in February 1899, having joined a collection of Norwegian Glaucous and Great Black-backed Gulls which Henry Pearson kept on his lawn.

In 1897 Henry Pearson and Henry Feilden were in the field again. This time the goal was the "ornithologically unknown" country (p. 58) around the

---

Plate 26 (*a*) Incubating King Eider Duck. Typically the King Eider breeds near freshwater lakes and further inland than the Eider. This bird was sitting, or rather lying, on her nest with outstretched head, on an island in one of the many lakes near Cambridge Bay.

(*b*) This King Eider's nest was on an islet in a shallow tundra pool. On 1 July she had not begun to sit and her three eggs were covered with grass and sedge and willow leaves. On 3 July she was incubating, did not cover the eggs when leaving, and had added some down. Cambridge Bay, 1986.

(a)

(b)

shores of Khaypudyrskaya Bay, that is the northeastern part of the already-mentioned Great Land Tundra. Chartering a Tromsø sealer, the *Laura*, they steamed north round the Norwegian coast. A brief detour was made east of North Cape, when they "steered close under the cliffs" of "the great bird-rock of Sværholtklubben", "slowed down, and sounded the steam whistle" (p. 67). "Thousands and thousands of birds rose into the air like swarms of bees. . . . It was a grand sight, this great nursery of Kittiwakes . . .". A week later, in the Barents Sea, a flock of King Eider on the sea were fired into at long range with both barrels of an eight-bore: five drakes were secured. But the route towards the mainland Russian coast in Khaypudyrskaya Bay was barred by ice. They landed instead at Dolgaya Bay on the northern coast of Vaygach Island (ostrov Vaygach), and explored the surrounding country using the ship as a base and its launch as a means of transport.

Pearson was fortunate, on his excursions ashore, to have the Norwegian sailor Hansen as his assistant, for it was Hansen's job to carry Pearson's camera, weighing 17 lb. (8 kg). Moreover, Pearson lent him a "pair of fishing waders with macintosh tops reaching well up the thigh" so that, when the water was too deep for Pearson's shooting boots, Hansen could carry Pearson over. A family of three Glaucous Gulls in down captured here were not so fortunate as those found in 1895; "the whole family now repose in a glass case in the Bird Gallery at the Natural History Museum". So many clutches of Little Stint's eggs were found on Vaygach Island that some were sucked for refreshment instead of being blown for egg-collections. But a fine series of 183 eggs (Newton 1905: 208–209) was collected and three clutches are beautifully illustrated in the colour plate which forms the frontispiece of Pearson's book. Two young Rough-legged Buzzards in down were taken aboard here and are figured in a photograph reproduced in the book. "I grieve to say", writes Pearson (1899: 92–93) that "the piece of flesh they are depicted fighting over is part of their mother, whom they lived on for a day or two after her skin was removed". After leaving Vaygach Island, its way south still blocked by ice, the expedition visited Khabarovo, landed on Dolgiy Island (ostrov Dolgiy) to study the birds there, and ended up by penetrating into the Kara Sea through Matochkin Shar Strait (proliv Matochkin Shar), the narrow strait that divides Novaya Zemlya into two.

Pearson's book, which recounts his ornithological adventures in 1895 and 1897 in the European Russian Arctic, was published in 1899, entitled "*Beyond Petsora eastward*". *Two summer voyages to Novaya Zemlya and the islands of the Barents Sea*. He himself contributed valuable notes on the avifauna of the places visited, while Feilden's important papers on the flora of Novaya Zemlya and the glacial geology of Arctic Europe and its islands were reprinted as appendices. Pearson, like Trevor-Battye, was a serious ornithologist who, in spite of the emphasis on collecting in his book, was making a real contribution to knowledge. A sequel was published in 1904 called *Three summers among the birds of Russian Lapland* describing the author's visits to the Murman Coast of the Kola Peninsula in 1899, 1901 and 1903. On these later expeditions egg-blowing and skinning evidently took up as much time as before, but the rather fine illustrations show that Pearson was also devoting time to photography,

Plate 27 Near Cambridge Bay, Mount Pelly's scree slopes held three Rough-legged Buzzard's nests containing eggs early in July 1986. Clearly they were rebuilt year after year and each had given rise to an "island" of vegetation among the boulders. This nest-site is typical for treeless tundra areas. In Scandinavia trees are more often used.

152 *In search of Arctic birds*

though not of birds. On one of these expeditions he had the good luck to collect two six-egg clutches of Rough-legged Buzzard's eggs in a single day (p. 123). On another he "had the satisfaction" of poking three young Hooded Crows out of their nest. "I never spare the nest of this bird, for, as in the case of our House-Sparrow, it is difficult to find much evidence of its usefulness" (p. 17). He was disappointed not to be able to collect eggs of the Spotted Redshank, though he took home at least two broods of young in down of this species. Explaining why he once carried a brood of loudly peeping young Wood Sandpipers back to base in his pocket, followed by the adult bird calling loudly the whole way, he writes (p. 160): "Young in down should never be killed when just found; they keep much better if carried home alive and then plunged into a jar of methylated spirit—the quickest and easiest death possible". Pearson's closing remarks on his last expedition may show that the Norwegian authorities at least were beginning to fear the effects of indiscriminate egg collecting (Pearson 1904: 169):

> Norway is now closed to those who wish to study birds in their breeding haunts and obtain specimens of eggs of British species, only breeding in the north, for their collections. Licences to take eggs are only granted by the Government to Norwegians! I understand it is probable that hotel proprietors on the Dovrefjeld will shortly have notice that they must refuse to allow all persons to remain in their hotels whom they have reason to think are collecting birds, eggs, or plants. But Russia is still open, and the genuine ornithologist who collects from scientific interest and not for trade purposes receives every assistance from the authorities. I can only conclude with grateful thanks for the many kindnesses experienced from all officials especially from the Governor of the Province.

On the other hand, Pearson's hint that the Norwegian government's measure was more xenophobic than conservationist may have been justified. In 1903 a Norwegian government expedition was overwintering at Matochkin Shar to study the aurora borealis or Northern Lights. Two members of it, J. Koren and H. T. L. Schaanning, took 17 clutches of four Little Stint's eggs, shooting the adult on each nest, on a single day, 10 July 1903 (Newton 1905: 205–209).

A WOMAN TAKES THE FIELD

In the spring of 1914 "the well-known Polish anthropologist of Oxford University", Miss M. A. Czaplicka, organised an expedition to study the native peoples of Siberia. With her went an "American gentleman", H. U. Hall, representing the Philadelphia University Museum, who was also interested in native peoples, an artist, Dora Curtis, and an ornithologist, Maud Doria Haviland. All four planned to travel down the Yenisey and spend the summer together at Gol'chikha on the river mouth, then the Misses Curtis and Haviland would return by sea to Britain while Miss Czaplicka and Mr Hall hoped to travel southwards up the Yenisey and overwinter with the natives in the taiga. Twenty-five-year-old Maud Haviland (Palmer *et al.* 1954), great-granddaughter of Dr John Haviland, lord of the manor of Fen Ditton, Cambridgeshire, and Fellow of St John's College, grew up in southeast Ireland, where she learned to handle a shotgun and developed a passionate interest in birds. Her experiences on the Yenisey and at Gol'chikha are described in her most readable book *A summer on the Yenesei*, published in 1915.

Seebohm's brief stay at Gol'chikha in 1877 had begun only on about 20 July; Maud Haviland arrived there on 29 June in 1914 and stayed for 2 months. She certainly made the most of her better opportunities and she benefited from the considerable additions to knowledge of the breeding birds of the area made in the meantime by another English ornithologist, Hugh Leyborne Popham, who had been at Gol'chikha in 1895 and 1897 and had published three papers in the *Ibis* on the birds he found there (1897a, 1898, 1901).

Maud Haviland was just as enthusiastic an egg collector, and just as good a shot, as any of her precursors so far mentioned. She was after the same birds, too. She took the eggs of Temminck's and Little Stint, Grey and Pacific Golden Plover, Grey and Red-necked Phalarope, and even those of the Curlew Sandpiper, which Popham (1897b) had been the first to collect, in 1897. She shot birds at the nest to collect their skins and to determine their sex. And she collected young in down too. The two young Curlew Sandpipers she collected were "the first that, to my knowledge, have been taken by an English ornithologist" (Haviland 1915a: 225). Among her disappointments were the failure

Plate 28 Curlew Sandpiper on its nest on the north coast of the Taimyr Peninsula, photographed by Pavel Tomkovich on 9 July 1984 near the River Uboynaya.

154 *In search of Arctic birds*

(a)

(b)

Plate 29 (a) Male Grey Plover in breeding plumage near its nest. In spite of its striking black-and-white plumage, this bird was often hard to see on the stony ground. (b) The mate of the bird figured in (a) was greyer than average, with little or no black on her underparts. In many Grey Plover pairs the sexes are not so easily distinguishable as in this Cambridge Bay pair.

to find a Bar-tailed Godwit's nest and the impossibility of properly blowing a clutch of Grey Plover's eggs she had carried back to base in a kettle: they turned out to be too much incubated.

As well as going after birds with her gun, Maud Haviland went after them with a "Birdland" quarter-plate reflex camera "built for me by Mr Armytage Sanders of London", armed with a 14-inch focal length lens. She was the first English ornithologist in Russia to use a hide, which she refers to as her "small green hiding-tent" (pp. 140–141, 132). This was still something new on a collecting expedition such as hers, though Oliver Pike (*In bird-land with field-glass and camera*, new edition 1900) and R. B. Lodge, not to mention the Kearton brothers, had already shown the way, and H. Witherby and Co. had published the first book on how to photograph birds, entitled *Photography for bird-lovers. A practical guide*, in 1911. Its author was Bentley Beetham. Maud Haviland's book is adorned with photographs of birds at the nest: Pacific Golden Plover, Little Stint, Temminck's Stint, Grey Phalarope and Willow Grouse, in several cases taken from her hide. She must surely have read Lodge's exclamation in the chapter "Photography for naturalists" of his *Pictures of bird life* (1903: 15), "How interesting to future generations of ornithologists to see permanent photographic records of the first recorded nests of the Little Stint and Grey Plover, with the birds themselves photographed on or near the nests, as first discovered by Seebohm; or some of Wolley's Lappland discoveries!"

Besides her gun and her camera, Maud Haviland put her pen to excellent use. She had already written and published a collection of nature stories entitled *The wood-people: and others* (1914). Though her approach would now be condemned as anthropocentric and over-sentimental, yet she portrays natural scenery and the activities of birds with perception and freshness, witness her description of the tundra at Gol'chikha (1915a: 219–220):

In the afternoon the skies cleared, and as I started on a solitary ramble up the valley, I saw the tundra under another guise. Last night we saw its dour side, its greyness, its loneliness, and seemingly under the scourge of the wind and the rain, its hopelessness. The frame of the land was just as the ice had left it. Its horizons lay in long, open curves, all angles planed away by the firm hand of the glaciers. Most likely the form of the swamps and the rivers had not changed since the mammoth lumbered over the frozen mudhills. But to-day I felt more clearly the promise of the tundra—its huge fertility, its immensity, its strange, indefinable magic. Nowhere, except in the Alps, may be seen such a profusion of flowers—forget-me-nots, lupins, saxifrage, pedicularias, and poppies—purple, blue, crimson, and orange—and in the hollows the willows were fragrant with bloom. On every hillock stood a plover in gold-studded livery, playing on his wild pipe, or malingering piteously to lead me from his hidden nursery. Down in the hollow, a pair of godwits whistled to one another in notes like the striking of flint on steel, and red-throated pipits dropped carolling among the flowers. As I walked quickly beside the river-bank, little waders ran before me down the sandy spits, too busy to be afraid, and a fine willow-grouse rose with a *whirr* and boomed away over the hill. Beside the ford, the gulls flew to and fro, and stopped at their own purple shadows on the sandbanks. And yesterday there had not been a bird to be seen, and all the flowers had hidden their rain-drenched heads! All this transformation had been caused by a little sunshine.

Nor was her writing limited to books. She contributed a fine series of papers in 1915–1917 to the *Ibis*, the *Zoologist* and *British Birds* giving detailed accounts of the breeding behaviour of some of the more interesting birds she had met with: the Curlew Sandpiper, Little Stint, Grey Phalarope and others.

After her adventurous trip to the tundras of the lower Yenisey, Maud Haviland (Palmer *et al.* 1954) was caught up in the First World War, working as a chauffeuse for the Scottish Women's Hospital in Romania and for the French Red Cross in Paris and Soissons. After the war she took a zoology degree at Cambridge and did research on insect damage to vegetation in Guyana (formerly British Guiana). In 1922 she married H. H. Brindley, a Fellow of St John's College like her great-grandfather. Her interest in birds proved lifelong: she was a founder of the Cambridge Sanctuary Club, which bought and managed a small bird reserve on the outskirts of the city, and when David Lack's *The birds of Cambridgeshire* was published in 1934 it incorporated her notes of migrant waders seen at Cambridge Sewage Farm, including a Dotterel, a Grey Phalarope and two Black-tailed Godwits. She died on 3 April 1941.

CHAPTER 7

Ornithological treasure trove

During the nineteenth century certain birds or their eggs became highly prized as trophies. To qualify as treasure trove of this sort they had to be extremely rare in themselves, as nearly as possible unknown in collections, extraordinarily beautiful and, if possible, only to be found in remote areas in very high latitudes. The birds which attracted most attention of this sort were the Arctic gulls, Ross's Gull and, to a lesser extent, the Ivory Gull and Sabine's Gull. The eggs which were most sought after, partly because of the great beauty and variety of their markings, were those of waders. Among these were several species which were familiar in western Europe or the United States of America in winter or on passage, but which were unknown on their High Arctic breeding grounds. The obsession of men and women like Harvie-Brown, Seebohm and Maud Haviland with the eggs of the Little Stint and Grey Plover has been touched on in the previous chapter. These collectors and others vied with each other to be the first to discover and collect the eggs of each species. But their claims did not always stand up to scrutiny. "The results of our somewhat adventurous journey exceeded our most sanguine hopes," wrote Seebohm of his 1875 expedition with Harvie-Brown (1901: 235). "Of the half-dozen British birds, the discovery of whose breeding grounds had baffled the efforts of our ornithologists for so long, we succeeded in bringing home identified eggs of three—the grey plover, the Little stint and Bewick's swan." As to the efforts of ornithologists being baffled for so long, as a matter of fact only one had been to look for birds in the Siberian Arctic and he had found nests and eggs both of the Grey Plover and Little Stint! This was the Baltic German A.F. Middendorf (1853: 209, 221), who had travelled from south to north through the central Taimyr Peninsula as long ago as 1843 on behalf of the Russian Imperial Academy of Sciences at St. Petersburg.

THE EGGS OF THE KNOT

Of all the Arctic breeding waders whose eggs were sought and prized by collectors, the Knot took pride of place. Although often very numerous on migration in European and American estuaries, its breeding places in the Arctic are widely separated from each other, mostly very inaccessible, and for the most part in very high latitudes. It breeds in Greenland but only in the northern half; it has occasionally bred in the Spitsbergen Archipelago; on the mainland of the Soviet Union it breeds only in northern Taimyr; it breeds only on some of the Soviet Arctic archipelagos and islands, namely Severnaya Zemlya, the New Siberian Islands (Novosibirskiye ostrova) and Wrangel Island (ostrov Vrangelya); and its breeding range in the North American Arctic is nearly restricted to the more northern and central parts of the Canadian Arctic archipelago. Moreover it often breeds on high inland plateaux where it is hard to locate. But what made the eggs of the Knot even more special was that the nest with eggs really is exceedingly difficult to find because the superbly camouflaged bird sits tight, indeed unless virtually trodden on she remains on the eggs.

In the December 1918 issue of the *Wilson Bulletin*, the much-respected American quarterly journal of ornithology, an article appeared by W. Elmer Ekblaw entitled "Finding the nest of the knot". With hyperbole similar to that employed by Seebohm, Ekblaw's opening paragraphs ran as follows:

> To ornithologists and bird lovers the world over the most important result obtained by the recent Crocker Land Expedition to the Arctic regions was undoubtedly the discovery of the nest and eggs of the knot (*Tringa canutus*). Two full clutches of eggs, the nests in which they were laid, and the sitting birds upon them, were brought back to the American Museum of Natural History of New York.
>
> Few eggs have been so eagerly sought as those of the knot; for a hundred years or more the nesting-places of this bird, so common on our shores in migration time, had been known to be far arctic and probably circumpolar; almost every expedition to the North for the last century has been definitely instructed to seek the nest and eggs; yet until this latest American expedition, the knots had foiled all explorers and successfully guarded the secret of their nests.

The Crocker Land Expedition was based at Iita (formerly Etah) in the district of Avanersuaq in northwest Greenland in 1913–1917, and of the Knots' nests described by Ekblaw, one was found on 28 June and the other a few days later, in 1916. The nests were on a dry stony plateau called North

Mountain just north of Thule Air Base (Pituffik; Map 28) and only a mile or two east of the former Eskimo settlement of Uummannaq. Ekblaw's colleague and friend Dr Harrison J. Hunt (Hunt and Thompson 1980: 77), the expedition's surgeon, was the actual finder of both these nests and he describes the event as follows:

> Ekblaw was with us a good part of that summer, too. There was one thing Ek wanted most of all, and that was to find the nest and eggs of the knot (*Tringa canutus*), which had never been found, although ornithologists had searched for it for many years, all over northern Asia and America. Many days Ek and I had hunted for it from Etah. We would take two sticks with a rope between them and go carefully over the ground to raise a bird, but we had no luck. One day at Thule I was out alone and all of a sudden nearly stepped on one. There the bird was, sitting on the nest with three eggs in it, just in a hollow in a grassy spot among the rocks. The bird and eggs, both the color of an old shingle mottled with brown, were almost invisible. I had to poke the bird to make her move off the nest. I marked the nest with a long line of stones and went for Ek. When we came back, we photographed the bird, nest, and eggs, and then, although we hated to do it we shot the bird, carefully cut the nest out of the ground, and packed everything for the Museum. Later I came across another nest with four eggs in it. Both nests were on an upland a mile from the shore and not far from the icecap.

Although these two clutches were, as Ekblaw says, brought back to the American Museum of Natural History in New York, the museum did not keep both. In 1918 it exchanged the three-egg clutch for something else it wanted (American Museum of Natural History, Reference Library Archive no. 1016, Crocker Land Expedition file no. 5) and this clutch found its way into the collection of John E. Thayer (Bent 1927: 136).

Convinced as these Americans were that their Knot's nests were "the only nests yet definitely reported with eggs" (Ekblaw 1918: 99), they could easily have discovered the falsity of their claim. For the leader of the Crocker Land Expedition, D. B. MacMillan, had been present in July 1909 near the northern tip of Ellesmere Island with Robert Peary, of North Pole fame, when two Knot's nests containing eggs were found! Twenty-three-year-old George Borup (1911: 277–278), youngest member of the expedition, describes the scene in his usual lively fashion:

> The Commander [Peary], who was very much interested in ornithology, was out for the scalp of every bird's egg available, and all the Eskimos were put at this work. Whenever they'd report a nest with eggs, I'd visit and photo it, and then the egg would be brought to the ship and Mac[Millan] would operate and blow it. As a rule, these aquatic birds took special pains to construct their nests on young islands in the middle of lakes, surrounded by the deepest water, to be secure from foxes.
>
> Now if you are suffering from ennui and need diversion, strip off your clothes and wade out waist-deep in ice-cold water, stepping on sharp stones every once in a while, holding a kodak in one paw, and praying to all that's holy that you will not slip. Then take a photo of an egg in its native habitat, grasp it in your free hand, and paddle shoreward.
>
> We got two sets of eggs of the Tringa Carnatus [sic]—the knot—the eggs of which had never been obtained in the Western Hemisphere before. This was a trophy, for fair.

(a)

(b)

D. B. MacMillan (1934: 257) adds that one of the Knot's nests was discovered by "Mr. Scott, our second engineer" and the other by one of the Eskimos. And he states that "The eggs of this bird had never been found previously". From expedition surgeon John W. Goodsell's journal (1983: 147) we learn that Oodaaq had found the first nest on 25 June and Scott had found his nest on 26 June. Peary himself, in a letter to Colonel Feilden, gives the dates as 26 and 27 June (Feilden 1920: 282); he seems never to have mentioned this important find in print.

What is the truth about the first discovery of the eggs of the Knot? To find it we have to go back to the nineteenth century. Discounting some earlier claims of Knot's nests containing eggs because they were never properly substantiated and are not now verifiable (Yarrell 1882–1884: 415), we come to the earliest incontrovertible discovery of breeding Knots made in July 1876 in northern Ellesmere Island by the two naturalists of the British Arctic Expedition, Chichester Hart and Henry Feilden. They and their colleagues searched in vain for nests with eggs, but on 11, 12 and 30 July obtained broods of young and, in one case at least, the adult bird with them (Feilden 1878: 163; Yarrell 1882–1884: 416).

So, in 1876, the Knot's breeding haunts had been found near the northern tip of Ellesmere Island, but what of the egg? In the summer of 1883 an American expedition led by Lieutenant Adolphus W. Greely, US Army, was based on the northern Ellesmere Island coast. Knots arrived on 3 June on their breeding grounds but Greely and his companions failed to find a nest in spite of careful searching and patient watching of the birds. On 9 July Greely (1886b: 377) "directed that a few knots be killed for specimens, having before ordered that they be undisturbed until the nest was found". A female was shot that day which contained "a completely-formed hard-shelled egg ready to be laid. A shot had broken in the shell at one point, but it was not sufficiently injured to prevent measurement". This, the first fully-authenticated Knot's egg, was lost with much of that ill-starred expedition's material and never reached a museum or egg collection. It was found in the same general area as all the nests so far mentioned, namely northwest Greenland and northeast Canada.

For 10 years there was no further news of Knot's eggs. But then, in 1893, Alfred Newton, surprisingly, was able to add one to his collection (Newton 1905: 207–208). It was given him by his friend Lord Lilford. An entry in Lord Lilford's "everyday book on the events of his aviaries" for 1893 runs thus (Trevor-Battye 1903: 283):

June 14th. Knot (*Tringa canutus*). An egg that I am convinced is of this species laid in aviary.

Plate 30 (a) A Knot with chicks crouching somewhere nearby permits close approach on the summit plateau of North Mountain near Thule Air Base in northwest Greenland. This is where Hunt found Knot's nests in 1916, and British egg-collector William Hobson (1972) took three clutches in 1967.

(b) Knot brooding newly-hatched downy young among Arctic Bell Heather flowers a few miles from Thule Air Base. 14 July 1984.

But was this egg really laid by one of the several Knots kept in the aviaries at Lilford Hall in Northamptonshire? Newton thought so, and Lord Lilford confirmed that, when one of his knots died in the following year, it showed signs of having laid an egg (Newton 1905: 207–208).

The ornithological world may have been surprised in 1896, when Henry Seebohm's *Coloured figures of the eggs of British birds* was posthumously published, to discover that he too could have boasted of a Knot's egg in his collection. The chapter on the Knot contains the following paragraph (Seebohm 1896: 144):

> The egg which I have ventured to figure is one of a clutch of four sent, with the parent bird shot on the nest, to me by Mr. Verslev, the chief tenor of the opera in Copenhagen, who received it from Coloniforsteher Bolbre, who procured it in 1875 on Disco, in Greenland, near Godhaven, in lat. 69°.

The coloured figure referred to does look very like a Knot's egg; it could also represent that of a Turnstone, a bird which does breed at Flakkerhuk on Disco. In his *Birds of Greenland* Salomonsen (1950: 226) dismisses this and another record of the Knot having apparently bred at Disco as "due to misidentification", apparently on the rather flimsy grounds that no other evidence was forthcoming of this bird breeding so far south in west Greenland.

When that stickler for authenticity Alfred Newton brought out his *Dictionary of birds* in 1896 he discounted both his own and Seebohm's Knot's eggs: "Much curiosity has long existed among zoologists as to the egg of the Knot, of which not a single identified or authenticated specimen is known to exist in collections", he wrote (Newton 1896: 299). This curiosity was not to be satisfied for some years, although, exactly 10 years before Newton wrote these words, and unknown to him, the nest and eggs of the Knot really had been discovered and fully authenticated.

The finder of this first scientifically authenticated Knot's nest and eggs was a young Baltic aristocrat of German descent but Russian adoption, known variously as Baron Eduard von Toll or Eduard Vasil'evich Toll. He found the Knot's nest while exploring the west coast of Kotel'nyy Island, the largest of the New Siberian Islands (Map 5), on either 21 June or 2 July 1886 (Pleske 1928: 266). With Aleksandr Aleksandrovich Bunge, he was investigating the geology of the mainland coast and offshore islands between the Kolyma and Lena rivers on behalf of the Imperial Academy of Sciences at St Petersburg (Barr 1981). But his previous Knot's eggs were in such damaged condition when they arrived at the Academy's Zoological Museum that he never published his discovery (Toll 1909: 323).

It was finally during Eduard Toll's last expedition of 1900–1903, described above at the close of Chapter 3, that the nest and eggs of the Knot were really and truly discovered, this time to be brought back to St Petersburg in good condition and preserved in the Academy's Zoological Museum there. Among the personnel of this expedition were two competent ornithologists, the German Hermann Walter and the Russian Aleksey Andreyevich Byalynitsky-Birulya, alias A. Birula. The expedition ship *Zarya (Dawn)* overwintered in 1900–1901 about half-way along the north coast of the Taimyr Peninsula, just in West Taimyr and not far from the Taimyr River (Map 40). In the early summer of 1901 the expedition members dispersed in various directions from the ship to explore the area and search for zoological and botanical specimens. Great excitement was caused on 23 June, when one of the ship's officers and a sailor returned to the ship from Bonevi Island (ostrov Bonevi), with a Knot's egg. On 2 July, Petr Strizhev, who was one of two crewmen in charge of the expedition's dogs, brought in two Knot's eggs and the adult bird, which was shot to pieces. Hermann Walter was furious. He had given instructions that only complete clutches were to be taken and that only geese, ducks and grouse were to be shot (Toll 1909: 323, 327). Then, one of these two precious eggs was broken by crewman Sergei Tolstov. Walter was so upset that he lost a game of chess to Toll. Finally on 3 July, Birulya found a Knot's nest with three eggs on Bonevi Island, shot the adult bird, and photographed the nest with and without the shot bird lying beside it (Birulya 1907, Plate 5; and Pleske 1928, Plate 25, Fig. 1). Subsequently, when the same expedition had moved to the New Siberian Islands, the Knot was found to be a common breeder and more clutches of eggs were taken, Birulya himself obtaining a series of adults, eggs, and downy young in summer 1902 on New Siberia Island (ostrov Novaya Sibir), the most easterly in the archipelago.

Ekblaw and MacMillan's failure to mention these Russian discoveries is hard to explain. A paper by Hermann Walter describing the finding of breeding Knots, Sanderlings and Curlew Sandpipers in summer 1901 in West Taimyr had been published in the *Annuaire du Musée Zoologique de l'Académie Impériale des Sciences de St. Pétersbourg* in 1902. Admittedly it was in German, but it appeared in the *Ibis* translated into English in 1904 (Dresser 1904). There were delays in the further publication of these important finds. Birulya's detailed report of the expedition's ornithological work came out in 1907 in the *Transactions of the Imperial Academy of Sciences*, but it was in Russian with no English or French summary. Toll's widow, Emmy von Toll, published his journal, with mentions of the finds of Knot's eggs in 1886 and 1901–1902, in 1909, in German. Eventually, in 1928, most of Birulya's above-mentioned paper together with much other material was published in English in Theodore or Fedor Pleske's *Birds of the Eurasian tundra* as Volume 6, Number 3 of the *Memoirs of the Boston Society of Natural History*.

ARCTIC BIRDS' EGGS IN NEWTON'S COLLECTION

There was a strong commercial element in nineteenth-century egg-collecting. Rare eggs were just as often bought and sold as exchanged or given away. But these financial transactions are usually hidden from our view. None the less, a glance at some of the entries in the printed catalogue of the superb egg-collection begun by John Wolley and continued by Alfred Newton, the famous *Ootheca Wolleyana*, referred to at the start of Chapter 6, provides some insight into the workings of the ornithological treasure trove. For Newton was at the centre of a sort of international egg-mart.

Take, for example, the Sanderling's eggs in this collection (Newton 1905: 204–207). On a trip to Iceland in 1858 Edward Newton, Alfred's brother, and John Wolley obtained a single egg from a local collector. The Sanderling was and is not known to breed in Iceland, but though no locality was given, Newton thought this very probably was a genuine Sanderling's egg. At any rate, it represented this species in the collection until 1870, when he obtained from Professor Baird of the Smithsonian Institute in Washington D.C. one of the clutch of four Sanderling's eggs—the first ever discovered—collected by Roderick MacFarlane in Arctic Canada and mentioned in Chapter 6.

Of the prized Arctic waders, the Little Stint was perhaps the best represented in Newton's collection (Newton 1905: 208–209). He does not tell us how, but he managed to obtain one egg of the clutch of four Little Stint's eggs—the first ever discovered—taken by A. F. Middendorf in the Taimyr Peninsula on 1 July 1843. Then he had an egg taken by Harvie–Brown on 22 July 1875 near the mouth of the Pechora and others taken by Arnold Pike in Novaya Zemlya on 8 July 1894, by C. B. Hill at Gol'chikha on the Yenisey on 10 July 1895 and by H. J. Pearson on Vaygach Island on 6 July 1897. Finally he possessed two clutches of Little Stint's eggs purchased from a Mr Marsden. They were among the 17 clutches taken on 10 July 1903 by the Norwegian expedition to Novaya Zemlya mentioned in Chapter 6. All these eggs were meticulously authenticated in the printed catalogue as to date and locality; it would be interesting to know what was paid or exchanged for them.

That museums and private collectors were all in the game together there can be no doubt. This is well illustrated by another series of scarce eggs from the High Arctic in the Newton collection, the eggs of the Ivory Gull. Here we see examples of both exchange and purchase (Newton 1905: 341). In July 1867 Edward and Alfred Newton, on a visit to Stockholm, showed the eminent ornithologist C. J. Sundevall, who was director of the Swedish Natural History Museum, some Dodo's bones "we had with us, for which he was good enough to say he would let us have" an Ivory Gull's egg in exchange. Sundevall had in his possesion two Ivory Gull's eggs which had been taken on 30 July at Murchison Fjord, in Northeast Land (Nordaustlandet) in the Spitsbergen Archipelago by Anders J. Malmgren during the 1861 Swedish Spitsbergen expedition (Map 19). He had obtained them from another Swedish ornithologist, W. Meves, who had unfortunately dropped both eggs. However, one was not too badly damaged, and became no. 4668 in Newton's collection. Some time later he was able to purchase two Ivory Gull's eggs from the curator of the museum at Tromsø in Norway, which had also been taken in Northeast Land, on 3 August 1887. Finally he obtained through his friend Henry Dresser an Ivory Gull's egg taken on 7 August 1897 in Franz Josef Land (Zemlya Frantsa Iosifa) by English explorer Frederick G. Jackson.

ROSS'S GULL

The ultimate in Arctic ornithological treasures was a small gull which was and is universally credited with astonishing beauty and extreme rarity. The adult in summer has pink underparts and a narrow black neck ring. Nineteenth-century Scottish naturalist William MacGillivray called it 'Ross's Rosy Gull', because James Clark Ross was credited with shooting the first known specimen of it and because of the beautiful rosy tint of its underparts. Afterwards this name was split between different countries: while the Americans and the English call the bird 'Ross's Gull' and the French 'Mouette de Ross', the Danes, Dutch, Germans, Russians, Swedes and others use their own different versions of 'Rosy Gull'. While Ross's Gull thus has two different vernacular names, it had the unusual distinction of acquiring three different Latin names, at a time when only three specimens had been collected.

The first known specimen of Ross's Gull was collected near Godhavn (now Qeqertarsuaq) in west Greenland by the German-speaking mineralogist Karl Ludwig Giesecke, probably on 2 May 1813, and donated by him to the Hof-museum in Vienna. There, curator Carl Schreiber catalogued it as *L.* (for *Larus*) *collaris*, the "Collared Gull", because of the black neck ring, and left it at that, without publishing any description of the bird (Hjort 1985). Then, in summer 1823, two specimens were collected near Igloolik off the northeast coast of the Melville Peninsula, Canada, during Captain W. E. Parry's "second voyage for the discovery of the Northwest Passage". The first of them was shot by James Clark Ross on 23 June. The task of publishing "the

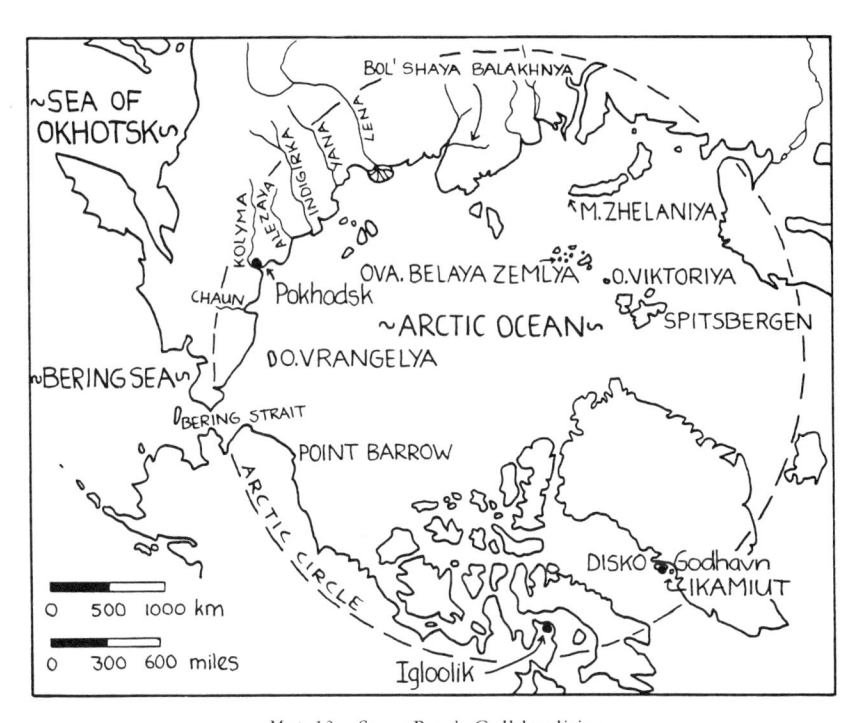

Map 13 Some Ross's Gull localities.

specimens of zoology obtained on Sir Edward Parry's second voyage" was given to John Richardson, and in the zoological appendix to Parry's (1825) *Journal of a second voyage for the discovery of a Northwest Passage* he named the supposedly new gull *Larus Rossii* after J. C. Ross, "as a tribute for his unwearied exertions in the promotion of natural history on the late Arctic voyages". His English name for *Larus Rossii* was the "Cuneate-tailed Gull". Although a specimen was exhibited and Richardson's description of the bird read out at a meeting of the Wernerian Society in January 1824, the zoological appendix was not published until 1825. After the meeting Richardson, "by direction of the Lords Commissioners of the Admiralty", handed over the Ross's Gull specimen he had described to the Edinburgh University Museum. Here it naturally came to the notice of Assistant Keeper William MacGillivray, just launching himself on a brilliant career as an ornithologist, who at that moment was working on a paper on gulls. In this paper, which was published in 1824 before Richardson's zoological appendix, MacGillivray innocently referred *pro tempore* (for the time being) to the new gull as *Larus roseus*, explaining that it "is to receive its proper designation from Dr. Richardson". But this temporary name was taken up and used in subsequent ornithological publications because, according to accepted rules of zoological nomenclature, MacGillivray had given the first published description and binomial designation of the new gull. It made no difference that MacGillivray, when he subsequently (1842) placed the gull in a genus of its own called *Rhodostethia*, gave it Richardson's specific name of *Rossii*. *Roseus*, altered to *rosea* to agree with *Rhodostethia*, remained and still remains the accepted specific name. Thus, at a time when only three specimens existed in museums, Ross's Gull had been given three different Latin specific names: *collaris, roseus* and *Rossii* (Swainson and Richardson 1831: 427–428; Murdoch 1899; Densley 1988).

For many years after the first specimen was procured in Greenland Ross's Gull remained a rarity. An occasional expedition recorded the odd sighting and that was all. In 1875, when Howard Saunders made an inventory of the existing specimens in museums, there were fewer than a dozen, and four of them came from Disko, west Greenland (Saunders 1875). In 1879, when Raymond Newcomb, naturalist of the *Jeannette*, saw and collected his first Ross's Gulls on 8 October not far north of Wrangel Island (ostrov Vrangelya), the ship's captain and expedition commander G. W. De Long (1884: 151) commented in his journal that this was "a most valuable prize and rare beyond calculation. In all Europe there is but one (at the museum in Mainz), and there is no record of one of the United States". The *Jeannette*'s drift in the Polar ice pack and her naturalist's ornithological work have already been described in Chapter 3. Newcomb and his colleagues eventually secured seven or eight Ross's Gulls and Newcomb managed to bring the three best skins home, part of the time wrapped in a piece of canvas under his jacket, to be preserved in the Smithsonian Institution. "This species is the loveliest I ever saw", he wrote (1888: 282).

During the autumn of 1881, whilst Commander De Long and his men were struggling over the ice and across stormy seas to the Siberian mainland, and then fighting for life and dying in the Lena Delta, another American Arctic expedition was establishing itself at Point Barrow on Alaska's north coast. Here, the expedition's ornithologist John Murdoch (1885 and 1899) suddenly struck gold. On 28 September, while he and his companions were busy unpacking, he noticed flocks of gulls passing along the coast. His sight identification of Ross's Gull was confirmed when he shot one. But the arsenic and tools for skinning birds were not yet unpacked. He laid the precious specimen down on a barrel in the cold store tent, only to find later that it had vanished—the dogs had eaten it! He need not have worried. More and more Ross's Gulls passed by the point flying in a northeasterly direction during the next few weeks. On some days they were "exceedingly abundant". More specimens were obtained as the birds came in from the sea and circled around the expedition's station and over the lagoons. On 22 October the movement stopped and no more Ross's Gulls were seen until the next autumn, when they again became abundant on and after 21 September 1882, and continued so until 9 October. Just how many Ross's Gulls Murdoch shot was not revealed in print. Professor Baird "expressly forbade our making public the numbers collected" for fear of being overwhelmed with requests for gifts or exchanges (Murdoch 1899: 152). Elliot Coues took Murdoch jokingly to task for "vulgarizing this beautiful bird". According to Murdoch "Our expedition succeeded in obtaining a large series of this rare and beautiful bird—more, in fact, than there were before in all the museums of the world put together—and a still larger series might have been obtained had the weather and other conditions been favourable" (Murdoch 1885: 123). Fortunately for Ross's Gull Murdoch was suffering from a bad cold in fall 1882 and, on his own admission, did not do much shooting. The shot birds were put out in the "block house" or cold store at the time and skinned later during the winter. Murdoch describes how he would fetch in half-a-dozen birds one evening, representing a morning's work skinning, and place them on a rack behind the stove in the living quarters overnight to thaw them out.

Murdoch may have flooded the market with Ross's Gull specimens, but he had not answered the question where his gulls came from and where they were going. He assumed that the thousands that passed Barrow must retrace their steps farther out to sea later in the same autumn because they "must of necessity seek lower latitudes" in which to winter, inferring that they wintered in the Bering Sea. In 1885 (p. 124) he speculated "that some land yet to be discovered, and north of Wrangel Island, will one day yield a glorious harvest of the eggs of this splendid species".

In the very same year in which Murdoch wrote those words Ross's Gull was discovered breeding. American ornithologists were informed of this exciting event one year later in a brief note in the *Auk* (Dalgleish 1886). Paul Müller, the Danish governor at Christianshåb (now Qasigiannguit), had discovered a nest of this rare species on 15 June 1885 on an islet off the island of Ikamiut in Disko Bay, west Greenland. The Ross's Gulls were nesting in the middle of a colony of Arctic Terns and both adult birds were at or near the nest, which contained two eggs. One of them was broken; the other went to Copenhagen's Zoological Museum, while the skin of one of the old birds found its way into Seebohm's collection (Salomonsen 1950: 332). Ikamiut was only about 150 miles (250 km) north of the Arctic Circle, further south than most Ross's Gull sightings up to then, and the event was universally regarded as exceptional. Indeed some ornithologists, notable among them Howard Saunders, refused to accept it as authentic. He maintained that the egg was that of a Sabine's Gull. A century later Yorkshire ornithologist Mike Densley saw the egg and expressed doubt in a letter to me "about it being that of Ross's Gull. The shape of the egg and its background colour do not appear right to me." In any case, this discovery had not satisfactorily answered the question, which many were asking, as to the whereabouts of the bird's main breeding area. In 1899 Murdoch was still speculating about Ross's Gull's breeding haunts. He thought Norwegian explorer Fridtjof Nansen was right in postulating a breeding colony in the far northeast of Franz Josef Land, namely in the group of islands forming Hvidtland or Hvidtenland, now ostrova Belaya Zemlya. But he believed that the "main breeding ground" was on a large hitherto undiscovered island north of Alaska—Keenan Island of American geographers.

It was a Russian civil servant and High Court judge, Sergei Aleksandrovich Buturlin, who first solved the mystery of the regular breedings haunts of Ross's Gull. This amateur ornithologist was sent off from St Petersburg in 1904 to the Kolyma via Yakutsk to arrange food supplies for the people of northern Siberia after the Japanese blockade of Vladivostok had interrupted the normal lines of communication (Potapov 1990). In summer 1905, in the delta of the River Kolyma, the most easterly of the great north-flowing rivers of the Soviet Union, he found Ross's Gull breeding "quite commonly". It was the most numerous gull there. Members of the British Ornithologists' Union must have been excited to read in their copies of the *Ibis* early in 1906 a letter sent from Pokhodsk in the Kolyma Delta by Buturlin (1906: 131–139) on 30 June 1905 announcing his find. "Up to the time of writing (June 23rd)", he wrote, "I have collected 38 skins and 36 eggs of this pretty bird, though I have spared large numbers of adults and their nests expressly to enable me to acquire a sufficient quantity of the young in down and of specimens in the immature plumage" (p. 131). Buturlin's birds were nesting on islands in lakes or in swamps in an area of marshy tundra some 200 miles north of the Arctic Circle, in scattered colonies of 10 or 15 pairs in company with Arctic Terns. He gives a charming description of their behaviour soon after their arrival in the area on 30 May (p. 134):

> Rosy Gulls hovered over this lake catching flies and other insects, or swam upon the surface, though they more often sat on the snow and ice in the vicinity. Both birds of a pair usually sat close together; and if the male, easily distinguishable even at a distance by his much more intense coloration, thought that others came too close, he actually tried to push his mate to one side; or if a male attempted to approach a second time (some of the younger, paler-coloured birds not having yet paired) he would engage in a short fight with the intruder—in which he was sometimes aided by the female—with angry cries of "miaw-miaw-miaw", to which the retreating culprit responded with a "á-dac, á-dac, á-dac", repreated with differet intonations. Every now and then the male tried to express his feelings to his mate by pecking her curiously, as if trying to kiss her, with his open beak on her head or neck, or made a few steps round her to one side or the other, shewing off as some Pigeons do; then

Plate 31 Adult Ross's Gull in breeding plumage photographed by Pavel Tomkovich in July 1972 near the River Indigirka in Siberia.

with a sound like t r r r r r r lowered his neck and breast to the ground, and in this position, with all the hinder part of the body, the tail, and the ends of the folded wings high up in the air, continued for some seconds his little promenade before the female, who very rarely engaged in such antics.

Buturlin was able to describe the breeding biology of Ross's Gull in some detail, and he established that it also nested on the lower reaches of the two rivers to the west of the Kolyma, the Alezaya and the Indigirka. Many years elapsed after this before new discoveries extended the known breeding range of Ross's Gull in the Soviet Union. In 1959 the geologist Oleg Kuvaev found a small colony at the mouth of the River Chaun in the Chukchi Peninsula. Then, in 1971 and 1972, a group of Soviet ornithologists, including Savva Mikhailovich Uspensky, a leading expert on Arctic birds whose book on the birds of the Soviet Arctic is discussed in Chapter 9, and A. A. Kishchinsky, whose researches will be discussed in Chapter 10, found and photographed Ross's Gulls breeding in the delta of the River Yana, some 300 miles west of the Indigirka, as well as on the Indigirka itself (Uspensky 1977; Kishchinsky 1988). In the following year, 1973, about 50 pairs of Ross's Gull were found breeding in the eastern part of the Taimyr Peninsula along the lower reaches of the Bol'shaya Balakhnya River (Map 40). Later they were discovered nesting in the Lena Delta (Orlov 1988: 158). Thus Ross's Gull is now known to have bred in every important river delta debouching into the Arctic Ocean between the Taimyr and Chukchi peninsulas. In the second half of the present century odd pairs of

Ross's Gull have bred, usually unsuccessfully, in Spitsbergen (1955), Greenland (1979) and Canada (1976 on) (Densley 1988).

Although the main breeding area of Ross's Gull now seems to have been accurately described, doubt still shrouds its winter quarters. Few ornithologists have taken seriously the notion based on the autumn flight at Barrow in a northeasterly direction, and still subscribed to by some authors (for example Dymond *et al.* 1989: 148), that Ross's Gull winters in the Arctic Ocean. As early as 1957 it had become clear to Russian ornithologists from observations made from the first four Soviet floating ice stations that, while Ross's Gulls, evidently non-breeding birds, were fairly numerous in summer in the pack ice of the northern Arctic Ocean, they were more or less absent in winter (Rutilevsky and Uspensky 1957). This was confirmed by V. Butev (1959), who saw none during two winters at Cape Zhelaniya, the northern tip of Novaya Zemlya, and by Valery Orlov (1988: 157), who spent a winter on Victoria Island (ostrov Viktoriya) between the Spitsbergen and Franz Josef Land archipelagos, and never saw Ross's Gulls after October. Recent research at the Alaska Fish and Wildlife Research Centre in Anchorage has confirmed what Murdoch had supposed, namely that the birds passing Point Barrow eastwards in early autumn return westwards later in the autumn, often further out to sea (Divoky *et al.* 1988), presumably making for Bering Strait. Moreover, winter sightings from Japanese waters (Orlov 1988: 158) bear out Murdoch's suggestion that the main wintering area of Ross's Gull lies south of Bering Strait. Wintering Ross's Gulls probably spend their time far out at sea both in the Bering Sea and in the Sea of Okhotsk (Potapov 1990 and see now Il'ičev and Flint 1990), up to a thousand miles south of their breeding haunts. If this is the case, the account of Ross's Gull in *The birds of the Western Palearctic* (Cramp and Simmons 1983: 861) will have to be revised. Under the heading "Movements" the following in any case somewhat dubious statement occurs: "Winter distribution unconfirmed; presumed to be along ice edge of Arctic Ocean, thus in latitudes north of principal breeding grounds."

Ross's Gulls were still being shot in the twentieth century for collections and for food. In a note published in the journal *Condor* in 1929, entitled "Ross Gulls for dinner", Clinton D. Abbott, of the San Diego Society of Natural History, stated that his society had received "many valuable consignments of Arctic birds and mammals, including what may be the largest series of Ross Gulls in the country" from Charles D. Brower. "In connection with his latest gift, which included twenty-five Ross Gull skins" Brower wrote:

> I did get a good crack at the Ross Gulls again this fall [1928]. One day, the 26th of September, they were around in thousands. If I could have had the time, I could have had several hundred birds to skin. The eskimo shoot them for food, and they are mighty good at that. I have eaten them many times, and this fall I had them fried and roasted until I almost turned into a Ross Gull myself. They taste just as do the Golden Plover, and are just as fat in the fall.

Another opinion about the culinary value of Ross's Gull was volunteered to me in a letter from Mike Densley, which he has allowed me to quote:

During my two-month stay at Barrow, in September–October 1975 in pursuit of passage Ross's Gulls, I spoke to a number of local Eskimos, young and old, and was told by them all that the shooting of Ross's Gulls there for food had ceased many years ago. A biologist there in 1975 was collecting specimens of Ross's Gull for the University of Alaska Museum in Fairbanks. Some of the carcasses from the skin mounts he prepared on the spot were 'microwaved' by us to try as food. Not a lot of meat, edible, not fishy, a little like chicken!

An occasion when Ross's Gulls could have been collected for scientific purposes, but were not, occurred during the Arctic expedition of the Swedish icebreaker *Ymer*. At least 1000 individuals, many in summer plumage, were seen in July–September 1980 in the pack ice west of Spitsbergen and north of Spitsbergen and Franz Josef Land. These must have been non-breeding birds summering in the pack ice of the Arctic Ocean. This remarkable experience was described in English in an article in *British Birds*, and in Swedish in a beautifully illustrated book by artist of the expedition Gunnar Brusewitz (Meltofte *et al*. 1981; Brusewitz 1981). He is also the illustrator of this book. Although he was able to make *post mortem* portraits during the expedition of several species that the expedition's biologists collected for scientific purposes, not a single Ross's Gull was shot.

CHAPTER 8

Goose puzzles in North America

SAMUEL HEARNE'S GEESE

It took ornithologists some time to sort out the six species of goose inhabiting the North American continent, all of which breed in the Arctic. They are the White-fronted, Snow, Ross's, Emperor, Canada and Brent Goose. Instead of the usual increase in the number of species as new ones were discovered, with American geese it was a question of reducing the number. The pioneer Samuel Hearne, some of whose birds have been discussed in Chapter 5, started with too many: he had "no less than ten" (Glover 1958: 281–287):

1. *The Common Grey Goose*. It bred "in great numbers in the plains and marshes near Churchill River". On 9 August 1781 Hearne and some Indian assistants managed to drive a flock of 41 half-grown young and moulting adults "into the stockade which encloses" Fort Prince of Wales on the Churchill River, opposite modern Churchill (Map 8). They were

"fed and fattened for winter use". This goose was one of the many races of the Canada Goose.
2. *The Canada Goose*. "Does not differ in plumage" from no. 1 above, but is smaller. They "generally fly far North to breed". Evidently one of the smaller races of the Canada Goose.
3. *The White, or Snow Goose*. Under this head Hearne includes the white form of the Lesser Snow Goose, and the Greater Snow Goose, which has a white form only.
4. *The Blue Goose*. This is the blue or grey form of the Lesser Snow Goose.
5. *The Horned Wavey* is Ross's Goose. This and the above two geese in Hearne's list have been mentioned in Chapter 5 and will be discussed below.
6. *The Laughing Goose*. From his description of the plumage, Hearne is here describing the White-fronted Goose. He did not know where it bred.
7. *The Barren Goose*. The largest goose known to Hearne, differing "from the Common Grey Goose in nothing but size". No doubt one of the large races of the Canada Goose, probably the Giant Canada Goose *Branta canadensis maxima*, once thought to be extinct.
8. *The Brent Goose*. This is the Brant of Americans, the same species as our Brent Goose *Branta bernicla*. Hearne says that it breeds "in the remotest parts of the North", its breeding places never having "been discovered by any Indian in Hudson's Bay".
9. *The Dunter Goose*, "as it is called in Hudson's Bay, but which is certainly the Eider Duck".
10. *The Bean Goose*. Hearne claims to have "seen three that were killed" but was probably mistaken as there have been no subsequent records of this Eurasian species in the Hudson Bay area.

HUTCHINS'S GOOSE

Hearne's goose muddle was exacerbated by the learned and famous authors of the *Fauna Boreali-Americana*, William Swainson and John Richardson (1831: 470–471). To honour the eighteenth-century Hudson's Bay Company Chief Factor at Fort Albany, Thomas Hutchins, who had devoted much time to studying and collecting the birds he came across, they named a strange goose after him: *Anser hutchinsii*, and called it in English "Hutchins's Barnacle Goose". According to Richardson, the residents of the Hudson Bay area considered this bird "a small kind of the Canada goose". They were right. It was a small, pale-breasted form of the Canada Goose breeding in the High Arctic coastal tundra. Nevertheless, Swainson and Richardson did enjoy some support for the treatment of it as a separate species and for the name "Hutchins's Goose". A. C. Bent, writing in 1925, calls it "Hutchins Goose", and accepts it as a race of the Canada Goose, but quotes (p. 224) a letter of Roderick MacFarlane to the Smithsonian Institution in which MacFarlane states: "I have no doubt about Hutchins goose being a good species; its mode of nesting alone would go far to prove it distinct from the Canada goose . . .".

174   *In search of Arctic birds*

(a)

(b)

Plate 32   (a) Hutchins's Goose, later called Richardson's Goose, is now recognised as a form of the Canada Goose. This bird is standing at its nest on an islet in one of the Cambridge Bay area's tundra lakes. (b) When sitting, the bird figured above holds its head down low in the normal posture adopted by incubating Canada Geese.

More recent writers have confusingly abandoned the name "Hutchins's Goose" and have called the bird instead "Richardson's Canada Goose", or just "Richardson's Goose" (Owen 1980: 54; Madge and Burn 1988: 147), but nobody nowadays regards it as a distinct species.

BLUE GEESE AND SNOW GEESE

Although the Canada Goose has been generally accepted for the last hundred or more years as a single species, ornithologists found it hard to agree on the status of the snow geese and blue geese. Hearne considered them to be two separate species. Some later writers believed the blue geese to be merely the young of the snow goose. When Arthur Cleveland Bent came to write the relevant volume of his *Life histories of North American wild fowl*, he opted for two distinct species, *Chen hyperborea*, the Snow Goose, in which the adult was white all over except for black wing tips, and *Chen caerulescens*, the Blue Goose, in which the adult was mainly dark grey in colour except for a white head and a blue-grey wing patch. He divided his snow geese into two subspecies, the Greater Snow Goose, which was larger and had a more northerly and westerly breeding range, and the Snow Goose or Lesser Snow Goose. Surprisingly, though he was writing as late as 1925, Bent (pp. 179–180) had to admit that "there seems to be no authentic record of the finding of a nest of the blue goose and, so far as I know, the nest and eggs in a wild state are unknown to science". "To find the breeding resorts of the blue goose is one of the most alluring of the unsolved problems in American ornithology. It is really surprising that such a large and conspicuous species, which is numerically so abundant, can disappear so completely during the breeding season." And again, "No one knows where the blue goose goes to spend the summer . . ."

The challenge was taken up by the Canadian government's Wildlife Service, which instructed one of its officials, J. Dewey Soper, to search for breeding Blue Geese, using aircraft, kayak or any other suitable mode of transport. In 1929 he was successful. He saw Blue Geese accompanied by goslings on the mudflats of western Baffin Island and discovered a large breeding colony which he described in the *Bulletin of the Department of the Interior of the Dominion of Canada* for 1930 (Eifert 1962: 232–234; Bannerman 1957: 255 also see Soper 1942). It is now within the bird sanctuary named after him (Map 38). Here the geese were nearly all Blue and Soper saw no sign of interbreeding with the very few white birds or Lesser Snow Geese he noted. Then, in the summer of 1930, the American ornithologist and bird painter G. M. Sutton (1932) found other breeding colonies of Blue Geese on Southampton Island (Map 8), but here the white Lesser Snow Goose was a common breeder and Blues were even occasionally found interbreeding with Snows. Nevertheless, Sutton agreed with Soper that the Blue Goose and Lesser Snow were two distinct species. Although he saw some geese whose plumage appeared to be intermediate between the Blue and Snow Goose, he supposed them to be hybrids. It looked as though Bent's division of the birds into two separate species had been confirmed by expert field work, but in 1942, in a paper in the *Auk*, Tom H. Manning, who had camped near the centre of a breeding colony of geese on Southampton Island from 25 June to 7 August 1934, argued that the Blue and Lesser Snow Geese ought to be treated as a single species. They were identical in every respect except plumage colour. His conclusion soon found general acceptance, and the Snow Goose, now usually called *Anser caerulescens*, is agreed to have two colour phases, a light (snow) and a dark (blue) form. On top of this, two races or subspecies are recognised—the Greater Snow Goose, in which the blue phase is extremely rare, and the Lesser Snow Goose.

Even though most Snow Geese breed in remote areas inaccessible to biologists except in expensive and specially equipped expeditions, an astonishingly detailed knowledge of Snow Goose biology has been acquired in the 60 years since the first discovery of the breeding haunts of what was formerly called the Blue Goose. It is now known that a large proportion of the Lesser Snow Goose population is concentrated in eight huge colonies, each numbering 30 000 pairs or more (Owen 1980: 42), extending from Wrangel Island in the Soviet Union in the west to three neighbouring colonies in Baffin Island. The total wintering population in the United States of America was put at over 2 million in 1987–1988 by the Canadian Wildlife Service in a survey prepared for the North American Waterfowl Management Plan (Hugh Boyd, personal communication). The Greater Snow Goose breeds in northern Baffin Island, Ellesmere Island, and northwest Greenland and numbered more than 400 000 individuals in the autumn of 1984. The population of this beautiful snow-white goose with black wing tips is one of the few that it is possible to census, because the entire breeding population gathers at staging areas along the St Lawrence River every spring before continuing on northwards. At the beginning of this century the Greater Snow Goose population was down to a few thousand individuals, so it has made a remarkable recovery (Owen 1980: 430); indeed, the bird may well be more numerous now than ever before.

Besides monitoring the size of Snow Goose populations, the Canadian Wildlife Service keeps a watchful eye on the proportion of blue morphs in them. Variations in this proportion had puzzled earlier naturalists. Already in the previous century John Rae, naturalist, Hudson's Bay Company surgeon and Arctic explorer, had noted that the proportions of blue and white geese varied from place to place and from time to time. In 1888, writing of three places in James Bay, at the southern end of Hudson Bay, over which the geese passed on migration, he stated (Rae 1890: 138; see Map 8):

> I may mention that 45 years ago the blue-winged and white-wavy geese visited Moose in about equal numbers, as they still do; whereas at Albany, 100 miles to the north, there were great numbers of the white birds and scarcely a blue-wing to be seen. Now the two kinds are about equally abundant there, whilst at Rupert's River, 100 miles east of Moose, now, as formerly, the blue-winged birds are alone met with.

Goose puzzles in North America 177

(a)

(b)

Plate 33 (a) This Greater Snow Goose pair were nesting on the bank of a shallow pool in northwest Greenland, where the bird has increased substantially in number in the last 50 years. (b) Greater Snow Goose near its nest in northwest Greenland. The Sun Glacier, which debouches into the head of McCormick Fjord, north of Qaanaaq, is visible in the background at least 5 miles away.

ROSS'S GOOSE

Although Hearne's "Horned Wavey" was undoubtedly Ross's Goose, it was another Hudson's Bay Company official, Bernard Ross, who in 1861, having sent the skin of a goose new to him to the Smithsonian Institution, had the species named after him by John Cassin. At that time and for a long time afterwards, the breeding haunts of this smaller version of the Lesser Snow Goose in its white phase were entirely unknown. That assiduous collector Roderick MacFarlane (1891), whose ornithological activities have been described in Chapter 5, never succeeded in finding Ross's Goose breeding in the Anderson River area which he worked in the 1860s. He regarded it as the rarest goose there and collected only one specimen, a male shot at Fort Anderson on 25 May 1865. As with the so-called Blue Goose, so with Ross's Goose, A. C. Bent was forced to admit in 1925 (pp. 185–188) that "Absolutely nothing seems to be known about its breeding habits in a wild state." Although "no one knows . . . whither it goes when it wings its long flight northeastward across the Rocky Mountains in the early spring", its winter quarters in California were familiar to many wildfowlers and ornithologists.

But, as A. C. Bent pointed out, the nesting habits of Ross's Goose had been studied in captivity 20 years or more before he wrote. At a meeting of the British Ornithologists' Club on 20 March 1901 a Dutchman, Frans Ernst Blaauw, exhibited the first Ross's Goose's egg ever seen by a white man. It had been laid at his place Gooilust in Holland by a solitary female, and was of course infertile. Early in the following year Blaauw was lucky enough to obtain a male Ross's Goose, "through the courtesy of Doctor Heck, of Berlin". "The birds soon paired, and in the beginning of May, 1902, the female made a nest under a bush in her enclosure". Five eggs were laid and she began sitting on 30 May, having first "abundantly lined her nest with down from her own breast". Twenty-one days after the start of incubation all five of the eggs hatched but, sadly, the young died while still in down. A year or two later a clutch of three Ross's Goose eggs was transferred to a hen and one gosling was successfully raised so that Blaauw could communicate to the *Ibis* in 1905 a detailed account of the development of its plumage during the first few months of its life (Bent 1925: 186–187).

It was not until June 1938 that the first Ross's Goose nest in the wild was finally discovered. Angus Gavin, manager of the Hudson's Bay Company's Perry River Post, which was situated on an island called Flagstaff Island off the river's mouth, found a colony of about 50 pairs breeding on islets in a lake some "12 miles up the Perry River and 8 miles southeast along a tributary" (Gavin 1947; see Map 18). The Canadians were only just the first with this important discovery. Since 1935, Charles E. Gilham, of the Fish and Wildlife Service of the United States government's Department of the Interior, had been surveying Arctic wildfowl and searching for breeding Ross's Geese. In 1939 he flew over the Perry River and saw numerous white birds on the lakes and marshes, but weather conditions precluded his landing to inspect them closely. In 1940 he had chartered a plane for an on-the-ground investigation of the Perry River, but had to cancel his trip at the last moment. In June of that year Angus Gavin and a Hudson's Bay Company colleague, Ernest Donovan, canoed upstream from the Perry River Post. They set out at 4.0 p.m. Two Eskimos seated fore and aft of the company officials paddled away all night in bright sunlight so that they could reach and inspect the Ross's Goose colony Gavin had discovered 2 years before (Eifert 1962: 235–237).

The mystery of the breeding Ross's Goose had been solved on the eve of the Second World War. It was only natural that, after the war, efforts would be made to rediscover the colony found by Gavin and to search for others, especially because the species was believed by some to be "in danger of extinction" (Scott 1951: 12). In 1949 Peter Scott took part in an expedition which was flown to the Perry River by air and remained there throughout the goose breeding season. Consisting of himself and two Americans, Paul Queneau, a geologist, and Harold Hanson, a waterfowl biologist of the Illinois Natural History Survey, the expedition was financially supported by the Arctic Institute of North America. Its aims were to study the bird life of the Perry River area, including the breeding geese, and especially Ross's Goose, to catch and ring geese and ducks, to collect specimens of birds, and to capture Ross's and other geese alive to add to the wildfowl collection of the Severn Wildfowl Trust (now the Wildfowl and Wetlands Trust), which Peter Scott had founded in 1946 at Slimbridge in Gloucestershire, on the Severn Estuary.

The success of the Perry River Expedition (Scott 1951), like that of so many other Arctic expeditions, was in large measure due to assistance gladly given by local Eskimos. In particular, against a payment of $12 and 15 packets of cigarettes, two of them took Scott and Hanson to a breeding colony of Ross's Goose not known to Gavin, which contained about 350 pairs of birds. The geese were sitting on their nests scattered over six small islands in a large lake, each attended by her gander. The Eskimos acted not only as guides but also as porters on the 3-day walk to this colony and back to the expedition's base camp. In all, 1951 Ross's Geese were counted by the expedition; nine of them, together with two White-fronted Geese, were captured and taken home alive. Other Ross's Geese were shot for specimens and some were eaten. On 21 June an immature shot by Queneau for the skin was cooked for supper. Scott's diary recorded that "we had been forced to shoot a Ross's goose for food" (Scott 1951: 93); and on 18 July "We ate Ross's goose for supper." They also dined on King Eider and on 20 June Scott records that he was snug in his sleeping-bag "after a huge dinner of white-front and lesser Canada". This must have been one of the last ornithological expeditions to the Arctic whose members dined on the very birds they were studying, though the frequent descriptions of meals—typical of many an Arctic explorer's diary—show that on Scott's expedition this was the exception rather than the rule.

Although Peter Scott and his colleagues of the Perry River Expedition made no general survey of breeding Ross's Geese in the coastal lowlands south of the Queen Maud Gulf (see Map 38), they evidently believed the near 2000 geese they had counted represented almost half the total world population, for Scott wrote in the forward to his book that this was "probably no more than 5000, possibly much less". He may not have been far off the mark but fortunately since that time the Ross's Goose population has increased substantially (Ryder 1969). Winter counts made in the San Joaquin Valley, California, from 1956 on, showed an average population, over the nearly 20-year period 1956–1974, of over 23 400 birds. At the beginning of this period (1956–1960) the count averaged 13 000 birds; at the end (1966–1974) the average was 27 000 (Bellrose 1976: 131–132). Since then the Ross's Goose population has gone from strength to strength. Canadian Wildlife Service figures showed that, by the early 1980s, there were over 100 000 individuals. Now, in 1991, there could even be 200 000 Ross's Geese in existence. It is hard to be exact because of the difficulty of separating them in the field from the white-phase Lesser Snow Geese that share their breeding and wintering areas. If Ross's Goose ever was in danger of extinction, this is certainly not the case now.

CHAPTER 9

Arctic ornithologies and their authors

THE ARCTIC AS A WHOLE

Attempts have been made in the past, and will no doubt be repeated in the future, to write an Arctic ornithology describing the avifauna of the Arctic as a whole. The task should not be too difficult, for many Arctic species are circumpolar in their distribution and, as mentioned in Chapter 1, the longest possible list of Arctic breeding birds could not much exceed 150 species. One such attempt was made in the eighteenth century by Danish naturalist Morten Thrane Brünnich (Mearns and Mearns 1988). Although he began by studying oriental languages and theology, this near-genius soon turned to natural history, becoming professor in that subject at Copenhagen University at the age of 32. By that time he had published books on entomology, on Mediterranean fishes, and on the Eider. His *Ornithologia borealis* or *Northern ornithology* was published in 1764, when he was 26. It was based on specimens from the Faeroes, Iceland, Greenland and Norway—territories then forming part of Denmark's colonial empire—in the natural history collections put together by two Danish civil servants, of which Brünnich had been appointed curator. In this book he published the first description of the Great Northern Diver, as well as of the (non-Arctic) Manx Shearwater and Jack Snipe. He also distinguished clearly between the two European species of guillemot, which had up to then always been confused. The more northerly species, the Thick-billed Murre of Americans, became known in Britain as Brünnich's Guillemot after Edward Sabine proposed the specific name *Brunnichi* in 1818.

Like Brünnich, Thomas Pennant (1726–1798) was hampered in his attempt to describe the northern avifauna by lack of knowledge: both wrote before the great exploring voyages of the nineteenth century. Although he is perhaps best known for his correspondence with Gilbert White, Pennant was also a prolific writer in zoology. His *British zoology*, published in 1766, went through four editions in his lifetime, but only in the fifth edition published in 1812, fourteen years after his death, did his name appear on the title-page (Mullens and Swann 1917: 464–469; and see Mullens 1909). His *Arctic zoology* came out in two volumes in 1784 and 1785, the introduction and quadrupeds in Vol. 1 and the birds in Vol. 2. Originally intended as a zoology of British North America, the "fatal and humiliating" Declaration of American Independence of July 1776 caused Pennant to alter its character to a circumpolar zoology of northern regions: he was unwilling to become the zoologist of the new, independent North America and no longer even willing to travel there, which he had planned to do (Pennant 1784: Introduction). Even though Vol. 2 of the *Arctic zoology* contained a quite remarkable amount of information about northern American birds, mostly obtained by Pennant from his friend Joseph Banks and officials of the Hudson's Bay Company, huge gaps naturally remained. Pennant was well aware of this fault. At the end of his book (1785: 585) he wrote: "I have done as much as the lights of my days have furnished me with." He surely had. A curiously prescient discourse, which demands quotation in full though it has little to do with birds, follows this sentence (Pennant 1785: 585–586):

> In some remote age, when the *British* offspring will have pervaded the whole of their vast continent, or the descendants of the hardy *Russians* colonized the western parts from their distant *Kamschatka*, the road in future time to new conquests: after, perhaps, bloody contests between the progeny of *Britons* and *Russians*, about countries to which neither have any right; after the death of thousands of claimants, and the extirpation of the poor natives by the sword, and new-imported diseases, a quiet settlement may take place, civilization ensue, and the arts of peace be cultivated. . . .

Early in the twentieth century the Berliner Herman Schalow produced the first comprehensive work on the avifauna of the Arctic. Entitled *Die Vögel der Arktis*, it was published in 1904. At age 19, Schalow had both joined the German Ornithological Association and obtained a post in a Berlin bank. Thereafter he was part-time ornithologist and full-time banker (Palmer *et al.* 1954: 502–503). His book on the birds of the Arctic is especially valuable for its very full bibliography of books and papers containing references to Arctic birds, arranged chronologically. It contains noteworthy discussions of problems still of interest to present-day researchers, such as the boundaries of the Arctic, the origins of Arctic birds, breeding habits of Arctic birds, the most northerly birds, and the relationship of the Arctic and Antarctic avifaunas. The main part of the book is devoted to species accounts, which concentrate on geographical distribution, and are of course now hopelessly out of date. Schalow included 144 species and subspecies as having bred in the Arctic.

For half a century Schalow's *Birds of the Arctic* remained the standard, and indeed only, work on the Arctic avifauna. Then in 1956–1958, Danish ornithologist Hans Johansen brought out a new version or replacement rather forbiddingly entitled *Revision und Entstehung der arktischen Vogelfauna* or *Revision and origin of the Arctic avifauna*, also published in German, but with an English summary. It appeared as fascicules 8 and 9 of a series called *Acta Arctica* published at Copenhagen by a group of Scandinavian Arctic experts called the Societas Arctica Scandinavica. Johansen's was the first ornithological work in this series; earlier volumes had dealt with Eskimo culture, the microfauna of northern Canada, and Arctic botany. Johansen criticised Schalow for defining the Arctic too narrowly, so as to include only the Arctic Ocean and islands bordering it, and, even so, admitting too many birds to the Arctic avifauna. He reckoned that only 80 species could be considered truly Arctic. The bulk of his book is taken up with species accounts which attempt to elucidate the geographical distribution of each species in the past and present, and of its various subspecies or races, in the light of the greatly increased knowledge of recent events in geological time gained in the half-century before he wrote. Thus he was able to show how, as the world climate cooled at the start of the last ice age, the birds then living in the Arctic were forced to adapt. They did this in two ways. Some learned to cope with severe cold and winter darkness and remained in the Arctic: for example, the Ptarmigan, Ivory Gull and Snowy Owl. Others, like the *Calidris* sandpipers and the geese, evolved migratory powers which enabled them to flee the Arctic altogether during the winter. Many of Johansen's hypotheses or conclusions are still accepted today, for example the splitting of the Arctic Zone between the two great zoogeographical regions of the Nearctic and Palearctic.

Recently, oil industry ecologist Bryan Sage has attempted to put the Arctic avifauna in a nutshell, that is to describe it in a single chapter of his book *The Arctic and its wildlife*, published in 1986. With the help of superb photographs, some excellent diagrams, a first-hand knowledge of the birds of northern Alaska and a thorough study of the literature, he succeeds very well indeed. Set out in the tables are estimated breeding densities of some bird species in selected parts of the Arctic, population estimates of Arctic breeding geese, and a comparison of breeding strategies and habitats of some Arctic nesting sandpipers.

REGIONAL ORNITHOLOGIES

Since the close of the Second World War earlier accounts of the avifaunas of different Arctic regions have mostly been replaced by scientific, more or less definitive, works based on the study of museum collections, expeditions in the field, and very thorough reviews of the literature. Those discussed in this chapter are listed in Table 14.

Table 14 Some regional Arctic ornithologies published since the Second World War.

Author	Date of publication	Title	Where published
Bailey, A. M.	1948	*Birds of Arctic Alaska*	Denver
Snyder, L. L.	1957	*Arctic birds of Canada*	Toronto
Todd, W. E. C.	1963	*Birds of the Labrador Peninsula*	Toronto
Salomonsen, F.	1950	*The birds of Greenland*	Copenhagen
Løvenskiold, H. L.	1964	*Avifauna Svalbardensis*	Oslo
Uspensky, S. M.	1958	*Birds of the Soviet Arctic*	Moscow

BAILEY ON THE BIRDS OF ARCTIC ALASKA

What makes Alfred Bailey's book especially interesting is that, while half of it is devoted to a species-by-species account of the birds, the other half describes the author's field work and the exploration history of the North Slope or northern coastal strip of Alaska. The field work he later described again in a separate book (Bailey 1971): *Field work of a museum naturalist. Alaska-southeast; Alaska-far north. 1919–1922.* Both books are illustrated with the author's own photographs.

Born in 1894 in Iowa City, Alfred Bailey (Phillips 1981) studied at the University of Iowa at that place. In 1912–1913 he voyaged to Laysan Island in the Hawaiian Islands, sponsored by the United States Bureau of Biological Survey. His first important post was at the Louisiana State Museum in New Orleans, where he was Curator of Birds and Mammals between 1916 and 1919. While there he had organised a cruise among the bird islands off the Louisiana coast that had a decisive impact on his future career. Among the passengers were two ornithologists. One of these was the director of the Colorado Museum of Natural History, now the Denver Museum of Natural History, J. D. Figgins. He had been to northwest Greenland with Robert Peary in 1896 as the expedition's taxidermist, and collected bird specimens for the American Museum of Natural History in New York, which was supporting the expedition (Chapman 1899). The other was E. R. Kalmbach, of the U.S. Bureau of Biological Survey, whose chief was the celebrated Arctic ornithologist E. W. Nelson. Both Nelson and Figgins offered Bailey positions, he accepted them one after another. First Nelson's offer took him to southeast Alaska in 1919–1921 to represent the Biological Survey there; then, in 1921, Bailey accepted Figgins's offer of a curatorship at the Denver Museum of Natural History with the chance of doing field work "far north of the Arctic Circle". In particular, he was to head a museum expedition to collect specimens of polar bears, caribou, seals and walruses, as well as Arctic birds and plants "to be exhibited in large ecological displays in the new north wing of the museum". With Bailey, as assistant, went Russell W. Hendee, who had just graduated from the University of Iowa.

Although the Expedition of the Colorado Museum of Natural History to Arctic Alaska took place in 1921–1922, Bailey did not bring its results together into a single report until 1948. On the expedition itself he travelled along the entire thousand-mile length of the north coast of Alaska from Wales on Bering Strait to the Canadian border at Demarcation Point (Map 9). Between them he and Hendee collected 2000 bird skins. But Bailey did much more than this: he taught two American traders and two native Eskimos, based at strategic points along the coast, to collect and skin birds and encouraged them to send specimens to his and other museums. And this they did, keeping up the good work year after year. From Wales the two Eskimos Dwight Tevuk and Arthur Nagozruk sent specimens of many rare visitors from Asia, including the first ever Wryneck for North America. At Wainwright A. J. Allen ranged up to 100 miles inland to collect where naturalists had never travelled before. But it was at Barrow that Bailey enjoyed his greatest success, for his student there was the famous Charles D. Brower, all of whose five sons joined their father in what became one of his favourite pursuits late in his life: collecting birds. They sent hundreds of skins to Denver and other museums and Bailey (1948: 40) claimed that "through their efforts more has been added to the knowledge of Arctic birds than through the combined work of all other collectors who have visited the far north". In fact the Browers added 63 species and subspecies to the list of the birds of Barrow, including rare stragglers from much further south in North America like the Killdeer, Brewer's Blackbird, and the Chipping Sparrow. None of these birds had any business to be in the Arctic. It was only after the Arctic bird-collecting system he had set up collapsed with the deaths of Allen and Brower in 1944 and 1945 respectively that Bailey at last decided to publish the full account of his field work and of the birds of the North Slope, which came out in 1948.

Bailey and Hendee arrived at the Eskimo village of Wainwright in the U.S. Coast Guard Cutter *Bear* on 6 August 1921. While Hendee settled in at Wainwright, Bailey continued along the coast eastwards in the *Bear* as far as the United States' frontier with Canada at Demarcation Point. He was back at Wainwright before the end of August, where the expedition established its winter quarters in the attic of the school house. Within days of his arrival there, on 2 September, Bailey collected a small duck along the shore that was heading his way. It proved to be the first record of the Baikal Teal for the American continent. During the rest of September and October, while trips were made in either direction along the coast, many migrants were seen and collected, including Ross's and Ivory Gulls. In November and December, with the onset of the "dark days" of the Polar winter, the two naturalists whiled away the time skinning the many birds which had been stored in the permafrost ice cellar, as well as photographing Eskimos in their igloos with flashlight and participating in their celebrations—which normally took place in the school house.

For the summer season's field work, beginning in March, the expedition again split up. While Hendee remained at Wainwright, Bailey sledged southwest down the coast to Wales, where he remained for some months. At Wainwright, Hendee camped out with the Eskimos on the edge of the shore ice by the open leads of water which form in the spring some 10 miles offshore. Here the Eskimos hunt the bowhead whales which migrate eastwards along the coast every spring and here Hendee was able to watch the spectacular spring migration of waterfowl and collect specimens. The main birds involved were divers, ducks and geese, prominent among them the eiders, for the Arctic coast of Alaska is one of the few places in the world where all four species of eider may be seen together, namely the Eider, King Eider, Spectacled Eider and Steller's Eider.

On 12 March Bailey had set out from Wainwright to sledge the 650 miles (1000 km) or so down the coast to Wales on Bering Strait, with an Eskimo companion, a heavily loaded sledge and eleven dogs. Unlike most Arctic explorers, whose narratives of sledge journeys invariably dwell on the hazards and discomforts they overcame, Bailey's radiate enthusiasm and enjoyment: for him, the weather always seems fine, the load light, and the trail excellent. On this particular trip he had little time for birdwatching, but two Snowy Owls, some Snow Buntings and redpolls, and many flocks of Willow Ptarmigan (Willow Grouse) were seen. At Wales the spring migration was soon under way, and Bailey divided his time between boat trips in an Eskimo umiak in the open water leads among the ice of Bering Strait, and hikes over the inland tundra and hills. These hikes were especially rewarding ornithologically. On 29 May he secured the first Great Knot to be recorded in North America. Later he became the first ornithologist to find the nest and eggs of the White-billed Diver or, as he knew it, the Yellow-billed Loon.

Alfred Bailey's contribution to Arctic ornithology, important though it was, was outshone by what many would regard as his *magnum opus: The birds of Colorado* published in 1965 in two volumes jointly with R. J. Niedrach. Although he worked in 1926–1927 in the Field Museum of Natural History in Chicago and was Director of the Chicago Academy of Sciences in 1927–1936, he returned in 1936 to the Denver Museum of Natural History, where he was Director for over 30 years until 1969. He died in 1981 (Phillips 1981).

SNYDER ON THE BIRDS OF ARCTIC CANADA

Although he spent most of his life in Canada, Lester L. Snyder (Long and Barlow 1986) was an American who was elected life fellow of the American Ornithologists' Union in 1947, having joined in 1919. Born at Panora in Iowa in 1894 in the same year and state as Alfred Bailey, he went to the State University of Iowa at Iowa City in 1913 and was probably a classmate of Bailey there, for besides studying general zoology, ornithology and entomology, he also took a class in museum techniques. However, he had not formally completed his university education when he took up his first position, which was on the curatorial staff of the Royal Ontario Museum in Toronto, Ontario. For the next 46 years, until his retirement in 1963, Snyder worked at the Royal Ontario Museum, first as technologist in gallery work and ornithology, and then, from 1935, as curator of birds and as assistant and associate director of zoology. His contributions to ornithology included the building up of a splendid ornithological library at the museum and the organisation of systematic field work in Ontario. He increased the museum's research collection of bird specimens from 5000 to 100 000 skins, eggs and nests, published 183 papers, and wrote *Ontario's birds* which came out in 1951 and was chosen as book of the year by the *Toronto Daily Star*.

Arctic birds of Canada (1957) is based on an extensive search through the literature, the personal observations of ornithologists working in the Arctic, unpublished reports of the Canadian Wildlife Service, and the research collection of skins at the Royal Ontario Museum, rather than on the author's own field work, which was limited. The book is adorned with a series of admirable line drawings by Terence Michael Shortt showing each species in its habitat. He was the finest Canadian bird artist of the day, and had first-hand knowledge of many of the species figured. Snyder's book, which is prefaced with a concise analysis of Canada's Arctic avifauna, is designed in the first place as an identification guide. Its distribution maps show the probable summer range of each species and the places where breeding has been established. Under the heading "Characteristics", descriptions are given of adults and young, both in the field and in the hand, as well as of downy young. The book is small enough to take on excursions into the tundra, where the descriptions of the field characters of the Arctic breeding waders will be found useful. Seventy birds "which habitually frequent the Canadian Arctic" are fully described in the text and an appendix mentions a further 92 which have occurred there. These include a Lapwing collected in October 1926 in Pangnirtung Fjord, Cumberland Sound, Baffin Island (Map 38), and a Passenger Pigeon, a species now extinct, which came aboard Captain John Ross's ship the *Victory* on 31 July 1829 during a storm in Baffin Bay.

TODD ON LABRADOR BIRDS

Snyder omitted Labrador from his book on the grounds that Oliver L. Austin Jr had already published on "the birds of that region" and that "little of it is true Arctic". But Austin's book, *The birds of Newfoundland Labrador* (1932), covers only the Labrador coast and not the truly Arctic area in the northwest of the Labrador Peninsula bordering on Hudson Bay. The very reason that helped persuade Snyder to omit Labrador urges its inclusion here, for W. E. Clyde Todd's (1963) *Birds of the Labrador Peninsula and adjacent areas* is one of the great bird books of the twentieth century. And he does include the northwestern part of the peninsula, an area of treeless tundra that he and other writers rightly consider to belong to the Arctic "Life Zone". Among Arctic species that breed here are the Tundra Swan, the Eider and King Eider, the Ptarmigan, all three of the smaller skuas: Pomarine, Long-tailed and Arctic, the Snowy Owl and the Snow Bunting. But of Arctic waders, only the Purple Sandpiper and Grey Phalarope are found and the absence of other typical High Arctic waders such as the Sanderling or Baird's Sandpiper shows that only the tip of the Labrador Peninsula extends into the Arctic. None the less, that tip is about 250 miles wide and nearly 400 miles long.

At age 13, Walter Edmond Clyde Todd (Parkes 1970) published his first ornithological paper in the *Oologist*. It was a note on the birds of his then home county of Beaver, Pennsylvania. Born in 1874, he left school in 1891 to take up a government post in Washington with the U.S. Department of Agriculture's Division of Economic Ornithology and Mammalogy. His first task was to put in good order the division's collection of bird stomachs in alcohol, many of them obtained by himself. He was still with the division in 1896, when it was transformed into the Bureau of Biological Survey, but in 1899 he moved to the Carnegie Museum at Pittsburgh, Pennsylvania, where he soon became curator of birds. George Sutton (1980) describes how, in about 1917, when he was about 19 years old and still an undergraduate, his father took him to see Todd at the Carnegie Museum. The youthful Sutton asked Todd if he could have a job as an assistant and he was duly set to work colouring the black-and-white illustrations of the *Catalogue of birds in the British Museum* which curator Todd kept near his desk. Later, Sutton was placed in charge of the egg collection. Though he recognised Todd as a great ornithologist and a meticulous curator who kept everything in excellent order, he found him somewhat eccentric. Todd had never tasted coffee because his mother had told him not to drink it. For the same reason, he did not drink tea. He never travelled, worked or collected specimens on Sundays. He abhorred smoking, he did not believe in tipping, and he had no university degree. Even in print, Sutton always referred to Todd as "Mr Todd": everyone called him "Mr Todd"; even though he had three Christian names, none of them was ever used. One of Todd's eccentricities not mentioned by Sutton was his aversion to introduced species, and especially the House Sparrow. Entirely ignored in his classic work on the birds of Pennsylvania, though it had become abundant there, it was grudgingly included in small type in his book on Labrador birds, where it was said that, since its "unfortunate introduction" this "foreign interloper" had spread over much of the U.S.A. and Canada.

When Sutton first met him, Todd was already at work on his *magnum opus*, the *Birds of the Labrador Peninsula*. From 1901 onwards until 1958 he organised 25 field expeditions to Labrador on behalf of the Carnegie Museum, many of which he led in person. Sutton, who accompanied him on expeditions number nine, ten and eleven (1920–1926) complains that he and his colleagues had to vacate the tent from time to time while Mr Todd made use of the rubber bath tub he always took with him on camping expeditions. Well over 5000 specimens were collected for the Carnegie Museum on these expeditions, forming the basis for Todd's book, which was published in 1963, when he was 89 years old. One of the most lavishly-produced and elaborate regional avifaunas ever written, its 819 large quarto-size pages include sections on 338 species and subspecies recorded for Labrador, a gazetteer giving details of localities mentioned, a very full chronological bibliography with its own index of authors, descriptions of all 25 Carnegie Museum expeditions, historical and geographical introductions, and eight fine colour plates by George M. Sutton. It is not a book for the field, it weighs in at 7 pounds, nor is it in any sense analytical, but as a straightforward descriptive account of the birds of a single region, rich in detail meticulously gathered together, it is a *tour de force*.

SALOMONSEN AND THE BIRDS OF GREENLAND

Denmark's best known twentieth-century ornithologist, who died in 1983 aged 73, was another museum man, Finn Salomonsen (Ferdinand 1979a; Evans 1985). Appointed curator of birds at the University of Copenhagen's Zoological Museum in 1943, he was chief curator there from 1958 to 1979. Like Todd's, Salomonsen's passion for birds began early in life. At age 16, in 1925, he was lucky enough to be able to join an ornithological expedition to west Greenland (Scheel 1927). It was led by the most famous Danish ornithologist of the day, the stockbroker E. Lehn Schiøler, who had befriended Salomonsen as a schoolboy. The object of Schiøler's expedition was to obtain specimens of birds and eggs for his private ornithological museum and to collect material for a book he was writing on the birds of Denmark, which was to contain a "conspectus" on the birds of Greenland. Salomonsen went as assistant to help prepare specimens. The expedition spent July and August exploring the bird life of the fjords and islands between Nuuk (Godthåb) and Qeqertarsuaq (Godhavn) (see Map 37) by boat, bringing back over 500 skins (Nielsen 1979). It gave rise to the first of Salomonsen's 200-odd scientific papers, published in the journal of the Danish Ornithologists' Union (*Dansk Ornithologisk Forenings Tidsskrift*) in 1925. It also established his interest in Greenland and its birds, which proved to be lifelong; he made nearly 20 more expeditions there in the decades that followed.

Towards the end of his life Finn Salomonsen became almost a god in ornithological circles. Besides being editor of the journal of the Danish Ornithologists' Union for nearly 20 years (1943–1961), he was its president from 1959 until 1971. Indeed he *was* the Danish Ornithologists' Union, and predictably, when he reached his seventieth year in 1979, that body took the unprecedented step of dedicating a special double issue of its journal to him. Forty years after his first communication to the International Ornithological Congress in 1930 he was elected president for the 1970 meeting at The Hague. On this occasion his presidential address was on the same subject as that earlier communication, namely zoogeographical and ecological problems in Arctic birds. Salomonsen was elected Honorary Member of the British Ornithologists' Union in 1947. Without ever holding a professorial chair, his scientific or academic reputation stood so high that it was often assumed that he did. His obituary notice in the journal of the British Ornithologists' Union the *Ibis* unofficially accorded him the title of "Professor" (Evans 1985).

After taking his degree in zoology at Copenhagen University in 1932, Salomonsen embarked on research there and was awarded a doctorate in 1939 for his thesis on the "moults and sequence of plumages" in the Ptarmigan. This important work was written in passable English, a language its author had learnt during a 2-year residence in London. It was based on over 1200 museum

(a)

(b)

skins of Ptarmigans and 36 fresh skins collected on an expedition to northwest Greenland in 1936. Salomonsen not only set out to describe in detail the course of the three annual moults of the Ptarmigan but sought to understand the factors that brought them about. Showing that the white winter plumage represented an "extra" moult peculiar to the more northerly members of the grouse family (Ptarmigan and Willow Grouse), he argued convincingly that the change of plumage into winter whiteness was a function of declining temperature. In all Ptarmigan populations that he found, the autumn moult began when the mean temperature dropped to around 2°C (36°F) and was complete by the time the temperature dropped to freezing point. Conversely, spring warmth triggered off the moult into summer plumage. Salomonsen was able to show that presence or absence of snow cover had nothing to do with the Ptarmigan's plumage changes. In the far north, for example, the birds became white in autumn *before* the first snowfall, and he cited a nineteenth-century Norwegian authority who recommended October as a good month for Ptarmigan shooting because the white birds then stood out against the bare fell.

Among the Greenlanders, Salomonsen (1967: 13) was known far and wide as *"tingmiarsiôq* (he who studies birds)", not because they were acquainted with his books and papers on the birds of Greenland, but rather because of his tireless efforts to persuade them to ring birds and to report recoveries of rings (Mattox 1970; Nielsen 1979). Not that Salomonsen was an innovator in this respect; he built on firm foundations. The schoolmaster from Viborg in Denmark, Hans Christian Mortensen—known as "Birds" Mortensen—was the first bird ringer; the founder of organised ringing in Greenland was the medical doctor at Uummannaq (Map 21), Alfred Bertelsen (1948). Enlisting the aid of the local population and the governor of Greenland, Knud Oldendow, who had been a pupil of Mortensen's at school in Viborg, Bertelsen ringed a total of 681 birds of nine species in 1926–1945. Of the 681 rings, a startling 17%—115 in all—were recovered from birds mostly reported "found dead", which usually meant shot. After a pause during the Second World War, Salomonsen went to Greenland in 1946 and set up a new scheme. Local people were recruited and paid to do the ringing and a reward was offered for the return of every recovered ring together with the leg and foot of the bird on which it was found. The scheme was organised by Salomonsen and the Zoological Museum of the University of Copenhagen; the Danish government's Ministry for Greenland paid the wages of the ringers and the rewards; the local authorities recruited the ringers; and the Carlsberg Foundation paid for many

---

Plate 34 (*a*) Ptarmigan pair photographed near Thule Air Base, northwest Greenland, on 29 June 1983. While the male's white winter plumage, not yet replaced by the brown summer plumage, makes him easily visible on grassy tundra, the female's moult has been completed, and she merges well with her background.

(*b*) The female Ptarmigan in summer plumage is almost invisible except when she moves. Her more rapid and complete moult in the early summer doubtless helps to camouflage her while she is incubating. The wings of both sexes remain white the year round.

of the scientific expenses. In 1946–1974 180 000 birds were ringed, and the work continues. Besides knowledge of the migration routes, dispersal and longevity of many species (Salomonsen 1967), it has been possible to use these ringing results to speculate on the origins of Greenland's avifauna. For example, Danish ornithologist Bent Nielsen (1979) has shown that Greenland's common passerines, the Wheatear, Redpoll, Lapland Bunting and Snow Bunting probably arrived in Greenland from different directions: the Wheatear and Redpoll from Europe and the Lapland Bunting from America. The Snow Bunting apparently colonised west Greenland from America and east Greenland from Europe.

Before Salomonsen came to the scene solid foundations had been laid for an ornithology of Greenland by Danish workers. Herluf Winge had published his *Birds of Greenland (Grønlands fugle)* as long ago as 1898; A. L. V. Manniche's work on the birds of northeast Greenland had been published in 1910 and 1911. He was the official ornithologist of the Danish expedition to northeast Greenland of 1906–1908, which was based on and named after the ship *Danmark*, and his paper gave a very full account of the birds of a hitherto ornithologically unknown area. He listed 38 species seen, of which 18 certainly bred, for he found their nests. Among the seven birds "which certainly breed, but whose nests were not found" was the Knot. Manniche's paper was beautifully illustrated with photographs, a map, and coloured plates of Ptarmigan in different plumages and of the eggs and downy young of the Sanderling as well as of the adult birds. An even better basis for Salomonsen's later work was laid down by Knud Oldendow, whose *Bird life in Greenland (Fugleliv i Grønland)* appeared in 1933. It made no claim to completeness, being mainly limited to birds collected or observed by the author, who, however, included an extensive bibliography of works on the birds of Greenland.

No one could have been better qualified than Finn Salomonsen to write *The birds of Greenland*. This massive de luxe folio-sized volume, printed in Danish and English in parallel texts, was published in 1950. The Danish government contributed to the publication costs and Danish Prime Minister Hans Hedtoft, who was also Minister for Greenland at the time, wrote the preface. The book's 608 pages are enlivened with 52 full-page reproductions of watercolour paintings of birds against a background of Greenlandic landscape by artist Gitz-Johansen. Although his landscapes are vividly evocative of Greenland's scenery and his use of colour is superb, Gitz-Johansen's birds are often disappointingly inaccurate. Still, the illustrations play their part in making this one of the most highly-prized books in any ornithological library. The extensive bibliography misses nothing and indeed every (often hard-won) scrap of knowledge of Greenland's birds is brought together here, species by species. While the detailed distribution of each bird in Greenland naturally takes up a fair amount of space in the species accounts, the author also gives information on habitat, food, exploitation by Greenlanders, field identification, breeding biology, migration and other topics. It is a very readable and often fascinating work written by an all-round ornithologist with wide and varied interests who treats his birds in many other contexts than a merely Greenlandic one and includes many interpretative and theoretical comments.

After 1950 Salomonsen wrote *Birds in Greenland*, published in Danish in 1967 with no English summaries, which is important for its species-by-species analysis of ringing results, and, again only in Danish, the section on birds in *Greenland's fauna*, published in 1981. A further major work in English was his *Ornithological and ecological studies in southwest Greenland* (1979), much of which is taken up with an account of the seabird colonies.

LØVENSKIOLD ON SPITSBERGEN BIRDS

To some people, the title of Herman Løvenskiold's (1964) book on the birds of Spitsbergen may seem cryptic. It is *Avifauna Svalbardensis*. The prospective reader may at once be reassured that *only* the title and the scientific names of birds are in Latin; the rest of the book is in English. Naturally, as a patriotic Norwegian, the author felt constrained to use the new-fangled Norwegian name Svalbard for what everyone else has always called Spitsbergen. If the reader gets as far as page 51 he or she may find enlightenment on this point: "To many readers the name Svalbard will be more or less unknown, but the correct signification is the following: Svalbard is a group name for all the islands in the Arctic Ocean placed under the sovereignty of Norway." The author goes on to explain that his book includes all birds found on Spitsbergen, the adjacent islands of Edge Island (Edgeøya) and Northeast Land (Nordaustlandet), and also Bear Island, which lies to the south (Maps 19 and 26).

Løvenskiold's reputation in ornithological circles was really made in 1947 when his *Handbook of Norwegian birds (Håndbok over Norges fugler)* was published in Oslo. In it he included a list of 42 species recorded in Spitsbergen and Bear Island. Thereafter, his growing interest in the birds of the archipelago led him to participate in eight expeditions there in the years 1948–1960. These expeditions, led by naval officers, were organised annually by the Norwegian government's Polar Institute (Norsk Polarinstitutt) in Oslo, founded in 1948 as part of a long-term project for scientific research in Spitsbergen. Taking with him at least one assistant and staying for several weeks each time, Løvenskiold was able to visit many different areas. In spite of some harassment by polar bears, this field work gave him an unrivalled first-hand knowledge of the Spitsbergen avifauna. But far more of his time was spent in the libraries of the Norwegian Polar Institute and Oslo University. His search of the printed literature on Spitsbergen birds was extremely thorough and he also made use of nearly 50 trappers' diaries dating from 1895 to 1940 which were preserved at the Polar Institute. His *Avifauna Svalbardensis* probably contains everything knowable about Spitsbergen birds up to 1958.

The author modestly describes his 460-page book as a "paper". In fact it is arguably the most comprehensive and scholarly of all the regional Arctic ornithologies described here. In an interesting account of the avifauna, Løvenskiold points out that Spitsbergen has only one common breeding passerine, the Snow Bunting, no breeding diurnal birds of prey or owls, six species of geese and ducks, one diver and two skuas. In Arctic breeding waders it is relatively poor, only the Grey Phalarope and the Purple Sandpiper being at all common; the Knot, Dunlin, Sanderling and Ringed plover have all bred but none is common or widespread. Løvenskiold reckoned that 93 species have probably occurred in Spitsbergen and Bear Island, 45 of which probably breed, but only 26 of these could be considered "common breeders". Distribution maps for over 30 species, each on a separate sheet, are provided in a wallet at the end of the book, and on them every case of a bird found breeding or just seen is marked with a red dot or circle. In his introduction, the author reviews climatic changes, discusses the problem of non-breeding among Arctic birds, and describes the "food of the sea birds and where they find it". The species accounts are very full, covering distribution in Spitsbergen in great detail, breeding biology, ringing results, and devoting several pages, in some cases, to a bird's "general habits". Much other information is given, some of it presented in tables, including the dates of arrival in the spring and the onset of egg-laying of the commoner species. This wealth of detail on each species takes up more than 10 pages in the case of the Little Auk and Ivory Gull, and about 7 pages for many birds. At the end of his book Løvenskiold provides a lengthy annotated chronological bibliography with 655 numbered items. This is indeed an ornithological classic which will surely remain for many years an invaluable book of reference.

USPENSKY ON THE BIRDS OF THE SOVIET ARCTIC

In turning from the massive tome of a Todd or a Salomonsen to Uspensky's slim 166-page paperback, which sold at 3 roubles when it came out 1958, it really is hard to say whether one is moving from the sublime to the ridiculous or from the ridiculous to the sublime. Certainly, among the books discussed in this chapter, Uspensky's is probably the only one that the man or woman in the street, or an ordinary birder, could afford to buy. Yet it is the only one that he or she certainly could not read, because *Birds of the Soviet Arctic* is only available in Russian. This is in fact somewhat surprising, because two other books by Uspensky *have* been among the very few Russian bird books translated into English, namely his *Bird bazaars of Novaya Zemlya* in 1958 and *Life in high latitudes. A study of bird life* in 1986.

Savva Mikhailovich Uspensky began the serious study of Arctic ecology and ornithology in the 1940s, when he spent some time in Novaya Zemlya carrying out research on colonially nesting seabirds. His thesis for the degree of candidate in biological sciences, for which he studied at the Moscow Municipal Pedagogical Institute, was published in 1951. Its subject was the ecology, conservation and economic value of some Novaya Zemlya colonially nesting seabirds. He had already published his first scientific paper on the adaptive traits of Brünnich's Guillemot's eggs. This research was brought together in the already-mentioned book on the bird bazaars of Novaya Zemlya. By the time the original Russian version of it was published in 1956, Uspensky was working at the Seven Islands (Sem' ostrovov) nature reserve on the Murman

Coast (Map 15) under L. O. Belopol'sky. Thereafter he continued his often lengthy field trips to different parts of the Soviet Arctic, working from west to east. Thus in 1957–1958 he was in the field on Vaygach Island (ostrov Vaygach), in the Yugorskiy poluostrov and in the Great Land Tundra (Bol'shezemel'skaya tundra), the European Russian Arctic area partly worked by Henry Pearson in 1897 (Map 12). He published papers on the breeding biology, density and distribution of 73 species found here. By the 1960s Uspensky had shifted his field work beyond the Urals to the far northeast of Soviet Asia, participating in a large-scale expedition to explore the natural history of the lower Indigirka from the southern fringe of the tundra to the sea and visiting Wrangel Island (ostrov Vrangelya) in 1960 and 1964 to count Snow Geese and polar bears on behalf of several nature conservation agencies. He was again in the field in the Arctic in 1972, when he studied and photographed Ross's Gull on the lower Yana river (Belopol'skii 1961: 8, 322; Il'ičev and Flint 1985: 36, 139; Portenko 1981; 39; Uspensky 1977).

Between 1955 and 1972 Uspensky wrote three books especially for the well-known series of illustrated monographs on different species and groups of birds called Die Neue Brehm-Bücherei published by A. Ziemsen, of Wittenberg in East Germany. These were written originally in Russian and the publisher had them translated into German; the Russian texts never appeared in print. The subjects of the three are the genus *Calidris* or Eurasian sandpipers, the eiders, and geese of northern Eurasia. Together they form a compact synthesis of knowledge of these three very important groups of Arctic birds. In 1964 Uspensky published a fascinating little book called *The Arctic through the eyes of a zoologist* which, unfortunately, was never translated from the Russian.

Birds of the Soviet Arctic is a handy little paperback, crammed with information, published by the Soviet Academy of Sciences. It aims to introduce the local bird life to people who live and work in the Arctic—personnel of expeditions and meteorological stations, crews of icebreakers and other ships, native inhabitants—who would hopefully, by reading and using it, be able to contribute to knowledge of the Arctic avifauna; 5000 copies were printed. The 100 or so main Arctic species found in the U.S.S.R. are each described and illustrated in a series of species accounts which take up two-thirds of the book. The illustrations are smaller and cruder than those in most field guides and there are only two colour plates, of waders in breeding plumage. Care is taken to explain the size of each bird, often relative to others, the field characters are well brought out, and the distribution and habitat of each species are indicated. In an excellent introductory sketch, the ecology and distribution of Arctic birds is described, their seasonal life, including breeding and migration, is outlined, and information is given on the exploitation of Arctic birds for food and clothing. The book concludes with an instructive and interesting account of the methods of study of birds, which deals with their observation, the making and preservation of skins and the collection of eggs, and photography and field sketches. Because of shortage of space, much is dealt with only briefly; for example, there is only one page on ringing. But the author does explain in some detail exactly what observations are most valuable, such as arrival and departure dates, date of egg-laying, weights of nestlings and so on, and how they should be logged in the field notebook. He even pinpoints particular species and localities about which very little is known and more information is required. Finally, addresses are given to which the amateur ornithologist can send observations and other material, namely the Zoological Museum of the author's own Moscow State University, and the Ornithological Institute of the Soviet Academy in Leningrad.

CHAPTER 10

Bird studies in the Russian and Siberian Arctic

Politically, until very recently, when the Yakut or Yakutskaya Autonomous Republic and the Chukchi or Chukotskiy Autonomous Region declared their independence from Russia, the whole of the Soviet Arctic lay within the borders of one of the 15 republics that made up the Soviet Union, namely the Russian Soviet Federated Socialist Republic or, more simply, Russia, which extended from the Kola (Kol'skiy) to the Chukchi or Chukotskiy Peninsula. Geographically, the Soviet Arctic is divided by the Ural Mountains (Ural'skiy khrebet) into the Russian Arctic within Europe, and the very much larger Siberian Arctic in Asia. The Russians themselves use the terms European Arctic, Western Siberian Arctic, Yakut(ian) Arctic, and the North-East.

ORNITHOLOGY IN THE SOVIET UNION

It may be true that there are no birdwatchers in Russia; certainly, in common with Mediterranean countries like Greece and Italy, it lacks the numerous

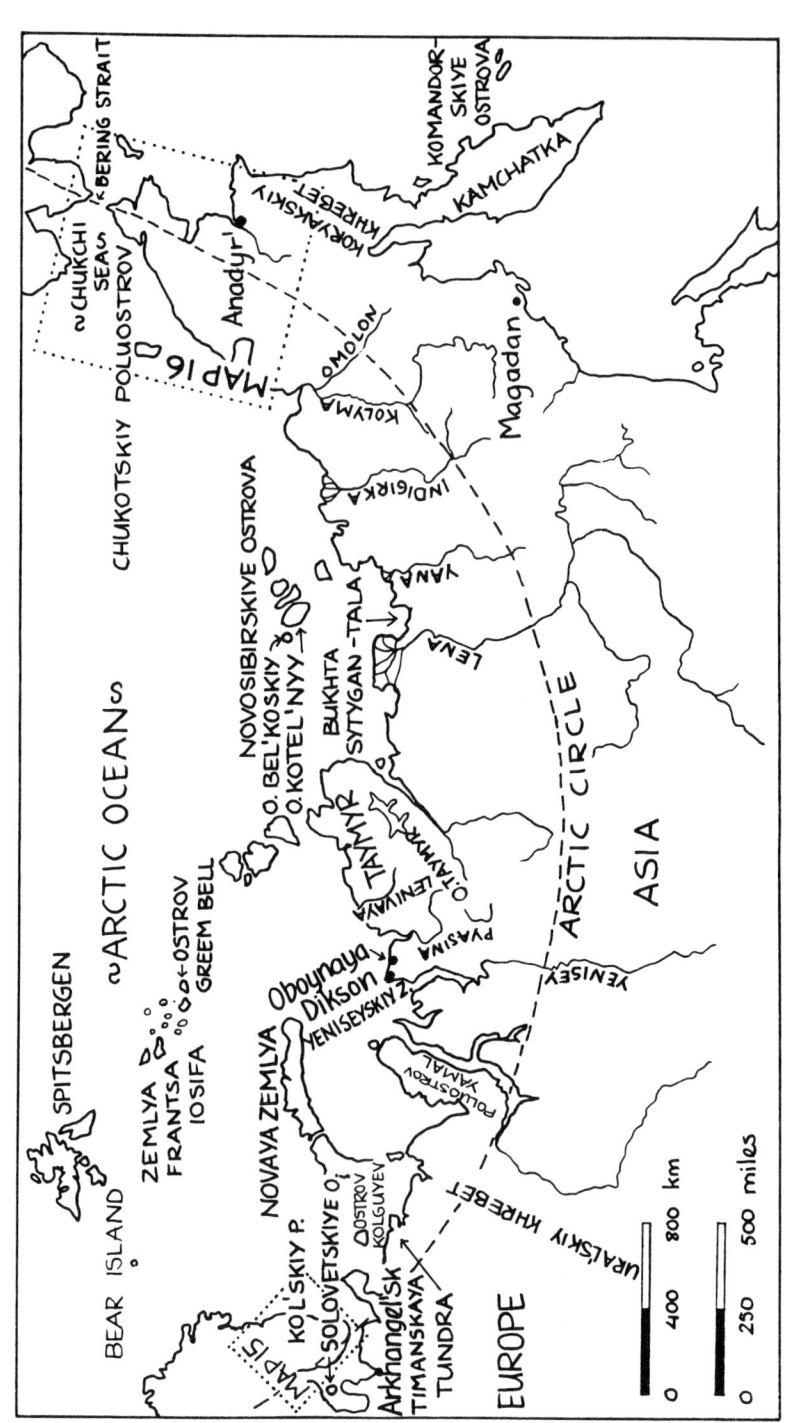

Map 14 The Russian and Siberian Arctic.

enthusiasts characteristic of England, Sweden or Holland. Yet the widespread interest in nature is evidenced by the print runs of journalist Valery Orlov's paperbacks: 50 000 copies were printed of his 1980 book on the wildlife of the Taimyr and 80 000 of his 1988 book on the far northeast of Siberia. Even V. Kryuchkov's much more serious work of 1979 on ecology and conservation in the Far North was issued in 21 100 copies. More recently, a popular, illustrated guide to Arctic birds came out in the series *Nature in Our Country* written by the highly respected professional ornithologist V. K. Ryabitsev (1986). If amateur birdwatchers are scarce in Russia, professionals have increased greatly in number, from the few dozen at the beginning of the Soviet period to about a thousand today (Il'ičev and Flint 1985: 18). Most are employed by the Academy of Sciences of the U.S.S.R. in its various biological research institutes, for example at Sverdlovsk, Yakutsk and Magadan; others work in universities. Still others staff the nature reserves or zapovedniks. Nor have Russian ornithologists been far behind in bringing together knowledge of the birds of their huge country. Between 1951 and 1955 G. P. Dement'ev, N. A. Gladkov and others produced a six-volume *Birds of the Soviet Union (Ptitsy Sovetskogo Soyuza)*, and this is currently being superseded by an entirely rewritten work called *Birds of the USSR (Ptitsy SSSR)* edited by V. D. Ilychev and V. E. Flint, Volume 1 of which was published in 1982 to coincide with the 18th International Ornithological Congress held in Moscow that year. The first of these works was translated into English in 1966–1970; the second (originally an East German enterprise) is being issued in a revised German edition by Aula Verlag, of Wiesbaden, publisher of the *Handbuch der Vögel Mitteleuropas* (Il'ičev and Flint 1985, 1989), and by A. Ziemsen of Wittenberg (Il'ičev and Flint 1990). Both the Russian handbooks are, of course, essential reading for the student of Arctic birds. The second, by Ilychev and Flint, is prefaced by an invaluable 174-page history of ornithological studies in the Soviet Union, arranged region by region, backed by a 16-page chronological list of the most important publications on the avifauna of the Soviet Union from 1832 to 1984. In 1968 Vladimir Flint collaborated with other Russian ornithologists to produce Russia's first popular field guide, published in a revised and updated English version in 1984 by Princeton University Press. The superb illustrations were specially painted for this work by Yu. V. Kostin, author of *The birds of the Crimea (Ptitsy Kryma)* published in 1983. Sadly, he only lived long enough to illustrate Volume 1 of the new *Birds of the USSR*. The sole Arctic birds portrayed there are the divers. Kostin's exquisite colour plate of them shows what are generally accepted as five species: Red-throated, Black-throated, White-necked or Pacific (or Arctic Loon in North America), Great Northern and White-billed.

Russian ornithology has undergone changes between the publications of Dement'ev and Gladkov in the 1950s and Ilychev and Flint in the 1980s and 1990s. The number of puzzle species about which little or nothing was known has been much diminished; the white specks on the map, representing areas with unknown avifaunas, have nearly disappeared. The list of birds of the Soviet Union has been lengthened to its present 800, partly through intensive study of peripheral areas and partly because taxonomists have divided into two several species which had previously been treated as a single one, for example the Lesser Golden Plover and Black-throated Diver. Other changes were methodological: counting birds from the air, colour marking to identify individual birds, the use of electronic gadgets to establish incubation parameters and sophisticated methods of researching a bird's diet. Russian ornithology has also become more institutionalised. Two veteran centres are outstanding: the Zoological Institute of the Academy of Sciences of the U.S.S.R. in Leningrad, and the Zoological Museum of Moscow State University. The former is the direct successor of the Zoological Museum of the Imperial Academy of Sciences of St Petersburg. Its collections originated at the beginning of the eighteenth century in Peter the Great's cabinet of curiosities and now include the largest assembly of bird skins in the country. The collections of Moscow University's Zoological Museum date from its foundation in 1791 but were completely destroyed during Napoleon's invasion of Russia in 1812 and had to be rebuilt thereafter. In terms of numbers of skins, Moscow is now second only to the Leningrad institute, and has the largest egg collection in the Soviet Union (Tomkovich, personal communication). Other ornithological institutions are much more recent. The All-Union Ornithological Conferences began in 1956 in Leningrad. The Ornithological Committee of the Soviet Union, which was set up by the Academy of Sciences, and was responsible for the publication of Ilychev and Flint's *Birds of the USSR*, is also a recent creation. The journal *Ornitologiya* was launched in 1958 by G. P. Dement'ev and V. F. Larionov, and published by Moscow State University in conjunction with the All-Union Ornithological Committee. The All-Union Ornithological Society was founded in the early 1980s with, as emblem, a Red-breasted Goose designed by Peter Scott.

THE BIRDS OF FRANZ JOSEF LAND

In the Soviet Union professional ornithologists have perforce had to spend much of their time in what amounted to exploration. Thus in 1980–1981, when the All-Union Scientific Research Institute for Nature Conservation and Nature Reserves mounted its expedition to Franz Josef Land to gather information in support of the proposal to make the archipelago into a zapovednik or nature reserve, a team of professional zoologists was sent there which included the author of the *Birds of the Soviet Arctic*, Savva Uspensky, and Pavel Tomkovich, curator of ornithology at Moscow State University's Zoological Museum and expert on sandpipers. They wrote a joint paper published in English in 1987 surveying the bird life of this little-known archipelago, the world's most northerly. After a thorough search of the literature, which consisted mainly of G. P. Gorbunov's 1932 study *The birds of Franz Josef Land*, they concluded that 37 or 38 species had been recorded on the archipelago. Among them were two vagrants from much further south, the Swift and the Swallow, each of which had been seen in two different years. The only land birds proved to breed were the Red-throated Diver, Brent Goose, Purple Sandpiper, Arctic Skua and Snow Bunting, but there were nine species of seabird, namely Fulmar, Eider, Glaucous Gull, Kittiwake, Ivory Gull, Arctic Tern, Little Auk, Black Guillemot and Brünnich's Guillemot. More than 60 different bird bazaars or seabird colonies were counted.

Pavel Tomkovich described his share of the expedition's field work in a separate paper (1984). He was allotted the most easterly island in the archipelago, named in 1899 by its American discoverers, of the Second Walter Wellman North Polar Expedition, after the then president of the National Geographic Society, Alexander Graham Bell. Only a quarter of 1708-km^2 (659-square-miles) ostrov Greem Bell (Graham Bell Island) is ice free. Tomkovich worked the northwest of the island, from 13 to 23 June and from 7 to 12 August 1981. Between those dates he was based on the east coast. The temperature in June averaged $-1 \cdot 1°C$ (30°F), and snow fell on 6 days of the month. Twenty-four species of bird were recorded, including three new for Franz Josef Land: the Black-throated Diver, Herring Gull and Sanderling. Tomkovich made interesting observations on one of the world's northernmost Kittiwake colonies. The first egg was laid on 24–25 June, but most birds did not lay until early July, and the last eggs were laid around 10 July. Clutches of one or two eggs were the only ones seen, the average clutch size working out at 1·42. The first chicks a few days old were seen on 20 July. Tomkovich saw a few Ross's Gulls, both in breeding and second-year plumage, in July, but no sign of breeding,

Plate 35 At this Kittiwake colony on an offshore island 50 miles west of Qaanaaq, northwest Greenland, at about 77°N, egg-laying had not yet begun on 19 June 1985 when the photograph was taken.

Plate 36 Ivory Gull in a breeding colony on Graham Bell Island in the Franz Josef Land archipelago, photographed by Pavel Tomkovich on 17 July 1981.

and he discovered several colonies of Ivory Gulls, some deserted ones of previous years, some newly-established that year, and some which had been occupied for several seasons. The seven occupied colonies contained in all 170–200 pairs, and they were up to 8 km (5 miles) from the coast. All were on the summits or slopes of low hills in flat country.

SANDPIPERS AND STINTS

Tomkovich must have felt more like an explorer than an ornithologist on Graham Bell Island, but already before those adventurous times he was an expert on sandpipers, especially of the genus *Calidris*. He had photographed breeding Sharp-tailed Sandpipers on the Indigirka tundra in 1973; Great Knots on dry Alpine tundra in the Koryak Mountains or Highland (Koryakskiy khrebet or Koryakskoe nagor'e) in 1976; Red-necked Stints, Spoon-billed Sandpipers, Baird's Sandpipers and Western Sandpipers in the Chukchi Peninsula around Uelen and Dezhnev in 1978 and 1979; and in Franz Josef Land he found time in 1981 to photograph a Purple Sandpiper on its nest. These photographs were published in *Ornis Fennica*, *British Birds* and *BTO News* (Prater and Grant 1982; Myers *et al*. 1982; Tomkovich 1988b).

During these years he and other Soviet ornithologists made important progress in wader research, and their studies soon went far beyond the descriptive regional ornithologies or distributional investigations of earlier workers. In 1973 Vladimir Flint organised the first All-Union Conference on Wader Biology, which showed that Russian ornithologists were keeping pace with others and which acted as a precedent for later wader conferences in 1979 and 1987. At the third conference it was decided to set up a Wader Study Group under the aegis of the All-Union Ornithological Society, and this planned to intensify wader research in the Soviet Union, co-ordinate the colour ringing of waders, and issue an information bulletin. These developments were described by Pavel Tomkovich (1988b) for the benefit of English-speaking readers in the November–December 1988 issue of *BTO News*, the Newsletter of the British Trust for Ornithology. The first two bulletins of the Soviet Wader Study Group were issued in 1989.

Tomkovich has a long series of papers on the breeding biology of Arctic waders to his credit. He contributed much to the second All-Union Wader Conference, the proceedings of which were edited by Flint in 1980. As sole author, he provided a study of the territorial behaviour and pair formation of the Little Stint on migration, and another on the breeding biology of the Western Sandpiper in the Chukchi Peninsula. Jointly with Flint, he described the breeding biology of the Great Knot and the Red-necked Stint. After the third Soviet wader conference, held in Moscow in October 1987, Tomkovich was co-editor of the 32 papers presented (Flint and Tomkovich 1988), among which was his report of the discovery of a new species of Arctic wader nesting in the Soviet Union, namely the Semipalmated Sandpiper, hitherto known only as a North American species breeding in the High Arctic between northwestern Alaska and Labrador.

Besides his work on little-known members of the genus *Calidris* like the Sharp-tailed Sandpiper, Great Knot and Red-necked Stint, Tomkovich has also studied the more familiar ones. While on Graham Bell Island in Franz Josef Land in the summer of 1981 he devoted much of his field work to the only breeding wader of the archipelago, the Purple Sandpiper, which here, at about 81° N, is at the extreme north of its range, for it does not breed in the far north of Greenland (Tomkovich 1985). The bird was scarce and patchily distributed on the ice-free northern peninsula of the island. No other places suitable for breeding Purple Sandpipers were found in the rest of the island, which is nearly all covered with a permanent ice-cap. Ten nests or broods were found and there were thought to be 12–14 breeding pairs in the 460-km^2 (176-square-mile) area of barren tundra. The birds only arrived on 15 and 19 June, when single Purple Sandpipers were seen on bare patches of raised ground, otherwise devoid of any kind of living thing, where wind had removed the snow. Later, nests were found to be concentrated in the few parts of this Polar desert where there was some vegetation. Tomkovich's field work was carried out on difficult terrain in around zero temperatures, but somehow he managed to measure the sandpiper's eggs, weigh their young, and observe and record their behaviour. The population was evidently only on the verge of existence in such extreme conditions and, in the course of a month, Tomkovich found five mummified sandpiper corpses, showing that in prolonged bad weather birds could easily come to grief. Food items were small, sparse and of limited variety, and it was no surprise that in this population of Purple Sandpipers both the birds themselves and their eggs were smaller in size than Purple Sandpipers elsewhere. In this way the birds' energy requirements had been reduced in the face of the restricted food supply. Other unusual features of the Graham Bell Island Purple Sandpipers were likewise adaptations to their harsh environment: they devoted most of their time in the pre-nesting period to feeding rather than to courtship or territorial activities; they did not concentrate into flocks at the end of the breeding season; and there was some evidence that the onset of moult was later, and its duration shorter, than in other Purple Sandpipers.

Before his summer's field work in Franz Josef Land, Pavel Tomkovich, with a colleague, had studied another abundant Arctic and sub-Arctic wader, namely Temminck's Stint (Tomkovitch and Fokin 1983). This was in 1977 in an area of well-vegetated alluvial sand around Sytygan-Tala Bay (bukhta Sytygan-Tala) not far east of the Lena Delta. Here the birds were breeding at optimum density in places where food was abundant. Although one clutch of eggs was found being incubated by two birds, most were products of the so-called double-clutch system of this species, in which a female lays two clutches within a few days of each other. While her first mate incubates the first clutch, a second male fertilizes her second clutch, which she then incubates (Cramp and Simmons 1983: 314). Sytygan-Tala Bay is well within the northern limits

(a)

(b)

Plate 37 (a) Semipalmated Sandpiper incubating in marshy, sedgy, tundra near Cambridge Bay, where it breeds as abundantly as the other common sandpiper—Baird's. (b) Close-up of the Semipalmated Sandpiper figured above, sitting on three eggs. These attractive stint-sized waders fly around low over water or marsh in pairs or small groups constantly uttering their reeling calls.

of Temminck's Stint's range, which, east of the Taimyr Peninsula, is formed by the shore of the Arctic Ocean. But in 1982 Tomkovich (1988a) found Temminck's Stint breeding far north of its usual haunts, in the neighbourhood of Dikson, in the true Arctic tundra where the species had never before been found. In the following summer he several times saw a male in song flight even further north near the lower Lenivaya River on the north coast of the Taimyr Peninsula. Here was another wader population struggling to survive on the northern fringe of its range. The birds were nesting in association with human settlements and weather stations both on the island of Dikson and on the mainland around Dikson, but Tomkovich also found them far from human habitation on the coastal tundra some 60 km (37 miles) east of Dikson around the mouth of the River Uboynaya. Here there were at most 15 nests in an area of 80 km^2 (31 square miles), giving a density, where the birds were most crowded, of one nest in every 2 hectares (5 acres), as opposed to two nests per hectare further south in the bird's range. Another striking difference between these far northern Temminck's Stints and those nesting further south was their arrival 2–3 weeks later than in the south. The courtship period of the males

206   *In search of Arctic birds*

Plate 38   Incubating Purple Sandpiper on Graham Bell Island in the Franz Josef Land archipelago, photographed by Pavel Tomkovich on 19 July 1981.

Plate 39   Temminck's Stint in the Varanger Peninsula, north Norway. This bird's minute downy young were hiding in the dense marsh vegetation, and it could be photographed with a telephoto lens without a hide.

was shortened to less than 20 days, but even so these Taimyr Temminck's Stints remained far behind those breeding further south. Tomkovich reckoned that, while the main hatch in the south of the bird's range was in the first 10 days of July, and that at Sytygan-Tala Bay further north in the range in the second 10 days of July, the main hatch of the Taimyr birds was not till the last 10 days of July. Delays in egg-laying and incubation were accentuated by prolonged periods of bad weather. On 12 July 1984 when a snowfall with north wind lasted from 09.30 to 15.00 hours and covered the tundra, a sitting female Temminck's Stint was off the nest for 9 hours and 50 minutes. No wonder the eggs were found to be cold and lightly covered with snow when the nest was inspected. Tomkovich surmised that this most northerly of all breeding populations of Temminck's Stint consisted in the main of females that had already laid a clutch further south earlier in the breeding season, birds that had lost a nest further south, and inexperienced young birds.

The reader must not be misled into assuming that Tomkovich is the only Russian wader expert; others will be mentioned in the following pages. Tomkovich (1988b) himself has drawn attention to A. Ya. Kondrat'ev's important monograph on the biology of waders in the tundra of northeast Asia (1982). Kondrat'ev's field experience in the Arctic, amounting to 15 complete summer seasons, is second to none. Some of his many publications will be discussed below.

PIONEERS OF RUSSIAN ARCTIC ORNITHOLOGY

In spite of twentieth-century social and political changes in Russia, modern Russian ornithologists are very conscious of the debt they owe to the great naturalists of the nineteenth century. Uspensky dedicated his *Life in high latitudes. A study of bird life* (Uspenskii 1986) to the nineteenth-century traveller "Aleksandr Fedorovich Middendorf, outstanding naturalist and one of the earliest investigators of Arctic fauna". The first investigation of the birds of the Russian Arctic was undertaken by members of the famous 6-year expedition led in 1768–1774 by German naturalist Peter Simon Pallas, who had accepted an invitation from Catherine the Great to join the Imperial Academy of Sciences in St Petersburg (later Leningrad) (Mearns and Mearns 1988). Split into several sections, the expedition covered much of the Russian Empire, its aim being to make a complete systematic survey of the Russian fauna. I. I. Lepekhin and other expedition members travelled into the European Russian Arctic as far as the Timanskaya tundra, the Solovetskiye ostrova or Islands and the Kola Peninsula. Among the more than 70 bird species described for the first time from Russia in the expedition's publications were two Arctic breeding geese, the Snow and the Red-breasted (Il'ičev and Flint 1985: 15).

Of Baltic German extraction, A. F. Middendorf, alias A. T. von Middendorff, was born in St Petersburg in 1815. He was a fine all-round naturalist as well as a great traveller (Leonov 1967). He joined the Academy of Sciences at St Petersburg in 1850, having graduated in 1837 at the University of Dorpat, or Tartu as it now is, in Estonia. In 1840 Middenhorf (1843) travelled in the Kola Peninsula with K. E. von Baer and produced a first list of the birds of that area, numbering 80 species (Il'ičev and Flint 1985: 21). In 1843 he became the first naturalist to investigate the animal and plant life of the Taimyr Peninsula (Middendorff 1848: 17–23). Starting in May and travelling from south to north with the local nomads on their spring migration, in a party comprising nine sledges and 36 reindeer, he stayed for some time in the Lake Taimyr (ozero Taymyr) area, followed the Lower Taimyr River (Nizhnyaya Taymyra) northwards, and reached the shores of the Gulf of Taimyr (Taymyrskaya guba) (Map 40), where he saw Gyrfalcons, in August, "all in dark plumage" (Middendorff 1853: 127). Everywhere in Taimyr he saw Snowy Owls "Motionless, for hours at a time, spying out for lemmings". He noticed that the female Ptarmigans completed their moult into summer plumage long before the males; one was already in the brown plumage on 5 June while the males were still white at the end of the month (see Plate 34). Although Middendorf was disappointed by his failure to find a Knot's nest, he did find Little Stints and Grey Plovers breeding 32 years before British ornithologists Harvie-Brown and Seebohm claimed to have been the first to find them. He was the first naturalist to discover and describe in print the nest and eggs of the Grey Plover (Middendorff 1853: 209–210). He met with these birds both in the south of the peninsula and in the Byrranga Mountains (gory Byrranga). On 26 June he found a female on a nest containing four eggs. "Since nothing reliable appears to be known about these last" and he could find no picture of them in F. A. L. Thienemann's great work (1825–1838 and 1845–1856) on the breeding of European birds with illustrations of their eggs, he gave their measurements and description, and figured one in colour on his Plate 19. The day before, in level mossy tundra, he had found nests of Steller's Eider well-lined with down, each containing 7–9 fresh eggs. "The male stayed near the female, who only flew unwillingly from the nest with a cry recalling that of the Teal, but more rattling" (p. 235). In his Plate 23, Middendorf figured three of their beautiful greyish-blue eggs, showing the variation in their size, rightly commenting that they had never before been found.

Among Russian ornithologists of the late nineteenth century, Fedor Dmitrievich Pleske stands out because of his Arctic studies (Il'ičev and Flint 1985: 17, 21, 22, and Tomkovich, personal communication). He began his ornithological studies as member of an expedition to the Kola Peninsula launched by the St Petersburg Natural History Society in 1880. After his return he searched the literature thoroughly, obtained data on the western parts of the area from Finnish ornithologists, and published his *Critical review of the mammals and birds of the Kola Peninsula* in 1887. He included 235 bird species, 30 of which were not numbered because the author was not fully satisfied with the evidence for their occurrence. His career in the Zoological Museum of the Imperial and subsequently Soviet Academy of Sciences in Leningrad was divided between ornithology and entomology. After some vicissitudes, Pleske became one of the few Russian ornithologists whose work appeared in English, for in 1928 his classic *Birds of the Eurasian tundra*, which appeared under the name Theodore Pleske, was lavishly published by the

Plate 40 Steller's Eider does not breed nearer Britain than the Taimyr Peninsula but non-breeding birds occur regularly in the Varanger Fjord, Norway, where this group of drakes was photographed.

(a)

(b)

Boston Society of Natural History as Volume 6, Number 3 of its *Memoirs*. This book was originally intended to publish the ornithological findings of Eduard Toll's Russian Polar Expedition of 1900–1903, described in Chapter 3, but was extended into a general account of the avifauna of the tundra zone or High Arctic from Spitsbergen and Bear Island (Bjørnøya) to the north shore of the Chukchi Peninsula. But Pleske took care to incorporate most of the expedition's ornithological material, including splendid photographs and paintings of the birds and landscape of the Taimyr Peninsula and the New Siberian Islands, in his book. The most substantial part of it consists of a systematic list of Arctic birds. These species accounts are often very detailed. For example, for the Lapland Bunting Pleske lists all the specimens of birds, nests and eggs obtained by the Russian Polar Expedition, describing each of the seven nests or clutches in considerable detail, and provides an elaborate table compiled from all available sources giving the dates of clutch completion, hatching and fledging of the Lapland Bunting in different regions of the Russian Arctic. In some cases, too, for example in the account of the Knot, as mentioned in Chapter 7, he gives long extracts from the journal of A. Birulya, one of the expedition's naturalists, which had appeared in Russian in 1907 as *Sketches of the bird life of the Polar coast of Siberia*. The quality and interest of these may be judged from the following excerpt referring to the tundra just inland from the north coast of the Taimyr Peninsula about 80 miles (150 km) west of the estuary of the Lower Taimyr River (Nizhnyaya Taymyra) (pp. 262–263):

> To give an idea of the life of the Curlew Sandpiper during its nesting season I will reproduce an extract from my diary of 24 June 1901, showing the life of the tundra from first impressions. Lying on a dry hillock, I was able with the help of a Zeiss prismatic field-glass to study the tableau of tundra life unrolled before me. My attention was attracted first of all to the dozens of Curlew Sandpipers careering about with swift and impetuous flight, at times in pairs, at times in little flocks of two or three pairs. During such aërial performances these handsome and dainty Sandpipers reminded me very much of butterflies in temperate climes flitting about on a sunny June day. At times these masters of flight would rise to a considerable height, mingling in serried ranks in the air; again, they would join together and sweep close to the ground. Often an entire flock would alight upon the ground and make a loud clamor. Then abandoning their aërial games, the whole company would again seek the tundra and begin to run about among the dried herbage probing the soil with their curved beaks. From my observation point I soon saw a pair of Turnstones and a Curlew Sandpiper fly to a little hillock near me, showing signs of anxiety. On going to the spot I found a nest with four eggs, similar in color to those of the Curlew Sandpiper found previously. In order to be sure as to which of the two

---

Plate 41   (*a*)   Nest and eggs of Lapland Bunting in a relatively open situation near Cambridge Bay. The nest is lined with Ptarmigan feathers.

(*b*)   Female Lapland Bunting incubating in the same nest viewed from a different angle. In all 14 nests found near Cambridge Bay in summer 1986 the female was incubating.

Sandpipers this nest belonged, I withdrew to some distance and watched the birds with my field-glass. After about five minutes a Curlew Sandpiper approached the spot where the nest was, and taking a turn about it, disappeared from sight amongst the vegetation. The two Turnstones took posts near by on stones and uttered anxious cries. When I again flushed the Curlew Sandpiper from the nest the Turnstones joined it, in mock pursuit, but did not come near the spot where the nest was. I went toward my observation point and saw that the Sandpiper had come back to its nest. I could readily make out among the grass its head with curved bill, uneasily turned from side to side. Before shooting it I several times flushed it from the nest, but it returned each time with less delay, and the last time almost immediately. Near at hand, at a distance of five or ten paces, I found two nests of the previous year, belonging either to the Curlew Sandpiper or to the Turnstone. The nest of the Curlew Sandpiper was placed between two little bushy hillocks in a hollow in the dry ground, near a small pool. It was on a little tussock and consisted of a hemispherical depression in the moss and compacted grass, quite unlined except for the dried moss and a very small quantity of lichen. It contained four pear-shaped eggs of a light olive color with pale brown spots chiefly disposed about the large end. They were placed in the nest in the form of a rosette with their pointed ends inward. Dissection of Curlew Sandpipers shot near their nests showed that both sexes share equally in the incubation of the eggs. Nevertheless the Curlew Sandpiper leaves its nest readily, even during the last days of incubation. Throughout this period it flushes while one is still forty or fifty paces distant, swiftly circles about one with its short quick cry—*pit-pit*—and alights at some distance on a stone to await one's departure. Since all the lower parts of its nuptial plumage are of a dark maroon color, that appears almost black at a distance, its dark silhouette stands out clearly against the light background of the sky. In flight its down-curved bill and white under tail-coverts are obvious features. It never attempts the ruse of trying to attract the enemy away from its nest but evidently depends chiefly for protection upon the concealing tints of the surrounding tundra, a view that explains why it leaves its nest so soon; if it stayed for a longer time on the nest, as the Knot does, it would run the risk of betraying the spot where the nest is. If the nest contains well-incubated eggs, the bird returns to it very soon, approaches directly toward it without any precaution, stands on the ground at a little distance, looks about it and at once settles upon the eggs.

Besides the species accounts, Pleske gives a separate bird and literature list for each of the 21 sections into which he divides the Eurasian tundra, and his book also includes an account of the Russian Polar Expedition of 1900–1903 and an analysis of the composition of the avifauna of the Eurasian Arctic.

Another pioneer of Arctic ornithology active in Russia at the beginning of the twentieth century was Sergei Aleksandrovich Buturlin, already mentioned in Chapter 7 because of his discovery in June 1905 on the lower Kolyma River, almost a century after the first specimen of the bird was described and named, of the nest and eggs of Ross's Gull. Buturlin was born at Montreux, Switzerland, in 1872 and died in 1938. Interested in game birds and an inveterate traveller, he made a large personal collection of skins and eggs from many parts of the Soviet Union and published a fine series of avifaunistic studies. In the north, he travelled in the Archangel (Arkhangel'sk) area, to Kolguyev Island (ostrov Kolguyev), Novaya Zemlya and the Chukchi Peninsula, as well as to the Kolyma Delta. After his return from the 1905 journey he wrote a book on the birds of the lower Kolyma which was unfortunately never published, the manuscript having disappeared at the time of the 1917 October Revolution. All that was published on the famous expedition were his letters to the *Ibis* about the discovery of breeding Ross's Gulls quoted in Chapter 7, and a note on the game birds of the lower Kolyma. In 1911 he co-authored with A. Tugarinov a detailed avifauna of the Yenisey area, and his work on the birds of the Soviet Far North appeared in 1916–1917. In 1918 Buturlin accepted a post as assistant at the University of Moscow's Zoological Museum and remained there as a professional ornithologist for the rest of his working life. A prolific writer, it has been said that he contributed 40% of the contents of the journal *Messager Ornithologique (Ornitologicheskiy Vestnik)*, which was published in Moscow in 1910–1927 (Potapov 1990). Towards the end of his life, in 1933–1935, he co-authored a study of Russian birds called *Systema avium Rossicarum* with G. P. Dement'ev which was serialised in a French ornithological periodical. The two also combined to write the five-volume *Comprehensive key to the birds of the U.S.S.R.* which was published in Moscow and Leningrad in 1934–1941 (Il'ičev and Flint 1985: 17, 195, 196, 198, 338).

Buturlin's contribution to his joint work with A. Ya. Tugarinov on the birds of the Yenisey province was made in the museum at Krasnoyarsk and in his colleague's collection of more than 4000 skins, rather than in the field, where Tugarinov was the active partner. The area covered by Tugarinov's journeys in 1905–1910 included the lower Yenisey valley north of the taiga as far as the coastal tundras along the shores of the Gulf of Yenisey (Yeniseyskiy zaliv), where Maude Haviland was to work a few years later (Il'ičev and Flint 1985: 125, 126; Tugarinow and Buturlin 1925). The marshy tundras round the mouth of the Yenisey were alive with birds, and Tugarinov recorded Little and Temminck's Stints, Dunlins and Curlew Sandpipers, Red-necked and Grey Phalaropes, Ruffs in drier, grassier areas, as well as Golden and Grey Plovers. On 6 July 1908 he was fortunate to come across breeding Red-breasted Geese (Tugarinow and Buturlin 1925: 186; author's translation):

> Following along the shore of the Gulf of Yenisey we found a whole family—six goslings besides both old birds—which had settled among some driftwood washed ashore. Surprised suddenly, the old birds were completely bewildered; they just honked continually and, with far outstretched necks, repeatedly shifted the position of their feet. The first hasty shot, a miss, did not frighten them; they hurried to the driftwood to hide themselves there. The second shot killed the goose. The gander flew a short distance away, then returned again toward the goslings. The dog accompanying us went up to retrieve the shot bird: the goslings scattered and fled and the gander ran to meet the dog with a threatening posture. On 18 July more broods were encountered; the already tolerably well-grown downy goslings were from 27 to 32.5 cm [11–13 inches] long. The old birds were now moulting hard though some, whose moult had only just begun, could fly. This time the old birds were more cautious, and, followed in a boat, they tried to swim away while the goslings hurried off in various directions with long and adroit dives.

Tugarinov continued to make valuable contributions to knowledge of the birds of the Siberian Arctic. Another work of his on Yenisey birds appeared in 1927, and in 1934 his "Materials for an avifauna of eastern Taimyr" was

214 *In search of Arctic birds*

published jointly with A. I. Tolmachev, who had written a paper on the birds of Kolguyev Island in 1928 (Il'ičev and Flint 1985: 197, 198; and see Chapter 6 above).

SEABIRDS OF THE KOLA PENINSULA

In the extreme northwest of Russia, its coasts warmed by the Gulf Stream though most of it is north of the Arctic Circle, the Kola Peninsula separates the Barents Sea from the White Sea (Beloye more). The seabird colonies along the shores of both seas, on the Murmansk Coast (Murmanskiy bereg) and on the Kandalaksha Coast (Kandalakshskiy bereg), have been intensively studied by Russian ornithologists, and two outstanding works, by L. O. Belopol'sky and V. V. Bianki, have been translated into English. Belopol'sky's *Ecology of sea colony birds of the Barents Sea*, published in Russian in 1957 and in English in 1961, has become a much admired classic. Bianki's *Gulls, shorebirds and alcids of Kandalaksha Bay*, published at Murmansk in 1967, was not translated until 10 years later. This time-lag has flawed the reputation of his book, because rapid advances in ornithology in the 1970s made it to some extent out of date by the time it became known in the West (Evans 1980).

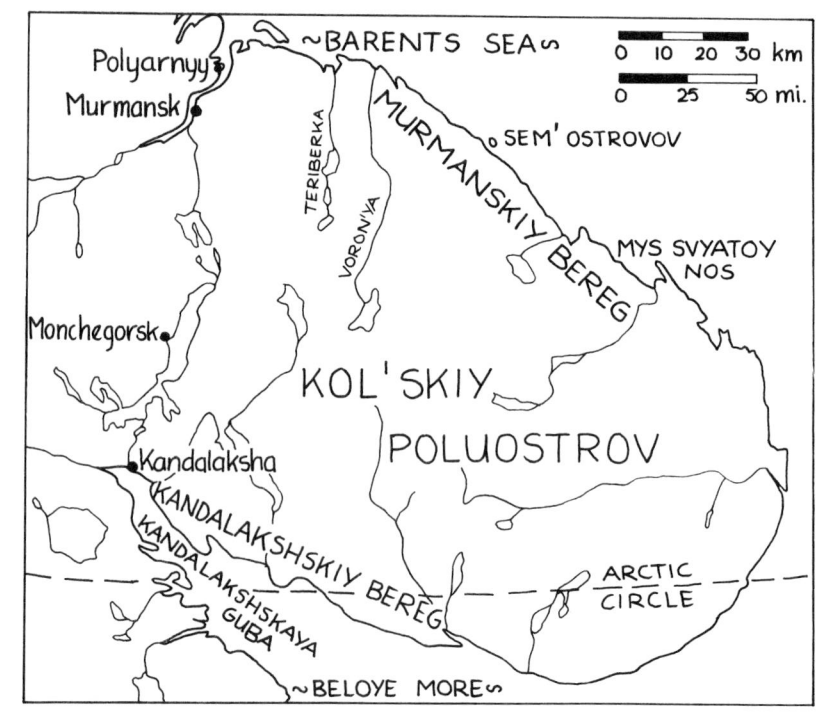

Map 15 The Kola Peninsula.

Leb Osipovich Belopol'sky joined the staff of the Murman Biological Station at Polyarnyy on the Barents Sea coast north of Murmansk in 1926. Concentrating on seabirds, he took part in three expeditions from there into the Barents Sea, in 1927, 1928 and 1929–30, and published a paper on the distribution of Fulmars and Kittiwakes at sea. In 1930–1931 he was the zoologist of an exploring expedition to the Chukchi Peninsula but spent most of his time there on the south coast, south of the Arctic Circle (Portenko 1981: 12). His stay resulted in a paper on the avifauna of the Anadyr' area (Il'ičev and Flint 1985: 141, 337). In the next 2 years he enjoyed an Arctic experience which must be nearly unique even among Russian ornithologists: he twice sailed along the Northern Sea Route from the Barents Sea to Bering Strait: in 1932 on the icebreaker *Sibiryakov* and in 1933 on the *Chelyuskin*. Both were Soviet government expeditions. The second ended in shipwreck. The *Chelyuskin* was crushed in the pack ice of the Chukchi Sea and Belopol'sky and his shipmates had to camp on the ice for 2 months in the late winter before being rescued by air. It was only after these adventures that he settled down to study seriously the breeding seabirds of the Seven Islands (Sem'ostrovov) where he was chiefly responsible for creating a zapovednik, or nature reserve, in 1938, complete with a well-equipped laboratory or field station. Here Belopol'sky was in charge of the scientific research of the reserve from 1938 until 1943, apart from an expedition to Novaya Zemlya in 1942, and again between 1946 and 1951. The Second World War hit Belopol'sky hard: his research was interrupted, three of his graduate students were killed, and many of his pre-war data were lost. He reckoned that in all he had completed a total of 97 months in the field when his book was written. Only a small part of the data he amassed on seabird ecology was published in the 13 papers by him appearing between 1930 and 1957; the bulk of it was reserved for use in his book (Belopol'skii 1961: 5). Belopol'sky by no means worked in isolation. At least four other ornithologists had studied the Seven Islands seabirds before 1938 and after then the number of researchers working under Belopol'sky's direction increased. Many of them were graduate and other students who often only came for the summer, some of whom assisted with the ringing programme initiated in 1937. Belopol'sky was in constant touch with other Russian ornithologists, especially G. P. Dement'ev, and one of the young researchers he helped to train was S. M. Uspensky (Il'ičev and Flint 1985: 23; Belopol'skii 1961: 7–8).

The *Ecology of sea colony birds of the Barents Sea* has taken on a new interest in the 30 years or more since it was published because of the increasing evidence, even in remote Arctic areas, of the negative effects on seabird populations of pollution and over-exploitation of fish stocks by humans. These came to a head in the late 1980s. Belopol'sky studied a flourishing community of 18 species of colonially-breeding seabird before any of these man-made disasters were detected. He even measured the thickness of the egg shells of 11 species and found that Brünnich's Guillemot's eggshells were thicker (0.50 mm) in Novaya Zemlya than on the Murman Coast (0.30 mm). The suggested explanation was that the cliff ledges there were of jagged slates while on the Murman Coast they were of smooth granite. Belopol'sky found little evidence of food shortages seriously affecting seabird populations. In 1940 shortages of

216   *In search of Arctic birds*

suitable fish round the Seven Islands delayed egg-laying by the local Herring and Great Black-backed Gulls for about a month, but the Kittiwakes and guillemots were unaffected. Then, on 17 May, a fish processing vessel anchored at the islands to load and process the local fishing catch, and the resultant waste fish brought on mass egg-laying by the large gulls (p. 176). In 1947 stocks of herrings and sand-eels were low but birds dependent on them were able to switch to other prey. For example, Kittiwakes made good the sand-eel scarcity by taking more capelins than usual (pp. 101–103). Only in 1950 did real disasters occur, when Arctic Terns abandoned incubation en masse along the Murman Coast (p. 18) and on Kharlov Island, one of the Seven Islands, where the younger chick in some Black Guillemot nests died, as did most of the youngest nestlings in Kittiwake's nests which had clutches of three eggs. Examination of stomach contents showed that in both cases local shortages of fish had forced the birds to bring less nutritious food items, mainly molluscs (Kittiwakes) or crustaceans (Black Guillemots) to their young (p. 262).

The two main sections of Belopol'sky's book deal with food and reproduction. Analysis of stomach contents showed how Fulmars fed largely on pelagic gastropod molluscs, Razorbills and guillemots on fish, Little Auks on pelagic crustaceans, and Eiders on littoral crustaceans and molluscs. Was sufficient

Plate 42   Black Guillemot portrait. Unlike most photographs in this book, this one was not taken in the Arctic but on Mykines in the Faeroes.

food available? The 1500 Eiders breeding on the Seven Islands were credited with taking about 60 tons of molluscs each summer; careful calculations showed that the Kharlov Island littoral alone supported 380 tons of mussels. Belopol'sky studied the seasonal and geographical variations in the diet of each species and compared the diets of the two sexes. He found that male Kittiwakes fed on fish mainly in April and May, on molluscs early in June, on crustaceans in April, May and early June, and on insects in July; the females ate more fish than the males in June, more molluscs in April and May, and more crustaceans in July (p. 68). In the section on reproduction, Belopol'sky gives information on times of arrival, dates of egg-laying, lengths of incubation, rate of weight increase in young and fledging dates for most of the 18 species studied. He discusses such questions as whether or not the incubating duck Eider ever leaves her nest. He asks why Eiders breeding on the Seven Islands use a new nest every year? Is it because the old nest usually harbours fleas?

One thorny question finally resolved by Belopol'sky was the connection between the elongated, pointed or pyriform shape of the guillemot's egg and the likelihood of it tumbling off the bird's often narrow breeding ledge. Yu. M. Kaftanovsky, author of at least five papers on the guillemots and other birds of the Seven Islands, but one of the graduate student victims of the Second World War, had claimed in a 1941 paper that, though the guillemot's

Plate 43 Three drake and two duck Eiders feeding among rotting seaweed on the shore of the Varanger Peninsula, north Norway. Store Ekkerøya, 12 June 1972.

eggs did not spin round like a top when inadvertently pushed by the bird or when blown by the wind, as had been popularly reported, none the less the pyriform shape ensured that they fell less often than a round egg would. In 1947 Belopol'sky noticed at Bezymyannaya Bay, Novaya Zemlya (Map 35), that most of the Brünnich's Guillemot's eggs that did fall off the cliffs were newly-laid or very slightly incubated. At his request in 1948 S. M. Uspensky carried out an experiment at this bird bazaar. Soon after the birds had laid he fired a shot from a boat below the cliffs. "An immense crowd of murres simultaneously rose from the cliff, and the eggs started 'pouring'" (p. 132). Just before the eggs were due to hatch the experiment was repeated, but this time, though a swarm of birds took wing, "not a single egg tumbled off the ledge". Belopol'sky was able to show that, as the embryo begins to develop inside the egg, the initially very small air space at the larger end gradually expands, shifting the egg's centre of gravity toward the pointed end. This causes the partly or well-incubated egg to lie on its pointed end so that it spins round on the ledge rather than rolling off it. This makes biological sense: if a female losses an egg soon after it is laid, she will replace it by re-laying; but if it is lost later on, ovary degeneration may be so far advanced that she is unable to lay again that year.

Vitaly Bianki's (1977) book on the *Gulls, shorebirds and alcids of Kandalaksha Bay* is another of the very few Russian ornithological studies to have been translated into English. Though most of the White Sea (Beloye more) is just south of the Arctic Circle, the northern end of Kandalaksha Bay (Kandalakshskaya guba) extends to the north of it. The climate in the bay is more continental than that of the Murman Coast and winters are cooler; the sea ice does not break up until late April to early June. Bianki describes the breeding habitats and biology of the predominant species only. For these he gives detailed data on habitat selection, breeding distribution and behaviour, nest sites, egg and clutch size, length of incubation and fledging, and so on. Three shorebirds are dealt with (the Ringed Plover, Oystercatcher and Turnstone), two gulls (the Common and Herring), the Arctic Tern, and two auks (the Razorbill and Black Guillemot). The Razorbill, here near the eastern edge of its range, breeds on low rocky islets, devoid of trees, far out in the bay. The nests are in jumbled boulders or even piles of driftwood and are often only a few feet above sea level. Even though they sometimes arrive in the bay before the ice has gone, the Kandalaksha Bay Razorbills are only present for 3 months of the year. The author found only two small breeding colonies, containing fewer than 100 pairs in all. He brings his book to a close with a warning that human pressures are adversely affecting the bird populations in the bay: a swarm of motor boats in summer lands people on islands not within the Kandalaksha Nature Reserve and these visitors plunder the nests of gulls and Eiders and to a lesser extent those of Arctic Terns and Oystercatchers. The bay has a total of 860 islands, but up to 1967, the year Bianki's book was published, only 130 of them were inside the nature reserve. Since then many more have been added to it.

Although the Kandalaksha Zapovednik, or nature reserve, was created in 1934, it was not until 1946 that an expedition was sent there from the University of Moscow to make a thorough study of the birds. Up till the 1950s it was the breeding Eiders of Kandalaksha Bay, because of their economic value, that received most attention from researchers. Between 1958 and 1963 the National Ornithological Station was based at the Kandalaksha Zapovednik, and it was at this time that Bianki, having joined the staff of the zapovednik, began his studies of the common marine birds other than the Eider which were reported in full in his book. Published in Murmansk, it was Volume 6 of the *Transactions of the Kandalaksha State Nature Reserve* (Il'ičev and Flint 1985: 23–24; Bianki 1977: v–vi, 1).

THE CHUKCHI PENINSULA AND WRANGEL ISLAND

The far northeastern tip of Asia with the islands lying to its north was one of the last regions of the Soviet Union to be investigated systematically by Soviet ornithologists. Field work in the northern or Arctic part of the Chukchi Peninsula, namely the area between the Chukchi Mountains or Highland (Chukotskiy khrebet or Chukotskoe nagor'e) and the coast extending from Chaun Bay (Chaunskaya guba) in the west to Uelen in the east, was only begun in 1932. The pioneer was Leonid Aleksandrovich Portenko, born in Kiev, who became one of the most distinguished twentieth-century Russian ornithologists. He died in 1972 at the age of 75 having spent a great deal of his working life as curator of birds in the Ornithological Laboratory at the Zoological Institute of the Soviet Academy of Sciences in Leningrad. One of his earliest expeditions to the Arctic was in 1930, when he was sent by a government agency to investigate the economic value and rational means of exploiting the bird bazaars of western Novaya Zemlya (Krasovsky 1937: 35). In the following year he travelled to the Anadyr' region in the south of the Chukchi Peninsula on the lookout for American species that might occur in Siberia. Finally, in 1932, having reached Anadyr', again as zoologist on the Chukchi-Anadyr' Expedition of the All-Union Arctic Institute at Leningrad, he arranged a late summer voyage round the southern and eastern coasts of the Chukchi Peninsula for himself and his assistant P. T. Butenko. Stopping at various points en route, they went ashore to collect birds as often as possible, afterwards making use of the skinning laboratory on board ship. When the ship reached its furthest point, the settlement of Dezhnev, they explored the surrounding tundra, collected Baird's Sandpipers and Sabine's Gulls, and saw a pair of White-billed Divers. Then the two ornithologists took a dog sledge to Uelen, the iron runners sliding easily over the wet grass in the rainy weather. On the return voyage more birds were collected, including Spoon-billed Sandpipers, and Lesser Snow Geese were seen (Portenko 1981: 12–14).

Determined to explore the Chukchi Peninsula and its birds much more thoroughly, in 1933 Portenko managed to have himself and Butenko posted by the Arctic Institute to the weather station at Uelen, where they would be based for a year while studying the local avifauna. They achieved a great deal, though their field work was curtailed in the autumn by the need to help

220   *In search of Arctic birds*

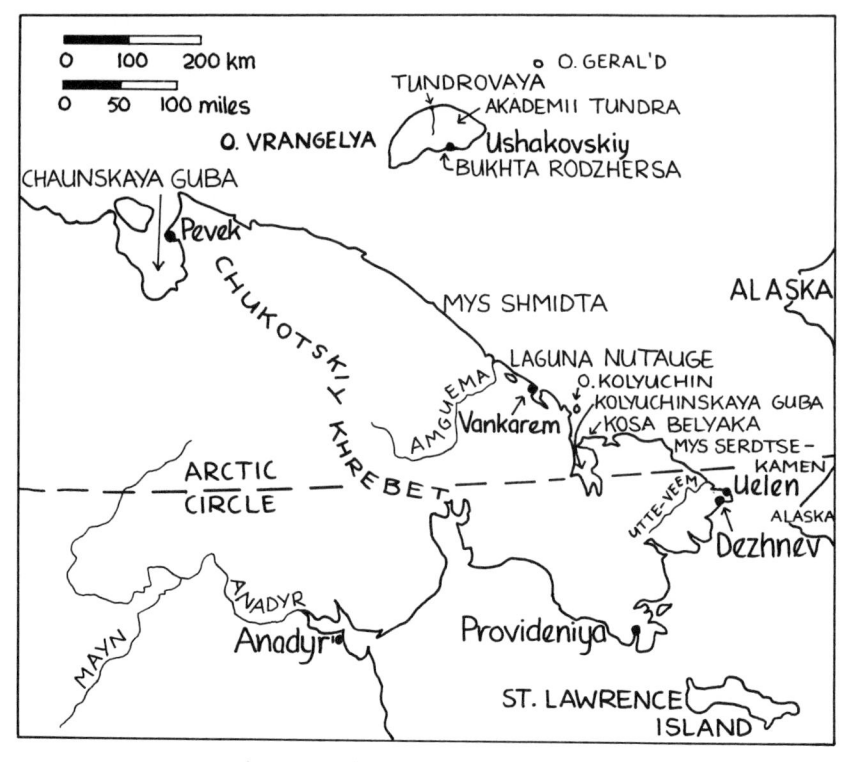

Map 16   The Chukchi Peninsula and Wrangel Island.

transfer the weather station from its previous site at Dezhnev and put up the new buildings at Uelen. Then in the spring of 1934 the entire staff of the station, not excepting visiting ornithologists, took part in the rescue of the passengers and crew of the *Chelyuskin*, which was crushed by the ice in the Chukchi Sea and sank on 13 February, leaving them stranded on the ice in a makeshift camp—among them of course Belopol'sky.

Arriving at Uelen on 15 August, Portenko had soon seen Sandhill Cranes and Emperor Geese in flight. On 22 August he climbed to the summit of a neighbouring hill from which Alaska could be seen in fine weather. In the rays of the setting sun, even the greenish tint on the slopes of the nearest mountains could be made out. Behind them on other mountains remnants of snow showed up here and there, and a distant range of higher mountains was visible in the blue haze. Flocks of migrating geese, Emperors, Brents and Lesser Snows, sometimes attended by a Peregrine and Long-tailed Ducks, were a feature of the autumn at Uelen. September brought a Spectacled Eider and Gyrfalcons; in October, while resting walruses were seen on ice floes, roaring like lions, Snowy Owls and Ross's Gull were seen. In April Portenko was able to take advantage of the air-lift organised to rescue the shipwrecked passengers of the *Chelyuskin*. He persuaded pilot Ivan Doronin to fly him and a supply of dog food from Uelen to Vankarem. From there he explored the coast northwestward by dog sledge as far as Cape Schmidt (mys Shmidta) weather station, returning in May via Kolyuchin Island and Cape Serdtse-Kamen'. On this adventurous trip newly-arrived flocks of Eiders and Glaucous Gulls were seen, the first Sandhill Crane was sighted and the mating call of the male Ptarmigan was heard. Snow Buntings were occupying nest sites and cormorants were seen carrying nest material in their claws. Thereafter, Portenko and Butenko spent the summer making trips, often with a walrus-skin canoe made for them at Uelen, so that they could explore the local lagoons and rivers. A journey up the Utte-Veem River in July enabled Portenko to collect some of the passerine birds of the willow scrub which grew in the river valleys of the interior up to the height of a person: pipits, warblers, thrushes and wagtails. By the end of August 1934 they were once again wearing fur coats and the first migrating birds of the autumn were seen. Their year in the Arctic came to a close when they left Uelen on 3 September (Portenko 1981: 14–20).

Wrangel Island, part mountain part tundra, about 150 km (95 miles) long and lying about the same distance north of the Chukchi Peninsula coast, was more or less uninhabited until 1921. Since 1926 there has been a Soviet weather station on the island and a permanent settlement at Ushakovskiy on the south coast. Portenko was determined to study the island's little-known bird life, and was lucky enough to be appointed zoologist on an expedition there organised by the Soviet Academy of Sciences in 1938 to investigate reports of the finding of a mammoth corpse. He arrived early in August, too late for most of the summer birds: the young Sabine's Gulls were flying strongly and the Knots had already migrated. There was only one thing for it: Portenko decided to overwinter in a hut he built for himself at the weather station in Rodgers Bay (bukhta Rodzhersa) so that he could study the island's breeding birds in the following summer. On excursions from the station on foot in September and October he saw Long-tailed Ducks, King Eiders, Eiders, Glaucous Gulls and Ravens and once had to beat a hasty retreat when confronted by a large male polar bear while armed only with "two half-charges of small shot for buntings". On 4 November he saw "a massive flight of Ivory Gulls". By the time the collecting season came round in June Portenko had an expert taxidermist on hand to do the skinning: he had whiled away the long winter hours running a single-student class in taxidermy for S. I. Korobko, wife of a wintering explorer. Although he occasionally had the use of a vehicle, Portenko's summer excursions into the tundra and mountains inland, and along the coast in either direction from his base in Rodger's Bay, mostly had to be conducted on foot after the first week of June, when dog sledging became impossible. Rivers swollen with meltwater so as to be uncrossable, and repeated fogs, restricted the area he could cover. Warm sunny days were few and far between; mostly it was cold and wet. He never reached the main breeding grounds of the Lesser Snow Geese, finding only abandoned nests with eggs apparently sucked by gulls. On 5 June he shot Eiders and a single Spectacled Eider, and 6 July he devoted to studying and photographing nests of Snowy Owls. By early August he was confined to the immediate surroundings of Rodgers Bay, awaiting the annual ship to take him off the island, which it eventually did on 28 August, after he had filled in time shooting walrus and carefully packing his bird skins in solidly made wooden boxes. Each skin was individually wrapped "in stiff, thick paper and sealed with wax". All reached Leningrad in good condition. It was 20 years before another ornithologist wintered on Wrangel Island (Portenko 1981: 31–39).

The fruit of all these labours in the field was a detailed and very readable two-part monograph entitled *Birds of the Chukchi Peninsula and Wrangel Island*. Unfortunately, its publication was long delayed. In 1967 Portenko expressed his frustration with this delay to the then curator of birds at the Smithsonian Institution, George E. Watson, his opposite number in America, who was visiting him in Leningrad, showing Watson "the manuscript stacked high on his desk" (Portenko 1981: vii). Five more years elapsed before the first volume of the Russian edition came out in 1972, followed closely by the second. An English translation of Volume 1 was eventually published in India by the Amerind Publishing Co., nearly a decade after the author's death, but for some unexplained reason, Volume 2 never appeared. For all that, Volume 1 alone is a most worthwhile book. It covers the divers, shearwaters, waterfowl, raptors and waders and contains the fascinating introduction describing the author's own field work which has been drawn on largely in the foregoing pages. It is only fair to say that by 1980 later works, to be mentioned below, had so transformed our knowledge of the birds of the Chukchi Peninsula and Wrangel Island that, in scientific but not literary terms, Portenko's classic has been largely superseded.

ALEKSANDR ALEKSANDROVICH KISHCHINSKY

There is no very special reason for giving Sasha Kishchinsky a section all to himself except that he was an outstanding ornithologist cut off by an untimely death in his forty-third year in 1980 (Flint *et al.* 1982). Sasha is of course short for Aleksandr; Kishchinsky is often spelt Kistchinski. In April 1990 *British Birds* figured two splendid photographs of a duck Spectacled Eider taken by "Sacha Kistchinsky" (Ogilvie 1990). He was one of the few Russian ornithologists to have published a paper in English in the *Ibis* (Kistchinski 1975), and, though his interests ranged widely, one has only to glance at a list of his publications to appreciate the extent of his contribution to knowledge of Arctic birds.

Born in Moscow in 1937, Kishchinsky's ornithological interests developed early in life: he was reading bird books at the age of 5 and wrote and illustrated his own *Birds of the Soviet Union* when he was 8 years old. The same was true of his intellectual gifts: at school he was awarded a special scholarship and a gold medal. Before entering the Faculty of Biological Sciences of Moscow State University as a student in 1953 he had become a member of the youth section of the All-Russian Society for the Protection of Nature. At Moscow University he was inspired and encouraged by one of the leading ornithologists and teachers of the day, Professor G. P. Dement'ev, and by the crowd of young ornithologists working in the university's Zoological Museum. His practical work as a student took him on expeditions to the north coast of the Kola Peninsula in 1955 and 1956 and resulted in the publication of his first scientific paper, on the biology of the Gyrfalcon, which came out in 1958 in the first volume of *Ornitologiya*, a year after he had completed his studies at Moscow by obtaining his diploma or first degree.

Kishchinsky's three breeding pairs of Gyrfalcons were nesting on the rocky banks of the rivers Teriberka and Voron'ya and of a tributary of the Teriberka between 6 and 25 km (4–15 miles) from the Barents Sea (Map 15). Each of the three pairs had two or more nests which they used in different years. The 20-year-old student did everything possible in the field during his three summer visits in 1955–1957, and must have had his work cut out when he got home to analyse the contents of the prey remains and the Gyrfalcon pellets he collected, which contained fragments of 702 different individual animals. He turned up a few surprises here because, though the expected birds and small mammals predominated, he also found examples of the falcon's diet in hard times: reindeer, showing that carrion was resorted to, fish and 24 beetles. A surprisingly high 40% of the prey items found were of mammals. Lemmings and voles were favourite prey among mammals; Willow Grouse and Ptarmigan among birds. Other species of bird taken included Red and Black-throated Diver, Long-tailed Duck, Red-breasted Merganser and Golden Plover; but only one Snow Bunting, two Meadow Pipits, three Redwings and two redpolls among the more common passerines. In June 1956 in a 12-day period the pair of Gyrfalcons breeding nearest the coast brought to the nest a shrew, a vole, an unidentified small mammal and 25 birds—two Willow Grouse, one Ptarmigan, one unidentified *Lagopus* species, a skua, an Arctic Tern, a Herring Gull, a Glaucous Gull, ten Kittiwakes, two guillemots, an unidentified auk, a duck Teal, a Golden Plover and a Snow Bunting. Food was not always abundant for these falcons: in 1957 none of the three pairs nested. In conclusion, Kishchinsky (1958) thought that Gyrfalcon distribution north of the tree line was dependent on the availability of suitably sheltered cliff ledges for nest sites and on sufficient food early in the season. In the north of the Kola Peninsula, at any rate, Gyrfalcons could be divided into two "ecological groups": those breeding near the coast within reach of bird bazaars, and those breeding on the inland tundra or forest tundra, which were more or less dependent for survival on the presence of Willow Grouse in early spring.

After his student years in Moscow, Kishchinsky worked for a time on the birds of the Black Sea area. Then, in 1959, he joined an elaborate Academy of Sciences expedition to Kamchatka and found himself in a detached section exploring the Koryak Mountains in the company of Leonid Portenko who, thereafter, became a friend, guide and source of inspiration to the younger man. Portenko invited him to enrol as his research student at the Zoological Institute of the Academy of Sciences in Leningrad. In the years after 1962 Kishchinsky worked there for his second or candidate's degree, then returned to Moscow in 1965, and served for 3 years on a government scientific committee before joining the Central Laboratory for the Protection of Nature of

224 *In search of Arctic birds*

what is now the All-Union Scientific Research Institute for Nature Conservation and Nature Reserves. In the next few years his post in this organisation enabled him to make a series of ornithological expeditions into the Arctic and Soviet far east. In March–May 1968 he was on ostrov Kotel'nyy and ostrov Bel'kovskiy in the New Siberian Islands (Kishchinsky 1988: 14). In the spring of 1969 he was studying Polar bears on Wrangel Island. In 1970 he led an expedition to the north shore of the Chukchi Peninsula which included two students of Moscow University. They stayed at Cape Schmidt (mys Shmidta) from 31 May to 5 June, then flew by helicopter to the east shore of Nutauge Lagoon, 25 km (16 miles) east of the Amguema estuary (Map 16). Field work in the surrounding tundra occupied them here until early in July when the expedition explored the lower reaches of the Amguema by boat. Then from 13 to 21 July the coast between Nutauge Lagoon and Vankarem was traversed before the return trip was made to Cape Schmidt on 26 July. Kishchinsky presented the 65 bird specimens collected to the Moscow University Zoological Museum and gave his preliminary data to a grateful Portenko for use in his book (Portenko 1981: 27). Some results of this expedition, namely notes on the breeding biology of the Emperor Goose and the Spectacled Eider, were published in English in 1971 and 1974 in the journal *Wildfowl*. In 1971 Kishchinsky spent 4 months exploring the Indigirka Delta with V. E. Flint, and in 1972 he was with S. M. Uspensky on the lower Yana River (Kishchinsky 1988: 14).

(*a*)

(b)

Plate 44   (a) A female Grey Phalarope on the nest. This bird is not incubating, which in this species is done by the male, but laying. Cambridge Bay, 23 June 1986. (b) Grey Phalarope pair on a tundra pool near Cambridge Bay. The more brightly coloured bird on the left is the female.

It was on these expeditions to the Chukchi Peninsula and the Indigirka and Yana deltas that Kishchinsky collected the data for his paper on the "Breeding biology and behaviour of the Grey Phalarope *Phalaropus fulicarius* in east Siberia" which was published in the *Ibis* in 1975 with his name spelt Kistchinski. During the three expeditions, 93 Grey Phalarope's nests were found and 126 birds were shot and dissected. Although the Grey Phalarope had been studied in Spitsbergen and in North America, little was known about this interesting bird in Siberia where, at least on the northern Chukchi coast and in the Indigirka Delta, it and the Lapland Bunting are the commonest species. It bred in areas of low, flat, moss-sedge or polygonal tundra flooded by the spring thaw, where there are numerous temporary and permanent pools and lakes. Kishchinsky gives details of the breeding chronology. In 1971 he arrived in the Indigirka Delta during the spring thaw before the Grey Phalaropes, which began to come in on 4 June, reaching a peak on 8 June, in pairs and flocks, flying low and invariably coming from the east. No wonder they started feeding as soon as they alighted: those collected had empty stomachs and must have flown non-stop for long distances. Kishchinsky found that after egg-laying, the pairs, which may only be formed temporarily to provide a male with a clutch, break up, and the females, leaving the males to incubate, depart altogether from the nesting area and gather in the lakes and coastal lagoons with non-breeding and unsuccessful males, moving to the seashore by the end of June. He described the Grey Phalarope's courtship displays, all of them initiated by the female, and was able to confirm that the species, like other phalaropes, lacks any true territorial behaviour. Siberian nests were not found in clusters of 5–10 as recorded elsewhere, but were for the most part distributed evenly over the tundra, except for an occasional example of two nests in close proximity.

In 1973, after an expedition to the Yamal Peninsula (poluostrov Yamal), Kishchinsky joined the staff of the Institute of Evolutionary Animal Morphology and Ecology of the Soviet Academy of Sciences in Moscow. Continuing his work on the birds of the Soviet far east, he worked in the field there in the summers of 1974–1976, and in the summer of 1977 he spent 3 months in Alaska, studying waterfowl especially with W. J. L. Sladen and other American ornithologists. The climax of this work came in 1977–1978 when he wrote up his doctoral dissertation on the distribution and ecology of the birds of northeast Asia, with which in 1979 he gained his third and last degree, that of doctor. Parts of this dissertation were eventually published in Russian in book form in 1988, for the book he was planning at the time of his death had unfortunately remained unwritten. Even though much important work had been done in the 10 years that elapsed between its completion and its publication, Kishchinsky's *Avifauna of northeast Asia, History and present status* is one of the most important and interesting Russian ornithological publications of the second half of the twentieth century. Roughly half the book is devoted to species accounts, among which Arctic or sub-Arctic birds are naturally very well represented. Of particular interest are the accounts of the Spectacled Eider, Siberian White Crane, Spoon-billed Sandpiper, Great Knot and Ross's Gull.

THE MAGADAN TEAM

A big step forward in zoological research in northeast Asia was taken in 1968 when the Institute of Biological Problems of the North was set up at the Soviet Academy of Sciences' Far Eastern Scientific Centre in the former gold-rush town of Magadan, now a city with a population of 150 000. Zoologists working there established permanent field stations on Wrangel Island and in Chaun Bay (Chaunskaya guba) in 1969 and in 1972–1974 temporary stations were set up in Kolyuchin Bay (Kolyuchinskaya guba) and on the middle course of the River Omolon, a tributary of the Kolyma, and in the Kolyma Delta. The institute soon possessed a collection of over 1000 bird skins. Its team of zoologists busied themselves through the 1970s in different parts of the region, not all of them Arctic (Kishchinsky 1988: 15–16; Il'ičev and Flint 1985: 143–146), and knowledge of the animals, birds and plants of the region was soon transformed. A major research programme to study the region's avifauna and the adaptations of birds to life in the Arctic was initiated. The institute's three ornithologists, A. V. Krechmar, A. Ya. Kondrat'ev and A. V. Andreev were more particularly concerned with northern birds and regions. In 1978 the three of them brought out a book on the ecology and distribution of the birds of the far northeast of the Soviet Union, in which Kondrat'ev wrote on the birds of Kolyuchin Bay and Kolyuchin Island, and Krechmar and Andreev dealt with the birds of the middle reaches of the Kolyma and Omolon rivers (Krechmar, Andreev and Kondrat'ev 1978). Kondrat'ev, who comes from Irkutsk and prefers to spell his name Kondratyev (but the transcription conventions used here do not allow it), followed this up in 1982 with an important monograph on the biology of waders in the tundra of northeast Asia. His name became known to British birdwatchers in this same year, when his splendid photographs of a sitting Long-billed Dowitcher and a male Grey Phalarope were published in *British Birds* in an article on "Waders in Siberia", though no mention was made of his ornithological work (Prater and Grant 1982), which included papers on the breeding biology of the Spoon-billed and Western Sandpipers both published in 1974 (Kishchinsky 1988: 272). Of these three Magadan researchers, Arseny Krechmar, a student of Portenko, was the veteran. He had already in 1958–1964 investigated the birds of western Taimyr, including the Red-breasted Goose, in a series of expeditions and, in 1966, published a very thorough account of his researches (Il'ičev and Flint 1985: 125, 342).

In 1972–1974 Krechmar and Kondrat'ev (1982) turned their attention to the Emperor Geese breeding in the remote uninhabited coastal tundras of

Plate 45 Emperor Goose at nest on the Belyaka Spit, on the north coast of the Chukchi Peninsula. The photograph was taken by Pavel Tomkovich on 9 July 1987.

Kolyuchin Bay on the north coast of the Chukchi Peninsula (Map 16). Their work was based on different temporary field stations from which they could explore the tundra round the shores of the bay and both east and west of it; one of these was at the mouth of the bay on the low-lying promontory called the Belyaka Spit (kosa Belyaka). Lying just on and very slightly north of the Arctic Circle, Kolyuchin Bay is at the extreme northern limit of this goose's breeding range, which straddles Bering Strait, the biggest concentration of breeding birds being on the American side, in the deltas of the Yukon and Kuskokwim rivers some 500 miles (880 km) further south. After an aerial reconnaissance of the goose's nesting grounds round the bay, the Soviet ornithologists made use in 1973 and 1974 of automatic recording devices constructed in their institute's laboratory and first developed by the senior author. These had already been used to study the Snow Geese on Wrangel Island. They were successfully installed in two Emperor Goose nests in 1973, and in three nests the following year, in one of which continuous data from egg-laying to hatching were obtained. The work was made more difficult because of the rather sparse distribution of breeding geese—less than one pair per 10 km^2 (4 square miles). Arriving at the end of May already paired, the geese built and laid around 15–20 June, almost invariably choosing nest sites by water. While other workers had reckoned the clutch size at from 2 to 11 eggs, usually 5 or 6, in Kolyuchin Bay 12 nests containing full clutches contained only 2–5 eggs each (average 4.17 ± 1.02). Krechmar and Kondrat'ev did find one nest containing 7 eggs but were sure that this resulted from two geese laying in the same nest, something which frequently happened in Snow Goose's nests on Wrangel Island. The automatic recording device showed that during egg-laying the goose gradually increased the frequency and duration of her vists to the nest. The temperature in the artificial egg placed in the clutch by the researchers, after varying wildly in the first few days, gradually stabilized at around 24°C (75°F) and remained very stable at between 25 and 30°C throughout incubation, when the goose spent 95–98% of her time sitting on the nest, leaving only once or twice a day for a short period. Clutch temperature was unrelated to the temperature of the surroundings, even when the goose was absent.

In the later 1970s Kondrat'ev shifted his field work from Kolyuchin Bay to the Chaun Bay lowlands, where a permanent biological station was sited. Using colour neck-collars, he here investigated the ecology and migration of the breeding swans. These yellow-billed Bewick's Swans migrate in winter to Japan, and on one occasion, at least, have successfully interbred in the Chukchi Peninsula with the black-billed Tundra Swan of North America (Flint *et al*. 1984: 28)—a good reason for regarding them as conspecific, that is belonging to one and the same species. Kondrat'ev also studied Sabine's Gulls breeding in colonies and single pairs in 1974–1977 and, together with his wife, he published a paper in 1984 in the journal *Ornitologiya* on the growth and development of nestling Sabine's Gulls. The Kondrat'evs followed up these researches with another paper in the same journal, published in 1987, in which the breeding ecologies of Sabine's and Ross's Gulls were compared.

A. Ya. Kondrat'ev should not be confused with another member of the group of ornithologists based at Magadan, A. V. Kondrat'ev, who hails from

Leningrad and has been working chiefly on ducks (Potapov, pers. comm.). A. Ya. Kondrat'ev also collaborated closely with his colleague at the Magadan institute, the Leningrader Aleksandr Vladimirovich Andreev. Appointed head of the institute's laboratory of ornithology in 1986, Andreev's scientific reputation, his research in extremely rigorous conditions, and his flair for popularization have made him a well-known figure in Russia. His candidate's thesis on the adaptations of birds to winter conditions in the sub-Arctic, which was published in book form in 1980, earned him an invitation to Moscow to receive the prize of the Young Communist League (Komsomol). A documentary film made for television by a professional team from Novosibirsk was much admired. Entitled "From the field notebook of ornithologist Andreev" it opened with a pair of swans in flight over the green tundras of the lower Kolyma, studded with blue lakes, and followed with views of the mock battles of Ruffs lekking on the lake shore. Ornithologist Andreev was shown providing the commentary while sitting in his tent, wading waist-deep into the marsh, camera held on high, to photograph a Ross's Gull, and surrounded by the electronic equipment devised by his friend Krechmar and further perfected by himself (Orlov 1988: 141–165).

At the end of May 1978 Andreev and A. Ya. Kondrat'ev (1981) contrived to have a breeding colony of Ross's Gulls within walking distance of their field station in the Kolyma Delta. On the left bank of the great river about 60 km (37 miles) north of the mouth of the River Omolon (Map 14), it was in what was locally known as the Khalerchinskaya Tundra, named after the Yukagir word for a gull—*khalarcha*. It was visited weekly, and then daily, through the breeding season. In the following year a hide was built on the edge of the colony to minimise disturbance. Over the two seasons data on a total of 52 nests, some of them apart from the main colony, were obtained. The birds' activities at some nests were registered automatically and temperatures in different parts of the nests were recorded, with devices adapted by Andreev from meteorological instruments or borrowed from his friend Krechmar. The flat tundra country was usually free of snow in the first days of June and it was then that the gulls began to appear at their nesting places, either on the banks of large shallow lakes or in eroded marshy depressions among sedge hummocks. Eggs were laid from early June on and the young hatched, after about 20 days of incubation, early in July. In late July and early August first the adults and then the young gulls left the tundra for the ocean where most of their lives are spent. Clutches were of two or three eggs. In 1978, out of 60 eggs that hatched, 54 downy young were produced, but only 22 of these actually fledged; and the figures for 1979 were no better. This mortality of about two-thirds of the young birds was attributed to heavy predation by large gulls, falcons, and pike. It will not necessarily affect the size of the population of Ross's Gull if the species is well enough adapted to life at sea or among the ice floes during the rest of the year.

From the late 1960s, Andreev had been interested in the biology of the Ptarmigan and Willow Grouse; an early paper of his in the *Zoologicheskiy Zhurnal* for 1971 described peculiarities of the ecology of the Ptarmigan on the Commander Islands (Komandorskiye ostrova; Map 14) (Il'ičev and Flint

1989: 394). Moving to Magadan, he worked out the winter energy requirements of the Black-billed Capercaillie in the northern taiga and studied its courtship, and in 1977 authored a paper on temperatures inside the winter snow-holes of the Hazel Grouse. The question that Andreev sought to answer, in the first place in his 1980 book, was: how do the 20-odd non-migratory bird species inhabiting the north Siberian taiga manage to survive the winter? Though this region is more sub-Arctic than Arctic, nevertheless winter temperatures of −60°C (−76°F) are encountered. In a fascinating article in English published in *Science in the U.S.S.R.* in 1986 Andreev described some of the highlights of his and his Magadan colleagues' 15 years of research into these problems. The tiny Siberian Tit survives winter life in the northern forests with only 4 or 5 hours of daylight to find the 7 or 8 grams of food, consisting of frozen insects, their larvae and eggs, and larch seeds that it needs for survival during the long cold nights. To save heat while roosting, it pulls in its legs and head, fluffs out its specially adapted plumage (over 100 barbs and barbules per cubic millimetre), and seeks refuge in a hole in a tree. This is fine at −45°C (−49°F), but what if the temperature drops to −60°C? Then, as the Norwegian ornithologist Svein Haftorn (1971) had already shown, the tit can lower its body temperature by 5 or 10 degrees Celsius and pass the night in semi-hibernation. The Russian workers were able to show, by placing a thermosensor under a sleeping bird's wing, that members of the genus *Corvus* could survive in the same way.

At their field station on the lower reaches of the Kolyma, more than 250 km (150 miles) north of the Arctic Circle, Andreev and his colleagues studied the winter life of the Willow Grouse. The population reached a peak in 1979–1980, then crashed. Andreev was able to explain why this happened (Andreyev 1986: 51–53; also see Andreev 1991):

> During the short winter day the bird manages to peck up between three and five thousand willow buds and twigs, after which it digs itself into the snow, where for 18 to 20 hours it slowly swallows and "burns" its wood "fuel". This is rough food, containing as it does about 30 percent cellulose, but all the *Tetraonidae* are well-adapted to digesting it. Their strong gizzard is full of gravel collected in autumn along rivers and streams, and their extended intestines have well-developed caeca. The gravel in the gizzard grinds the twigs, separating the bark from the pulp, and then the mixture is pushed through the small intestine, with up to 65 percent of it being utilized in the process. The rest of the stuff, which cannot be absorbed during the first stage because of the speed of digestion, ends up in the caeca where the digestive process is completed and the fats and a quantity of cellulose are utilized.
>
> All would be fine were not willow twigs, unlike other food of this kind, so poor in proteins (4–6 percent) and lipids (3–4 percent). This is why willow grouse begin to suffer from protein deficiency by the middle of winter when they lose weight rapidly, a fact that may cause their death in spring. Why does this happen?
>
> In autumn, when the day is still long and there is little snow, the birds easily find shoots rich in protein, but a little while later the weaker individuals have to collect whatever was left behind by stronger and more experienced birds. Most of the shoots are either eaten up or buried under snow in the middle of winter, the day is shorter, and there is hardly any time to look for anything; the store of gravel in the gizzard dwindles and it is hard to replenish it because of the snow. When spring

comes the shoots that had not been covered with snow become so thick that it is hard to peck anything off from them. Some birds even damage their bills as they try to remove the bark from the shoots. When there are no stones left in the gizzard digestion is 5–10 percent less efficient. Protein deficiency becomes increasingly pronounced. While in winter the birds lose weight owing to the depletion of the fat reserves, in spring the mass of their muscles reduces. The loss of weight can be reversed up to a point and is generally not dangerous, as the birds restore their form as soon as the first patches of snow-free ground can be found. But they fail to get strong in the years when their population is large, and especially if spring is late and the snow cover is thick.

The result is that the females lay no eggs and those that do are very late, and by the time the young are hatched the males lose all interest in them and take no part in raising them up. This is what the various predators make use of and the bird population is decimated.

Although still in his early forties, Andreev has already launched several research students on ornithological careers. Among them is Eugene R. Potapov, who began by working on the biology of tundra raptors while a student at Leningrad University and joined the staff of the Institute of Biological Problems of the North at Magadan in 1984. In 1990–1991, while researching for his final doctor's degree, he holds a research scholarship at the University of Oxford's Edward Grey Institute of Field Ornithology (Andreev, pers. comm.).

WRANGEL ISLAND AGAIN

Since the pioneering work of Portenko and others the avifauna of Wrangel Island (Map 16), a remote outpost consisting of 7000 km^2 (2700 square miles) of tundra and mountain which was never even thought of as part of Russia until the present century, has been very intensively studied. Partly because of the establishment of the Wrangel Island zapovednik, or nature reserve, in 1976, some of this research has been carried out by the reserve's own staff. The centre of interest, the main reason for the zapovednik, was the presence on the island of the only surviving sizeable colony of breeding Lesser Snow Geese in the Soviet Union. The man chiefly responsible for Russian Snow Geese studies is Evgeny (or Eugene) Viktorovich Syroechkovsky. Originally a member of the Magadan team, he began work on the Wrangel Island Snow Geese under Krechmar's supervision. Later he moved to the U.S.S.R. Academy of Science's Institute of Evolutionary Animal Morphology and Ecology in Moscow, and has since moved to the bird-ringing centre there. His series of publications on the ecology and behaviour of the Snow Goose goes back to the early 1970s, and in 1981 he published a summary of his 10 years of research on the structure of the Snow Goose colony on the island. He found that the size of the colony, measured both by the number of birds and by the area they covered, varied markedly from year to year, depending on the amount of snow cover or, otherwise put, the date of the spring thaw. In some years at the start of the breeding season almost the entire area used by the geese on the banks of the upper reaches of the Tundra River (Tundrovaya) and the slopes above was free of snow; in other years geese could only breed here and there. Because of this, the colony varied in size from 200 to 2600 hectares (500 to 6400 acres) and in number of breeding geese from 10 000 to 120 000. Syroechkovsky was able to divide it into several zones, based on a central area which was suitable for nesting geese virtually every year and which allowed room for 10 000–12 000 birds. Around it in a peripheral zone a further 24 000 geese could breed in 6 or 7 years out of 10. Outside this zone still further space, constituting a sort of reserve zone, became available in 5 or fewer years out of 10, when the snow-melt was exceptionally early. Productivity in the central nucleus of the colony was higher than in the outer areas and exceeded mortality, making up for the low productivity of the outlying areas, where it did not compensate for mortality. This subdivision of the colony into productive and reserve areas was thought to play an important role in stabilizing goose numbers in a regime of varying environmental conditions.

The history of the Wrangel Island Snow Geese was also described by Syroechkovsky (Bousefield and Syroechkovskiy 1985). Before the settlement of Ushakovskiy was founded in 1926, geese nested in nearly all the valleys in the mountainous central part of the island. In the years that followed, the gathering of large numbers of goose eggs annually not only by local inhabitants but also by scientific expeditions, and a rapidly increasing arctic fox population sustained by waste from the inhabitants' sea mammal exploitation, caused a rapid diminution in goose numbers. Breeding in the southern part of the island ceased altogether, and by the early 1940s only three or four large nesting grounds remained. No wonder Portenko in 1939, working out of Rodgers Bay on foot, never reached breeding Snow Geese in any number. Then, in 1948 and 1954 about 150 reindeer were introduced and soon increased in number; by 1978 the herd was 7000 strong. Further disastrous consequences for the Snow Geese followed: while the foxes increased still further in number through feeding on the waste from the annual autumn reindeer harvest, thousands of reindeer now migrated every spring through the upland valleys occupied by breeding geese, to reach their summer grazing in the Academy Tundra. They walked through the goose colonies trampling on nests and browsing on the eggs, which they ate in quantity. Already by 1960 the Snow Geese were reduced to the one substantial colony studied in the 1970s by Syroechkovsky in the upper Tundra River valley. Since the creation of the zapovednik the situation has been much improved. Human habitation of the island, research workers excepted, has been brought to an end, and the reindeer are being brought under control (Ovsyanikov 1990). Continuing research will in future ensure the maintenance of a viable population of Snow Geese on the island.

Wrangel Island not only has breeding Snow Geese, it is also a classic locality for breeding Snowy Owls, which prey on a vigorous population of two species of lemming, the collared and Siberian. Like other lemming populations, this one is subject to cycles of scarcity and abundance, and Litvin and Baranyuk (1989), following Portenko, worked on the island from the early 1980s and

Plate 46   The Snowy Owl and the lemming. Barrow, Alaska, June 1988.

used other data from 1969–1987 as well to study the relationship between the lemming and Snowy Owl populations. They were helped by the staff of the zapovednik. To establish lemming abundance, Litvin and Baranyuk did not have to trap lemmings systematically themselves; colleagues did that for them. Two of them, Chernyavsky and Tkachev from Magadan, published a book in 1982 on their Wrangel Island lemming studies called *Population cycles of lemmings in the Arctic*. In all, Litvin and Baranyuk collected data on 200 Snowy Owl's nests. The correlation between numbers of owls and lemmings was remarkable. In their 100-km^2 (39-square-mile) study area between 1969 and 1987, they found 15 nests in 1970, 11 in 1986 and 10 in 1981—all of them "good" lemming years. But in the "poor" lemming years of 1969, 1973, 1974, 1977 and 1984, no owls bred in their area, indeed in 1973, 1974 and 1977 no Snowy Owls could be found breeding anywhere on Wrangel Island. Litvin and Baranyuk were also able to confirm what others had already established, namely that the clutch size and productivity of the owls depended on the lemming population. The mean clutch size varied annually from 5·43 to 9·43 and full clutches varied in size from 3 to 14 eggs. I. V. Dorogoy, a member of the Magadan team, has also published studies on the relations between Snowy Owls and lemmings on Wrangel Island.

THE GEESE AND THE FOX

Naturally, Russian ornithologists have studied the well-known stratagem used by Arctic breeding geese, and other birds, of nesting in close proximity to avian predators in order to gain protection from their arch-enemy the Arctic fox. Syroechkovsky (1977) described a nice example of this from Wrangel Island. Besides the single large goose colony he studied intensively, other small colonies of Snow Geese were found scattered about the island in association with Snowy Owl's nests. Such colonies also included Brent Geese and common Eiders. In 1970–1971 Syroechkovsky found 14 of these colonies, each containing up to 50 nests, and geologists working on the island reported 4 more to him. Of these 18 colonies of anserine birds associated with Snowy Owls, 8 were of Snow Geese only, 2 were of Snow Geese and Eiders, 4 were of Snow and Brent Geese, and 4 were of Snow and Brent Geese and Eiders. These colonies did not usually survive from one year to another but two relatively large ones, each in a place where a Snowy Owl nested almost every year, were more or less permanent. In two cases, when a Snowy Owl pair shifted their nest site 200–300 metres from one year to the next, the geese and Eiders moved with them. In years when Snowy Owls did not breed on the island, these colonies disappeared altogether. The mutual benefits gained by the owls and

Plate 47 Arctic fox, ubiquitous predator of breeding birds on the tundra, photographed at Thule Air Base, northwest Greenland, where for a time the animals would beg for scraps outside the back door of the dining hall.

waterfowl in this breeding system were transparent: the geese warned the owl of an approaching fox, and the owl would not tolerate an Arctic fox within 200–250 metres of its nest. Somewhat harder to explain was the arrangement of anserine nests within the colonies: the Brent Geese nested nearest the owl, within 1–15 metres; the Snow Geese nested a little further away, and the Eiders furthest of all. Syroechkovsky thought the Brent Geese breeding on Wrangel Island were entirely dependent on the Snowy Owl and could only nest successfully when, in a good lemming year, the owls were abundant.

While he was exploring western Taimyr in the 1960s, A. V. Krechmar was able to investigate the anti-fox breeding stratagems of the Red-breasted Goose, as described in his joint paper with V. V. Leonovich which appeared in 1967 in the journal *Problems of the North (Problemy Severa)*. The two researchers explored some 500 km (300 miles) of mostly unexplored country along the River Pyasina and its tributaries in western Taimyr (Map 14). They found that the geese frequently nested near to a Peregrine falcon's nest. Indeed, out of 22 Peregrine's nests investigated, 19 were in close proximity to colonies of breeding Red-breasted Geese, and three other Peregrines nests had concentrations of non-breeding geese near them. The closest goose's nest was only 1·5 metres (5 feet) away from the falcon's nest. Krechmar and Leonovich also found Red-breasted Geese nesting among Herring Gulls and once they discovered two previous year's goose nests close to a Rough-legged Buzzard's nest. After the breeding season they frequently came across concentrations of moulting Red-breasted Geese near Peregrine's nests, and their conclusion was that the tendency of this goose to nest near a falcon's nest has evolved into one of the most characteristic traits of its breeding biology so that, at the present time, its survival depends on that of the Peregrine.

BIRDS OF THE YAMAL PENINSULA

A model avifauna of the birds of an Arctic region was published in Moscow in 1984. In the *Birds of Yamal*, Nikolay Nikolaevich Danilov and two colleagues, one of them the V. K. Ryabitsev mentioned at the beginning of this chapter, produced an annotated list, in substantial detail for the more interesting species, of all the peninsula's birds, in a nicely printed book of over 300 pages. It has a map, descriptions and photographs of the different habitats, and discussions of numerical changes in bird population, conservation measures and other problems. It is prefaced by a history of ornithological work in Yamal and a physical description of the area and was put together at Sverdlovsk in the Urals where Danilov and his colleagues are based. It will be splendid if comparable studies are published for other Arctic areas but it will not be easy to improve on the standards laid down by this one.

THE BETTER-KNOWN NORTH?

The reader of the foregoing pages may well have the impression that the birds of the Soviet Arctic, and especially of the far northeast, are relatively well known. Such a notion would not be altogether far from the truth. The authors of the *Birds of the USSR* (Il'ičev and Flint 1985: 175–189) included in their first volume a geographical assessment of the level of avifaunistic research in the different parts of the Soviet Union compiled by Yu. A. Isakov. He compared the different regions according to different criteria. On the criterion of modern publications on the avifauna, assessed on a scale of 1 to 4, 1 for the fewest and 4

Table 15 Some natural regions of the Soviet Arctic graded according to published knowledge of their birds (after Il'ičev and Flint 1985: 173–189).

Natural region	Points out of a possible 40
Kola Peninsula	26
Bol'shezemel'skaya tundra (Great Land Tundra)	19
European Arctic Islands	16
Yamal Peninsula	21
Gyda Peninsula	10
Taimyr Peninsula	19
Upper Olenek	0
Lower Lena Valley	16
Yana-Indigirka-Kolyma basin	23
New Siberian Islands	13
Chukchi Peninsula	18
Wrangel Island	26

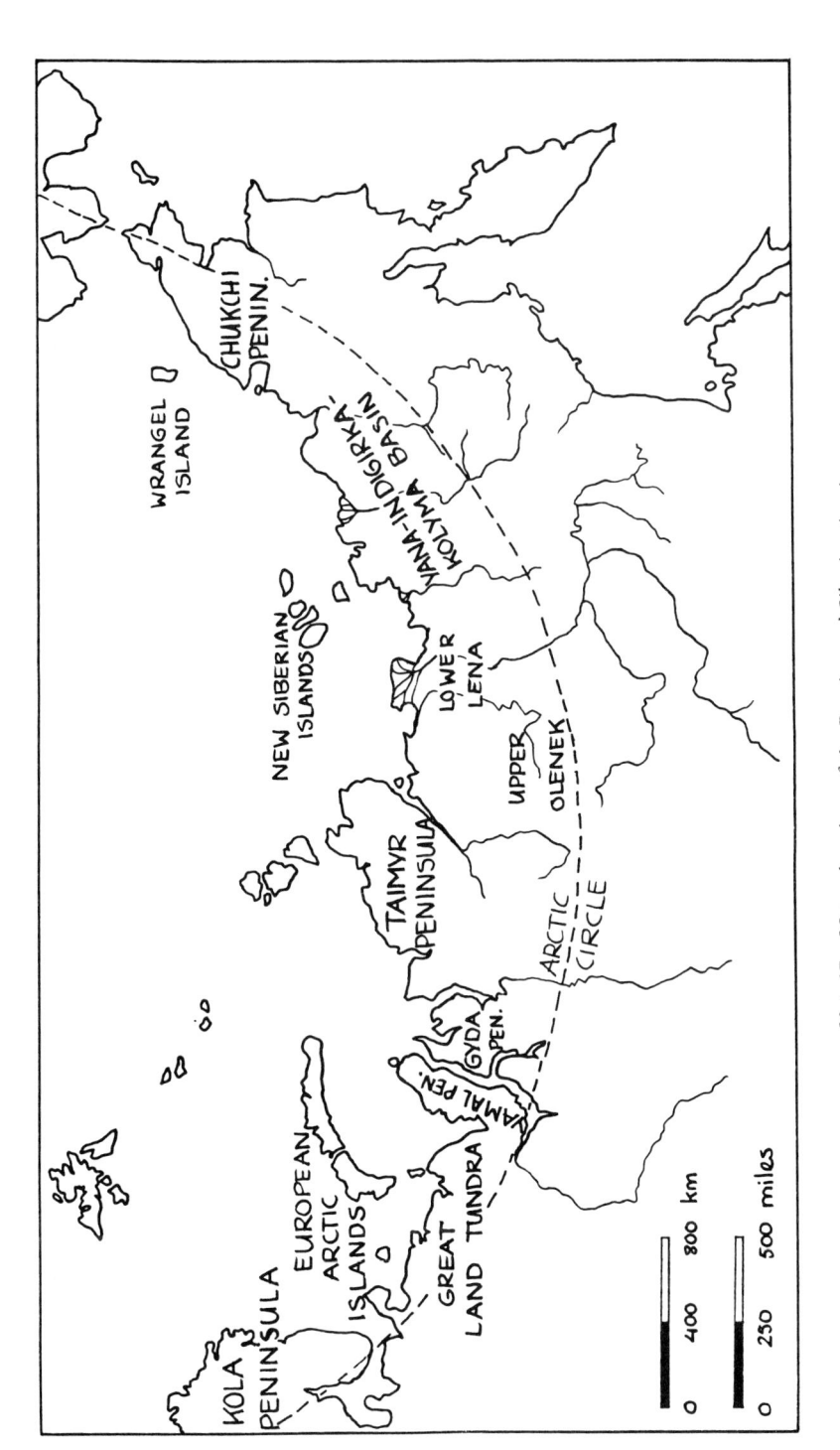

Map 17 Natural regions of the Russian and Siberian Arctic.

for the most, publications, Taimyr and the Yana-Indigirka-Kolyma basin scored a healthy 3. Even the Chukchi Peninsula scored 2, while large parts of the country, especially western Siberia, scored 1 only. On the criterion of publications about birds in general, most of the far northern areas scored 2 as against numerous 3s in European Russia and the south. Taking all criteria together, these maps show clearly that the ornithologically least-studied part of the true Arctic was the Gyda Peninsula (Gydanskiy poluostrov). Wrangel Island with its total of 26 points on all counts, ranked high—as high for instance as the southern Urals or the Kola Peninsula, and much higher than most other parts of the Soviet Arctic. Table 15 tells the tale; the upper Olenek area has been included because though barely Arctic it is north of the Arctic Circle. It is the only part of the Soviet Union scoring nothing. It goes without saying that, in the 10 years since Isakov made the assessment on which Table 15 is based, a spate of publications on Soviet Arctic birds, many of them mentioned in this chapter, have made Russia's Arctic avifauna much better known than it was then.

It looks as though Arctic ornithological studies in the Soviet Union are going from strength to strength. Pavel Tomkovich (pers. comm.) completed 3 years' research on the breeding biology of the Spoon-billed Sandpiper in the Chukchi Peninsula in 1988. Evgeny (or Eugene) Eugen'evich Syroechkovsky, uncle of E. V. Syroechkovsky, recently launched an international expedition designed to span several years, to the Taimyr Peninsula. In the first year, 1989, Polish and German ornithologists took part; in 1990 Dutch and French experts joined in and Tomkovich led a detachment of the expedition to the Lower Taimyr River (Nizhnyaya Taymyra) to begin a study of four of his beloved *Calidris* sandpipers: the Knot, Sanderling, Little Stint, and Curlew Sandpiper. Preparations for British participation in 1991 were being made at the time of writing this (December 1990). All in all, Arctic ornithology in Russia is in such a flourishing state at the moment that one might conclude that it is playing a dominant role there in ornithology as a whole.

CHAPTER 11

Ornithological research from Alaska eastwards to Spitsbergen

There need be nothing very special about ornithological research in the Arctic. The ranges of many Arctic birds extend southwards into the sub-Arctic or further and a species may be studied in any part of its range. For example the Eider, widespread and abundant throughout the Arctic, is also a common breeding bird in Scotland. The English biologist Peter M. Driver, who wished to study the display of the Eider, only went to the Belcher Islands in Hudson Bay, in the Canadian Arctic, because he could thereby obtain financial support from the Arctic Institute of North America and from McGill University, Montreal. Such finance had not been forthcoming in Britain (Driver 1974). On the other hand, the Arctic does pose special problems which tend to bring different research disciplines together, and this has promoted the establishment of special Arctic research institutes. The all-pervasive snow, ice and cold make special clothing and equipment for field work necessary whatever the nature of the research. The short summer, the continuous winter darkness, and the extremely low winter temperatures all pose problems to animals, birds

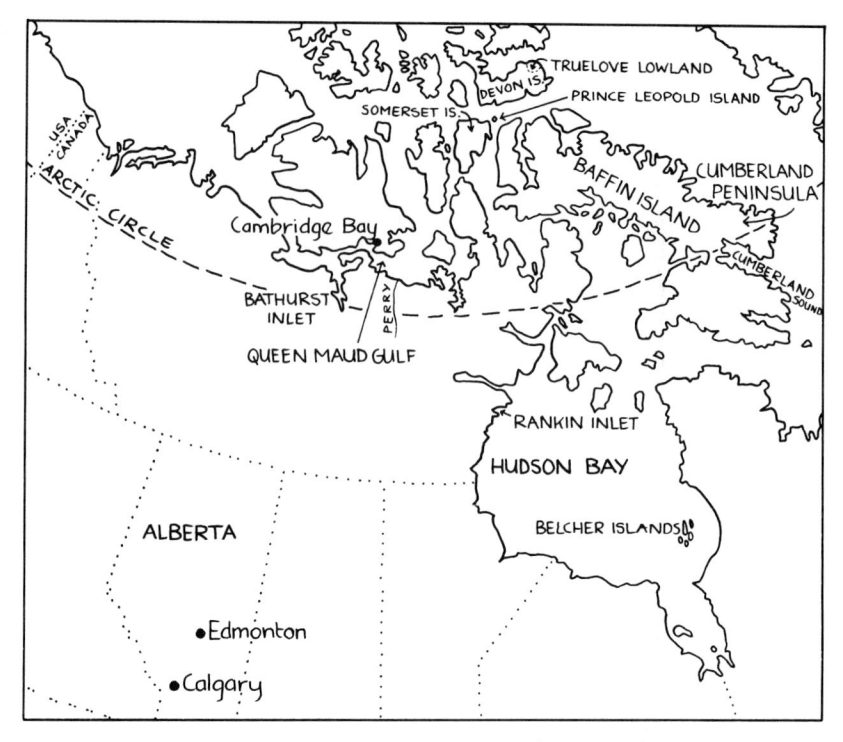

Map 18 Research in Canada.

and plants alike. Even for summer field work the ornithologist must needs join an expedition or organise one. Thus one can meaningfully consider Arctic ornithological research as something apart.

ARCTIC RESEARCH INSTITUTES

Limited space allows mention here of only a few of the research institutes that have actively promoted and organised ornithological research in the Arctic, in every case as part of a broad programme of biological and other scientific studies. Pride of place in the English-speaking world must surely go to the Arctic Institute of North America, since 1979 based at the University of Calgary in the Canadian province of Alberta, partly because its journal *Arctic* has published so many important papers on Arctic ornithology in the last 25 years. For example, the March 1988 number carried studies on the breeding biology of the Peregrine Falcon and Gyrfalcon in the Canadian Arctic. It has also funded ornithological research projects and since 1960 has maintained a summer field station on the north coast of Devon Island that is available to independent researchers. Here an area of lakes and meadows called the Truelove Lowland was the scene of intensive ornithological investigation in the 1970s and 1980s when a complete census of birds was made in mid-July every year for nine successive years (Pattie 1990).

Another institute which has consistently encouraged ornithological research in the Arctic is the Oslo-based Norwegian Polar Institute (Norsk Polarinstitutt) which is an agency of the government's Department of the Environment. It was responsible for publishing Løvenskiold's book on Spitsbergen birds and for enabling him to undertake several seasons' field work in the archipelago. In spite of the thoroughness of Løvenskiold's book, which came out in 1964, much new information about Spitsbergen's birds has been obtained and published since then by the Institute, which has had a full-time biologist on its staff since 1967, and has continued to organise or support annual ornithological expeditions to Spitsbergen. In 1983 it hosted an international symposium on "current research on Arctic geese" in Oslo and printed the 18 papers presented in its series of publications (Mehlum and Ogilvie 1984). Some of this goose research will be described later in the chapter.

While research in the North American Arctic is catered for by the Arctic Institute of North America and that in the Spitsbergen Archipelago by the Norsk Polarinstitutt, work on the natural sciences in Greenland has been the speciality since its foundation in 1878 of the Commission for Scientific Research in Greenland. Its publication *Meddelelser om Grønland* is the oldest serial publication of scientific research in the Arctic. After appearing since its inception in 1879 as a single series for a hundred years, it was split in 1979 into three separate series: *Bioscience, Geoscience*, and *Man and Society*. Although the Commission was originally called the Commission for Geological and Geographical Research in Greenland, it published numerous papers on botany and ornithology in its *Meddelelser*, including A. L. V. Manniche's important work on "The terrestrial mammals and birds of northeast Greenland", which appeared in 1910. Since the Second World War it has published reports on bird ringing in Greenland, information on new or rare species recorded in Greenland, and papers on ornithological research in Greenland, some of which will come under notice later in this chapter. The Commission for Scientific Research in Greenland, which used to form part of the Danish government's Ministry for Greenland, was transferred on 1 January 1989 to the newly-established Danish Polar Centre in Copenhagen.

The Arctic Centre of the University of Groningen in the Netherlands is relatively newly-established and comparatively modest. Founded in 1969 by a group of professors, it has no government support although it is the only Arctic institute in the country. In the first 10 years of its existence its main research project was an archaeological and historical investigation of seventeenth-century Dutch whaling in Spitsbergen under the direction of Louwrens Hacquebord. Since then it has taken ornithology under its wing. One of its periodical international symposia, held in October 1987, was devoted to Polar seabirds and included papers on seabird research in Spitsbergen and Canada. It has published some papers on Arctic birds in its *Circumpolar Journal*, and, as will be described later, its senior research assistant, Ko de

Korte, has worked and published on the breeding biology of the Long-tailed Skua in Greenland.

An Arctic research institute of a quite different kind was the United States Naval Arctic Research Laboratory (known as NARL) at Point Barrow, Alaska. Situated on the coast 3 miles (5 km) northeast of the village of Barrow, at the northern most tip of United States territory, it was founded by the then newly-created Office of Naval Research in 1947 (Map 31). The site, at about 71° N, was already a busy one, for since 1944 it had been the main supply base for the exploration for oil in the United States Naval Petroleum Reserve No. 4. This had been established in 1923 and covered some 35 000 square miles (90 650 km^2) of northern Alaska (Map 39). The Naval Arctic Research Laboratory was well placed for ornithological studies because of the proximity of the tundra with its breeding birds and of the coast with its migrant gulls and wildfowl. Its first scientific director was Laurence Irving, a professor at the Edward Martin Biological Laboratory, Swarthmore College, Pennsylvania, whose later study of Eskimo bird names in Arctic Alaska was described in Chapter 2. He was the author of an interesting book published in 1972 entitled *Arctic life of birds and mammals including man*, and thanks to his efforts research at Point Barrow continued to include important programmes in biology and ornithology. His own initial research programme consisted of measuring the metabolic activities of various birds and animals in relation to Arctic temperatures; it will find a place later in this chapter. Ornithological research in the early years included the investigation of the birds of the Colville River basin, inland from Barrow (Map 20), begun in the summer of 1952, by which time researchers from 13 different universities were busy at the Laboratory. This institution was not run directly by the United States Office of Naval Research, but its administration was contracted out successively to various universities. Its research role was, primarily, to provide accommodation, transport and laboratory facilities for the different teams working in the field. Planning, advice and supervision were provided by the Arctic Institute of North America, already mentioned.

It would require at least an entire chapter of this book to review all the ornithological research carried out at the Point Barrow Naval Arctic Research Laboratory over the years by individual ornithologists. But, by way of example, mention may be made here of British ornithologist Mike Densley's work on the fall migration of Ross's Gulls in Alaska. In August and September 1975 he stayed at the laboratory while watching Ross's Gulls offshore and was able to complement his field observations by examining the series of Ross's Gull skins at the Laboratory, many of which had been collected along the coast early in October 1958 (Densley 1979).

HUXLEY'S RED-THROATED DIVERS

One of the earliest pieces of ornithological research in the Arctic still frequently referred to today is Julian Huxley's study of the courtship of the Red-throated Diver. It was carried out in 1921 during the first of three expeditions to Spitsbergen organised by a group of members of Oxford University, most of them ornithologists. Indeed, the 1921 Oxford University Spitsbergen Expedition, to give it its official title, could boast an unsurpassed galaxy of ornithological talent. It was led by one of the great ornithologists of the day, F. C. R. Jourdain, and among the 12 other members were several who subsequently became well known, namely the Scottish bird photographer and writer Seton Gordon, who wrote a book about the expedition (1922) as well as others about birds in Scotland; the founder of the present-day science of ecology, Charles Elton, later director of the Bureau of Animal Population at Oxford; and Julian Huxley (Binney 1925; 257–258; Hayes 1934: 202–218), whose brilliant public career included the holding of chairs of zoology in both Britain and the United States of America. He was also at different times secretary of the Zoological Society of London and general secretary of UNESCO (United Nations Educational Scientific and Cultural Organisation). The primary purpose of the expedition was ornithological, and at least nine papers on Spitsbergen birds by expedition members appeared in scientific periodicals in the ensuing years and were reprinted in Vol. 1 of *Spitsbergen papers* by the Oxford University Press in 1925. One of them was a major review of the Spitsbergen avifauna by Jourdain, which appeared in the *Ibis* for 1922. Jourdain also published a paper in the *Auk* on the breeding habits of the Barnacle Goose in Spitsbergen, and his colleague on the expedition, A. H. Paget Wilkes (1922), wrote in a more light-hearted vein "On the nesting of the Barnacle Goose in Spitsbergen" for another American bird periodical, the *Journal of the Museum of Comparative Oology, Santa Barbara, California*. He described how he obtained two clutches of Barnacle Goose's eggs in the precipitous valley inland from Longyearbyen called Longyeardalen, where the species has not bred since about 1930:

> As soon as we arrived, Messrs. Jourdain, Gordon and myself started out. After walking along the flat for a certain distance we found ourselves in a long valley with steep mountain-sides on which the snow still lay about in great patches. A snow-stream ran down the centre and filled the lower level of the valley. The slopes on either side consisted of big boulders which had come down from the rocky bastions and towering cliffs above us, the bastions forming a ridge along the valley. It was here that we looked with care and eagerness. Walking on near the stream, we were soon delighted to see a pair of Barnacle Geese fly out from the mountain-side. They flew round us somewhat excitedly and finally pitched into some rocks about half a mile further on. Though we could not see the exact spot, this gave us a valuable clue . . . I scrambled up the screes and boulder-strewn side of the valley, and on drawing nearer, saw the goose quite distinctly, sitting on its nest, with the gander standing on guard. After a stiff bit of climbing I found myself within 50 yards of the pair and stopped a moment to notice the beauty of the plumage and characteristic actions. The steel-grey and black feathers of this handsome brant [*sic*] were truly remarkable, at close quarters, standing out against the dull-coloured rock; but perhaps its most distinctive point is the white cheek-patch which gives it the name '*leucopsis*'. Another steep 'pitch' of rocks and I was on the bastion. In a second or two the gander was off, flying away with his deep double-syllabled cry and the remarkable creaking of wings characteristic of the geese.

The next moment the goose, which had been sitting upright on her eggs, stretched out her neck to its fullest extent on the ground and lay motionless. I stood some twenty paces from her. Then, as I advanced, she got up and flew off. Both birds circled round, with creaking wings and deep cries, and finally settled on a neighboring bastion some hundred yards away. The nest . . . consisted of a hollow in the soil liberally lined with light-coloured down and containing 5 eggs. A large amount of hard, dried excrement was mixed with the down and the hollow had evidently been used for several seasons, since the remains of old nests were seen below the top layer of down. The eggs are small for the size of the bird, and in colour and texture do not vary much from those of the Brent Goose (*Branta bernicla*). . . .

After carefully packing up eggs and down, I was about to descend into the valley again when I saw two geese on the next bastion, about 100 yards away. The pair whose nest was at my feet were on the wing. I scrambled round to the next bastion as quickly as possible and on the way both birds got up. Climbing out on to the bastion I saw another nest with four eggs in a very similar position to the first, out on the point of the rock, on a space some 15 square feet in area, with a sheer drop on three sides. Having no other receptacle I was forced to pack this second clutch in my pocket, with the down, and so began an anxious journey into the valley and back to the sloop.

Before he joined the 1921 Oxford University Expedition to Spitsbergen, Huxley had already published important papers on the sexual behaviour of two species of grebe, the Great Crested Grebe and the Dabchick or Little Grebe. His study of the Red-throated Diver was a logical continuation of the earlier ones. The bird breeds all round the Arctic and in all suitable parts of Spitsbergen. Huxley and his companions camped for 11 days starting on 30 June on Prince Charles Foreland (so named in 1612 after the son of King James VI of Scotland, but now Norwegianised to Prins Karls Forland), off the west coast, where a large tract of low marshy country dotted with small lakes provided an ideal habitat. The party, which consisted, besides Huxley himself, of Charles Elton, botanist V. S. Summerhayes, and R. W. Segnit, a geologist, camped on the shore of Richard Lagoon (Barentsfjorden), a 5-mile long sheet of brackish water joined to the sea through one narrow opening, at the northern tip of the Foreland. This was an ideal place for studying divers, which bred on shallow pools in the tundra, fed on the open sea and socialised and courted on the lagoon. On one occasion eight were seen together there.

Huxley began his paper by describing the main courtship actions of the Red-throated Diver and gave his interpretations of their significance. In the "Plesiosaur race" two or three birds swam fast through the water close together with their necks extended forwards rather stiffly and heads and beaks somewhat down-pointing, so that they resembled miniature Plesiosaurs. They would swim in one direction, then turn and swim in the opposite direction, after apparently changing the leader. Sometimes the wings were stretched out at right angles to the body. A special note which Huxley called a "rolling growl" always accompanied this ceremony. Under the influence of sexual excitement, as Huxley surmised, the divers sometimes dipped their heads and beaks into the water as if looking into it, shook their heads, or dived with a splash instead of their usual quiet submergence. On a few occasions one bird was seen to dive with a splash and at once the other bird spread its wings and

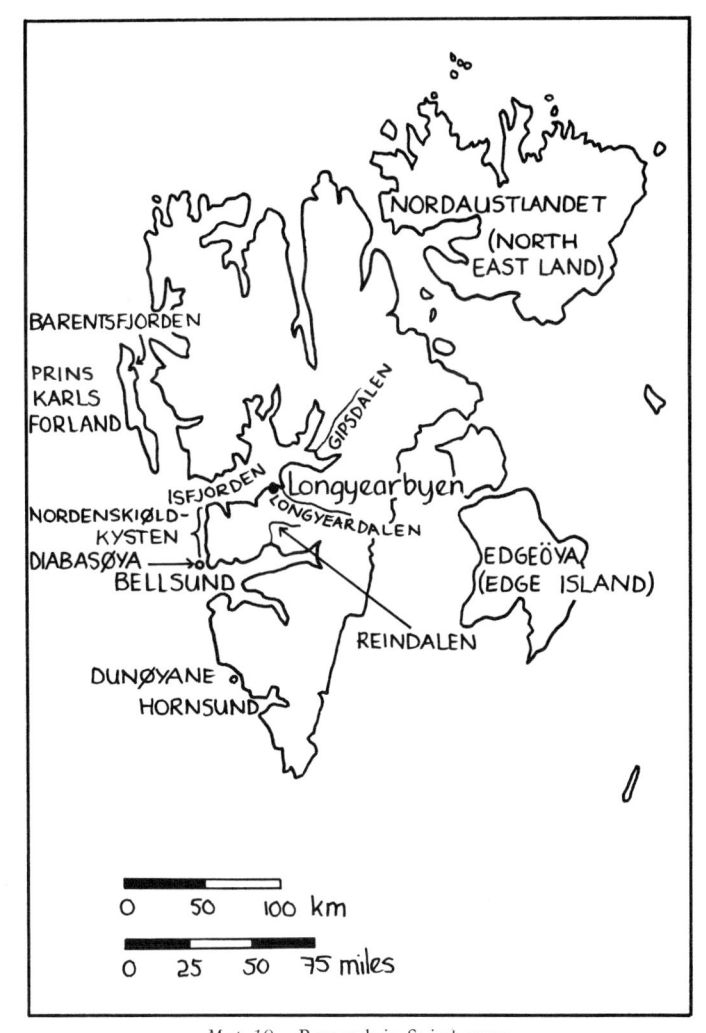

Map 19 Research in Spitsbergen.

flapped along the surface for a few yards. Then the bird that had dived came up out of the water vertically, the neck stretched up but the bill down-pointing, and remained in that position for several seconds. Again, in what Huxley termed the "snake ceremony", two mated birds would swim one behind the other "with their necks arched so that the tip of the bill is submerged".

To watch his divers more closely, Huxley set up a hide on the edge of a small lake a mile from camp where a pair was established. The next day he spent 9 hours in the hide, keeping up a continuous watch on the Red-throated Diver pair. His paper contains detailed descriptions of the "activities of a particular

246 *In search of Arctic birds*

(a)

mated pair" based on his notebooks written in the hide. Here is a typical extract (Huxley 1923: 258–260):

> Shortly after, the male arrived. The female speedily joined him, and the pair swam towards me, the female leading, with bill slightly open. When close to the bank, the female suddenly performed a remarkable action, obviously of a sexually stimulating nature. She stretched her neck forwards at an angle of perhaps 30° with the horizontal, the head and beak also pointing up but at a less angle. The neck itself was straight and rigid. In this pose she swam close up to the male, crossing his bows, so to speak. Whatever its significance, the performance was exciting to watch. There was a tenseness about the bird's attitude, a rigidity, which has been commented on by other writers, notably by E. Selous, in connection with various sexual ceremonies in other birds. I can best describe the impression it made on me by saying that it was like that apparently produced by certain sexual dances of savage tribes – the whole thing fraught with the significance of sexual emotion, and mysterious in the sense of being thus emotionally charged far beyond the level of

(b)

Plate 48 (a) Avanersuaq's coastal mountains form a backdrop to this incubating Red-throated Diver nesting in a shallow lake in Steensby Land, 50 miles southwest of Qaanaaq. (b) The Red-throated Diver sits somewhat awkwardly on its nest, which is a heap of moss and mud in shallow water presumably built up by the bird.

ordinary life, but completely natural and without restraint. It is fairly clear that, even in animals, the emotional tension during sexual excitement is far higher than at almost any other time, and that the impression given to the observer is, therefore, not wholly a subjective one.

On this occasion the male was absolutely unresponsive to the female's "stimulating" action. The pair went off together; after some time the female started to go by a long route across dry land to the nest, but soon gave it up. The pair then swam, the female again leading, to a little bar covered with moss on the far shore. This bar I afterwards examined, and found that on it there was a rudimentary nest, built almost entirely of the moss which was here abundant, both on the bar and under the surface of the shallow water near by. Another such "nest" was found on the shore of a second tarn on which a second pair was breeding. Very similar structures are built by the [Great] Crested Grebe; and in both cases their function appears to be the same—they are the places on (or near) which the act of coition occurs. Just before arriving here both birds simultaneously "looked into the water" . . . for about 20 seconds; there was also a good deal of beak-dipping by both male and female. On arriving at the bar the female turned and gave an energetic splash-dive. The male responded by a less energetic splash-dive. The male then crawled out on the mossy bar, onto what was later found to be the pairing-nest; there he stood upright, stamped alternately with his two feet several times, and then sank down as if brooding. He plucked small fragments of moss in his beak, and apparently placed them around himself as if adding to the nest. The female meanwhile was swimming close to the bar, in water scarcely deeper than her draught, giving repeated splash-dives. Her tail (which of course was very short, as in all Divers and Grebes) was repeatedly waggled up and down. The association with this motion with copulation in birds is obvious and well known. However, she did not land, but swam across to the right and back, giving several more splash-dives; the male got up and scrambled back into the water. The pair then went to the nearest point to the true nest, both diving twice *en route* (not splash-dives); the female led, at a fast pace. After a short abortive excursion overland towards the nest, followed by swimming off the shore, she went right up to the nest and settled down on it.

It would take us away from the Red-throated Diver and the Arctic to analyse the rest of Huxley's paper, which is in fact the greater part of it, comprising his fascinating "general discussion" of the significance of courtship ceremonies in the divers and grebes and other birds studied by him. It is this in particular which marks out his Arctic ornithological research from earlier work. There is meticulous field observation and vivid descriptive writing in the publications of such earlier ornithologists as E. W. Nelson, but the analysis, the theoretical discussion, provided by Huxley is new and exciting.

TINBERGEN ON SNOW BUNTINGS IN GREENLAND

Julian Huxley was not the only natural scientist in the years between the two World Wars whose research early in life took him to the Arctic, but who subsequently became famous in other fields. E. Max Nicholson, co-founder of the British Trust for Ornithology in 1933 and director-general of the Nature Conservancy in the 1950s, was another. For, during the Oxford University Greenland Expedition of 1928, he organised a "complete count of the bird population of 8.3 square miles" of tundra inland from Godthåb (Nuuk) and later published a paper in the *Ibis* on the birds seen in the area (Nicholson 1930; and 1931: 93). Nicholson's Arctic experiences seem to have helped persuade the young Dutch zoologist Nikolaas Tinbergen (Manning 1990), then a graduate student at Leiden University, to undertake ornithological research in east Greenland in 1932–1933. Thereafter Tinbergen, universally known as Niko, became famous in other fields, like Huxley and Nicholson never returning to work in the Arctic. Co-founder with Konrad Lorenz of modern ethology, the scientific study of behaviour, among his best-known books were *The study of instinct* (1951) and *The Herring Gull's world* (1953). He was Professor of Animal Behaviour at Oxford University in 1966–1974, having moved to Oxford in 1949 as a lecturer. Four years after his brother Jan, the economist, won a Nobel Prize, Niko gained one in medicine for his work on autistic children. He died at Oxford on 21 December 1988.

In 1931 Niko Tinbergen (1934, 1958) was offered the opportunity of joining a small Dutch expedition which, as part of the activities of the International Polar Year 1932–1933, was to carry out meteorological observations at Ammassalik in east Greenland. He would be attached to the group but would undertake his own programme of zoological research. To do this he had first to resolve two problems that might have stood in the way. First, he was in the middle of writing a doctor's thesis on the bee-killing digger wasp *Philanthus*. He persuaded the university authorities at Leiden to allow him to put the work on ice for a year. Second, he was about to be married to Elizabeth (Lies) Rutten. He persuaded her to defer the honeymoon and go with him to Greenland, and managed to arrange for this to be possible. Food and lodging were provided by the Dutch Polar Year Expedition; finance for the research was made available by several Dutch foundations, one of them belonging to his own university of Leiden.

Tinbergen had planned in advance to use most of his time in east Greenland to make a detailed study of the commonest Arctic passerine, the Snow Bunting. He had also agreed to obtain as complete a collection of Eskimo artefacts as possible for a museum in The Hague by bartering sugar, tobacco and other things brought for the purpose, and to make botanical and zoological collections as well. But the Snow Buntings had the first priority and, in February 1933, well before the return of these summer visitors, the Tinbergens had installed themselves at Kuummiut, 25 miles up Ammassalik Fjord from the outer coast (Map 21). They had selected this place in the previous autumn as a good Snow Bunting spot.

Kuummiut was just south of the Arctic Circle and the darkness of the Polar winter was tempered, even at midwinter, by at least a couple of hours at midday when there was enough light to read by indoors. But birds were few and far between. After November, when the last migrating Gyrfalcons were seen, the only land birds were Ptarmigans and Ravens. Tinbergen whiled away the time studying the Eskimos' dogs with a zoologist's eye. Much influenced by Eliot Howard's studies of territory in birds, he soon hit on the interesting fact that the huskies held group territories. The more than 100 huskies in Kuummiut were divided into teams or packs of 6–10 dogs each and each team defended a territory around their owner's house where their food was put out for them. Strangers were driven off by the entire pack in concert. The individual dogs did not hold territories but often quarrelled among themselves over social rank, and each pack had its leader. The first Snow Buntings arrived at Kuummiut on 22 March and for the next few weeks they and other summer visitors, Wheatears, Lapland Buntings, and Purple Sandpipers, were concentrated on the only snow-free area around, namely the village. Here, with the help of local Eskimo boys, the Tinbergens trapped and ringed these migrants with clap-nets.

From their first arrival until the end of the summer the Snow Buntings breeding near Kuummiut were under almost daily observation, and Tinbergen was able to put together a detailed and comprehensive account of their breeding biology, which was published in 1939 in the *Transactions of the Linnaean Society of New York* under the title "The behaviour of the Snow Bunting in

Plate 49 Snow Bunting pair in northwest Greenland. The Snow Bunting is the sparrow of the Arctic, being the most abundant passerine bird and one which quickly colonises human settlements.

spring". The males arrived first, some still not in full summer plumage, and fed in flocks on fallen grass seeds in Kuummiut or in house ruins, using the *pee* call and the trill of wintering Snow Bunting flocks. A month after their first arrival, birds would begin to leave these flocks in the early morning, perch on a rock in the open country, and sing for the first time. At first the flocks would be formed again later in the day, but soon the males, now dispersed over the snow-covered tundra and hillsides, would hold territory by singing from certain boulders in a certain area. Intruding males were driven off and territorial fights between rival males were often seen. Eventually, some weeks after the males, the female Snow Buntings arrived, at first in flocks. But they quickly dispersed and each would sooner or later attach itself to a male. At first the male would treat the female with hostility, but one that insisted on remaining would soon be accepted. If the male attempted coition too soon, the female would fly off, and a sexual chase would ensue. Soon the two birds would forage together and inspect possible nest holes, while the female would defend the territory against other females. Tinbergen's account of the Snow Bunting's nest-building (by the female, the male accompanying her only), egg-laying and incubation, and the rearing of the young, was pieced together from many hours' continuous observation in the field. He showed convincingly, contrary to what was believed by some at the time, that the Snow Bunting was just as territorial as any other bunting, though birds would range far and wide outside their territory when feeding young, so that good insect places would be visited by birds from different territories. The theoretical part of the paper, which followed this admirable description of the Snow Bunting's breeding habits, included discussions of how the sexes recognise one another, of the function of fighting and of territory, and of the significance of song. In this Snow Bunting paper Tinbergen opened up several lines of enquiry which were subsequently developed by himself and others. For example, he described how male Snow Buntings engaged in combat would make pecking movements, as if feeding, during lulls in fighting, and this was the start of a series of studies of what later came to be known as displacement activities.

In early July the Tinbergens took a brief break from watching Snow Buntings and paddled their canoe out to an offshore island to collect a few hundred Arctic Tern's eggs for their larder. While there, they came across some Red-necked Phalaropes and Tinbergen at once began studying these birds, somehow fitting them in between spells of Snow Bunting watching. An important paper resulted, published in the journal of the Dutch Ornithologists' Union *Ardea* in 1935, on "The behaviour of the Red-necked Phalarope in spring".

THE PTARMIGAN'S DIET

Paul Gelting, author of *Studies on the food of the east Greenland Ptarmigan*, which was published in Copenhagen in 1937 in *Meddelelser om Grønland*, was, naturally enough, not an ornithologist but a botanist, for the adult Ptarmigan feeds exclusively on plants, and only a botanist could study its diet effectively. When Gelting set out for Greenland he had no plans to study Ptarmigans; he was intent on botanical work. In the summer of 1931 the Danish explorer and geologist Lauge Koch took his fifth expedition to Greenland. This was the Three-Year Danish East Greenland Expedition, 1931–1934, said in 1934 to have been "the greatest scientific expedition that has ever wintered in either the Arctic or Antarctic". It was credited with "3 ships, 12 motorboats, 2 seaplanes, a total personnel of 68, which included 36 scientists and students; 4 wintering stations were maintained complete with wireless" (Hayes 1934: 191). Among this army of researchers, Gelting was one of the expedition's botanists, and he later contributed importantly to knowledge of the Greenland flora with his *Studies on the vascular plants of east Greenland between Franz Joseph Fjord and Dove Bay*, which was published in 1934, also in the *Meddelelser om Grønland*.

The possibilities of identifying the plant remains found in the crops of freshly-killed Ptarmigans dawned on Gelting on the autumn of 1931 and he had soon mapped out a research programme which was eventually, in 1937, to lead to the successful public defence of his thesis and the acquisition of his doctorate in the Faculty of Mathematics and Natural Science of the University of Copenhagen. One of the aims of his botanical work had been to study the

(a)

(b)

relationship between vegetation and snow cover, and Gelting soon found that it was snow cover that determined the Ptarmigan's access to food. For, unlike the lemming, which lives in burrows under the snow and can find its plant food there throughout the winter, the Ptarmigan, though it does habitually dig into the snow to feed, cannot survive through the winter in this way. As darkness increases late in the autumn, and snowdrifts, hardened by the wind, become deeper, digging in the snow to find snow-covered vegetation becomes uneconomical and the Ptarmigan is forced to move to snow-free areas.

Soon after his arrival in Greenland, Gelting set about the task of measuring snow cover along the shores of the outer fjords of the east Greenland coast in the neighbourhood of his base camp at Eskimonæs on Clavering Island (Map 22). Fifty bamboo canes were stuck into the ground, 5 in each of 10 different plant communities. To measure the thickness of snow cover, notches were cut on the sticks at 50-cm (20-inch) intervals and the sticks were examined on various dates through the winter. The very strong winds that occurred throughout the winter tended to keep some raised or exposed areas more or less snow-free while thick drifts accumulated in the hollows, and Gelting was able to classify plants as growing either in snow patches or in snow-free areas. In particular, a dwarf Arctic willow *Salix arctica* and plants associated with it, was found to grow in areas covered by up to a metre (3 feet) of snow by the end of the winter, whereas purple saxifrage and mountain avens remained virtually snow free.

As the long winter nights set in, Gelting found the analysis of the contents of Ptarmigan's crops an excellent indoor occupation. The birds were shot by himself and his colleagues on the expedition and in the end he had 78 crops for study. These were dried out for a few days or longer before being opened, and the contents were then emptied out onto a sheet of white paper for sorting and identification. The individual plant fragments were picked up with tweezers and sorted into separate piles, and these were then stored in match boxes and tobacco tins. Among the photographs illustrating Gelting's thesis is one showing the entire contents of the crop of a young Ptarmigan arranged on a sheet of paper. About 5000 of the tiny leaves of the mountain avens are spread out in a rectangle. Below them are much smaller heaps, neatly spread out, of 13 fragments of twigs of Arctic bell heather, two leaves of bilberry and 74 buds of purple saxifrage. When they had been fully dried, the materials from the crops were carefully weighed so that the author could make a detailed quantitative study of the Ptarmigan's diet.

Altogether, remains of 52 out of a possible total of 163 plant species growing in the area were found in the Ptarmigan's crops. Of these, only four were

---

Plate 50 (a) A rocky knoll on the shore of a temporary lake formed by meltwater was the favourite perch of a male Ptarmigan when my wife and I camped here in the last week of June 1986, within a few miles of Cambridge Bay.

(b) Apart from a few dark feathers on the nape, this male Ptarmigan, which seemed to use the boulder as a look-out post for hours at a time, is still in its white winter plumage.

254 *In search of Arctic birds*

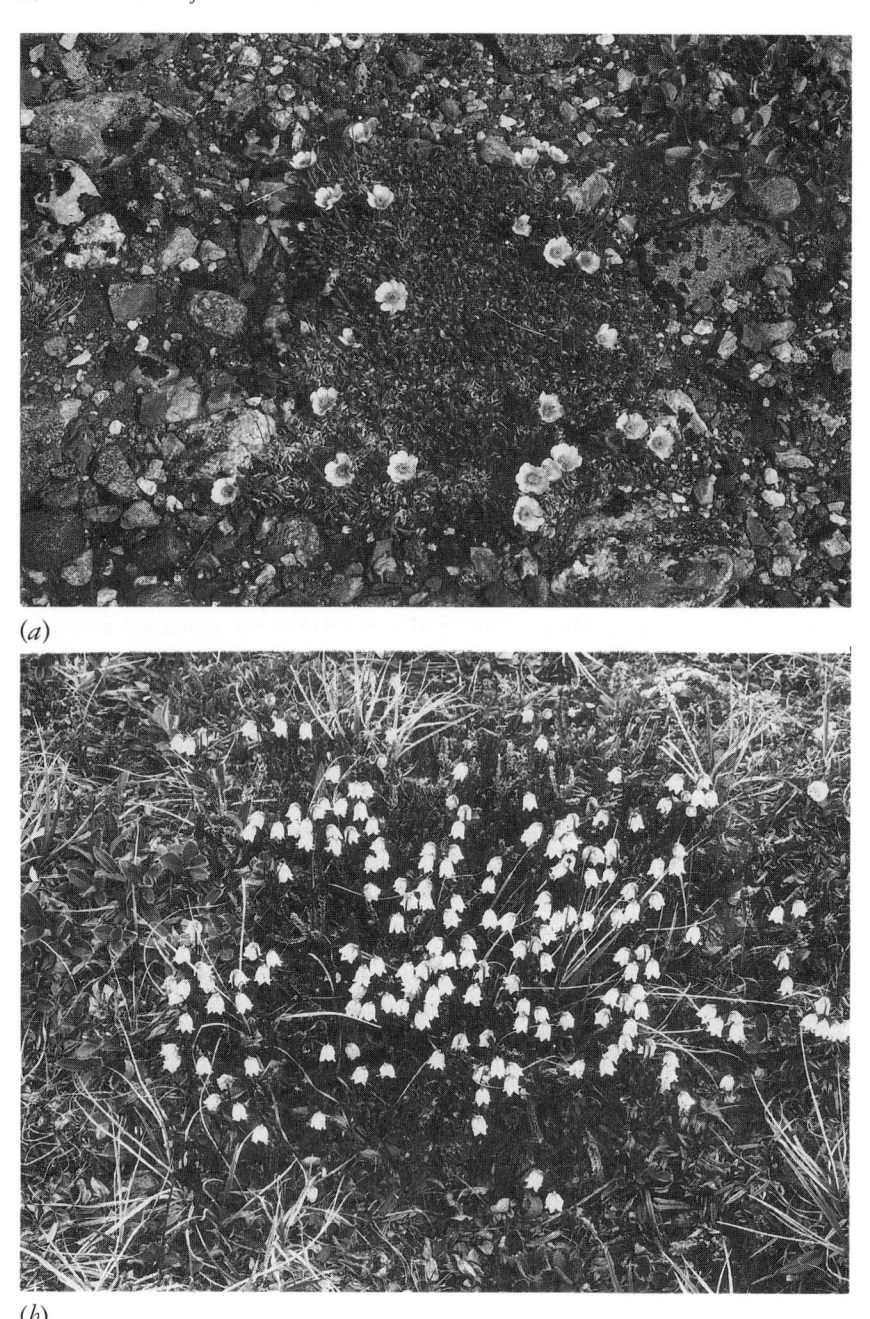

Plate 51 (*a*) Food plants of the Ptarmigan: Mountain Avens. (*b*) Food plants of the Ptarmigan: Arctic Bell Heather.

quantitatively important, namely Arctic willow, mountain avens, purple saxifrage and the alpine knotgrass or bistort *Polygonum viviparum*, all the rest put together amounting to less than 10% of the whole. The Ptarmigan was found to be choosy about which parts of the plant it ate; it went for buds and twigs of Arctic willow, leaves of mountain avens, shoots of purple saxifrage, and bulbils of alpine knotgrass. Analysis of the crop contents revealed very striking seasonal variations in the Ptarmigan's diet. It lived on Arctic willow, mountain avens and purple saxifrage in winter and early spring, and in the summer switched to a diet of alpine knotgrass. What the Ptarmigans ate between late May and early July Gelting did not discover, for the very good reason that, at the expedition's headquarters at Eskimonæs, Ptarmigan shooting was banned during the bird's breeding season, so no crops were available for that period. The probability was that until the alpine knotgrass emerged above ground in June and its much-favoured bulbous roots became available, the Ptarmigan's diet went through a more varied spell in May–June when a large variety of buds and shoots was probably taken.

Gelting's Ptarmigans migrated annually between the inner fjords and the outer coast. At the end of the summer, as soon as the snow covered the knotgrass and Arctic willow, they left the inner fjords and made for the open plains and flat plateaux of the coastal headlands and islands. Here, for a time, arctic willow would still be available; after the end of the year, when the willow was covered with snow, the Ptarmigans fed in bare wind-swept areas where mountain avens and purple saxifrage were available. As soon as the spring snowmelt allowed, they reverted to Arctic willow. What made things most difficult for the Ptarmigan in winter was not the thickness of the snow cover, which reached its maximum in March–April. This was even then such that windswept areas where purple saxifrage grew could be found. Rather it was the darkness of the Arctic night, peaking in January, which made it hardest for the Ptarmigan to find food. For whatever reason, it does not normally feed at night-time, and Gelting was able to show, theoretically, that by leaving the sheltered fjord shores and valleys and moving to the high plateaux of the outer coast at a height of 400 m (1312 feet), the Ptarmigan was able to avoid altogether the 24 darkest days of the year. Even so, he found that it had evolved special winter feeding habits. It gobbles its food down faster than in summer in order to fill its crop as quickly as possible. It fills its crop fuller in winter, so that a visible bulge appears in the bird's breast. And feeding Ptarmigans in winter are so busy nibbling vegetation that they can be approached much more closely than in summer.

These fascinating studies of the Ptarmigan's feeding behaviour round the year were supplemented by an investigation of the food value of the Ptarmigan's food plants. It turned out that the alpine knotgrass it ate in summer was rich in proteins while the Arctic willow and other constituents of its winter diet were rich in fats. Thus providentially for the Ptarmigan, it obtained body-building proteins when they were needed in and after the breeding season, and fats with a high calorific value to keep it warm in winter.

256 *In search of Arctic birds*

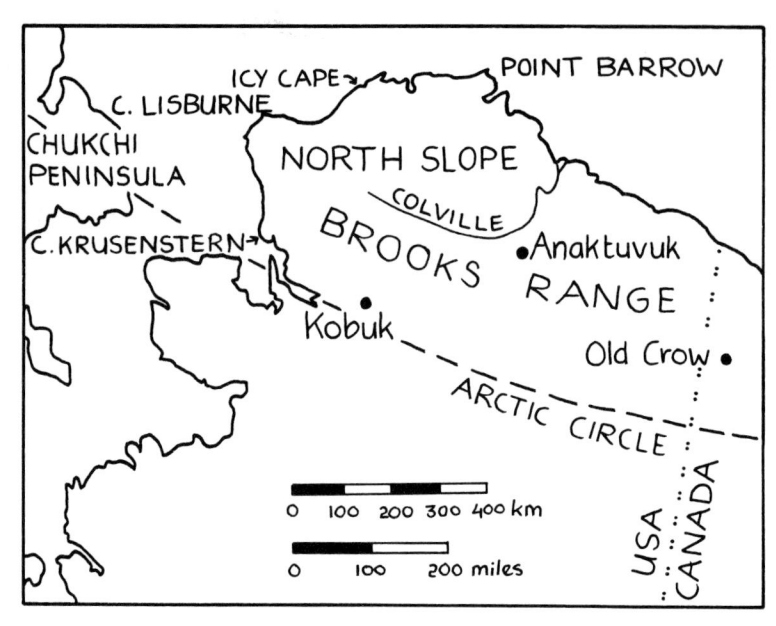

Map 20 Research in Alaska.

BIRD STUDIES IN THE BROOKS RANGE, ALASKA

Laurence Irving, the first scientific director of the Naval Arctic Research Laboratory at Point Barrow, soon turned his attention to the birds of the little-known interior of Alaska. He was a physiologist particularly interested in the adaptation of birds to the Arctic climate, who had served in the U.S. Air Force in the Second World War helping to study techniques for human survival in the Arctic. After his tour of duty at Point Barrow, he joined the Arctic Health Research Centre at Anchorage in 1949 and continued his studies of adaptation to Arctic cold. The Brooks Range, which stretches east–west along the 68th parallel from Cape Lisburne to the Mackenzie Delta, forming the southern boundary of Arctic Alaska, or the North Slope, had never been ornithologically explored. Between 1947 and 1957 Irving made repeated visits to Anaktuvuk Pass, in the centre of the range, flying to and fro from Barrow, 250 miles (400 km) to the northwest. His bird studies there were greatly facilitated by the local mountain Eskimos, or Nunamiut, and by colleagues and friends at Barrow, including the most ornithologically knowledgeable of Charles Brower's sons, Tom Brower. Among other research at Anaktuvuk, Irving and his helpers studied the temperatures of birds and mammals in early spring and weighed over 500 recently-arrived spring migrants.

In May 1952 Irving made the first of five flights to the village of Kobuk in the southern foothills of the Brooks Range. During these visits he gathered information about the local birds from the resident Eskimos to supplement his own observations and specimens. Finally, in 1957 he flew to Old Crow at the eastern end of the Brooks Range and remained there from early April to late in June. He and his helpers compiled a list of 107 species of Old Crow birds and obtained Indian names for 99 of them from a 75-year-old ex-chief called Joe Kay. Some hearts of newly-arrived and recently settled migrant birds and of resident birds were sent for expert examination to see whether the sudden transition from migrating to sedentary habits had affected their coronary structures, and 361 bird skins were deposited in the United States National Museum at Washington.

Annotated systematic lists of the birds of the three places in the Brooks Range worked by Irving in the 1950s were brought together in his book published in 1960 by the Smithsonian Institution in Washington D.C. entitled *Birds of Anaktuvuk Pass, Kobuk, and Old Crow. A study in Arctic adaptation*. These lists for the first time documented the migration and migratory routes of birds moving over the Brooks Range passes into the Arctic or North Slope, and this subject was discussed by Irving in a separate chapter of his book devoted to migration in Alaska. He showed how some of these migrants winter in Asia, others along the Pacific coast of North America, and still others winter east of the Rocky Mountains. At Anaktuvuk all five of the Asiatic wintering migrants regularly occurring in Alaska were recorded: namely the Bar-tailed Godwit, Wheatear, Bluethroat, Arctic Warbler and Yellow Wagtail. At least two of these species, the Wheatear and Arctic Warbler, migrate across about 100 degrees of longitude from China or the Philippines respectively to reach central Alaska. Irving also collected, compared and analysed data on the arrival and departure dates of migrants in the Brooks Range and on the date of the laying of the first egg. His discussion of the metabolism of warm-blooded animals and birds in the Arctic and their insulation, based on his own researches at Anaktuvuk and elsewhere, is fascinating but technical. Even he seems a little surprised at the way the tiny Black-capped Chickadee—American equivalent of the Willow Tit—can survive during the extreme cold of midwinter in the Brooks Range, where it feeds among the willow branches blown clear of snow miles to the north of its summer breeding habitat in the forest.

SNOWY OWLS ON BAFFIN ISLAND

When in 1953 P. D. Baird, director of the Arctic Institute of North America, then based at McGill University, Montreal, Canada, was organising another of his expeditions to Baffin Island, he invited a young Scottish naturalist, Adam Watson, to join him. Watson subsequently became a well-known figure both as a university teacher at Aberdeen and in the Nature Conservancy. He also

Plate 52 (*a*) Snowy Owl pair at a nest near Barrow, Alaska. The invariably whiter male bird has brought a lemming to the incubating female. (*b*) The sitting female Snowy Owl frequently changed her position and seemed often to stare suspiciously at the lens protruding from the front of my hide.

published important papers on the behaviour of the Red Grouse and the Ptarmigan. But on Baffin Island he chose to study the breeding biology of the Snowy Owl, and the results of a summer of hard work in the field were published in the *Ibis* in 1957. This detailed study, which also brought together information culled from the extensive literature on the Snowy Owl, added much to the hitherto somewhat fragmentary knowledge of that conspicuous but often elusive bird.

Watson's study area, which naturally became known as "Owl Valley", formed part of the 50-mile-long Pangnirtung Pass which cuts through the highlands of the Cumberland Peninsula (Map 18). He was fortunate in being able to spend an entire summer, from May until September, in the field, and even more so in the fact that 1953 was a good lemming year. "Owl Valley", in which the expedition's base camp was situated, was a 2-mile-wide strip of marshy tundra along the banks of a river. In one 5-mile-long section of this tundra six Snowy Owls were nesting. Their territories were carefully mapped out and found to average about a square mile (2.6 km^2) in size. Watson soon found that owl density coincided with lemming density. In two neighbouring valleys, where the tundra was drier and lemmings were scarce, not a single pair of breeding Snowy Owls was found. Trapping near the base camp showed that, in this owl-rich area, lemming density was about 20 animals per acre (0.404 hectares) in June–July and approached 130 per acre in mid-August, though the young lemmings mostly caught then were much smaller than the June adults, weighing only half as much.

The Snowy Owls studied by Watson fed exclusively on lemmings, nearly all of them brown lemmings. In all the owls' prey, which was investigated in pellets or found at nests and perches, he only identified the remains of one Snow Bunting and one Lapland Bunting. The owls seemed to have no trouble at all in catching enough lemmings to feed their large families, and food depots of up to five animals were formed at nests and perches, especially soon after the first young hatched. One bigamous male owl, which was feeding two females and 12 or 13 young, was still able to spend hours each day on a favourite perch doing nothing. He would normally catch a lemming within 5 minutes of beginning to hunt and once caught three within the space of 5 minutes. Since the young owls regularly regurgitate pellets at the nest, their food intake can readily be assessed by analysing the contents of the pellets, and Watson found that they often consumed a fifth or a quarter of their own body weight every day in food. In one territory, he reckoned that about 440–460 lemmings were consumed by the young owls, working out at an average of nearly two lemmings per young per day. From hatching to fledging, a brood of nine young Snowy Owls probably consumed around 1500 lemmings. By watching the adult owls for hours at a time, Watson was able to establish the amount of food they ate. On one occasion a male bird was watched for 24 hours continuously and seen to eat four well-grown lemmings himself and give his two females three each. An adult Snowy Owl was thought to require between 600 and 1600 full-grown lemmings per annum. Watson estimated that, since the total number of lemmings eaten in one (the bigamous male's) territory of about a square mile in size was 3000 to 4000 full-grown animals, about 5–7

(a)

(b)

were removed during the summer from each acre. Expressed in other terms, the proportion of lemmings taken by the owls in this territory during the summer was estimated at 8–20% of the biomass of lemmings in mid-August and about 20–31% of the numbers in June–July. Even in August, the 3 adult and 15 young owls in the territory were only consuming 0.2% of the local lemming population.

The average clutch size of six of Watson's Baffin Island Snowy Owl's nests was 8 eggs: in two nests 9 eggs were laid, two had clutches of 8 eggs and two had only 7 eggs. Clutches of 12 eggs and even 14 or 15 have been recorded elsewhere in good lemming years. Watson assumed that incubation began with the first egg, otherwise it would have been frozen; and that the spread of hatching over up to 10 days was due to the eggs having been laid at intervals of a day or two. He discounted claims that incubation of the last eggs to hatch is done by the young already hatched and that the first egg often hatches before the last is laid. Visiting Snowy Owl nests to ascertain and study their contents was sometimes hazardous. Once the eggs had hatched the male bird, recognisable because it was invariably smaller and whiter than the female, would repeatedly dive at Watson, usually from behind, when he approached a nest. Often arm-waving or shouting would cause the bird to sheer off without striking him, but he was sometimes hit on the head by the bird's wing or foot. He found that the only effective remedy "was to fire a shotgun at very close range", but even this "had to be done repeatedly".

LITTLE AUKS OF HORSE HEAD ISLAND

The Little Auk, or Dovekie as it is now called in North America, must be one of the least known of the auks breeding around the shores of the North Atlantic. Yet it may well be the most numerous. Concentrated in a few remote and widely separated colonies in the High Arctic, it is a bird whose breeding biology can only be effectively studied by mounting a special expedition. Apart from Norwegian ornithologist Magnar Norderhaug's studies of Spitsbergen Little Auks from 1968 onwards, which were supported by the Norwegian Polar Institute, no special attention had been given to Little Auk breeding biology until, in the summer of 1974, Peter G. H. Evans with three colleagues from the University of Aberdeen, all of them members of the Aberdeen University Expedition to Northwest Greenland of that year, set out to study the Little Auks breeding on Horse Head Island or Appalersalik, to give it its Greenlandic name. Too small to be marked on the map of Greenland, this

---

Plate 53   (*a*) The female Snowy Owl has arrived at the nest near Barrow and is about to resume incubation of her five eggs. June 1989. (*b*) The Snowy Owls pictured here were nesting on a knoll just a mile from the former United States Naval Arctic Research Laboratory, the buildings and other installations of which can be seen here in the background.

mile-long island with a plateau top 950 feet (290 metres) high, one of thousands of offshore islets along Greenland's northwest coast, is 60 miles (95 km) northnorthwest of the settlement of Upernavik (Map 21). Peter Evans was no stranger to this sort of undertaking: he had already counted seabirds on two of the most remote islets in Britain, namely North Rona and Sula Sgeir (Map 25). By the time his paper on the "Ecology and behaviour of the Little Auk *Alle alle* in west Greenland" was published in the *Ibis* in 1981, he had joined the staff of the Edward Grey Institute of Field Ornithology in Oxford, for a number of years.

Although Evans and his three companions were only on Horse Head Island from 9 July until 12 August, and the island was sometimes fog-bound or storm-bound, they accomplished a great deal, evidently making full use of the 24 hours of continuous daylight available to them. To investigate the patterns of attendance of the Little Auks at their colony, which was estimated to comprise 6000 pairs, counts were made every 2 hours of all birds visible either on the sea, in the air, or on the steeply-sloping ground where the auks nested. With each count, note was taken of the weather at the time: light intensity, cloud cover, temperature, precipitation, visibility, state of the sea and wind force and direction were all recorded. Thirty-five nest sites were marked with a number and inspected at regular intervals until the eggs hatched. The rate of growth of 26 chicks was elucidated by means of periodic measurements and chicks were also weighed every 2 hours through 24-hour periods. Selected nest sites were watched continuously for 24 hours at a time by a hidden observer so that the visits of the adult birds to a chick with food could be related to weight changes in the chick. To investigate the food being brought to the young, Evans and his companions collected 460 food samples dropped by the adult birds either naturally or as a result of their being purposely frightened. To study the behaviour of adult birds, a group was watched continuously for 10 minutes at a time; there were 230 of these 10-minute watches. The researchers also counted nests along three different transects through parts of the colony to discover the number of nests per square metre and relate nest density to the different habitats and plant communities. Finally, many adult birds were caught and ringed.

In the second and third weeks of July, when most pairs of Little Auks were incubating eggs, birds assembled on the sea in rafts during the late evening. After midnight, spectacular massed flights occurred in the air above the

---

Plate 54 (a) A massed flight of Little Auks at a colony some 18 miles southwest of Thule Air Base in northwest Greenland. Such flights zoom to and fro over the sea shore and penetrate up to half a mile inland. The valley slopes—incuding one's tent—are sprayed with pinkish droppings which, over the years, have a remarkable manuring effect on the lichens and other vegetation. In these places the rocks are plastered with thickly-growing soft black lichen, while between them is a luxuriant growth of bright golden-green moss.

(b) Little Auks perched in characteristic fashion on a boulder under which some of them probably have their nests. 17 July 1984.

(a)

(b)

colony, and more and more auks began to alight there until the maximum number of birds ashore was reached at about 06.00 h. The afternoon was the period of minimum colony attendance, only the incubating birds remaining.

The Horse Head Island Little Auks fed their young almost exclusively on a copepod, a marine crustacean called *Calanus finmarchicus*, which they carried in their well-developed throat pouches, and the Aberdeen University researchers thought that most of these were caught within a few miles of the colony. On average, food was brought to the chick just over five times every 24 hours, most of it during the night and only a very small proportion between 12.00 and 18.00. This feeding pattern was found to fit in well with the up and down movements of *Calanus finmarchicus*, which lives near the surface of the sea at night and descends to lower depths during the day.

The breeding success of these Little Auks was surprisingly low. Only 10 out of 20 selected eggs produced young that survived 20 days. Eggs were lost through desertion or infertility (4), disappearance for unknown reasons (2) or breakage (1). Of the 13 chicks which hatched, 3 were taken by Arctic foxes. It seemed possible that this low success rate was partly due to human disturbance and was partly a result of the accessibility of those nests that were discovered. The Little Auk lays its single egg in rock crevices underground and the nests actually found were likely to be those nearest the surface and most obvious to predators.

The behaviour of Little Auks above ground was studied quantitatively by means of many 10-minute watches of groups of birds. It was found that they spent three-quarters of their time ashore resting, either "with body hunched low to the ground" or "with body upright but resting on the tarsi". Often they were in pairs and from time to time these pairs were seen to interact. Up to the end of July two birds were occasionally seen facing one another and bowing their heads several times in rapid succession. Later, this head bowing was replaced by "head-wagging", when the two members of a pair would move their heads rapidly from side to side while calling and touching bills, sometimes holding their wings out slightly and fluttering them. When a Little Auk landed or moved towards a group of auks on a rock it would invariably walk with body arched upwards and bill pointed downward towards the ground. Sometimes when one member of a pair flew off it would use a special type of flight with slow wing-beats and then return to its mate. Most of these and other behaviour patterns recorded by Evans in the Little Auk were closely comparable to those of Guillemots and Razorbills. He also recorded two sorts of communal flight around the colony. In one, flocks of several hundred or more birds would circle round calling loudly as if singing. In the other, smaller groups of up to 50 or so, individuals would fly around silently except for the rushing noise made by their wings in rapid flight.

THE THICK-BILLED MURRES OF PRINCE LEOPOLD ISLAND

In 1971, Hugh Boyd, a former colleague of Peter Scott's at the Wildfowl Trust, whose ornithological career has taken him to Canada's Wildlife Service,

Plate 55   Arctic fox searching for Little Auk's eggs under boulders on the steep scree slope where they nest. 17 July 1984, in northwest Greenland.

masterminded for that organisation a research programme into the distribution, breeding biology and ecology of the seabirds of the eastern Canadian Arctic. Special attention was to be given to two abundant species, the Fulmar and Brünnich's Guillemot, known in North America respectively as the Northern Fulmar and the Thick-billed Murre. The eventual aim of this research was to improve understanding of the working of marine ecosystems in the far north which were threatened by plans to search for oil and gas there. Prince Leopold Island, at the western end of Lancaster Sound off the northeastern corner of Somerset Island, was seen from the start as a focal point for these studies (Map 18). It was in the very region threatened by oil exploration and transport, and its 250-metre (820-foot) high cliffs harboured tens of thousands of seabirds, including over 50 000 pairs each of Fulmars and Brünnich's Guillemots. Its incongruous name had been given to it by its discoverer, William Edward Parry, in honour of King George IV's son-in-law Leopold, the duke of Saxe-Coburg-Saalfeld, on 4 August 1891. Already that day, Brünnich's Guillemots, Fulmars and Kittiwakes were recorded as numerous in the vicinity (Parry 1821: 38).

In 1972 David Nettleship, co-author with his English colleague at the Canadian Wildlife Service Tony Gaston, of *The Thick-billed Murres of Prince Leopold Island* (Gaston and Nettleship 1981), designed and initiated a 3-year (1975–1977) research programme into the Prince Leopold Island Brünnich's Guillemots to form part of the Wildlife Service's larger seabird programme. Gaston's previous experience was in Africa and Asia, and his research for an Oxford University Ph.D. at the Edward Grey Institute of Field Ornithology had been on co-operative breeding in Indian birds. Now, he took charge of the 3-year programme of field work on Prince Leopold Island. Nettleship, who had studied biology in three different Canadian universities before also undertaking research at the Edward Grey Institute in Oxford, already had Arctic experience. Invited to study the relationship of breeding waders and their insect prey in the High Arctic by the Entomology Research Institute of the Canadian Department of Agriculture, he spent the summer of 1966 at Hazen Camp on Lake Hazen in Ellesmere Island (Map 21). Two important papers, on the breeding biology and ecology of the Knot (1974) and Turnstone (1973) respectively, resulted from this work. Nettleship's Knot paper, based on the discovery of seven nests containing eggs after prolonged searching, and on analyses of the stomach contents of 19 adult and young birds collected, provided the first published systematic data on the breeding biology and summer diet of this little-known species. These studies were complemented by a description of the area's avifauna which he wrote in conjunction with W. J. Maher.

It must have been a welcome change for Nettleship to switch his research from wading birds, whose nests are sparsely distributed over the tundra and exceedingly difficult to find, to a species like Brünnich's Guillemot, breeding in densely-packed masses on cliff ledges where hundreds of nesting birds can be watched simultaneously from above. On the other hand, the summer climate on the high plateau of Prince Leopold Island at 74° N was much less pleasant than that of the Lake Hazen area at 81° 49' N. On the island, cloud, fog and high winds were normal; rain, freezing rain or snow occurred frequently, and the mean summer temperature was around the freezing point. Hides made of plywood set up on the cliff top were essential to afford observers some protection from the weather.

The Thick-billed Murres of Prince Leopold Island is a readable, well-organised and nicely produced book, prefaced by 42 fine colour photographs. In many ways it is a model of how ornithological research ought to be presented, namely in a manner acceptable both to the specialist and to the general reader.

Plate 56 (*a*) Portrait of a Brünnich's Guillemot taken on 13 July 1972 on Hornøya off the Varanger Peninsula. The white mark on the bill serves to distinguish it at once from the common Guillemot.

(*b*) A stack on Hakluyt Island off the northwest Greenland coast, some 50 miles west of Qaanaaq, crowded with Brünnich's Guillemots on 19 June 1985.

(a)

(b)

It describes in detail how a huge breeding colony of seabirds can be effectively studied and the size of the population ascertained, by means of carefully selected study plots. Each plot consisted of a section of cliff easily visible from above inhabited by between 50 and 100 pairs of birds. A cairn marked the position of the observer and enlarged black-and-white photographs, some of which are published in the book, were used to mark and number the individual breeding sites. Variations in seasonal and daily attendance at the colony were investigated by counts of the study plots, in each of which the total number of nesting birds was known. Because one member of a breeding pair was always present at the breeding site, the total count of birds was never less than the number of breeding pairs. The study plots were used to investigate many aspects of the Brünnich's Guillemot's breeding biology. The roles of the male and female in incubation and brooding were found to be similar, the parents taking turns to look after the single egg or chick, working in approximately 12-hour shifts. The incubation period was 31–34 days, and the chick-rearing period varied from 15 to 30 days. Detecting the presence of an egg under a sitting bird was not easy; indeed it was found that during 6 hours of watching at a study plot, an observer would detect only just over half the eggs known to be present. Chicks were fed on whole fish, usually Arctic cod, at the rate of between four and eight per day. A survey from the air on one particular day suggested that some birds were catching fish up to 70 miles (112 km) from the breeding colony, and that the mean distance of feeding birds from the colony was about 35 miles (56 km). The authors reckoned that a bird could need nearly an hour to catch a fish for its chick, and might take over 2 hours on the round trip to feed it. They found some evidence that the adult guillemots fed themselves on the smaller prey items and took the larger fish to feed their chicks. Among the many intriguing questions posed by this study is that of the size of the colony: why does it have to be so large? What advantage accrues to the species from the concentration of 80 000 pairs of Brünnich's Guillemots on the cliffs of a single island? A suggested answer is that the colony's large size enhances its function as an information centre, enabling outgoing birds to find good foraging areas more easily by taking their cue from the direction of flight of incoming birds carrying fish.

BIRDS OF PREY IN WEST GREENLAND

All three of Greenland's breeding raptors have recently been intensively studied in west Greenland. They are the White-tailed Eagle, the Gyrfalcon and the Peregrine falcon. The first of these is confined to the Low Arctic part of west Greenland between Disko Bay and Cape Farewell (Kap Farvel). The Peregrine Falcon breeds as far north as Upernavik on the west coast and Ammassalik on the east, and the Gyrfalcon breeds all round the coast of Greenland including Peary Land in the far north.

The White-tailed Eagle is thought to have colonised Greenland after the last ice age. Since then the birds have been isolated from other White-tailed

Map 21 Research in Greenland.

Eagle populations, the nearest of which is in Iceland, and their larger size has tempted some authorities to regard them as forming a separate subspecies. They and the Norwegian White-tailed Eagles represent the only remaining reasonably stable and viable Western populations of a magnificent bird which was probably once found throughout Europe. In Asia this eagle is still fairly widespread and it breeds along the Arctic coast of Siberia as far as 72° or even 75° N. The attention of Danish ornithologists was first brought to bear on the Greenlandic White-tailed Eagles in 1947, when sheep-farming interests in south Greenland tried to have a system of awards for the killing of eagles introduced. In the following year the Zoological Museum of the University of Copenhagen carried out an investigation to discover how many eagles there were and what they lived on. This revealed that eagle numbers in the sheep-farming districts were declining rapidly and that, although the eagles exceptionally brought a new-born lamb to the eyrie, these were almost certainly already dead when taken. Fish and birds formed the major part of their diet.

In 1971 Danish raptor expert Benny Gensbøl initiated a research project into the distribution and breeding biology of the White-tailed Eagle in Greenland which was organised and supervised by the Danish Ornithologists' Union with the support of the World Wildlife Fund and the Carlsberg Foundation. The main part of this project consisted of three successive field expeditions to west Greenland in 1972–1974 to study the numbers and distribution of breeding eagles. These were supplemented by three further expeditions in 1976–1978 to study the eagles' breeding biology. The earlier expeditions were made up of two-man parties which searched the fjords and outlying coastal skerries in locally hired motor-boats. Later a fibreglass hide was used, which was set up 100 yards from the eyrie, constructed so that the observer could remain in it for 2 or 3 days at a time. To record and identify prey items brought by the adult eagles, two motorised Leicaflex miniature cameras were placed near the eyrie in a soundproof box. They were loaded with colour film and worked from the hide by remote control.

The findings of this Danish research effort of the 1970s were published in 1979 in a special issue of the journal of the Danish Ornithologists' Union (*Dansk Ornithologisk Forenings Tidsskrift*) devoted to Arctic ornithology and dedicated to Finn Salomonsen (Ferdinand 1979a and 1979b; Hansen 1979; Christensen 1979; Kampp and Wille 1979; Wille 1979). The breeding population of White-tailed Eagles in Greenland was estimated to be between 85 and 101 pairs and this was thought to represent between a half and a quarter of the population at the beginning of the century, before the decline set in. The production of young birds was reckoned to be one almost fully-fledged young bird per "occupied breeding locality", a figure similar to that calculated for the near relative of the White-tailed Eagle in Alaska, namely the Bald Eagle, which research has credited with one young per breeding attempt. The eagles used traditional breeding sites year after year and over half of the eyries in one area were found to be "easily accessible". Many of them were on low outlying skerries, often on the summit with a view all round. Nest material was birch and willow branches and twigs, with moss, lichen and grass for the lining. The biggest ever Greenland eyrie was nearly 6 feet (180 cm) high. The eagles' summer diet varied a great deal from nest to nest but was made up everywhere mainly of fish, with a few birds and an occasional Arctic fox. It was thought that, overall, at the southern end of the range about 90% of the birds' summer diet consisted of fish.

Along with this wealth of interesting new information about Greenland's White-tailed Eagles came new concerns about their conservation. Though they were granted the full protection of the law in 1973, the shooting of eagles continued thereafter. In 1974, twelve recently-shot eagles came to the notice of the researchers, who knew of three pairs shot near the eyrie in 1976. In spite of a ban on the export of eagles dating from 1933, birds were still being sent to Denmark to be stuffed and mounted as specimens, either in a ship's freezer or by airmail. A more serious threat than shooting was thought to be posed by the steel fox traps in use since 1967, in which eagles were being killed or maimed every year. Finally, the researchers drew attention to the recent increase in the construction of summer holiday cabins along the fjord shores and on the coastal skerries, which had negatively affected the breeding eagles in some places.

In the same years when Danish ornithologists were investigating the breeding biology of White-tailed Eagles in southwest Greenland, a team of American workers began a long-term study of Peregrine falcons and Gyrfalcons further north in west Greenland which was still continuing in 1990. Their researches are being carried out in the inland ice-free area on either side of Søndre Strømfjord and on the coast around Disko Bay. The programme was initiated by a geographer at Dartmouth College, Hanover, New Hampshire, called William Mattox, who had been in west Greenland in 1967 to trap and ring migrating Gyrfalcons. He and his friend William A. Burnham, who, in the late 1970s and early 1980s, was running a captive breeding facility for Peregrines in Colorado, were the authors of a paper published in 1984 in *Meddelelser om Grønland* which summarised some of his research (Burnham and Mattox 1984). Some time before then, in 1979, one of the party of five who carried out the first season's field work, James T. Harris, published a book describing his experiences entitled *The Peregrine Falcon in Greenland. Observing an endangered species*. He was a newcomer both to the Arctic and to Peregrine falcon studies and his book is fresh and original as well as informative.

This American research into Greenland's falcons was started in 1972 against a background of impending disaster, at least for the Peregrine, which had declined drastically in number on both sides of the Atlantic during the late 1950s and 1960s. In Britain first, and then in America, it was shown that this decline was caused by the poisoning of the birds through the ingestion of pesticides present in the bodies of their prey. These pesticides, notably DDT, aldrin, dieldrin and heptachlor, all of them chlorinated hydrocarbons, were widely used for controlling argricultural pests from soon after the end of the Second World War. They did eliminate pests, but they also eliminated birds and animals that preyed on the pests, for these lethal chemicals became concentrated in the body tissues of birds and animals at higher levels in the food chain. The Peregrine, at the top of such a chain, was particularly vulnerable. Birds were either killed outright by the accumulated poisons or suffered long-term effects which were equally disastrous to the species, namely abnormal breeding behaviour including egg-eating, the laying of infertile eggs or eggs with thin shells, and the death of embryos in the egg. Breeding failure brought about by these effects was the immediate cause of declining Peregrine populations, and one of the prime aims of the American research programme in Greenland was to discover to what extent the Greenland population was affected. In the year before its start, Tom Cade and others had reported thin-shelled and breaking eggs associated with high concentrations of DDT from Peregrine nests along the Colville River in Arctic Alaska. By 1975 this remote population had been reduced to a mere one-third of its original numbers, and the remaining pairs were producing less than one young bird each year. Clearly, Alaska's Arctic-breeding Peregrines were accumulating pesticides in their body tissues. These must have derived either from the migratory birds on which they fed in the summer, which could have ingested pesticides in their winter quarters far to the south, or from prey eaten by the Peregrines themselves in their winter quarters. Would the same be true of the Greenland Peregrines?

The first season's work allayed the worst fears for the health and viability of the Peregrines breeding inland around Søndre Strømfjord. In an area of approximately 800 square miles (2072 km^2), surveyed on foot, eight breeding pairs of Peregrines were found. Their eyries produced a total of 18 chicks or eyasses, which means that on average each pair produced over two young. It was thought that most of these young would fledge successfully. These figures for breeding density and success showed that the Greenland Peregrines had not yet suffered significant ill-effects from ingested pesticides. On the other hand, the researchers found evidence which proved conclusively that the falcons had been ingesting pesticides and other toxic chemicals. The contents of two addled eggs analysed in the United States contained concentrations of two of these substances, denoted by the letters DDE and PCB, which exceeded 200 parts per million and could only be regarded as dangerously high. Eggshell fragments from seven eggs of four different females were collected and their thickness was carefully measured. On average, it was found that eggshells in 1972 were 14% less thick than eggshells of Peregrine's eggs collected in Greenland before 1947. This was not far off the 20% reduction in shell thickness which elsewhere had caused Peregrine's eggs to break in the nest, and it was concluded that the Greenland population was very near a contamination level that would endanger it. Fortunately, there has been no evidence since then of an increase in contamination, and the breeding density and success of the Peregrines have continued to be satisfactory.

How did the Greenland Peregrines become contaminated with toxic chemical residues? To discover what they ate in the breeding season, Harris and his colleagues in 1972 collected pellets at eyries and made a census of the birds breeding in the tundra where the Peregrines hunted for their prey. One might assume that a large powerful falcon like the Peregrine would feed on birds the size of pigeons or grouse, such as the Long-tailed Duck and Ptarmigan, both of which bred commonly in the study area. But in the pellets Harris and his colleagues collected in 1972, though a single Ptarmigan and a single Red-necked Phalarope were represented, all the other prey remains were of small passerines: Snow Buntings and Lapland Buntings were the most numerous, and Redpolls and Wheatears were also found. These four small passerines were shown by the census to be the commonest birds in this part of Greenland. At the end of the first 10 years of the project Burnham and Mattox concluded that these four small birds accounted for 90% of the breeding falcons' prey, and that 70% of the falcons' diet was made up of Lapland Buntings. Of these prey species the Wheatear alone winters in the Palearctic, crossing the Atlantic to Europe and Africa. The other three, together making up the vast bulk of the falcons' diet, winter in Canada and the United States, where the use of pesticides was widespread. Analysis of samples of body tissues from these prey species collected in Greenland showed that all carried low levels of DDE and PCB, but the amounts were thought to be well below the level likely to cause egg-shell thinning in the Peregrine Falcons preying on them. The conclusion must be that the Greenland Peregrines also ingested these poisons in the prey they took while migrating through North America or wintering in South America.

Burnham and Mattox included some mention of Gyrfalcon biology in their Peregrine research. They thought that there were probably as many Gyrfalcons as Peregrine falcons breeding in their Søndre Strømfjord study area. Although the Gyrfalcons also depended in large measure on the same small passerines as the Peregrines for their summer diet, they showed no trace of egg-shell thinning or other ill-effects from pesticides. The probable reason for this was that these Gryfalcons only moved south within Greenland, or perhaps as far as Iceland, to winter, thus avoiding areas of intensive pesticide use. Another marked difference between the two species of falcon was that Gryfalcons made much use of old Raven's nests for their eyries while the Peregrines laid on cliff ledges. The Gyrfalcon also bred earlier in the year, and the Arctic hare probably formed a more important part of its diet than was the case with the Peregrine.

Latest news of this long-standing project, which in 1991 celebrates its twentieth consecutive summer's field work, is that 1140 Peregrines have now been ringed in west Greenland and experiments have been started with a radio transmitter attached to an adult Gyrfalcon: Polar-orbiting Tyros-N satellites picked up over 250 locations from a Gyrfalcon for 8 weeks after its capture and release in early July 1990 (Mattox 1990).

PEREGRINES AND GYRFALCONS IN ARCTIC CANADA

As mentioned earlier in this chapter, the March 1988 issue of the journal *Arctic* contained two papers on the natural history respectively of the Peregrine Falcon and the Gyrfalcon in the Canadian Arctic. The field work for this falcon research was carried out in 1980–1986 by two different teams of Canadian biologists, five in all, three of whom had worked or were working in the Department of Zoology of the University of Alberta at Edmonton, Alberta, and three of whom were currently employed by the Department of Renewable Resources of the government of the Northwest Territories. Court, Gates and Boag studied the Peregrines breeding in about 174 square miles (450 km^2) of coastal tundra in the northwest corner of Hudson Bay around Rankin Inlet, while Poole and Bromley's Gyrfalcons were nesting in a 772-square-mile (2000-km^2) strip of mainland coastal tundra southwest of Cambridge Bay, between Bathurst Inlet and Queen Maud Gulf (Map 18). Both falcon populations were thought to be at around the maximum density for the species, but the Gyrfalcons were more widely spaced out than the Peregrines. The average distance between nests was about 2 miles (3·3 km) for the Peregrines and about 6½ miles (10·6 km) for the Gyrfalcons. In spite of their apparent healthy state and high productivity, both falcon populations suffered losses of eggs and nestlings. The loss of one clutch and two broods of Gyrfalcons was directly attributable to weather. The clutch of eggs was lost after freezing rain and snow covered the nest. One brood loss was caused by meltwater saturating the chicks so that they died of exposure; the other resulted from the old Raven's nest used by the falcon having been built on snow, which subsequently melted so that the nest fell. There was no evidence that the Gyrfalcons had been affected by the low concentrations of pesticides found in addled eggs collected in 1983. On the other hand, the authors of the Peregrine paper found that "a certain proportion of nesting failures at Rankin Inlet each year were attributable directly to pesticide pollution".

Court *et al.*'s research on the Rankin Inlet Peregrines was certainly intensive. The authors were in the field during five successive summers from May, when the Peregrines arrived, until they left in September. In all, 47 adult and 202 young Peregrines were ringed with a standard United States Fish and Wildlife ring on one leg and a numbered colour ring on the other. This enabled the authors to ascertain that although the same nest was never used in successive years, fidelity to territories was high. It turned out that, although it began very much later, the breeding season of these Arctic Peregrines did not differ very significantly in length from that of more southerly populations, although the pre-laying courtship stage was shorter and incubation began before clutch completion instead of, as in southerly populations, with the laying of the last egg. Broadly speaking, Rankin Inlet Peregrines arrived in the second half of May and laid in the first half of June. The eggs hatched in the first half of July and the young fledged in the second half of August. The diet of these Canadian Peregrines was normally not very different from that of Burnham and Mattox's Greenland Peregrines. Early in the season Ptarmigan on spring migration were taken; later the four locally breeding small passerines, namely Snow Bunting, Lapland Bunting, Shore Lark and Water Pipit (now called Buff-breasted Pipit) were the Peregrines' most important prey species. But the Canadian Peregrines could and did do something that their west Greenland relatives could not, namely, switch to lemmings in a "good" lemming year. In 1985, when lemmings were numerous at Rankin Inlet, at least half the Peregrines brought them to their eyries to feed the young and in that year nearly twice as many young Peregrines were produced in the study area as in any of the previous 4 years. This was surprising, for the Peregrine had normally been considered a bird predator which remained unaffected by lemming abundance.

The area chosen by Poole and Bromley to study breeding Gyrfalcons had no permanent human inhabitants, but Eskimos hunted there at times in winter. Both Ravens and Golden Eagles were common, and the Gyrfalcons almost invariably took over the old stick nest of one or other of these birds for their eyrie. Occasionally they usurped a freshly-built Raven's nest, driving off the owners. Old nests of Rough-legged Buzzards were never used, perhaps because they were on lower cliffs and were more open to the weather. The 91 rings placed on young Gyrfalcons during this study had provided no evidence of the whereabouts of the birds' winter quarters when the paper was published, but the authors had reason to believe that the Gyrfalcons probably moved further south during the winter months. They began to return to take up their territories as early as February, and most pairs began laying in the first half of May—a month earlier than the Rankin Inlet Peregrines. The white and grey colour-phases in this population of Gyrfalcons were found to occur in almost equal numbers.

BREEDING LONG-TAILED SKUAS

While he was still a 25-year-old biology student at the University of Amsterdam, Ko (short for Jacobus) de Korte became one of the relatively few ornithologists to have wintered in the Arctic. He spent a year in 1968–1969 on Edge Island in the Spitsbergen Archipelago with three other young Dutchmen. The aim was to study Polar bears, but de Korte also devoted time and energy to observing and collecting birds, and subsequently described the avifauna of this little-known part of the archipelago in a detailed publication (de Korte 1972). Inspired by an abiding love of the Arctic, he next turned his attention to Greenland, and the summers of 1973, 1974 and 1975 found him at the mouth of Scoresby Sound in east Greenland studying Long-tailed Skuas and the breeding waders of the tundra. In each year he took with him a single companion and, in the usual grandiose manner, referred to this field work as the Dutch Greenland Expedition to Scoresby Sound, 1973, 1974 and 1975. Much of it was written up in a series of five papers on the ecology of the Long-tailed Skua published in the journal of the Institute of Taxonomic Zoology at the University of Amsterdam's Zoological Museum, where de Korte worked at the time, and elsewhere. To earn his doctorate in 1986 he submitted in place of a thesis these five papers together with two others on the Scoresby Sound waders. The whole was entitled *Aspects of breeding success in tundra birds. Studies on Long-tailed Skuas and waders at Scoresby Sound, East Greenland*. After being awarded his doctorate de Korte, in collaboration with his Amsterdam colleague Jan Wattel, published a revised version of his earlier unpublished paper on the food and feeding habits of the Long-tailed Skua. This appeared in the journal of the Dutch Ornithologists' Union (Nederlandse Ornithologische Unie), *Ardea*, in 1988, by which time de Korte was working part-time at the Arctic Centre in the University of Groningen.

The study area de Korte chose for his Long-tailed Skua research was on the north side of the entrance to Scoresby Sound in east Greenland not far from the Inuit settlement of Scoresbysund or Ittoqqortoormiit (Map 22), where provisions and other necessaries could be obtained and transport arranged with the help of the local people. From the point of view of investigating the Long-tailed Skua this area had two special advantages. One was the close proximity of the tundra here to the open sea as opposed to the pack ice, which would facilitate the study of the Long-tailed Skua's spring change from a marine life to a terrestrial one. The second was the fact that Scoresby Sound is at the southern edge of the species's breeding distribution in eastern Greenland,

Plate 57 Long-tailed Skua in flight over the Canadian tundra near Cambridge Bay.

which meant that the factors controlling its breeding success might more easily be studied here than in the centre of its range. The field work left de Korte in no doubt that the skuas' breeding success *was* limited. In 1973 eggs were laid but none hatched. This egg loss was attributed to Arctic foxes, which were seen in the area almost daily and were breeding there. In 1974 no Long-

---

Plate 58 (*a*) Breeding Long-tailed Skuas are often very approachable. The pair figured here were photographed near Cambridge Bay after we sat down quietly near their nest. One bird perched on my head more than once, and even pecked my ear.

(*b*) Long-tailed Skua perched on a dry stony tundra ridge a few yards from the nest: just two eggs of a deep olive-green colour on the bare ground in a shallow scrape. 5 July 1986.

(a)

(b)

tailed Skua's nests at all could be found. The birds held territories but did not breed because of shortage of food on the tundra. This was partly due to the very late spring snowmelt and partly to the scarcity of lemmings that year. In de Korte's third year in the field (1975), 11 nests were found and chicks were reared but not a single one fledged. Again the Arctic fox was the culprit.

At Scoresby Sound de Korte collected data on the density and territorial behaviour of Long-tailed Skuas, on the dates of their arrival, their fidelity to a particular site, and their general behaviour. He studied their clutch size and other aspects of their breeding biology, and analysed the relationship between their energy reserves and breeding behaviour. He also measured their breeding success and the growth of the young. To obtain information about their physical condition, the development of their sexual organs, and their food and plumage, 82 Long-tailed Skuas were collected. Some of these were shot by Greenlandic hunters to feed their dogs, others were collected by de Korte himself. His collecting was limited in the study area to non-breeding immature birds; other birds were shot outside the study area. From an analysis of the stomach contents of these birds and from some pellets he collected, de Korte found that the breeding Long-tailed Skua's summer diet comprised four main foods: terrestrial vertebrates, which made up almost three-quarters of the whole, berries and marine animals, comprising about 10% each, and insects (7%). Of the terrestrial vertebrates, two were found to be most important. One of these was the collared lemming, which runs about a great deal on the tundra surface after the snowmelt has deprived it of its winter nests in the snow. These little animals fall an easy prey to the Long-tailed Skuas, which hunt them either by hovering over the ground and diving down onto them or by swooping from a perch, usually a boulder. The other was the Snow Bunting. In mid-July, when the fledgling buntings leave their nests and flutter uncertainly over the tundra, large numbers are taken by the skuas, which were watched pursuing the Snow Buntings a few feet or more above the ground, forcing them to alight, and then killing them. Throughout the summer the Long-tailed Skuas were found to eat a few berries—bilberries, bearberries and crowberries. At first these were from the previous year: the frost had kept them in good condition through the winter. Later, unripe berries were found in the skuas' stomachs until late July and August, when they were replaced by ripe ones. As to the insects, the great bulk of those taken by the Long-tailed Skuas were in the form of larvae. The almost total change of diet among skuas, from fish and other marine animals in winter to the summer diet described above, was discussed by de Korte, who also drew attention to the fact that a Long-tailed Skua, after leaving the tundra soon after fledging, does not return to land until its third summer, when it has to learn again to find and exploit terrestrial food sources.

SPITSBERGEN GOOSE STUDIES

At the start of this chapter mention was made of an international symposium organised by the Norwegian Polar Institute in Oslo in 1983 on the subject of

Arctic geese. The 18 papers published in 1984 (Mehlum and Ogilvie 1984) summed up more than 20 years of work in the field which has made the three populations of geese breeding in Spitsbergen among the most studied in the world. The largest of the three is that of the Pink-footed Goose, which tends to nest on grassy slopes inland. It was reckoned in the mid-1980s at over 28 000 birds (Madge and Burn 1988). These Spitsbergen Pinkfeet winter along the coastal lowlands of continental Europe between Denmark and Belgium. The Brent Goose, the light-bellied form of which occurs in Spitsbergen, nests on offshore islets and was thought to number 7000 birds in the early 1980s. It winters in northeast England and Denmark. Thirdly, the Barnacle Goose, numbering over 10 000 in the mid-1980s, breeds on coastal islets and on rocks projecting above inland valleys. The Spitsbergen population winters round the Solway Firth in southwest Scotland and northwest England.

Modern research on Spitsbergen's geese really began in 1952 when a Sherborne School expedition comprising four Englishmen and a Norwegian caught and ringed 42 Pink-footed goslings in Gipsdalen (Map 19). In 1953 three of the Englishmen, James Goodhart, Russell Webbe, and Thomas Wright, returned to Spitsbergen and, mainly in Reindalen, caught and ringed over 600 flightless adult geese during their moult, most of which were Pink-footed Geese. While recoveries of these rings from migrating and wintering birds were beginning to provide valuable information, in summer 1962 the Norwegian Ornithological Spitsbergen Expedition, comprising seven students from the University of Oslo, ringed a total of 685 flightless Barnacle Geese on a group of offshore islets north of Horn Sound called Dunøyane. Later in the same year, on 26 October, Hugh Boyd, of the Wildfowl Trust, saw at least 46 of these ringed birds in the Caerlaverock National Nature Reserve near Dumfries in southwest Scotland. Then in February 1963, a Wildfowl Trust rocket-netting team caught 316 Barnacle Geese at Caerlaverock. Among them were 94 of the 685 birds ringed on Dunøyane in the previous summer but not a single one of the 609 Barnacles ringed in east Greenland in 1961, nor any of those that had been ringed in the Netherlands. From 1970 onwards British and Norwegian ornithologists from the Wildfowl Trust and the Norwegian Polar Institute developed a joint research programme to study the Spitsbergen Barnacle Geese. This international co-operation was extended in 1975 when ornithologists from the Zoological Laboratory of the University of Groningen in the Netherlands, led by Rudi Drent, currently chairman of the Arctic Centre there, joined the project. It was also supported by the Department of the Environment of the government of Norway and later financed in part by the North Atlantic Treaty Organisation and the Royal Society (Norderhaug 1984; Owen 1986).

In the early 1960s it became clear that a discrete population of 3000–4000 Barnacle Geese was breeding in Spitsbergen and wintering on the Solway Firth. The introduction in the 1950s of rocket-propelled nets and large individually numbered coloured plastic leg rings which could be read at a distance with a telescope, had made it possible to study this population in detail. After the original catch of 316 birds on the Solway in February 1963, other catches were made there in 1966 and annually from 1975 into the 1980s. Over the

280 *In search of Arctic birds*

same period, in the birds' summer quarters in Spitsbergen, moulting Barnacle Geese continued to be rounded up while flightless and driven into pens for ringing. Since 1973 every goose caught and ringed has also been aged, sexed, weighed and measured. As a result of this work, from 1977 onwards, about a fifth or a quarter of the population of Spitsbergen Barnacle Geese has become individually identifiable in the field. In their contribution to the 1983 symposium, Malcolm Ogilvie and Myrfyn Owen, of the Wildfowl Trust, reported that in all 4522 different geese had been ringed. By that time, with the help of Norwegian ornithologist Nils Gullestad, the migration routes of these Barnacle Geese had been elucidated. In spring they stop off for a few weeks at a staging area on the Norwegian coast in the district called Helgeland, consisting of tens of thousands of small offshore islands (Gullestad, Owen and

Map 22 Research in the North Atlantic area.

Nugent 1984). In autumn, while some are thought to make the journey from southern Spitsbergen to northern Britain in a single flight, many stop off at Bear Island. Even so the distance from Bear Island to the Solway in a straight line is 2400 km (nearly 1500 miles), so the geese probably have to fly non-stop for 30–40 hours (Owen and Gullestad 1984).

In 1984 Owen was able to use the results of the Spitsbergen Barnacle Goose research in 1970–1983 to analyse the dynamics and age structure of the population, which had increased from "just over 3000 in 1970 to 8000–9000 in the 1980s". This in spite of three disastrous breeding seasons in the Arctic in 1977, 1979 and 1981, when the proportion of young birds in autumn at the Caerlaverock refuge, where virtually the entire population assembles in October, was less than 5%. In an earlier paper, Owen and Norderhaug (1977) had shown that these poor breeding seasons were directly correlated to late springs and snowmelts in Spitsbergen. Owen was now able to conclude that the population increase had been due to a lowering of the mortality rate rather than to increased breeding success. This decrease in the death rate could be shown to be due neither to the prohibition of Barnacle Goose shooting in Norway in 1971, nor to the setting up of nature reserves in Spitsbergen in 1973 covering most of the islands where the geese breed. Rather it was a direct result of the creation by the Wildfowl Trust of its refuge at Eastpark, Caerlaverock, which comprised the main part of the National Nature Reserve there, besides additional undisturbed feeding areas on farmland.

In Britain, the shooting of Barnacle Geese has been prohibited since 1954, but several hundred were thought to be illegally shot every year. Following the extension of the refuge at Caerlaverock in 1970, the geese began to spend about half, instead of only about a fifth, of the winter there. In effect, they now spent most of the shooting season on the reserve and it was this that gave the population "breathing space" and enabled it to treble in numbers in the next decade or so. However, Owen was able to show that, as the population grew, so the proportion in it of mature geese that bred fell. Factors operating on the breeding grounds began to limit breeding success: competition for nest sites and competition for the available food, for example, were a probable cause of an increase in nest losses and loss of goslings. Thus a pattern emerges of the increasing population size being offset by decreasing fecundity.

This work of Owen and others on the Spitsbergen Barnacle Geese, which was continued and elaborated in the late 1980s, was complemented in a most important way by the group of Groningen-based Dutch ornithologists led by Rudi Drent. They studied breeding Barnacle Geese in relation to their food on the Nordenskiøld Coast (Nordenskiøldkysten) of Spitsbergen, between Ice Fjord and Bell Sound (Map 19), during five successive summers from 1977 to 1981, and their work was described by J. Prop, M. R. van Eerden and R. H. Drent in the most substantial single contribution to the 1983 Oslo goose symposium. A reconnaissance expedition in 1975 led to the choice of the Nordenskiøld Coast, where several colonies of Barnacle Geese bred on offshore islets, for the field work which followed. The 1977 season was devoted to establishing a depot of provisions, locating the permanent observation hut, which was lent by the Norwegian Polar Institute, and constructing an observation tower overlooking the breeding colony of Barnacle Geese on an islet called Diabasøya. To minimise disturbance of the geese, the 4-metre (13-foot) high tower was set up on the mainland 300 m (328 yards) from the islet, which was never visited while the geese were there. But the main task in 1977 was the rounding up and marking with individually numbered yellow plastic rings of all the geese that could be caught, which amounted to 1241 birds.

Three summers of intensive goose study by the Dutch workers on the Nordenskiøld Coast followed. In May and June observations were concentrated on the breeding islet and adjacent tundra. An astronomical telescope with a magnification of 80 times was found useful in the observation tower for reading ring numbers. While the parent geese were leaving the islet for the mainland in July it was possible to identify most of them individually and to count their goslings. Then, in July–August, the geese were carefully counted once more while they were moulting on the tundra lakes along the mainland coast. From the start of the operation observers manned small hides on the tundra to study the feeding behaviour of the geese. For example a 4·2-hectare (10-acre) study plot was divided into 25 × 25-metre sections with marker sticks, and a hide was set up on a neighbouring ridge. From this hide the behaviour of incubating female geese, which left their islet nests periodically to feed on the tundra, could be closely monitored. The density of their food plant here, a dwarf Arctic willow, *Salix polaris*, was measured by counting leaves in 3360 one-metre squares spread over the study plot, and the area covered by snow was carefully mapped daily or every other day. To measure the effect of grazing geese on this sparse tundra vegetation, small wire-mesh exclosures (as the researchers called them) were set up, inside which the geese could not graze. The peck-rates and food-preferences of individual geese were systematically recorded and it was found that 80% of the plant material seen to be ingested was made up of Arctic willow (*Salix*) buds, purple saxifrage flowers, and variegated horsetail shoots. Droppings collected and later analysed microscopically confirmed this. The researchers found that snowmelt was a vitally important phenomenon for the breeding geese because it was at the margin of the melting snow that *Salix* buds and other plant food became available to them. Counts showed that *Salix* buds were at their peak abundance a mere 5–6 days after the snowmelt had uncovered the plants. After another 5 days the buds opened and the leaves were not eaten by the geese, which then had to find another patch of *Salix* or switch to another plant species. One female goose spent 24 minutes foraging on a piece of tundra recently uncovered by the snow-melt and was seen to feed at the rate of 54 *Salix* buds per minute. During the same period she also ate 30% of all purple saxifrage flowers in the area.

Because the geese were recognisable by their ring numbers it was possible to compare the food intake rates of individual birds feeding on the dwarf willow and other plants. It was found that successful females had higher intake rates than birds which failed to hatch young. This difference could be explained on the hypothesis that some birds arrived in Spitsbergen in better condition to start with, were older and more experienced, and were dominant over their neighbours in the flock. It was also found from counting the dwarf willow buds

Plate 59 The dwarf or creeping Arctic willow, several species of which grow in the Arctic, is a food plant of the Barnacle Goose in Spitsbergen. This photograph was taken in northwest Greenland, near Thule Air Base.

inside and outside the exclosures and comparing the results, that the geese were causing a measurable decline in the quantity of food available. This reduced the food intake of some birds, and the authors were able to describe how nests come to be deserted: a given female, unable to find sufficient food quickly enough, was absent from the nest on its feeding expeditions to the mainland tundra for longer and longer periods until her eggs were eaten by a patrolling Glaucous Gull. Later in the season, when the geese were moulting and their goslings were feeding with them, it was found that large families were more successful at finding food than small ones, mainly because they were dominant over the small families, which were thus effectively excluded from the richest feeding sites. In this way food availability on the Spitsbergen breeding grounds was seen to control the size of the Barnacle Goose population by limiting the birds' breeding success, or to put it in another way, the geese in the study area were numerous enough to deplete existing food supplies so effectively that their reproductive output was thereby being reduced.

GREENLAND WHITE-FRONTS

The White-fronted Geese which breed in west Greenland and winter in Scotland and Ireland, and to some extent in Wales, were first identified as a distinct subspecies by Dalgety and Scott in 1948. They called this Greenland White-fronted Goose *Anser albifrons flavirostris*, after its mainly orange, instead of pink, bill. The total population was thought in 1979 to be around 15 000 birds, a decline, perhaps due to shooting in winter, having probably occurred since the 1950s. In the winter of 1977–1978 a group of enthusiasts at the University College of Wales at Aberystwyth, most of them students in their third year, conceived the idea of an expedition to west Greenland to study and ring Greenland White-fronted Geese, having become interested in the subspecies through watching the flock on the Dyfi estuary, on the Welsh coast north of Aberystwyth. During the planning of the expedition news came from Denmark of the discovery in Greenland of many flightless moulting White-fronted Geese in an area called Eqalummiut Nunaat, about 30–60 miles (50–100 km) north of Søndre Strømfjord Air Base (Map 21). They had been sighted from an aircraft that was surveying the local caribou herds. The area was upland tundra not far north of the Arctic Circle and near the edge of the Inland Ice, easily accessible by air from Søndre Strømfjord. Here eight of the members of the 12-person Greenland White-fronted Goose Study 1979 Expedition landed by helicopter and set up a base camp in the first week of May. They remained in the field until 20 August, covering the entire goose breeding season, being reinforced at the end of June by the remaining members of the expedition, constituting the ringing party.

Nearly every one of the 12 modestly self-styled "amateur naturalists" who spent the summer of 1979 in west Greenland studying White-fronted Geese, other birds, mammals, and plants and plant associations, made individual contributions to the closely-printed 319-page report which was published in 1981, edited by ornithologists Tony Fox, now of the Wildfowl Trust, and David Stroud, now of the Nature Conservancy. This admirable document, almost unique in the detailed information it contained, included a description of Eqalummiut Nunaat, detailed narratives of the expedition and its work, a summary of the "life history and ecology of the Greenland White-fronted Goose", an annotated systematic list of birds seen, and reports on mammals, fish, invertebrates, the flora, and meteorology, as well as a series of technical reports on the equipment used, the food consumed, the expedition's financial affairs, and much else besides. In 16 weeks, expedition members used 20 000 tea bags. When asked by their leader what they would like to eat the reply was invariably "As much as possible"; and indeed it did transpire that their daily ration of around 3000 calories left expedition members hungry, and that 4000 would have been better. As to finance, £2000 of the total costs of £10 000 were contributed by the expedition members themselves and over £3000 by the North Atlantic Treaty Organisation's Eco-sciences Panel and the Wildfowlers' Association of Great Britain and Ireland. The biggest single expense was £3000 for helicopter hire in Greenland. The expedition took with it at least 37 different "drugs and other preparations" but had no doctor; instead its members received instruction before departure from the Aberystwyth St John's Ambulance Brigade. The only serious mishap was the breakage or loss of a good deal of equipment and many packages of provisions in the air-drop, which was made by a Royal Air Force Hercules aircraft the day before the expedition arrived by helicopter.

The core of the Greenland White-fronted Goose Study Group's report was a series of papers on different aspects of the natural history of this species. Nearly all of the information was new, and it was the result of painstaking study in the field. For example, before the geese dispersed into the tundra to breed they fed during May in a marshy area not far from the expedition's base camp. Here they were carefully watched and their behaviour was recorded and quantified. The geese spent 68% of their time feeding, 11% of it roosting, 19% of it alert with heads up, and the remaining 10% or so was used for resting, preening and other activities. The researchers went further than this and, by watching individual pairs, established that, especially in pairs foraging alone, "the goose spends significantly less time alert and more time feeding than the gander" (Fox and Stroud 1981: 66). This behaviour helped maximise the goose's foraging opportunities so that she could build up her nutrient reserves prior to egg laying.

Another paper in the report dealt with predators and predation. Both Gyrfalcons and Arctic foxes were seen attacking geese, but these attacks were unsuccessful, though the remains of adult geese thought to have been killed by Arctic foxes were found on three occasions. Both Arctic foxes and Ravens were blamed for the disappearance from or breakage of eggs in four out of the seven nests found, but in two cases the nests were thought to have first been deserted in a snowstorm. It was concluded that the Raven, the breeding population of which had increased locally in the previous 20 years because of supplementary feeding, especially in winter, at the Søndre Strømfjord Air Base rubbish tip, might be contributing to the decline of the White-fronted Goose population in the area. Nests and nest-site selection was another subject covered: the nests were never grouped together but widely dispersed over the tundra, the smallest inter-nest distance being 1170 metres (1280 yards). A good all-round view from the nest and a marsh where the breeding pair could feed were thought essential. On the breeding behaviour of the geese new information was obtained by continuous watching of a nest from a plywood hide and by means of time-lapse photography. It was found that during much of the incubation period the gander grazed in a marsh some distance away, but at the beginning and end of incubation he spent more time within a few yards of the nest. The incubating goose left the nest roughly once very 24 hours for an average time of 24 minutes, most of which she spent feeding or drinking. She also collected droplets of water from her flanks while incubating in wet weather. The summer diet of the geese was studied by collecting and later microscopically examining the droppings of both adult and young geese, and cotton-grasses (*Eriophorum*) were found to be the most important food item. Other subjects dealt with included the behaviour of goose families, plumage variation and parasites.

The ringing programme was very successful. No fewer than 96 moulting geese were caught and ringed. Considering the difficulties, this was a noteworthy achievement. Flocks of the flightless geese were widely dispersed at the tundra lakes where they could seek safety from Arctic foxes by swimming out towards the centre of the lake. They were extremely shy of human disturbance and would leave the lake and run off quickly up the opposite hillside if anyone appeared in view, even as far as a mile away. The catching parties, each consisting of from six to eight expedition members, had to carry rings, nets, inflatable boats and food for 4 days. When a flock was located the party spread out to surround the lake while remaining out of sight of it, a process that could take up to 2 hours. Once the lake was surrounded, the catchers emerged and walked towards it, each varying the speed of his or her approach to keep the geese in the centre of the lake. For communication between catchers, which was absolutely essential to coordinate the various manoeuvres, the expedition had three "walkie-talkies" and also used hand signals, mirror flashes and a whistle. When the geese were safely in the middle of the lake three or four catchers round the shore could keep them there while the others inflated the boats and put up the nets in an arm of the lake where they could be set partly in the water, forming a funnel leading to the pen, which was on dry ground. Then the geese were chivvied into the funnel by moving the boats toward them while the catchers along the shores on either side prevented their escape to one side. Of these 96 geese ringed in Eqalummiut Nunaat in summer 1979, the surprisingly large total of 53 had been identified outside Greenland by the time the report went to press. Two were shot on passage in Iceland and 46 were identified by reading the rings in Scotland during the winter.

The Greenland White-fronted Goose Study Group did not rest on its laurels after the publication of the report of its 1979 expedition (Fox and Stroud 1981; see too Fox and Stroud 1988). It undertook another expedition in 1984, and in August 1988 and subsequent years the Fox–Stroud partnership was in action again in west Greenland. In 1988, partly funded by the government of Greenland, they carried out an aerial survey of their old study area and some other areas nearby. They also made further searches on foot in a new area north of Søndre Strømfjord Air Base. A fixed-wing low-flying aircraft was used for the aerial survey and nearly all the geese overflown were thought to have been visible from the air. The team's resolve to continue these goose researches by air and on the ground was reinforced by its 1988 experiences. Further work was also judged necessary to promote the conservation of a goose population which probably still only numbered between 20 000 and 30 000 birds. In their 1989 article reporting on the 1988 expedition, Fox and Stroud deplored the fact that, while the government of Greenland had recently taken important measures to protect these geese, the British government, represented by the Scottish Office, had permitted them to be shot under licence on Islay in 1987–1988!

LAPLAND BUNTINGS AND OTHER BIRDS AT BARROW, ALASKA

Among the ornithologists who used the Naval Arctic Research Laboratory at Point Barrow, Alaska, as a base for their research, one of the most regular summer visitors there was Frank Alois Pitelka. Arriving there in 1951 as leader of a University of California (Berkeley) team working on the population biology of Arctic land vertebrates, he returned year after year to study tundra ecology. In 1974 he published a detailed account of the avifauna of the Barrow region, by which time he was on the staff of the Museum of Vertebrate Zoology of the University of California at Berkeley and had published papers on lemming cycles, on the breeding biology of the Pectoral Sandpiper, on the ecology of the Arctic Slope of Alaska and other subjects, either alone or jointly with colleagues.

The Arctic Slope is the name given to the lowlands between the Brooks Range and the Arctic Ocean, and the Barrow region is the most northerly part of this area. Because it is only some 500 miles (800 km) from Asia as the crow flies and is on the mainland, Barrow has a longer bird list than most places in the North American Arctic, many of which are on islands, as well as a high proportion of stragglers from a long distance away. Thus Irving, in 1960 (p. 257), reckoned that the nearest regular breeding place of at least eight species recorded at Barrow was 1000 miles (1600 km) or more away. These are listed in Table 16. By the time Pitelka described the Barrow avifauna in 1974 the list of birds recorded there had been considerably lengthened, mainly because of the presence of ornithologists based at the Naval Arctic Research Laboratory. He reckoned that there were only 22 species regularly breeding at Barrow and 13 more that bred occasionally. Regular migrants numbered only 9 species. But a remarkable 44 species had strayed to Barrow from central Alaska, 11 more from Canada or southern Alaska, and 12 from Asia. So the Barrow list stood at 151 species when Pitelka wrote. He rightly predicted that more stragglers could easily be added. "Had my job been the discovery of stragglers, I would have long ago set up a small oasis of sham trees and shrubs on an area near the coast, clean of snow, in late May and early June. Then with an insulated deck chair and a hot toddy, one has merely to wait for the stragglers to come in!" (Pitelka 1974: 170). In winter, on the other hand, Pitelka recorded few or no birds at Barrow: "There are no winter residents and no regular permanent residents among either land or water birds. . ." (p. 171).

Pitelka's work on the demography of the Lapland Bunting or, to give it its American name, the Lapland Longspur, was carried out at Point Barrow between 1951 and 1973 and written up jointly in 1977 with his colleague

Table 16 Distance from their nearest regular breeding places of some birds recorded at Barrow (after Irving 1960).

Name of species	Scientific name	Miles from nearest nesting place
Scarlet Tanager	*Piranga olivacea*	2400
Eastern Kingbird	*Tyrannus tyrannus*	1800
Brewer's Blackbird	*Euphagus cyanocephalus*	1800
Wren	*Troglodytes troglodytes*	1200
Red-winged Blackbird	*Agelaius phoeniceus*	1100
Western Tanager	*Piranga ludoviciana*	1100
Killdeer	*Charadrius vociferus*	1000
Common Nighthawk	*Chordeiles minor*	1000

Thomas W. Custer. Custer and Pitelka also studied other breeding birds of the Barrow tundra. For example, in 1971 they collected information on 34 nests of the Pomarine Skua there (Custer and Pitelka 1987). Their Lapland Bunting study was based on nests found within 3 miles of the Naval Arctic Research Laboratory, and they concentrated their efforts on a 17-hectare (42-acre) study plot immediately southeast of the laboratory where 56 nests in all were found during the 7-year period 1967–1973 and where as many birds as possible were caught in sparrow traps and marked with U.S. Fish and Wildlife Service rings and with unique combinations of colour rings. During the course of the study most of the breeding females and all the males in the plot were so marked, and all nestlings were ringed in the same way before fledging. The number of nests in the plot varied from 15 in 1968, when six of them belonged to three polygamous males, each having two mates, to two in 1972. The date of the first egg laid was determined for a total of 227 nests in the 7 years and it was found that the first clutch was always begun in the first week of June and that all clutches in a given year were begun within 20–25 days of each other. Only two cases of attempts to raise a second brood in July were recorded. In one the female stopped feeding the young on the sixth day after hatching and they died; in the other, three of a brood of four were last seen in the nest ready to fledge but their fate was unknown. The authors thought it most unlikely that these rare attempts at a late brood ever succeeded. The adult Lapland Buntings at Barrow began their complete body moult around 1 July and this could only just be completed before the onset of cold weather and snow triggered off their departure on migration between mid-August and early September. Nor could young birds complete their growth and their post-juvenile moult in the time available if they left the nest much after mid-July, especially as the buntings' food supply declined abruptly after the emergence of adult crane flies at that very time.

The data presented by Custer and Pitelka were on breeding density, clutch size, hatching and fledging success, and survival of adult Lapland Buntings. They found that the onset of breeding depended on the snow melt, but it was also timed so that the peak abundance of larvae, mainly of craneflies, occurred soon after the young left the nest. Considering the severity and variability of the spring and summer climate in the Arctic, it was surprising to find that Lapland Buntings virtually stuck to their average clutch of five eggs whatever the weather. Even in the very late spring of 1969 their mean clutch size only dropped to 4·76. Nor did the Arctic climate seem to affect losses of eggs and young, which were comparable to those of open-nesting passerines in the temperate zone. The amount of predation varied from year to year, and the authors argued that a decline in Lapland Bunting numbers which they recorded during their study was probably due to the failure of the local brown lemmings to

---

Plate 60 (a) A female Lapland Bunting on the ground will sometimes allow close approach, as here. Crouched motionless it is all but invisible.

(b) Boulders are favourite song-posts for the male Lapland Bunting. Both photographs were taken in summer 1986 near Cambridge Bay.

(a)

(b)

reach their usual cyclic peak in those years: "If small mammals are scarce, predators rely more heavily on birds. . ." (Custer and Pitelka 1977: 523). The local predators that accounted for the bulk of Lapland Bunting nest failures were mostly skuas, both Pomarine and Arctic, but Snowy Owls, Arctic foxes and weasels were perhaps important in some years.

SOME ARCTIC RESEARCH PROBLEMS

While one does not have to travel to the Arctic to study geese or buntings, there are some ornithological puzzles that can only be solved there. Most of these problems involve more than a single species or group of species, and they are essentially products of the Arctic climate. Among them are non-breeding, recurring cycles of scarcity and abundance and their relationships with those of Arctic mammals, the way birds cope with continuous daylight, and the question of whether migrating birds can and do cross ice-caps. Some of these problems have already been briefly mentioned in the first chapter.

As a result of an expedition to northeast Greenland in 1936–1938, C. G. and E. G. Bird published a paper in the *Ibis* (1940) entitled "Some remarks on non-breeding in the Arctic, especially in north-east Greenland" in which they reported almost complete non-breeding in summer 1938 for all birds except passerines. Subsequent research has so far failed to document large-scale non-breeding affecting all species, but has confirmed that it does at times occur in certain places for certain species (Meltofte 1985; Sage 1986: 85–86). In the summer of 1947 A. J. Marshall joined the Oxford University Expedition to Jan Mayen at 71° N (Map 25) especially to study what his paper, published in the *Ibis* in 1952, described as "Non-breeding among Arctic birds". His conclusion, "based on the laboratory examination of 53 non-breeders" was that they were perfectly normal physiologically but were prevented from breeding for different reasons in different cases, such as lack of nest sites secure from predation by the numerous Arctic foxes, or food shortage, or the absence of suitable habitat. Since none of the phenomena usually associated with non-breeding in the Arctic—such as low temperature, persistent sea ice, late snowmelt and scarcity of lemmings—occurs on oceanic Jan Mayen Island, perhaps it was not the best place to study this subject. Some well-known examples of the phenomenon are the failure of Kittiwakes to breed in some years in the New Siberian Islands, the failure of divers and swans to breed in years when the tundra lakes remain frozen, the non-breeding of Snowy Owls and skuas in poor lemming years, and the failure of Brent Geese to breed in years of late snowmelt.

In 1986 R. W. Summers proposed in a brief paper published in *Bird Study* that the cycle of lemming abundance in the Taimyr Peninsula, of a peak year every 3 years or so, might be linked in some way with the breeding success of the Dark-bellied Brent Geese there, which he supposed showed a similar 3-year cycle. Perhaps Arctic foxes switched their prey to geese and goslings when lemmings were scarce? A series of papers followed which introduced

British birdwatchers to a central problem of ornithological research in the Arctic. In 1987 *Bird Study* carried no fewer than four papers in a single issue on the breeding performance of Brent Geese. Myrfyn Owen of the Wildfowl Trust denied the existence of a 3-year cycle in Brent Goose breeding success and thought the relationship between lemmings and Brent Geese "may well be spurious". André A. Dhondt of the University of Antwerp argued that goose reproduction would be poor in the summer *following* a lemming peak, because the many predators resulting from that successful breeding season, confronted with declining lemming numbers in the following summer, would then switch to birds. He also pointed out that goose reproduction would be adversely affected in any summer when the spring snowmelt came exceptionally late. Hugh Boyd, of the Canadian Wildlife Service, pointed out that there could be intrinsic reasons for a 3-year breeding cycle in Brent Geese. Since most Brent Geese do not breed until they are 3 years old, "high production in one year should be followed by an increase in breeders three years later". He also drew attention to the difficulty of obtaining adequate data on the Taimyr climate and on the present-day breeding distribution of the geese, which were 10 times as numerous in 1987 as they were in 1957. Nor was there any information available on the date of the Taimyr snowmelt in different years. Lastly, R. W. Summers, co-authoring with L. G. Underhill, restated his case for a 3-year breeding cycle in Brent Geese correlated with lemming abundance whether or not Dhondt's time-lag effect was taken into account.

In 1989 Dutch ornithologist Barwolt S. Ebbinge of the Research Institute for Nature Management (Rijksinstituut voor Natuurbeheer) at Arnhem, weighed in with a paper published in the *Ibis*. Going beyond mere speculation, he produced important new evidence in the shape of data on body-weights of Brent Geese caught with cannon nets in the Wadden Sea marshes just before they set off on their spring migration to the Taimyr Peninsula and on wind conditions during their flight. He showed that these parameters, as well as fox predation and snowmelt on the breeding grounds, need to be considered when explaining the ups and downs of Brent Goose breeding success, and he called this a "multifactorial explanation". Though Ebbinge had the good sense to ask Summers, Owen, Boyd, Dhondt and other goose experts like Rudi Drent to comment on his paper before publication, the 1990 volume of the *Ibis* contained a sharp attack on it by Underhill and Summers, their names now transposed, followed by a reply from Ebbinge and a brief rejoinder by Underhill and Summers. No doubt the debate will continue and our understanding of goose population cycles will continue to improve.

In the first issue of the *Ibis* for 1954 two papers appeared on the same subject, their differing titles reflecting differences in the mentalities of their authors. The more literary-orientated Cambridge clergyman and amateur ornithologist Edward A. Armstrong wrote on "The behaviour of birds in continuous daylight" while the zoologist from Oxford, J. M. Cullen, entitled his paper "The diurnal rhythm of birds in the Arctic summer". Both were trying to answer the same question, namely, how do birds react to the absence of the natural roosting time provided by some hours of darkness every night? Armstrong went to Abisko in Swedish Lappland at 68° N (Map 22) to obtain his data; Cullen spent the summer of 1950 on the island of Jan Mayen, where Marshall had studied non-breeding 3 years earlier, at 71° N, to study seabirds. Armstrong's observations of Bramblings, Willow Warblers and other song birds led him to believe that a lull in a passerine song occurred in the hours before midnight. Whereas Fieldfares he watched fed nestlings through the night, Willow Tits did not. Cullen's observations of Kittiwakes, Fulmars and Brünnich's Guillemots were likewise inconclusive. Much fuller data were obtained at Mesters Vig in east Greenland (72° N) (Map 22) in 1974 by two Danish students, Sten Asbirk and Niels-Erik Franzmann (1978), who accompanied an expedition there undertaken jointly by an English Wader Study Group and the Scottish university of Dundee (Green and Greenwood 1978). The two Danes between them maintained a 24-hour continuous watch on a Wheatear's nest with young on four different days in mid-July. They found that both male and female Wheatear went through a period of inactivity during the night of, on average, 4·75 hours' duration, which began some time between midnight and 03·45. They watched Snow Buntings in the same way, and found that their nocturnal period of inactivity began 2 or 3 hours earlier than the Wheatear's but was of a similar length. The stimuli that initiated these inactive periods and brought them to a close could not be identified.

Can migrating birds cross ice caps? Irving (1960: 279), when discussing the Wisconsin ice sheet which covered much of North American 10 000 years ago, stated that "land birds could not migrate over continental ice caps". But migration studies in Greenland have given him the lie. Already in 1967 Salomonsen (1979) concluded from ringing returns that there was a transglacial migration there, at least in spring, of White-fronted Geese, Brent Geese, Turnstones and Knots. Long before then, in 1949, a French Polar expedition (Expéditions Polaires Françaises) which set up a scientific station (Station Centrale) almost in the centre of the Inland Ice at 70° 45′ N, reported seeing geese, ducks, Ptarmigan, and a Snowy Owl, and several Turnstones were caught there in summer–autumn 1949 and spring 1950. The construction in 1960 of two radar stations called DYE 2 and DYE 3 belonging to the Distant Early Warning Line (DEW-Line), on the Inland Ice, turned out to be a crucial event in transglacial migration studies in Greenland (Map 21). On 18 May 1961 Willie Knutsen, a major in the Norwegian army staying at DYE 3, about 125 miles (200 km) northwest of Ammassalik, found a dead Wheatear. Then at DYE 2, some 186 miles (300 km) southwest of Søndre Strømfjord, on 21 July 1968, an exhausted Knot was picked up. A week later, a Turnstone was found dead there wearing a ring which had been put on its leg on St Agnes

---

Plate 61   (a)   On 8 July 1986 at Cambridge Bay this (probably female) Grey Plover gave a prolonged and elaborate distraction display after leaving its clutch of three eggs (figured in Plate 25). The "head down, wings out stance" shown here is also found in the Pacific Golden Plover.

(b)   The same bird lay flat on the ground for some time with wings outspread. It then clapped them noisily on the ground after raising and lowering one or both wings, conspicuously, deliberately and slowly. Here one wing only is being raised and lowered rapidly.

(a)

(b)

(a)

(b)

in the Scilly Isles, England, on 30 September 1960. In July 1971 Finn Salomonsen got permission from the Chief of Staff of the United States Air Force to stay for 5 days on DYE 2. He describes it succinctly (1979: 198) as being 84 metres (276 feet) high, weighing 3000 tons, having five floors, and being supported by eight steel pillars. He was disappointed not to see any migrating birds but consoled himself with the thought that the station was too far south and that his visit was too early in the season. He also found that the light was so intense in daytime that sunglasses were essential and these made it hard to see birds migrating overhead. He did find several dead birds: two Snow Buntings, two Wheatears, and a Red-necked Phalarope. He painstakingly interviewed the 15-man staff of the station about the birds they had seen. Considering that the crew usually worked inside and the structure had no windows, it is surprising how many birds the men *had* seen. Everyone had noticed Snow Buntings, and Mallards, Purple Sandpipers and probable Sanderlings were also reported. Particularly suggestive was H. E. Hunton's report of small flocks of White-fronted Geese flying toward Søndre Strømfjord on several days in the

---

Plate 62   (*a*)   The distraction display of a Knot in northwest Greenland on 3 July 1983, which was escorting three or four downy young when we came across it, consists of wing-raising, walking away with drooping fanned-out tail, and settling briefly on the ground as shown here.

        (*b*)   The Baird's Sandpiper has an elaborate distraction display during which it permits very close approach. The bird shown here led my wife and me across the Cambridge Bay tundra more than 100 yards from its nest, on 5 July 1986, in a fine example of the "rodent run". In the photograph this bird is fluffing out its feathers, depressing and fanning its tail, and "squealing" loudly.

spring of 1970. Later Hunton sent Salomonsen a Wheatear which actually flew into the station during the night of 21–22 August 1971 but died later.

In the springs of 1980 and 1982 a Scandinavian research team (Alerstam *et al.* 1986) was allowed to install automatic cameras to photograph the radar screen at DYE 4, which was on Kulusuk Island near Ammassalik, east Greenland (Map 21). To identify the migrating birds whose echoes were recorded on the screen, field observations were made near Ammassalik and on the shores of Disko Bay in west Greenland. As a result regular spring bird migration was demonstrated across the Inland Ice, in two directions, east or southeast and west and northwest. Divers, ducks and some seabirds were thought to cross the Inland Ice from west to east, while geese, High Arctic waders and probably Wheatears, were thought to be involved in the northwesterly flights. Iceland was the essential springboard for the northwesterly migration which passed the east coast of Greenland near Ammassalik, continued over the Inland Ice, and ended in northwest Greenland and the eastern Canadian Arctic. Typical of these migrants wintering in Europe and breeding in the general area of Ellesmere Island were thought to be Brent Geese, Ringed Plovers, Knots and Wheatears. These birds have to cross 450–700 km (280–435 miles) of a lifeless, waterless, windy, ice desert ascending to an altitude of 2500–2800 m (8000–9000 feet) with below-zero temperatures. Their total flight distance is 2300–3000 km (1430–1864 miles). Subsequently a group of Swedes from the Department of Ecology of the University of Lund, including three members of the 1980–1982 team, have confirmed that Brent Geese, Knots and Turnstones take off from northwest Iceland in spring to cross the Greenland ice cap to northwest Greenland and the eastern Canadian Arctic and that they follow a compass bearing of about 300°. Exactly how they do this is not yet understood (Alerstam *et al.* 1990).

Although it is a familiar phenomenon, the last word still has not been said on distraction behaviour, formerly known as injury feigning. In particular, it has not yet been satisfactorily explained why this seems to be more often found in the Arctic than elsewhere. It often takes the form of a so-called 'rodent run', when the adult bird runs away along the ground with fluffed-out feathers, uttering a squealing note so as to resemble a mammal running off. Or the bird retreating on the ground may appear to simulate injury—especially a broken wing—and inability to fly properly. This behaviour is very well developed among Arctic-breeding skuas and waders. The northern races of the Ringed and Golden Plover have much more elaborate distraction displays than western European birds of these same species, and many of the waders known for their distraction displays breed in the Arctic, for example the Sanderling and Purple Sandpiper. This behaviour pattern with its link to the Arctic and most probably to the Arctic fox, the predator it seems "intended" to deceive, deserves more attention than it has so far received (Duffey *et al.* 1950; Campbell and Lack 1985: 144–145).

CHAPTER 12

The birdwatcher's Arctic: where to go and how to get there

There is no intention here of providing an up-to-date travel guide for the Arctic birdwatcher. No details of costs or flight schedules will be set out, but rather a general idea will be given of what some of the more accessible places are like and of the ways and means of getting to them. Although some guided tours may be mentioned, names and addresses of tour operators will not be listed. Since the aim of most of these tours for birdwatchers, or birders as they are called in America, is to log as many species as possible, very few visit the Arctic. For, in a typical Arctic locality, a birdwatcher would be lucky to see 40 or 50 different species in all, or add more than a very few to his or her life list. Naturally, places like Florida, Israel or the Galapagos Islands are more attractive. Nevertheless, the Arctic has its share of "special" birds of limited geographical distribution, and of beautiful or spectacular ones.

THE ALLURE OF THE ARCTIC

The Arctic is also a very special place, with an elusive appeal that entices its visitors to return over and over again. It can be profoundly still and calm. The

298   *In search of Arctic birds*

occasional noises, of cracking ice, or perhaps, of a distant aircraft or vehicle, reverberate through the thin, calm air, seeming nearer than they are. Distances are foreshortened too. What looks like a few hundred yards across a gravelly plain to the foot of a glacier turns out to be more than a mile. The Arctic air is keen and invigorating. The winds that blow over the sea ice or from an inland ice cap are bitingly cold. Summer weather can be warm and balmy even in the High Arctic, but it can be foggy, grey and wet as well. Summer blizzards occur, though the fresh snow will not usually lie. The tundra landscape in summer is a mosaic of soft colours; shades of beige, brown and grey. Everywhere the ground is boulder-strewn, and the boulders are blotched and speckled with black, yellow, orange and grey lichens. Ground vegetation is short. Low, creeping or mat-forming plants predominate, some with white, yellow or purple flowers. Lakes are a feature of many an Arctic landscape. Often they are not ice free until July. The tinkling of pieces of melting ice as a breeze stirs the surface of the water is a characteristic summer sound on their shores. And of course, Arctic scenery in many areas is dominated throughout the summer by ice and snow. Snowdrifts often remain until July and the frozen sea, with its uneven surface pitted with innumerable pools of melt water and rain water, persists along many coasts. The extraordinary emptiness of the Arctic imparts a sense of wild, undisturbed, unspoilt nature, and this is enhanced in June when, to the visual beauty of the landscape is added the sound of bird song. The Snow Bunting sings like a lark high in the sky, as well

(a)

(b)

Plate 63 (a) An incubating Turnstone merges well into the background of the stony, grassy tundra of the Varanger Peninsula, north Norway. Breeding waders on the tundra can only survive if superbly camouflaged. (b) The beautifully marked eggs of the Stilt Sandpiper, just to the left of the boulder in the photograph, are unlikely to be seen by an Arctic fox or a skua and might even escape an Eskimo. Cambridge Bay, 30 June 1986

as from a boulder. The display flight of the Knot over the barren uplands is accompanied by a hauntingly melodious whistle. Perhaps most evocative of all is the cry of the Long-tailed Duck, which may be heard in almost every region of the Arctic. The summer traveller to the Arctic will experience the midnight sun, which is much more than a routine meteorological phenomenon; it is mysterious and exciting. A warm sunlight from low near the horizon casts long shadows and a soft glow. It is surprisingly difficult to get used to the fact that, at whatever time you wake at night, it is light.

For the birdwatcher who enjoys camping excursions into the wilderness the Arctic in summer is nearly ideal. It is cool enough to make backpacking pleasurable and, although the ground may be rocky and uneven or boggy, it is usually possible to cover 10 miles (16 km) or more in a day. Fuel for a camp fire in the form of Arctic bell heather or Arctic willow twigs (Plates 51 and 59) is often available, and driftwood occurs on many seashores. Best of all, excellent drinking water is to be found everywhere. Naturally one may camp anywhere one wishes and there is little likelihood of human disturbance. The weather will probably be mixed, but in July and August temperatures will scarcely fall

300 *In search of Arctic birds*

(a)

(b)

below freezing. In some areas, Greenland for example, high winds may be encountered; a top-quality lightweight tent is essential.

The bird life of the Arctic, though poor in variety of species and often in numbers of birds, has several attractive features, two of which perhaps deserve special mention. The Arctic offers many fine examples of camouflage, or what is scientifically termed adaptive coloration, in birds. These range from the pure white winter plumage of the Ptarmigan to the striking black-and-white pattern of the Turnstone's summer plumage. On the open tundra the waders' eggs merge just as well with their surroundings as the sitting bird does. The birdwatcher in the Arctic will also derive pleasure from the tameness of Arctic birds. Breeding waders often permit very close approach, phalaropes in particular, and the same is true of individuals of many other species.

ARCTIC VEXATIONS

However careful or fortunate the Arctic birdwatcher may be, he or she is bound, sooner or later, to encounter that ubiquitous pest, the mosquito. Although there are far fewer species of insect in the Arctic than in temperate zones, some of those that do occur are exceedingly abundant, and this applies to the two tundra species of mosquito. Although these can reproduce without the need of a meal of vertebrate blood, they will certainly take such a meal when opportunity offers. And it is not just their bites which annoy the birdwatcher. They often occur in such numbers that using binoculars can become difficult and photography almost impossible. Although the mosquito problem cannot be solved, it is possible to palliate or avoid it to some extent. Out on the open tundra, mosquitoes do not fly in very windy weather, but this is hardly helpful for the birdwatcher, who may find it difficult to find birds in such conditions. In a few places in extremely high latitudes, such as parts of north Greenland, mosquitoes may be rare or even absent. In some regions they are patchily distributed and it is possible to camp in places that are virtually mosquito-free. For example, around Cambridge Bay in Canada in 1960 they were not "uniformly abundant over the tundra, many places having few if any of them" (Parmelee *et al*. 1967: 14). In much of the Arctic mosquitoes are troublesome only for a few weeks, beginning in late June or early July. Birdwatchers would do well to plan their trips early in any case, to coincide with the spring return of birds to the tundra and the start of song, courtship and nesting activities. A 3-week visit starting about 10 June might altogether

---

Plate 64 (*a*) Watching Red-necked Phalaropes at close range in northwest Greenland. Near Thule Air Base, 27 June 1983.

(*b*) Although it scuttled away when first found on North Mountain near Thule Air Base, this male Ptarmigan allowed approach to within a yard or two. Demonstrating the "tameness" of Arctic birds in these photographs is the author's son John.

avoid serious trouble from mosquitoes. At Cambridge Bay in 1960 they were out in force only after 28 June. In 1986 my wife and I were camping 8 miles from there out in the tundra. Mosquitoes are not mentioned in her diary until 7 July when "today there were a few mosquitoes, but nothing much" and only on 9 July were they "out in force really for the first time". Of course the mosquito nuisance can be substantially reduced by protective clothing and insect repellants. The former is usually too warm, often gets in the way, and is hard to use with binoculars. The latter is smelly and unpleasant but effective for an hour or two. A combination of both is recommended, but not too much of either.

The Polar bear is in many ways a much more serious nuisance than the mosquito. Admittedly, if one chooses one's area carefully the chances of an encounter with one can be minimised. They are abundant around the shores of Spitsbergen but rarer inland. In Greenland they are absent from the west coast between Godhavn (Qeqertarsuaq) and Frederikshåb (Paamiut) (Map 28). There are some inland parts of the Canadian Arctic where they very seldom penetrate. Even when seen, a bear will usually not attack and can often be avoided. It is not the risk of meeting a bear, but the need to keep constantly on the lookout and perhaps to carry a rifle to save one's life by shooting it if this becomes really necessary which is annoying and frustrating for the birdwatcher. It is worth mentioning that an inexperienced person with a loaded rifle is probably more dangerous than a Polar bear, especially when one appreciates for how long one is likely to be exposed to each on an expedition. Things can be difficult for the bird photographer, who can hardly be expected to be on the *qui vive* for an approaching bear when sitting in a hide watching a Snowy Owl at the nest. The best advice one can give is not to wander about on one's own. If one is camping, then refuse should be burned and food kept well wrapped at least 100 yards from the tent, which should be surrounded by trip wires. If a rifle is taken, make sure you know how to use it. A Canadian leaflet advises against camping on beaches and suggests that scaring devices such as flares or propane cannon should be ready to hand. Members of at least one expedition frightened a polar bear away from their tent by rattling billy-cans, but the best precaution of all is to choose a locality where polar bears are scarce or non-existent. In any case the Arctic traveller should read carefully Danish biologist Henning Thing's (1990a) excellent booklet *Encounters with wildlife in Greenland*, available from the Greenland publishers Atuakkiorfik in Nuuk.

A TASTE OF THE ARCTIC: ICELAND AND NORTH NORWAY

For the European birdwatcher, the nearest and most easily accessible parts of the Arctic are east Greenland and Spitsbergen; for the American, Churchill on the west coast of Hudson Bay in the Canadian province of Manitoba is nearest. It is usually reckoned to be just in the Arctic and can be reached by rail. But in Europe at least some impression of the Arctic may be gained from a visit to Iceland, just south of the Arctic Circle, or to the Varanger Peninsula, in northeastern Norway, a little north of 70°N.

Iceland is only 2 or 3 hours flying time from Glasgow or London and there are daily flights throughout the year. It is not the true Arctic, but it has Arctic affinities, especially in its treeless landscape, its barren stony hilltops, and its glaciers and ice caps. Since British bird photographer G. K. Yeates published his book *The land of the loon* in 1951 describing two visits there in search of the Great Northern Diver, other Arctic enthusiasts who figure elsewhere in this book have followed. In 1951, Peter Scott, leading a Severn Wildfowl Trust Expedition there, took James Fisher along and they caught and ringed over a thousand Pink-footed Geese (Scott and Fisher 1953). In 1957 it was American bird painter George Miksch Sutton's turn to spend a summer birdwatching in Iceland and write a book about his experiences. Both of them acknowledge the help of the world-famous Icelandic ornithologist Finnur Guðmundsson, director of the Natural History Museum in Reykjavik from 1947 until he retired in 1977. Unfortunately he died before he could complete the definitive work on the birds of Iceland that he had long planned. Meanwhile his second cousin, Hjálmar Bárðarson, an enthusiastic bird photographer, had accompanied him on many expeditions with a view to illustrating his book with high-quality bird photographs taken in Iceland. After Guðmundsson's death, Bárðarson continued his bird photography and in 1986 published some 500 photographs accompanied by his own, more or less popular, account of Iceland's birds. This took the form of a de luxe volume, illustrated mainly in colour. Printed in Holland, it was privately published in Reykjavik simultaneously in Icelandic Danish, French, German and English. The ornithological visitor to Iceland might do well to look at it before setting out, even if he cannot afford to buy it (see too Breuil 1989).

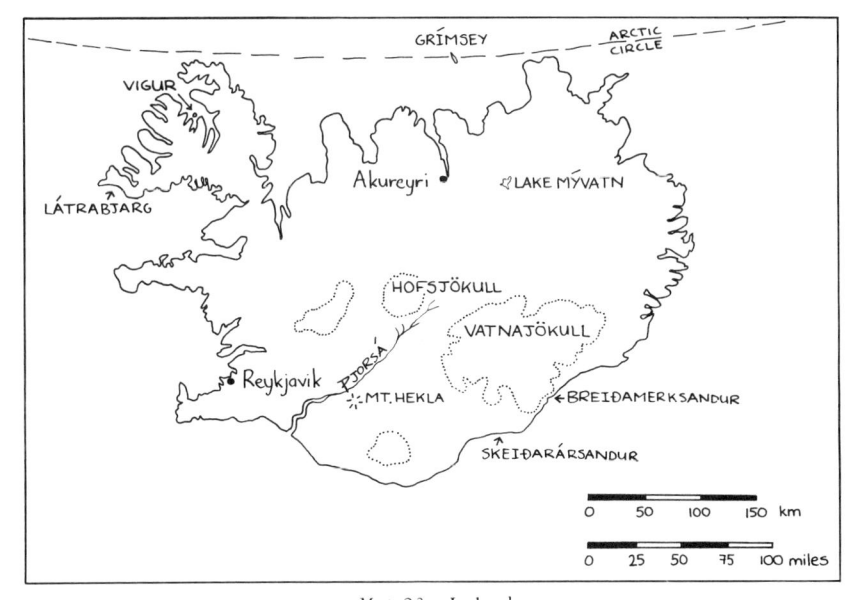

Map 23 Iceland.

Several truly Arctic birds reach the southernmost part of their breeding range in Iceland. The other breeding places of the Pink-footed Goose are in northeast Greenland and Spitsbergen, both in the High Arctic. The most southerly breeding colony of the Little Auk is or was on the island of Grímsey off the north coast: in 1983 only two pairs remained there. Two typical High Arctic waders, the Grey Phalarope and Purple Sandpiper, breed in Iceland; so do Gyrfalcons and Snowy Owls, usually associated with higher latitudes. In July 1949, Adam Watson, of subsequent Snowy Owl and Ptarmigan fame, was skiing in the mountains of northeast Iceland with two friends when he came across a sitting wader on a patch of snow-free ground at 2600 feet (792 m). It was a Knot brooding four newly-hatched chicks (Yeates 1951: 37). Was this an isolated case of a pair of Knots cutting their normal migration short by 500 miles, that is, failing to fly on across Denmark Strait to the Scoresby Sound area of Greenland? Or does a sparse population of breeding Knots inhabit the remote, inaccessible mountains of northern Iceland? They could be extraordinarily difficult to find. As well as some Arctic species like these, the summer birdwatcher in Iceland should find some 12 or 15 out of the 16 duck species which breed around Lake Mývatn. Along the coast of the northwestern peninsula he or she can see some of the finest seabird colonies in the world at Látrabjarg, including a quarter of a million Razorbills, more than half the world population (Grimmett and Jones 1989). And on the vast flat sandy and gravelly plains along the southeast coast—the Breiðamerkursandur and the Skeiðarársandur—are the world's largest colonies of breeding Great Skuas. Lastly the majestic Whooper Swan, a widespread breeder, will surely not be missed.

In the far northeast of Norway, the Varanger Peninsula (Vaughan 1979) reaches out into the Barents Sea toward Russia and the North Pole. Warmed by the Gulf Stream, it can scarcely be regarded as Arctic: there are even cows there, reputed the most northerly in the world. Yet it is further north than at least one of the places which will figure as undeniably Arctic in the pages that follow, namely Cambridge Bay in Canada. It has the advantage of being accessible by car: it is 1350 miles (2160 km) by road from the port of Göteborg in south Sweden. Or one can fly from Oslo to Kirkenes and then take the local bus. Though birch trees grow along the southern slopes of the peninsula, most of it is barren, rocky tundra with bare rounded fells rising to 2375 feet (724 m) at their highest point. A road runs along the coast to the only two towns, Vadsø and Vardø, and beyond; another crosses the northern part of the peninsula to Berlevåg and beyond. The sun is continuously above the horizon from 18 May to 26 July; the lighthouses are closed from 25 May to 11 August.

The Varanger Peninsula holds many of the typical European birds of the far north, including breeding waders like Bar-tailed Godwit, Temminck's Stint and Turnstone, and birds of the boreal forest like the Redwing, Brambling and Siberian Tit. At Syltefjordstauran is the world's northernmost Gannet colony, numbering 300 pairs, and in addition there are over 1000 pairs of Brünnich's Guillemot. This last also breeds on the islet of Hornøya, together with thousands of Guillemots, Puffins and other seabirds.

Arctic birds to look out for in the Varanger Peninsula in summer are the

Map 24 The Varanger Peninsula.

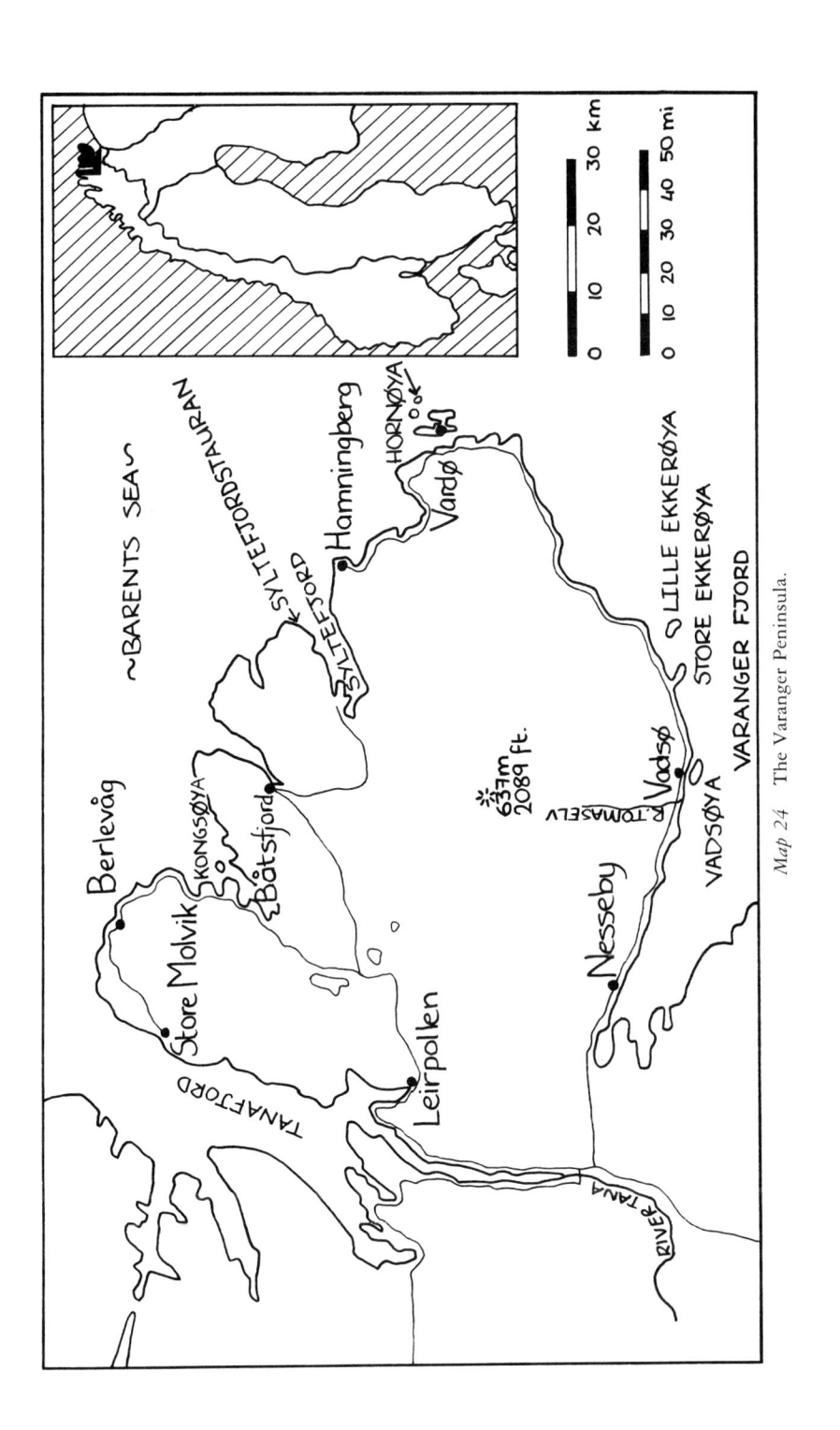

Long-tailed Duck, Little Stint, Long-tailed Skua and Snowy Owl; the Shore Lark and Snow Bunting are common. Late winter and early spring is the time to see that Varanger speciality Steller's Eider (Plate 40). Over 1000 have been counted in some years in the Varanger Fjord at Vadsø, together with larger numbers of King Eiders. For the European birdwatcher, the Varanger Peninsula is the nearest place to be sure of seeing either of these Arctic ducks. Otherwise he or she would have to travel to east Greenland or the Kanin Peninsula (poluostrov Kanin: Map 7) to see King Eiders. As for Steller's Eider, it breeds regularly only along Siberia's Arctic coast between the Taimyr and Chukchi peninsulas.

SVALBARD OR THE SPITSBERGEN ARCHIPELAGO

Spitsbergen lies between 76° and 82° N, and is well within the Arctic, but the Gulf Stream keeps the west coast free of ice in the summer months and helps to make the climate somewhat milder than it is in other places in the same latitude, such as Severnaya Zemlya. Although it is expensive to reach and difficult to travel around in, there is no need for the birdwatcher to go to the length and expense of responding to advertisements like the one that appeared in *The Times* on 27 November 1987:

> ARCTIC EXPEDITION sledging 80 days through Polar country, Spitsbergen to Edge Island, Feb. to March 1988, seeks fourth person. Joining fee £8000.

An international treaty was signed in Paris in 1920 granting Norway sovereignty over "the Archipelago of Spitsbergen". The signatories or contracting parties, namely Norway herself, the United States of America, Great Britain, Denmark, France, Italy, Japan, the Netherlands, Sweden and later the Soviet Union, granted their citizens equal rights to travel to Spitsbergen and fish and hunt there as well as to carry on "maritime, industrial, mining and commercial operations". Conservation was a prime consideration of the treaty, and Article 2 empowered Norway to "maintain, take or decree suitable measures to ensure the preservation and, if necessary, the re-constitution of the fauna and flora of the said regions, and their territorial waters" (Østreng 1977: 101–105). As a result of this treaty nationals of the above-named countries require no passport to visit Spitsbergen, provided of course that they do not travel via Norway, or indeed any other country. But they are subject to the rigorous Norwegian laws and regulations applying to Spitsbergen.

Apart from visiting it in one's own yacht or a chartered fishing boat, Spitsbergen cannot nowadays be reached by sea, except on one of the cruise ships sailing in Spitsbergen waters and calling at various places. In the 1980s the Dutch expedition ship *Plancius*, offering accommodation in two- and four-berth cabins, for 2 weeks at a cost of around £2000 per head, cruised during the summer months around the Scottish Islands, the Faeroes, Iceland, Jan Mayen, Spitsbergen and Bear Island. It was sometimes the only means of getting to those places. Passengers normally flew to and from Aberdeen or Tromsø to join and leave the ship. In charge of the arrangements was ornithologist Ko de Korte, of Groningen University's Arctic Centre. He saw to it that birdwatching opportunities were on offer: the places visited in 1988 included North Rona and Sula Sgeir, St Kilda, Mykines or Myggenæs in the Faeroes, and breeding colonies of Brent Geese and Ivory Gulls in Spitsbergen. A recent Spitsbergen cruise which made the headlines world-wide was that of the Soviet cruise liner *Maxim Gorky* with 575 German tourists aboard. Leaving Hamburg on 11 June 1989 the ship visited Iceland and was heading for Spitsbergen

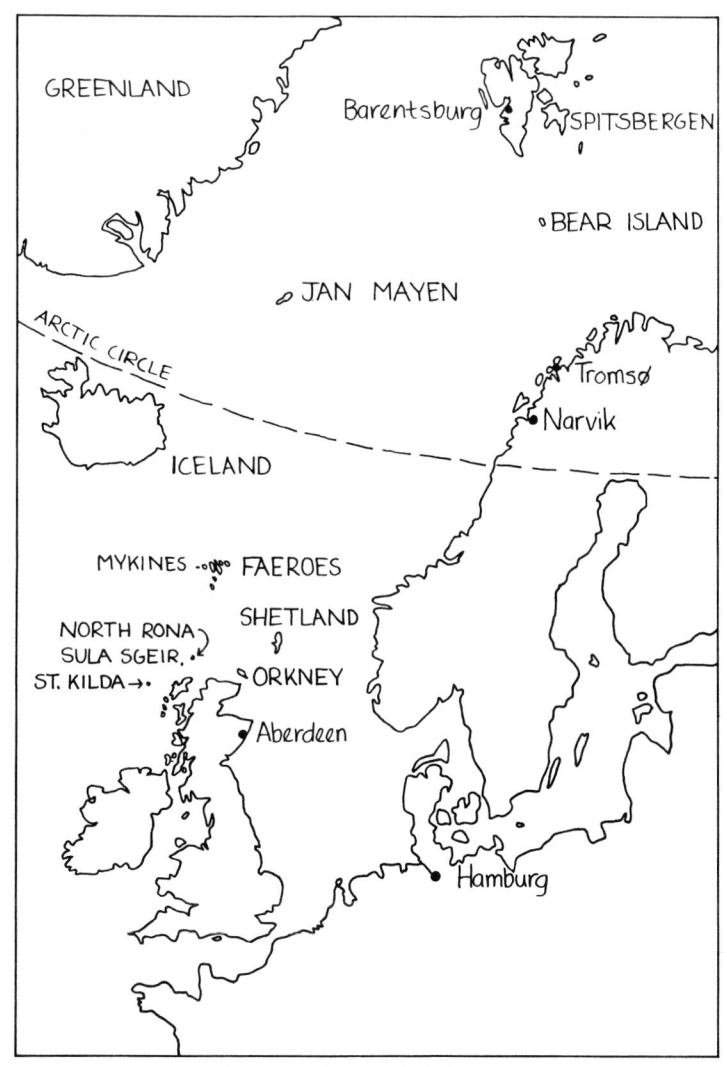

Map 25 Travel to Spitsbergen.

when she collided head-on with pack ice up to 5 metres (16·5 feet) thick while travelling at a speed of 18·5 knots. The passengers were all rescued and flown home from Barentsburg, the Russian mining settlement on Spitsbergen, where a 6-metre (20 foot) long gash in the ship's side was temporarily repaired. The captain was blamed, and suspended for 18 months (Barr 1990: 238–239).

Before the Second World War the Norwegian government subsidised a regular passenger ship service of five or six sailings each summer from Narvik or Tromsø in north Norway to Spitsbergen. After the war, sailings increased to up to 11 per summer, calling at Bear Island, Ice Fjord (Isfjorden) radio station, Longyearbyen and Ny Ålesund. But passenger sailings by sea were replaced in 1975 by up to four scheduled flights weekly from Tromsø to the new airport at Hotellneset near Longyearbyen. These Scandinavian Airlines System flights from Tromsø, which is easily reachable from Oslo, are the present-day visitor's best means of getting to Spitsbergen, though the more adventurous could try one of the once or twice monthly Aeroflot flights from Moscow.

It was probably the total protection of the reindeer by the Norwegian government in 1925, following the 1920 treaty, that saved the Spitsbergen subspecies of that animal from extinction. Like the Polar bear and the Arctic

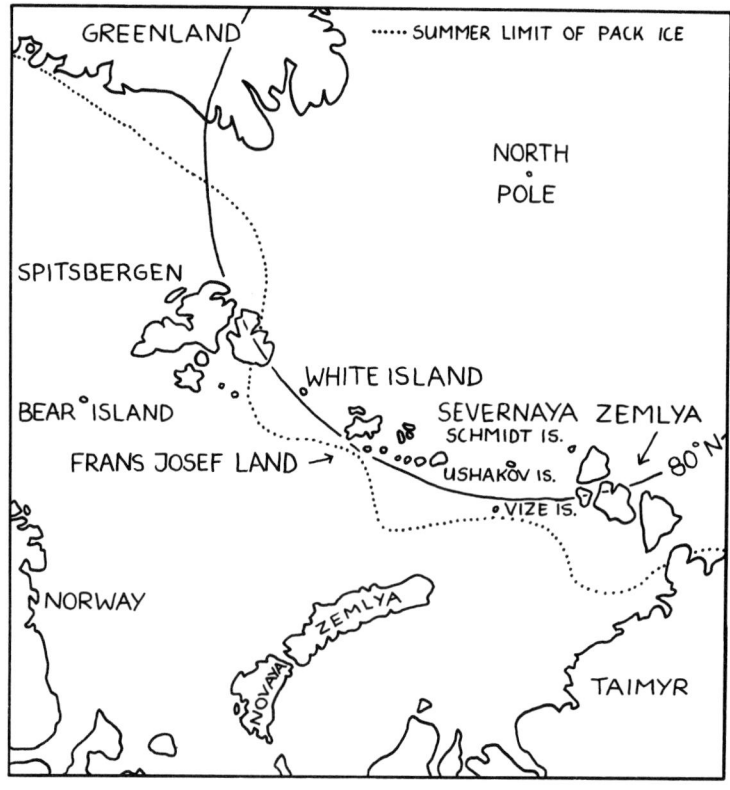

Map 26 The geographical position of Spitsbergen.

fox, it must have crossed to Spitsbergen over the ice—the archipelagos of Franz Josef Land and Severnaya Zemlya form the stepping stones along 80° N between it and the Taimyr Peninsula—and the longest gap being only about 200 miles (300 km). The introduced Arctic hare and muskox are the only other mammals in Spitsbergen. The archipelago's isolation is the probable cause of the absence of rodents there, and this in its turn may account for the absence of raptors and owls and the relative paucity of Arctic wader species and passerines. Yet there is much for the birdwatcher (Grimmett and Jones 1989). Auks (Alcidae) concentrate in large colonies mainly on the western cliffs which are washed by the open sea throughout the summer. There are thought to be over 1 million pairs each of Little Auk and Brünnich's Guillemot. In the north and east, Kongsøya has at least four colonies of Ivory Gulls, totalling nearly 100 pairs, and on islands in Liefdefjorden and Kongsfjorden thousands of pairs of Eiders breed. The Pink-footed Goose, which is said to be able to defend itself against the Arctic fox, breeds on mountain slopes and in valleys throughout the western part of Spitsbergen, while the Barnacle Goose breeds on flat offshore islets along the west coast. Among the common breeding birds found nearly everywhere are Red-throated Diver, Fulmar, Long-tailed Duck,

Map 27 Spitsbergen.

Ptarmigan, Purple Sandpiper, Red-necked Phalarope, Arctic Skua, Glaucous Gull, Arctic Tern and Snow Bunting.

Two things in particular make Spitsbergen a difficult country for the visitor. There is so much ice, snow, rock and mountain that travel from one area to another can only be accomplished by boat or aircraft. Secondly, of all Arctic lands, it is the most regulated. The Norwegians certainly take their treaty obligations seriously. They have in any case revelled in law-making ever since the Vikings. Before going into the field the visiting birdwatcher in Spitsbergen should be carrying a rifle, calibre 7.62 mm or larger, and must register himself and his companions, giving details of their names, means of transport, route to be followed and nationality, with the governor or *syssel-mann*. This all-powerful official may "visit and inspect most expeditions to distant places during their stay", and must be notified on the departure and return of the expedition. There is a campsite at the airport, a taxi in Longyearbyen but no roads elsewhere in Spitsbergen, and a boat or helicopter can usually be hired. There is no hotel in Longyearbyen and no shop worth the name. All this may sound forbidding, but should not deter the adventurous birdwatcher, who will surely receive a helpful response if he writes in advance to the governor of Svalbard at Longyearbyen explaining his plans and requirements—especially if he is a national of one of the signatory states of the 1920 treaty.

GREENLAND

For the birdwatcher in western Europe, east Greenland is the nearest part of the Arctic, but west and north Greenland are readily accessible too. Everywhere in Greenland the scenery is magnificent and, in the north and east, the visitor in summer can marvel at pack ice on the sea as well as admire the beauty of glaciers and of the Inland Ice, both of which can easily be reached from almost anywhere round Greenland, even in the far south. One can fly to Ammassalik or Kulusuk in east Greenland from Akureyri in Iceland, and to Nuuk (Godthåb) in west Greenland from Frobisher Bay (Iqaluit) in Canada. Or there are flights from Copenhagen to Narsarsuaq and Søndre Strømfjord (Kangerlussuaq). From this last one can fly to Kulusuk in east Greenland. From either Narsarsuaq or Søndre Strømfjord one can take a Grønlandsfly helicopter along Greenland's west coast north as far as Kullorsuaq or south to Nanortalik. Qaanaaq and other small places in northwest Greenland (Avanersuaq municipality) can be reached by a direct flight from Copenhagen to Thule Air Base (Pittufik) and then by helicopter. Places in east Greenland north of Ammasalik and Kulusuk are not so easy to reach but there are weekly Greenlandair flights from Reykjavik in Iceland to Constable Point near Scoresbysund (Ittoqqortoormiit). Destinations further north can only be reached by charter flight from Constable Point or Iceland. The Mesters Vig air strip now has no facilities or services of any kind: the place is uninhabited. Travel up and down Greenland's west coast between most of the inhabited places by Grønlandsfly helicopter is convenient but expensive. There are also regular passenger services by sea.

Since chartering a helicopter cost over £1000 per hour in the 1980s, most birdwatchers will have recourse to one of the other two means of local transport, namely hiring a motorboat or a dog sledge. These are also expensive and have the disadvantage of being available only at certain seasons. The sledge is not useable after the snowmelt in May in the south or June in the north. The motorboat is useless in the north and east until the winter ice has left the fjords, which may not be until July. The short breeding season of most Arctic birds in May–June thus comes at a time when neither boat nor sledge may be

Map 28   Greenland.

practicable. The isolation of many Greenlandic places, even some of those reachable by air, the cost of travel once there if a dog sledge or boat has to be chartered, and other practical difficulties, often make it worthwhile for several people to collaborate and mount an expedition.

The normal visitor or tourist was advised in the 1980s by the Danish government's Ministry for Greenland to contact the Greenland Travel Bureau in Copenhagen for advice and travel arrangements to and within Greenland. People organising expeditions had to obtain the Ministry's approval for their project. This, according to an official leaflet, was only granted if the expedition was properly planned, took all necessary safety measures, and carried adequate Polar equipment, and if its members were sufficiently experienced. Would-be expedition leaders were warned that the weather can be extremely nasty and is always unpredictable, that provisions cannot be obtained away from the larger settlements, that the official maps are inaccurate, and that the expedition would have to meet the full cost of any search and rescue operations. In spite of these dire warnings many a successful ornithological expedition to Greenland has been mounted in the last decade. In 1987 the Ministry gave permission for no less than eight such projects, including a survey of Brünnich's Guillemot colonies in northwest Greenland by Danish ornithologist Kaj Kampp, the ringing of Peregrines in south Greenland by Knud Falk for the University of Copenhagen's Zoological Museum, the recording of bird calls and songs, especially of the Knot and Sanderling, in Jameson Land, east Greenland, by English wildlife sound recordists Michael and Katherine Lea, and the ringing and measuring of Knots in northwest Greenland by Peter Prokosch from Husum in Schleswig Holstein (Commission for Scientific Research in Greenland, *Newsletter* 15, June 1987). Since 1 January 1989 the Ministry for Greenland's role has been taken over by the new Danish Polar Centre, which now deals with requests for permits for expeditions to Greenland and continues to lay down stringent regulations.

The ordinary birdwatcher intent on a visit to Greenland may well prefer to join one of the many guided tours there, or take a package holiday with one of the travel agents, or even book a holiday at one of the country's few hotels. Guided tours available in the 1980s included some arranged by a French firm. One of these consisted of 2 weeks hiking on foot in July in the magnificent country around Ammassalik; it was limited to 10 people. Another entailed flying from Paris to Iceland and then to Myggbukta and Mesters Vig, and making 2 or 3-day trips by canoe or on foot in the vicinity of both these places with possible visits elsewhere: 25 species of bird were thought likely to be seen. Scottish Polar explorer Angus B. Erskine, who first went to the Arctic in 1951 and 1952–1954 as Lieut. A. B. Erskine R.N., in charge of dogs and sledging for the British North Greenland Expedition, set up his own travel agency in 1979 specialising in Arctic tours, usually with ornithological opportunities. One of his oft-repeated Greenland tours, for groups of 6–8 people, was to Mesters Vig where the Gyrfalcon was a possibility, another was to Qaanaaq in the northwest where Snow Geese—the larger, all-white subspecies or Greater Snow Goose—were virtually guaranteed, and still another was to Ammassalik. These tours involved backpacking and camping, and both leaders and members have been the source of important records of Greenland birds. It was an Erskine Tour which spotted Greenland's first Sandhill Crane in 1985 not far north of Qaanaaq. As to hotels, why not use one of these as a base for birdwatching trips along fjords or out to the offshore islands? The Hotel Hans Egede in Nuuk (Godthåb), capital of Greenland and the country's largest town with more than 12 000 inhabitants, would be rewarding as a base for excursions. Or one could try the Arctic Hotel at Narsarsuaq near the southern tip of Greenland. From here boat trips along the fjords can be made to see the ruins of Eric the Red's farm and other Viking remains. It is easier to get to than Nuuk because there are direct flights from Copenhagen. For the more adventurous, the hotel at Qaanaaq might be worth a try.

A birdwatching visit to Greenland will never yield more than about 50 species. At Myggbukta, where the Swedish ornithologists comprising the Swedish Northeast Greenland Expedition—Myggbukta 1979, Magnus Elander and Sven Blomqvist (1986), studied the local avifauna from 17 May to 19 July, they identified 38 species. Rarities included a Snow Goose, two drake Pintails, Red-breasted Merganser, Pectoral Sandpiper, Ross's Gull and Greenland's first Wood Sandpiper. Myggbukta also proved an excellent base for seeing some typical High Arctic birds, such as Pink-footed and Barnacle Goose, King Eider, Long-tailed Duck, Knot, Sanderling, and Arctic and Long-tailed Skuas. The two Swedes compiled a complete list of birds seen at

Plate 65 A drake King Eider with two ducks swims across a tundra pool in Avanersuaq, northwest Greenland. The marshy ground now occupied by Thule Air Base once harboured this species, which still breeds not far off.

314　*In search of Arctic birds*

Myggbukta and this gave a grand total of 55 species. At Mesters Vig, also in east Greenland, in 1974, a more numerous English expedition (Green and Greenwood 1978) which covered a larger area, saw many of the same species. They found that Mesters Vig was a good starting-point from which to study breeding Barnacle and Pink-footed Geese, as well as breeding Gyrfalcons, Knots and Sanderlings, but no rarities were encountered.

The visitor to west or southwest Greenland will naturally see fewer High Arctic species, though Long-tailed Duck and Gyrfalcon will probably be encountered. Using Narsarsuaq as a base one can see breeding Red-throated

(a)

(b)

Plate 66 (*a*) and (*b*) Baird's Sandpiper is a North American species which has "overflowed" westward into the Chukchi Peninsula and eastward into Avanersuaq in northwest Greenland, where this incubating bird was photographed on 7 July 1984.

and Great Northern Divers, Harlequin Ducks, White-tailed Eagles, Peregrines, Purple Sandpipers, Red-necked Phalaropes, and Iceland and Glaucous Gulls. This should be a good area for rarities, and, besides the usual Arctic passerines seen almost everywhere in Greenland, namely Snow Bunting, Lapland Bunting, and redpoll of one or two species, Greenland's only breeding Redwings and Fieldfares, both of them recent colonists (since 1937), should be easy to find.

In the early 1980s I was lucky enough to obtain permission to stay with a companion at Thule Air Base (Pituffik) during three successive summers to watch birds in Avanersuaq or, as it used to be called, the Thule district, in northwest Greenland (Vaughan 1988). The United States authorities were most helpful, allowing us to stay at the North Star Inn, eat in the Dundas Dining Hall, and use other base facilities. Among these were a library, recreation centre, bowling alley, gymnasium, the Blue Nose and Top of the World Clubs, and the Base Exchange where one could buy food, clothing and much else. Most usefully for the birdwatcher, one could avail oneself free of charge of the base transport, consisting of shuttle buses and taxis, to travel from one part of the base to another, or to one of the outlying sites. During 2 weeks from 23 June 1983 my son John and I scoured the surrounding countryside for birds, but our list numbered only 20 species. Among these Baird's Sandpiper and Thayer's Gull were new to us. A Grey Plover was only the second record for northwest Greenland; but it was more exciting to hear the courtship song of the Knot and find these birds with young. Several pairs were nesting within a couple of miles of the air base.

In the next two summers, a Grønlandsfly helicopter was pressed into service and, with the help of some backpacking, half a dozen remote areas were visited and landings were made on some offshore islets. In the third season, 1985, my wife replaced my son as companion and field assistant. The bird list now grew more respectable, but still numbered only 32 species. Among them were Greater Snow Goose, which was photographed at the nest (Plate 33), King Eider, also found breeding, and Baird's Sandpiper, of which several nests were found. Pure white Gyrfalcons were seen several times. The breeding colonies of Fulmars, Brünnich's Guillemots and Little Auks were especially memorable. This is an exciting, still poorly-explored area which holds out the possibility of interesting vagrants from North America. Qaanaaq is the place to stay, though accommodation might be hard to find.

CANADA

The Canadian Arctic is so vast and varied that only a few places can be selected for mention here. Except for the extreme northern tip of the Yukon Territory, a coastal strip of Hudson Bay which is just in Manitoba, and the far north of Quebec Province, the Canadian Arctic is entirely within Canada's so-called Northwest Territories, or simply "the territories". It can be divided into two main regions: the Arctic islands, the largest of which are Banks, Victoria, Ellesmere and Baffin Island; and the mainland, where a wide strip of more or less coastal tundra extends from Tuktoyaktuk and Inuvik in the northwest to Churchill on the shore of Hudson Bay and the Ungava or Labrador Peninsula in the south and east. Along the Dempster Highway through the Yukon one can actually drive into the Arctic at Inuvik. Or one can take the train to Churchill. In the rest of the Canadian Arctic, scheduled air services will take you to almost any settlement from Montreal, Winnipeg or Edmonton, and there are numerous flights between settlements. Chartered aircraft, perhaps not prohibitively expensive for groups, are available at Iqaluit (Frobisher Bay), Resolute, Cambridge Bay and Inuvik to take you out into the vast wilderness. Travel Arctic, which is the tourist agency of the government of the Northwest Territories, based in Yellowknife, will provide detailed information. It issues an annual *Explorer's guide* and *Explorer's map*.

Because the Northwest Territories form part of Canada, birdwatchers in the Canadian Arctic have an excellent "where to find birds" book at their disposal in the shape of J. Cam Finlay's *A bird-finding guide to Canada* (1984). Thus at Inuvik they are advised to visit the rubbish tip to the east of town in May and

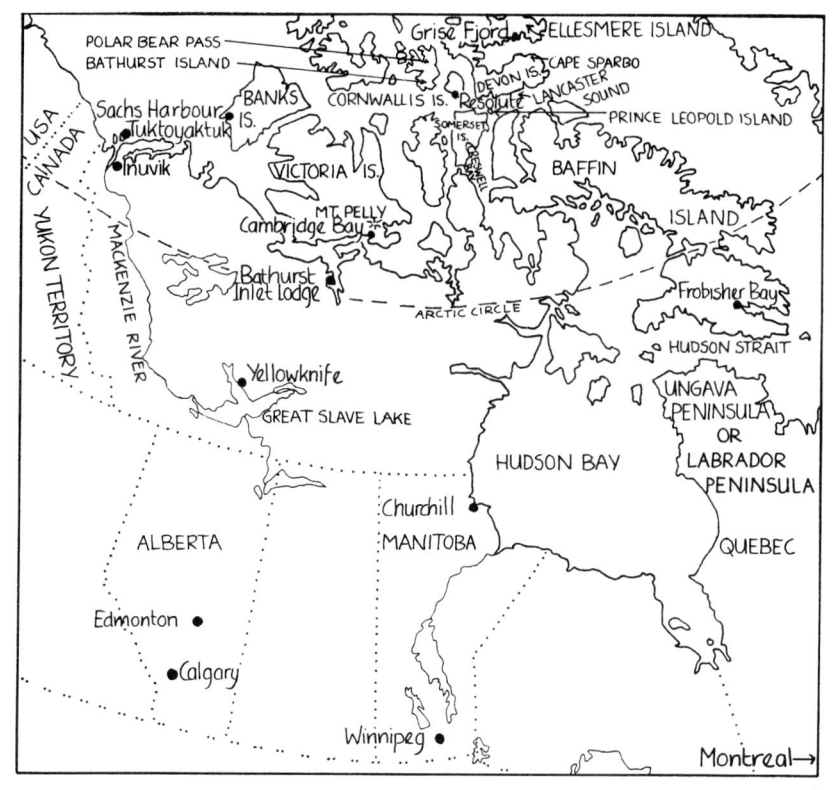

Map 29 The Canadian Arctic.

June "to check the six gull species: Glaucous, Herring, Thayer's, Mew [Common], Bonaparte's and Sabine's" (p. 345). To the west of town, up to 20 species of wader may be found on the sewage lagoons. A walk out into the tundra northeast of Inuvik should produce breeding Redpolls and Arctic Redpolls and Lapland Buntings. From Inuvik a flight to Tuktoyaktuk, or Tuk for short, is highly recommended. On the tundra there Semipalmated Plover, Whimbrel, Pectoral and Semipalmated Sandpipers and Hudsonian Godwits breed. Moreover, Tundra Swans and various scoter and scaup species are often present on lakes and lagoons in the vicinity. At opposite ends of the Canadian Arctic, Churchill is highly recommended for its list of 176 birds seen, among them Ross's Gull, which actually bred there in 1980–1982. It has been credited with 15 breeding wader species. The length of Churchill's bird list is due to its accessibility by rail and to its geographical location near the southern fringe of the Arctic. Thus it has High Arctic species like King Eider, American Golden Plover and Sabine's Gull, while more southerly breeding birds include Bonaparte's Gull and various species of hawk, owl and warbler. Among Churchill's rather special breeding birds are Harris's Sparrow and Smith's Longspur.

Another Canadian Arctic settlement with birdwatching possibilities is Resolute, on the south coast of Cornwallis Island (Finlay 1984: 351–352). Accessible by air from Edmonton or Montreal, it was founded as a joint United States–Canadian weather station and airstrip in 1947, and as the Inuit community of Kaujuitoq or "the land that does not melt" in 1953. Canadian biologist John Geale (1971: 53–59) was the first naturalist to spend an entire summer on Cornwallis Island. That was in 1969, and he saw a total of 30 species at Resolute, eight of which had not previously been recorded there. High Arctic species probably or certainly breeding were Long-tailed Duck, Eider, King Eider, Turnstone, Knot, Purple, White-rumped and Baird's Sandpipers, Sanderling, Grey Phalarope, Arctic and Long-tailed Skuas, Glaucous and Thayer's Gulls, Arctic Tern, Snowy Owl and others. If this is not enough, J. C. Finlay advises the well-heeled birdwatcher to charter a plane from Resolute for some really exciting but extremely expensive birdwatching. He or she could search for breeding Ross's Gulls on low islets not far from Resolute or for breeding Ivory Gulls on nunataks (ice-free rocks protruding from an ice sheet) on neighbouring Ellesmere, Devon, and Baffin Islands (Thomas and MacDonald 1987: 211–218). He or she could fly 100 miles (160 km) eastsoutheast to Prince Leopold Island off the northern tip of Somerset Island to see the teeming Brünnich's Guillemot colonies studied by David Nettleship and his colleagues, besides Fulmars, Kittiwakes, Black Guillemots and Glaucous and Thayer's Gulls. "Now fly south to Cresswell Bay on nearby Somerset Island for the Greater Snow Goose breeding grounds", Finlay (p. 352) recommends enthusiastically.

My own choice for a birdwatching visit to the Canadian Arctic was Cambridge Bay, on the south coast of Victoria Island. Its Inuit name is Ikaluktutiak, "the fair fishing place". Some good reasons for this choice were its long list of Arctic birds, already well known and written up by the American ornithologist David Parmelee and colleagues (Parmelee *et al.* 1967; see too Lok and Vink 1986, Vink *et al.* 1988), its unusually balmy summer climate, and the absence of both polar and grizzly bears. There is an excellent hotel and the mosquitoes are less troublesome than elsewhere.

It was 23 June 1986 when my wife and I arrived at Cambridge Bay after flying via Edmonton, where we spent a night, and Yellowknife, where we had time to find a White-crowned Sparrow's nest with four eggs. Taxi driver Harry took us from the airport to the hotel. The weather was so warm that we left our gloves and thick jerseys in two large suitcases containing reserve rations and films and other things there, and, after a visit to one of the two shops, trudged out of town with heavily loaded rucksacks. Two or three miles out into the tundra we camped by a river with a pair of Ptarmigan for company, having

---

Plate 67 (*a*) Incubating American Golden Plover photographed on its nest near Cambridge Bay in summer 1986, where my wife and I found six nests in all.

(*b*) American Golden Plover standing near its nest, Cambridge Bay. The plumage of this species is startlingly black-and-white. Even in paired birds, the sexes can be hard to distinguish.

(a)

(b)

already seen Grey Plovers, two sorts of phalarope, and Sabine's Gull. The next day's investigation of the marshy hollows and stony ridges around our campsite showed that this was, indeed, a birdwatcher's paradise. Whereas near Thule Air Base Baird's Sandpipers were thinly scattered over the tundra, here they were everywhere. We soon found that Semipalmated Sandpipers were just as common but preferred the wetter areas. We had good views of all three species of smaller skua. A Snow Goose flew over; a Snowy Owl was spotted sitting on a distant boulder. By the end of the first day we had found Snow and Lapland Bunting's nests and three wader's nests with eggs—those of the American Golden Plover, Baird's Sandpiper and Grey Phalarope.

A few days later, after spending a night in the Ikaluktutiak Hotel and doing some more shopping, we set out to walk the 8 miles or so (13 km) to Mount Pelly over undulating tundra, picking our way round innumerable lakes and marshes. Below the slopes of Mount Pelly we set up camp between two lakes on soft, mossy ground, near a herd of browsing muskoxen. Here we stayed for five idyllic days in virtually continuous sunshine and quite out of sight of the rest of mankind. We scarcely even saw a human artefact of any kind, except when looking at distant Cambridge Bay through binoculars from the summit plateau of Mount Pelly. Nowhere else have I seen Arctic birds so well or in such variety. A Canada Goose was incubating on the lake next to our tent; a White-fronted Goose had a nest on a dry ridge a few hundred yards away. To visit it we passed a pair of Long-tailed Skuas with two eggs on the bare ground and a Pectoral Sandpiper with four, well concealed in a marsh. On one lake Sabine's Gulls were nesting on an islet and up to 70 Tundra Swans were gathered, apparently not nesting because the ice on most of the lakes had not yet melted. On other lakes White-billed Divers and Arctic Loons (Pacific Divers) had taken up residence. Three pairs of Rough-legged Buzzards had nests with eggs on the scree slopes of Mount Pelly. Also within a few hundred yards of our lakeside camp were nests of Stilt Sandpiper containing marvellously beautiful greenish buff eggs mottled with grey and sepia and of that quintessentially Arctic breeding wader, the Grey Plover.

Our last night on Victoria Island was spent in the Ikaluktutiak Hotel. Even in and around the town itself (population 900), which is ringed by a seemingly endless expanse of tundra, there were birds to be seen. The American version of the Ringed Plover, the Semipalmated Plover, was breeding within the town limits, as were Snow and Lapland Buntings and Semipalmated and Baird's Sandpipers. Shore Larks, Grey Plovers, Stilt Sandpipers, phalaropes and probably King Eiders and Long-tailed Ducks were all nesting within a mile.

Several naturalists' lodges in the Canadian Arctic offer full board and special facilities for the wildlife enthusiast. Crediting the immediate vicinity with over 85 bird species, the management of one, Bathurst Inlet Lodge, open during the 1970s and 1980s, also issued lists of the local mammals, butterflies and plants. Some 40 or so bird species have been found breeding within reach of the lodge, which is about 60 miles north of the Arctic Circle. Among Arctic birds nesting here are Great Northern Diver, Tundra Swan, King Eider and Eider, Gyrfalcon and Peregrine falcon, Grey Phalarope, and Baird's and Pectoral Sandpiper. Daily boat trips were available from the lodge for the 18 guests, who enjoyed home cooking and individual attention. Such lodges are inevitably expensive because all supplies have to be flown in by chartered aircraft. Others are at Grise Fjord at the southern tip of Ellesmere Island and at Sachs Harbour on the west coast of Banks Island. According to Finlay (1984: 349) Sachs Harbour "is a perfect base to search for nesting Snow Geese, three jaeger [skua] species and Snowy Owl".

For the birdwatcher serious enough to pursue a research project and be classed as an ornithologist, there are summer field stations in the Canadian Arctic where permission might be given for a stay and some facilities provided. In the Truelove Lowland on the north coast of Devon Island the Arctic Institute of North America based at Calgary has a research station about 25 miles (40 km) southwest of Cape Sparbo. Here the regular breeding birds include Snow Goose, Eider and King Eider, Peregrine, Ptarmigan, White-rumped and Baird's Sandpipers and Grey Plover (Finlay 1984: 353). In 1968 the High Arctic Research Station of the Canadian National Museum of Natural Sciences in Ottawa was set up at Polar Bear Pass on Bathurst Island. Canadian ornithologists who worked there or at its satellite camps in the 1970s were Stewart D. MacDonald, a curator at the National Museum, who published on the breeding behaviour of the Ptarmigan and studied breeding Ivory and Ross's Gulls, and Philip S. Taylor, of the University of Alberta at Edmonton, who studied Snowy Owls and skuas. At Polar Bear Pass the Canadians hosted American ornithologist David F. Parmelee, who studied Sanderlings, and bird artist George M. Sutton (Sutton 1971; MacDonald 1981).

ALASKA

The far north of Alaska, beyond the Arctic Circle and the Brooks Range, is the only part of the state that really belongs to the Arctic, though the Seward Peninsula and the Yukon and Kuskokwim Deltas are classed as sub-Arctic. A relatively dense network of scheduled passenger services radiates out of Anchorage, which can itself be reached non-stop from major airports in the Lower 48, that is the rest of the United States, from Canada, and, in Europe from Amsterdam, Brussels, Copenhagen, Frankfurt, London and Paris. Within Alaska there are nearly 200 air taxi operators, and adventurous birdwatchers can easily have themselves put down anywhere in the wilderness and collected again later. Camping beside an inland lake or river can be serene, but mosquitoes may be present in force and the grizzly bear, just as dangerous as the Polar bear, takes the place of the Polar bear away from coastal areas. Although the places mentioned in what follows are relatively safe from bears, one can never be certain: a polar bear wandered into the town of Barrow in 1989 and was shot by one of the inhabitants.

Although there are excellent guided birdwatching tours to various points in Alaska, including Barrow, Nome and Gambell, it is probably no more expensive to go independently and be free to wander about on one's own. Barrow is surely the best base for birdwatching in Arctic Alaska. Excellent

Map 30 Alaska.

accommodation is available and the birdwatching season starts in May and continues into October: the spring migration of King Eiders and other waterfowl takes place in May and June, tundra breeding birds are active in June and July, and the autumn migration begins in August. In September and October large numbers of Ross's Gulls, and in some years, Ivory Gulls, pass along the coast.

The Top of the World Hotel in the town of Barrow, at 71° 18' N, 300 miles north of the Arctic Circle, is operated by Tundra Tours Inc., which also runs the Tundra Tours Bus. It is a subsidiary of the Inupiat-owned Arctic Slope Regional Corporation. The hotel is in the centre of Barrow, which is the largest Inupiat Eskimo community, numbering 3000 persons. To take the birdwatcher out of town into the tundra either southwards, or eastwards along Barrow's longest road (if it is open), which ends after a few miles at the town's gas well, or north along the coast to Point Barrow, there are at least four taxi firms, all appropriately named: Northern Taxis, Tundra Taxis, Polar Taxis

Map 31 Barrow and surroundings.

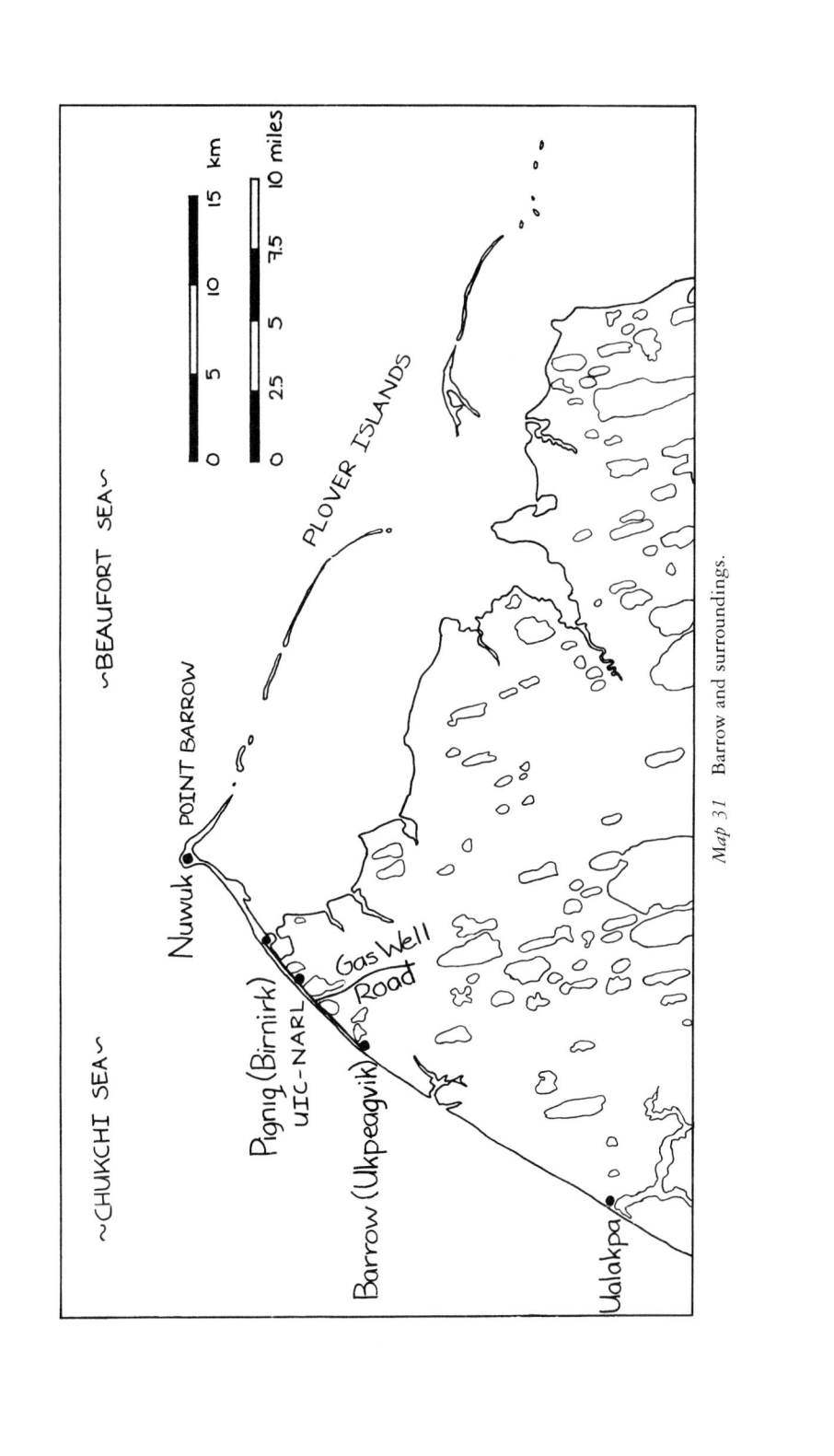

Plate 68 Near Barrow in 1989 Pomarine Skuas and Snowy Owls were next-door-neighbours on the tundra and often clashed.

and Arcticab. The Inupiat name for Barrow is also appropriate: it is Ukpeagvik, meaning "the place where owls are hunted", namely Snowy Owls, and this is probably one of the best places in the whole of the Arctic to see these magnificent birds. In 1989 my wife and I found a Snowy Owl's nest within a mile of the nearest buildings (Plates 52 and 53), and once seven Snowy Owls, perching on fence posts and boulders, were visible with binoculars from a single position. The visitor to Barrow can also stay at what used to be the United States Naval Arctic Research Laboratory (NARL) 3 miles from Barrow on the coast road which continues northeast towards Point Barrow. The laboratory's sprawling buildings have now been taken over by Barrow's municipality, the Ukpeagvik Inupiat Corporation (UIC), and the laboratory has been renamed the National Arctic Research Laboratory (still NARL, but now called UIC NARL). Also housed in the complex are the North Slope Borough's Wildlife Department and other offices. When my wife and I stayed there in 1989 the laboratory was indistinguishable from a rather good hotel, but somewhat cheaper, and indeed we discovered that the main laboratory is now being run as an 88-bed hotel, complete with conference rooms, recreation room, coffee and TV lounge and, of course, laboratories.

The birdwatcher's Arctic 325

(a)

(b)

Plate 69 (*a*) A Pomarine Skua arrives at its nest in marshy tundra near Barrow, Alaska, 27 June 1989. (*b*) The Pomarine Skua puffs out its breast and throws back its head as it settles onto its two eggs.

A stay at this hotel is warmly recommended for the birdwatcher. Situated among lagoons between the Chuckchi Sea and the tundra, it is surrounded by birds. Arctic Redpolls and Snow and Lapland Buntings visit the bird table outside the offices of the local Department of Wildlife, and they were joined in June 1989 by an unmistakable Pine Siskin, seldom seen in northern Alaska. Snow Buntings were nesting in a nestbox and Arctic Redpolls on an outside iron staircase (Plate 7). Semipalmated Plovers and Baird's Sandpipers were breeding in open spaces in the complex; Ravens had brought up a family presumably in a nest on a building; and four Red-necked Phalaropes were feeding on a tiny pool outside our front door on the day we arrived. Walks in the tundra and by the lagoons on the succeeding days revealed other birds, some new to us: nesting Pomarine Skuas and Snowy Owls were found, and Savannah Sparrows, Long-billed Dowitchers, Western Sandpipers and Red-necked Stint were all seen. We did not encounter any of the more unusual waders that have bred at Barrow, namely White-rumped, Curlew and Buff-breasted Sandpipers, and even Dotterel. Although we saw King Eiders we could find no Steller's, Spectacled or common Eiders, though all are regularly seen in the area (for Barrow's birds, see Johnson and Herter 1989).

Barrow would be an excellent base for further explorations in Arctic Alaska. A local airline has flights west along the coast to Wainwright, and MarkAir flights between Barrow and Anchorage stop regularly at Prudhoe Bay, along the coast to the east. In spite of the several thousand oil workers in residence at Prudhoe Bay, this is a well-known spot for displaying Buff-breasted Sandpipers in June, as well as for Spectacled and King Eiders and Sabine's Gulls. There is also a scheduled passenger service from Barrow 60 miles south to the tiny settlement of Atkasuk on the Meade River. Here the adventurous birdwatcher could camp on the lake-strewn tundra and enjoy a rather better summer climate than in often fogbound Barrow on the coast. Breeding birds to be looked for here should include White-billed Divers and Arctic Loons (Pacific Divers), Steller's Eider, Bar-tailed Godwit, and Red-spotted Bluethroat. Who knows what else remains to be discovered in such a seldom-visited corner of the Arctic!

The Brooks Range, which forms the divide between the treeless tundra of the Arctic and the boreal forest, offers many possibilities for an independent birdwatcher prepared to camp and backback. From Fairbanks one can fly into the Gates of the Arctic National Park to the Eskimo village of Anaktuvuk on the pass of that name at $68°$ N, where Laurence Irving (1960) did much of his field work on Alaskan bird migration. Arrive here in May on a flight from Fairbanks to watch the spring migration and stay to find breeding birds. Among the migrants are Tundra Swans, Brent Geese, Rough-legged Buzzard, Grey Plover, Turnstone, Whimbrel, Long-billed Dowitcher, Buff-breasted Sandpiper and Snow Bunting. Breeding birds likely to be encountered at Anaktuvuk include Semipalmated Plover, American Golden Plover, Wandering Tattler, Lesser Yellowlegs, Semipalmated Sandpiper, Arctic Redpoll and Redpoll.

Though it is south of the Arctic Circle, the coastal town of Nome, of Gold

Plate 70 Red-spotted Bluethroat male photographed in the Varanger Peninsula in 1972. In Alaska this Eurasian species is thought to be a recent colonist, having probably arrived in the last 10 000 years or so (Voous 1960: 220).

Rush fame, is a good headquarters for seeing Arctic birds. It can be reached by direct flight from Anchorage. Here one can stay at the Polar Arms, the Polaris Hotel, or the Nugget Inn. From this last, situated on the coast, the spring passage of divers, geese and ducks, and gulls can be watched. Also at Nome is the Aurora House Bed and Breakfast, where reindeer sausage features on the early morning menu. For the birdwatcher, the great advantage of Nome as a base is that it actually has roads; they are the only ones in the whole of northern and western Alaska, apart from the Dalton Highway from Fairbanks to Prudhoe Bay, which was built to service the pipeline and is not open to the public. Taken together, Nome's gravel highways total some 250 miles in length covering between them much of the inland tundra and mountain that makes up the Seward Peninsula (see Kessel 1989 for the birds of the peninsula). Look out for breeding Bristle-thighed Curlews, Long-billed Dowitchers, Bar-tailed Godwits and Wandering Tattlers. Look carefully at any golden plovers encountered. The Lesser Golden Plover has now been split into two species: the American Golden Plover is the common breeding bird of the North American Arctic, including Alaska, but the Pacific Golden Plover, breeding in Asia, also breeds in extreme western Alaska. Other special breeding birds of the Nome area are Red-spotted Bluethroat and Aleutian Tern. Quite a number of small American passerines reach as far north as Nome, for example White-crowned, Savannah and American Tree Sparrows.

From Nome one can fly to the Yupik-speaking Asiatic Eskimo village of Gambell, or Sivokak, at the northwestern tip of St Lawrence Island, from where the coastal mountains of the Chukchi Peninsula can be seen on a clear day. On St Lawrence Island the accommodation is poor, the weather is normally cold and foggy or wet and windy, and all birdwatching has to be done on foot. North American birdwatchers congregate here to see Asiatic rarities like the Terek Sandpiper, Great Knot, Red-breasted Flycatcher, Dusky Thrush and Fieldfare. Within a short walk of the village are thousands of breeding seabirds, and from a neighbouring headland migrant divers, geese including Emperor Geese, eiders of four species and many other birds can be watched. If Red-necked Stint was not found breeding in the Seward Peninsula, it might be discovered here. On St Lawrence Island, too, nests that extra-white version of the Snow Bunting, McKay's Bunting, a rarity which only breeds on some of the more northerly islands in the Bering Sea and is not universally accepted as a separate species.

Alaska is well provided with so-called wilderness lodges. A good one for the birdwatcher is the Iniakuk Lake Lodge in the Gates of the Arctic National Park, in the heart of the Brooks Range. It claims to offer comfortable accommodation and good food for up to 20 persons, and facilities for making camping trips into the mountains or along the river valleys. The birdwatcher can expect to see here a mix of birds like Arctic Loon (Pacific Diver), Long-tailed Duck, and American Golden Plover, whose breeding ranges extend well into the Arctic, and more southerly species here approaching their northern limits like Surf Scoter, Golden Eagle, Lesser Yellowlegs, Tree Swallow and American Dipper.

RUSSIA AND SIBERIA

One could argue that, because of the difficulties of getting there, and of travelling about once there, one might as well write off the Soviet Arctic; after all, nearly all its birds can be found breeding somewhere else. But in fact, there is a short list of rather spectacular or particularly interesting birds which can only be found nesting in the far north of Siberia. Among these Siberian Arctic endemics are the Red-breasted Goose and Curlew Sandpiper, which breed only in and a little to the west of Taimyr; and Sharp-tailed and Spoon-billed Sandpipers and Great Knot and Ross's Gull, all of which breed further east. The ranges of these species are widely spread, and it would be difficult to see all of them on any one visit to the Siberian Arctic.

Guided birdwatching tours to the Soviet Arctic seem unlikely to catch on for some time to come, though at least one was planned for summer 1991. Existing birdwatching holidays in the Soviet Union tend to divide their time between mountains, steppe and taiga, all far south of the Arctic. Undoubtedly such holidays are the easiest to organise in Russia and the only requirement is 12–15 like-minded Arctic enthusiasts. Once they are found, any travel agent specialising in birdwatching holidays would probably help, or contact could be made direct with Intourist.

Ever since the Russian cruise ship *Malygin* visited Franz Josef Land (Zemlya Frantsa-Iosifa) in 1931, exciting possibilities have existed of birdwatching in the Arctic pack ice north of the Soviet Union and of visiting some of the many islands there, from a cruise ship. However, since the Second World War most of the cruises have been for Soviet citizens only. Thus early in September 1987 200 Soviet citizens embarked at Murmansk on the luxury liner *Klavdiya Elenskaya* for a cruise to Franz Josef Land and the Taimyr port of Dikson in the Kara Sea (*Soviet Weekly*, 12 September 1987). Cruises in the Barents and Kara Seas were not new, but in 1989 for the first time a Soviet cruise ship "made a 20-day circuit of the Bering Sea and Strait" (Armstrong 1990: 128). In 1990 at least one Arctic cruise was open to anyone, provided he or she could afford it. For the birdwatcher it offered the chance to see Ross's and Ivory Gulls and other northern seabirds. "For around £10 000, you could go on the first ever two-week cruise to the North Pole this summer aboard the nuclear-powered *Rossia*" announced the *Soviet Weekly* proudly (10 May 1990). This was no cruise liner, but one of the Soviet Union's largest and newest icebreakers. The cruise was offered by the Murmansk Shipping Agency. The *Rossia* was to leave Murmansk on 30 July, planned to arrive at the Pole on 6 or 7 August, and would be back in Murmansk on 13 August after passing near Spitsbergen or Franz Josef Land.

During the 1970s and 1980s, only the favoured few made it to the Soviet Arctic, apart, perhaps, from Murmansk. To do so you had to be a Peter Scott or a Jeffery Boswall. One person who did was Gerald Durrell. His visit to the Taimyr Peninsula figured in the last two episodes of a 13-part television series he made in the Soviet Union. It was described in the book he wrote jointly with his wife Lee, which was beautifully illustrated in colour with photographs of Soviet landscapes, people, flowers, birds and animals, no less than 30 of which featured the author. One of the stars of the series was the muskox, and it was that recently-introduced animal which took Durrell to Taimyr in the first place (Map 40). He and his team flew from Moscow to Khatanga, and, after a lunch given in their honour at a local restaurant, were flown north by helicopter some 300 km (200 miles) to the valley of the lower Bikada River

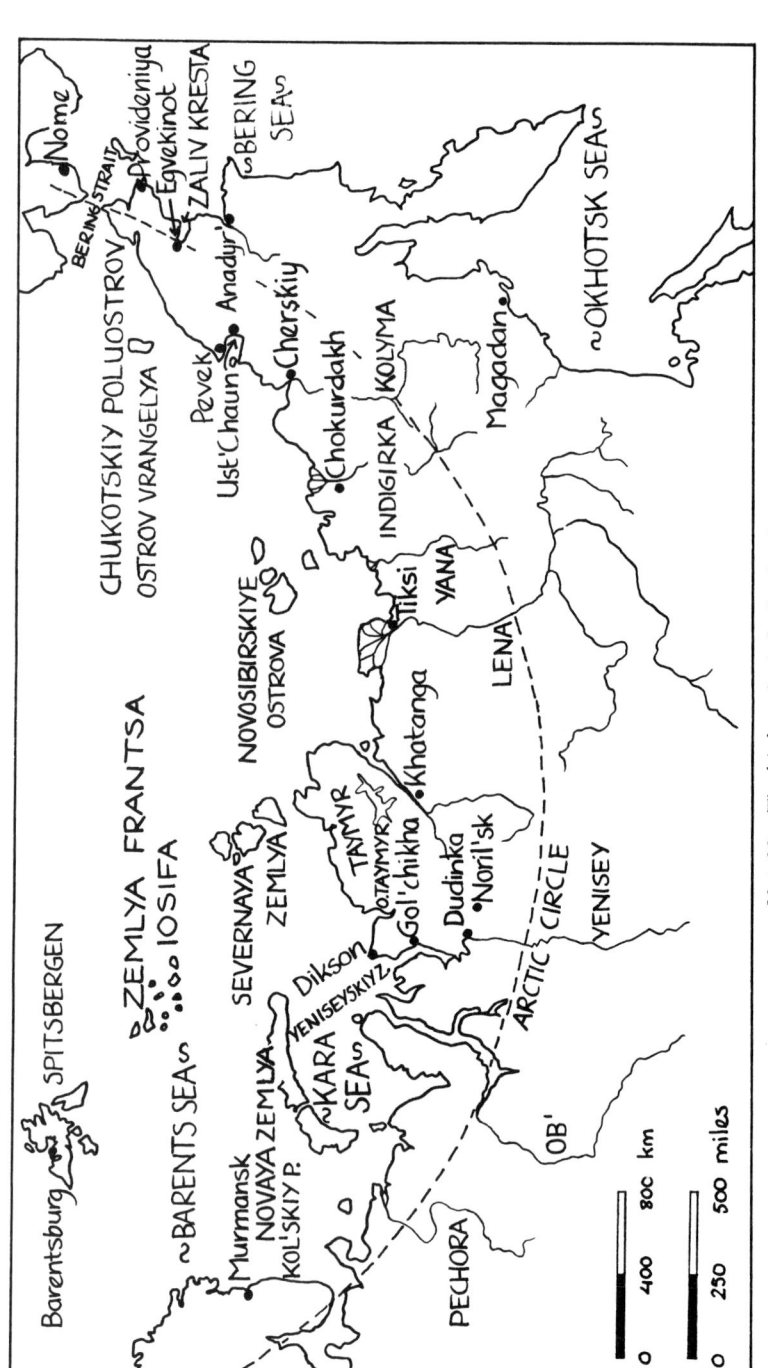

Map 32 The birdwatcher's Soviet Arctic.

near where it flows into Lake Taimyr (ozero Taymyr). Here Durrell and his party were billeted in buildings belonging to the Taimyr nature reserve. Later they were taken by helicopter along the winding course of the Malaya Logata River, and on another trip, to what is probably the world's most northerly forest, of larch trees, at Ary-Mas (*Soviet Weekly*, 21 December 1985). On an island off the mouth of the Bikada River they were shown a breeding colony of Red-breasted Geese (Durrell and Durrell 1986: 173). They

> cautiously approached their nests, some of which had young and some of which had eggs. They were extraordinary trustful and allowed us to approach within ten feet of them while they either manipulated their tiny young into the pools for swimming lessons or else snuggled more tightly over their eggs. What magnificent birds they were, as they wheeled across the sky, their red breasts glinting in the sun, uttering choruses of plaintive, honking cries like groups of wounded basoons.

Other breeding birds encountered were King Eiders, Dotterels, golden plovers of unstated species, probably Pacific, Bar-tailed Godwits and Sabine's Gulls.

Khatanga itself, at about 70° N, south of the Taimyr Peninsula and 600 km (400 miles) north of the Arctic Circle, is in the so-called forest tundra or transition zone between the tundra and the taiga. Pavel Tomkovich, curator of Ornithology at Moscow State University's Zoological Museum, has very kindly mentioned in a letter to me some of the birds which breed around Khatanga: Pectoral Sandpiper, Pintail Snipe, Bar-tailed Godwit, Spotted Redshank, Olive-backed Pipit, Siberian Accentor, Dusky Thrush, Arctic Warbler and Pallas's Reed Bunting. He adds that both Golden Plover and Pacific Golden Plover breed in the area.

It seems likely that independent travel in the Soviet Union will soon become easier than it has been, and since air travel is relatively cheap there, a birdwatching visit to the Soviet Arctic may not be prohibitively expensive. Khatanga is only one of several good birdwatching localities in the far north of the country reachable by air. Moving from west to east, from the nearest to the furthest away from western Europe, the Kola Peninsula (Kol'skiy poluostrov), though accessible by air and rail as far as Murmansk, provides no Arctic species not more easily to be found, because of the roads there, in the Varanger

Peninsula, north Norway. The nearest accessible point of the Soviet Arctic mainland with a substantially different bird list to that of the Varanger Peninsula is the lower Yenisey River. Gol'chikha, as well as other settlements along the banks of the river and the shores of the Gulf of Yenisey (Yeniseyskiy zaliv), can be much more easily reached now than when Maud Haviland watched birds on the Yenisey in 1914. There are airfields at Noril'sk and Dudinka, and even at Gol'chikha itself, as well as at Dikson, so there should be no problem in reaching the area by air from Moscow. Gol'chikha is probably the best birdwatching base. Here, on the western shore of the Taimyr Peninsula, in Asia rather than Europe and in Siberia, not Russia, European birdwatchers can hopefully add some exciting new breeding species to their lists: King Eider, Pacific Golden Plover, perhaps in some years Pectoral Sandpiper, Grey Plover, Curlew Sandpiper, Grey Phalarope and Pomarine Skua, and perhaps Red-breasted Goose.

Further east, a settlement reachable by air from Yakutsk or Moscow which gives access to the low-lying marshy tundras of the Siberian coast is Chokurdakh on the lower Indigirka River. Near here one should be able to find two of Siberia's Arctic endemics, Ross's Gull and the Sharp-tailed Sandpiper. Besides these, Tomkovich's bird list for the area includes Black-throated Diver, Broad-billed Sandpiper and Long-billed Dowitcher, which did not figure in the Khatanga list, as well as Pintail Snipe, Spotted Redshank, Siberian Accentor, Dusky Thrush and Pallas's Reed Bunting, which did. In suitable habitats near the River Indigirka Grey Plovers can be found nesting, and, near the river's mouth, Little Stint and Curlew Sandpiper. Finally the birdwatcher based at Chokurdakh should look out for King, Spectacled and Steller's Eider, Peregrine, Gyrfalcon, Siberian White Crane and Sabine's Gull.

Other places reachable by air from Moscow are Tiksi on the coast near the lower Lena River and Cherskiy on the Kolyma. Not far from Tiksi, in the Lena Delta, is a nature reserve which has breeding Brent Geese, Sharp-tailed and Curlew Sandpipers, and Ross's and Sabine's Gulls. In the mountains south of the delta Red-necked Stint and Rosy Finch breed.

In the far northeast of Siberia is the Chukchi Autonomous Okrug, or Region, which includes the Chukchi Peninsula, reaching out toward Alaska. The city of Magadan, administrative capital of the region, on the shore of the Okhotsk Sea, is the base for exploring this area. It is also the headquarters of the Institute of Biological Problems of the North of the Soviet Union's Academy of Sciences. At or near its field station near the settlement of Ust' Chaun on the south shore of Chaun Bay (Chaunskaya guba), according to the reports of people based there passed on to me by Pavel Tomkovich, breeding birds include White-billed Diver, Bewick's Swan, Spectacled Eider, Sandhill Crane, Grey Plover, Great Knot, Red-necked Stint, Little Stint, Sharp-tailed and Curlew Sandpipers, Long-billed Dowitcher and Ross's and Sabine's Gulls. Thus the southern shores of Chaun Bay hold no less than four Siberian Arctic endemics: the Great Knot, Sharp-tailed and Curlew Sandpipers, and Ross's Gull. Pevek, accessible by air from Moscow, is the biggest centre on the bay. From it both Ust' Chaun and Cape Schmidt (mys Shmidta) can be reached. All the birdwatching spots so far mentioned in the Soviet Union are well north of the Arctic Circle. Other places mentioned by Tomkovich are the town of Anadyr' and the settlement of Egvekinot at the head of Cross Bay (zaliv Kresta). These are just south of the Arctic Circle and many of the already-mentioned species can also be found here, as well as several not hitherto mentioned, namely the Emperor Goose, Baird's Sandpiper, Spoon-billed Sandpiper, Grey-rumped Sandpiper, Pechora Pipit and Siberian Rubythroat. Baird's Sandpiper is the only one of these birds that is typical of the High Arctic but the ranges of several others, including that of the Spoon-billed Sandpiper, extend north of the Arctic Circle. Tomkovich also recommends a visit to Provideniya, which has large colonies of seabirds in its bay and can be reached from Anadyr' or by charter flight from Nome, Alaska.

CHAPTER 13

Conservation

Although conservation has been a recurring theme throughout this book, the scattered comments and references already made need to be brought together, and some gaps need to be filled. The role of the native peoples of North America in the management of their own resource harvesting has been discussed in Chapter 2. Much of the research described in Chapter 11 was undertaken for conservation purposes by conservation organisations like the Canadian Wildlife Service or Britain's Wildfowl Trust. Even when this was not the case, the research often had important implications for conservation, such as the discovery of pesticide residues in Greenlandic and Alaskan Peregrine falcons.

Conservation, of course, can cover a multitude of sins. In Britain's northern uplands a great deal of money and effort is devoted to maintaining the Red Grouse population at an artificially high level so that as many as possible can be shot on and after 12 August. In North America a wildfowl refuge is not what it seems at first sight, namely a refuge where the birds are safe from disturbance or shooting, but rather a place where ducks and geese can be concentrated in large numbers and then shot by wildfowlers. Large areas can easily be declared national parks, that may look excellent on the map but mean little or nothing in practical terms. Both nature reserves and parks can be developed by their creators to attract more and more people, thus causing more and more disturbance. Conservation, too, may sometimes be used as an excuse to kill or control unpopular species, for example the larger gulls at seabird colonies. Most of these practices or abuses have little or no relevance for the Arctic, though one may justifiably be suspicious of the ease with which enormous national parks can be created in uninhabited tundra regions and inadequate conservation measures taken within them. On the other hand, now that most of the rain forest and much of the taiga have been destroyed, the tundra is the last great surviving natural ecosystem and, if this is not to be destroyed too, large sections of it must be set apart and conserved as parks or reserves. Conservation in the Arctic is indeed largely a matter of creating and maintaining large-scale tundra nature reserves, as well as establishing smaller ones to protect colonially breeding seabirds. The conservation of Arctic birds, most of which are migratory, depends also on regulating shooting and creating reserves for them far away from the Arctic.

INTERNATIONAL ACTION

What precisely does the Arctic need to be protected against? A number of large-scale threats to it and to Arctic bird life have been identified and discussed at international gatherings and their implications for bird life have also been described in the literature. It was these man-made threats to the environment and the need for a rapidly increasing world population to learn to manage natural resources rationally which inspired the initiation in 1964 of the International Biological Programme. In 1966 it was decided that certain "habitat groups" or "biomes" should be studied in depth and "Arctic and Subarctic Lands", later called the "Tundra Biome", was chosen as one of these. This research programme was well under way when an international Conference on Productivity and Conservation in Northern Circumpolar Lands was organised at Edmonton, Canada, in 1969 under the auspices of the International Union for the Conservation of Nature and Natural Resources (IUCN), then based at Morges in Switzerland. Many of the papers read at this conference, which were published in book form in 1970, were by ornithologists writing about conservation in particular Arctic countries—Norderhaug on Spitsbergen, Salomonsen on Greenland. Uspensky on the Soviet Arctic. For the IUCN, C. W. Holloway reviewed the status of those Arctic vertebrates which his organisation considered to be threatened with extinction or to be rare enough to require special monitoring. In 1975 the Convention on International Trade in Endangered Species of Wild Fauna and Flora (CITES) made a start in protecting these threatened species from commercial exploitation. Among the 800 animals and plants in Appendix 1, all of which were thought to be under threat of extinction, was the Gyrfalcon. No international trade in these species was allowed except under permit, but CITES permits were "frequently given for trade in Appendix 1 species for educational or scientific purposes and for hunting trophies" (Lapointe 1987: 480). In 1981, after research by American and Canadian biologists had shown that the north American Gyfalcon population "was not threatened with extinction and could sustain some exploitation", that population was taken out of Appendix 1. This was at the very time when, as mentioned in Chapter 4, there *was* an international trade in Gyrfalcons that could have threatened this rare and beautiful Arctic species of limited geographical distribution.

Another international conservation agreement which has repercussions for Arctic birds is the Convention on Wetlands of International Importance Especially as Waterfowl Habitat. This is usually called the Ramsar Convention, from the place in Iran on the southern shore of the Caspian Sea where it was adopted in 1971. It came into force in 1975, the same year as CITES. All Arctic states are parties to this convention, which establishes a List of Wetlands of International Importance to be designated by the signatory states. As far as the Arctic was concerned, initial response was poor: only 10% of sites listed in the first 10 years of the convention (1975–1985) were north of 60° N and most of these were in Finland and Sweden and not in the Arctic (Navid 1987: 491–493). But the Soviet Union had listed an 800-square-mile (2080-km^2) area centred on Kandalaksha Bay (Kandalakshskaya guba) in the White Sea (Knystautas 1987: 89), some of which is north of the Arctic Circle, and Canada had contributed half-a-dozen Arctic nature reserves including the Polar Bear Pass National Wildlife Area on Bathurst Island at 75° 45' N, and the Queen Maud Gulf Migratory Bird Sanctuary. In 1984 this last was thought to harbour some 9000 Snow and 30 000 Ross's Geese, many of them breeding (Finlay 1984: 350). These sites, registered in the convention's List of Wetlands of International Importance, are known as Ramsar Sites. In November 1988 the government of Greenland, which the Danes still prefer to call the Greenland Home Rule Authorities, declared no fewer than 11 Ramsar Sites, including the Greenland White-fronted Goose site Eqalummiut Nunaat, the western part of Jameson Land, and various tundra valleys, fjords, and offshore islands where seabirds breed (Grimmett and Jones 1989: 145–162).

In 1981, more than a decade after the inception of the International Biological Programme, its research into tundra ecosystems was published by Cambridge University Press in a massive 813-page volume, costing £90 (Bliss *et al.* 1981). Ten countries took part—Australia, Austria, Canada, Finland, Ireland, Norway, Sweden, Britain, the United States of America and the Soviet Union. A total of 24 Tundra Biome sites were investigated and compared. Only a few of these were in the High Arctic: Point Barrow and Prudhoe Bay in Alaska, Devon Island's Truelove Lowland in Canada, Disko in west Greenland, and four others in the Taimyr area of the Soviet Union. Of the 26 numbered sections in the book only one, on bird and mammal predators, was about birds. This book, edited by United States botanist L. C. Bliss, was only the tip of an iceberg. It was volume 25 of the International Biological Programme's research publications and it stimulated a great deal of other research, including ornithological work.

International interest in the Arctic environment continued through the 1980s, when at least three important international conferences brought new

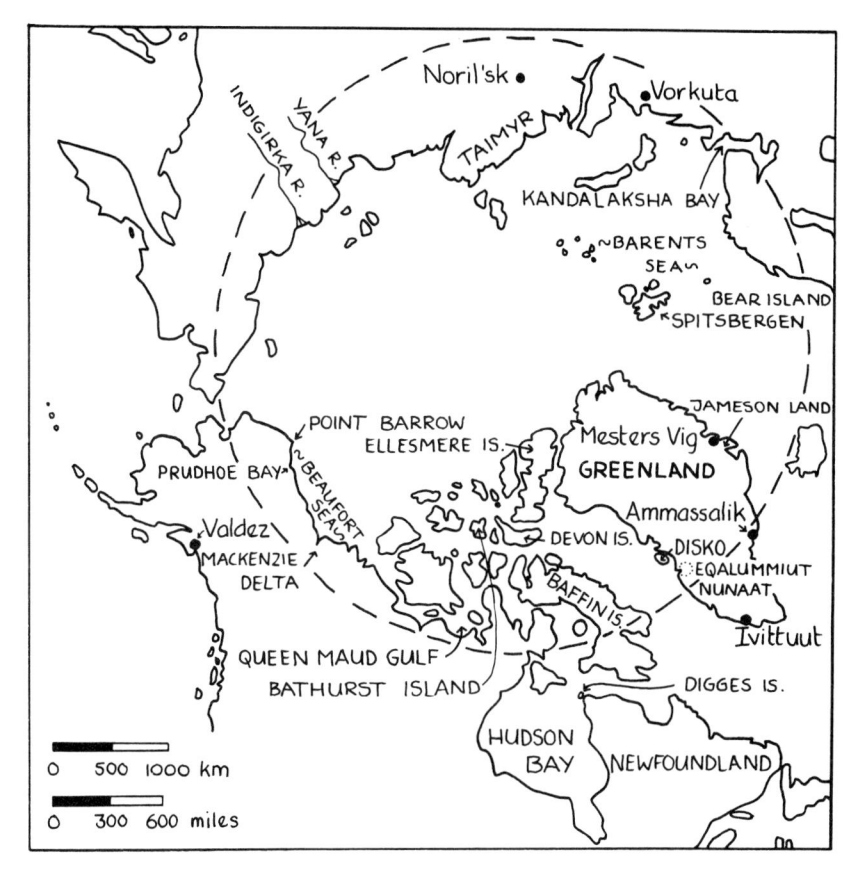

Map 33 Circumpolar conservation.

information to light and helped to focus world attention on the Arctic. In March 1980 the newly-formed Comité Arctique International based in Monaco held a conference at the Royal Geographical Society's London headquarters on The Arctic Ocean, The Hydrographic Environment and the Fate of Pollutants. The Swiss President of the committee, Louis Rey, edited the proceedings with Bernard Stonehouse's help and these appeared in 1982. Two international conferences on Arctic problems were held in 1985. At the Scott Polar Research Institute in Cambridge, England, the subject was Arctic Air Pollution, and 30-odd contributions, edited by Stonehouse of that institute, were published in 1986. At Banff in Alberta, Canada, the conference was titled Arctic Heritage, and attention was concentrated on land use and conservation. Its proceedings were published in 1987, edited by J. G. Nelson, R. Needham and L. Norton.

THREATS TO THE ENVIRONMENT

This international activity has helped to identify and describe the major threats to the Arctic environment (Holloway 1970; Norderhaug 1979; Sage 1981; Solomonov 1987; Stokke 1989; Kotlyakov and Sokolov 1990a: 203–278). Among them, those most important for birds are perhaps marine and air pollution, the search for oil and other non-renewable resources, and tourism. All of these could easily modify the environment in such a way as to affect birdlife adversely (Uspensky 1982). More directly, birds may be threatened by over-exploitation, either in the form of subsistence hunting, or shooting for sport, or by falling victim, like dolphins in fishing nets, to human exploitation of some other species. The actual and potential impact of this human activity in the Arctic, impinging directly or indirectly on Arctic birds, needs to be considered in detail.

MARINE AND AIR POLLUTION

The pollution of the seas by industrial and chemical wastes and pesticides has been less thoroughly documented in the Arctic Ocean than elsewhere, but pesticide residues and other poisonous substances have been detected in Arctic seabirds, often far away from the source of pollution. It is not always easy to decide whether the birds in question absorb these poisons in Arctic waters or bring them there from industrialised regions they may have entered on migration or in their winter quarters. Mercury levels in west Greenland Black Guillemots were shown in 1974 to have doubled in 20 years but were still comparatively low. Since many west Greenland Black Guillemots remain in the general area throughout the year, this mercury probably represents pollution in Arctic or sub-Arctic waters off the west Greenland coast. On the other hand, the pesticides DDE and aldrin, as well as polychlorinated biphenyls from industrial effluents, found in Greenlandic Brünnich's Guillemots, must have been absorbed in the birds' winter quarters around Newfoundland (data from Evans 1984: 76). Chlorinated hydrocarbons and polychlorinated biphenyls have also been found in seabirds and their eggs on Bear Island and elsewhere in the Barents Sea (Norderhaug et al. 1977: 101). Eider eggs from 70° N in north Norway were found to contain lower amounts of DDE than those from Spitsbergen in 78° N (Norderhaug 1979: 62), but none of these toxic residues in seabirds were at high enough levels to have deleteriously affected them. Nor so far is it likely that atmospheric contaminants like mercury, lead and zinc, which are undoubtedly being transferred via precipitation from the atmosphere into Arctic seas, as they are also by rivers, are as yet anything more than a potential threat to birds. A recent investigation of heavy metals in Greenland seabirds found that zinc, cadmium, mercury and selenium were all present in higher concentrations in birds from the northern than from the southern half of Greenland, but it was not suggested that these concentrations were affecting the health of the birds (Nielsen and Dietz 1989).

Pollution in the Arctic atmosphere is obvious and visible because it takes the form of so-called "Arctic haze" (Rahn 1982: 163–195). First observed in the 1950s, this became especially notorious at Barrow, Alaska, where it has been minutely studied. It is nothing less than industrial smog. It occurs especially during winter and spring and has also been noted in Spitsbergen and Greenland. Partly because of the large amounts of vanadium and sulphate in it, this Arctic haze or aerosol is though to derive from industrial pollution in middle latitudes. Its potentially damaging effect on the Arctic environment includes a warming of the atmosphere and ocean surface, an increase in precipitation or changes in precipitation patterns, and the deposition of noxious substances into the Arctic Ocean.

OIL AND MINERAL EXPLOITATION

The extraction of oil in the Arctic is almost bound, sooner or later, to lead to damaging oil spills, such as have already occurred in temperate waters. Serious exploration for oil in the North American Arctic began in the early 1970s after the discovery of huge oil and gas deposits at Prudhoe Bay in northern Alaska in 1968. Timely warnings of the risks to seabirds were given in a 1976 publication of the Canadian Arctic Resources Committee. This pointed out that concentrations of migratory ducks and geese in the area of the Mackenzie Delta, numbering up to 10 000 birds in one area, mainly of Snow Geese, could be decimated by an oil spill in the region. Also at risk in the southern Beaufort Sea, where oil exploration had already begun and where up to one-third of its total population at times gathered, was the Brent Goose. Enormous numbers of eiders, scaups, scoters and Long-tailed Ducks could be destroyed in the same area if an oil spill occurred when these birds were migrating. It was feared, too, that a single oil spill might obliterate entire colonies of Brünnich's Guillemots, some of which were huge. What if a spill occurred in the mouth of Hudson Bay, where in the late 1950s an estimated 2 or 3 million of these birds were nesting on the Digges Islands (Pimlott *et al.* 1976: 95–97; but see Bruemmer 1989, who estimated 800 000 birds on the eastern island and mainland opposite combined)? In 1986 the Arctic Institute of North America promoted the publication of a 212-page *Bibliography on the fate and effects of Arctic marine oil pollution* which contained 748 citations (Young 1986; see too Nelson-Smith 1982; Clark and Finley 1982).

So far there have been no major oil spills in the North American Arctic and oil exploration there has been more or less restrained. But the marshy lake-strewn tundra landscape round Prudhoe Bay has been laid waste. This "lonely and haunting wilderness", wrote oil ecologist Bryan Sage in 1982, is now "a great industrial complex, with mile after mile of gravel road network carrying an enormous volume of traffic, power lines, feeder pipelines, power stations, pump stations, a refinery, an airfield, dozens of miscellaneous buildings, completed oil wells and rigs drilling new wells" (Sage 1982). Oil first flowed south from here through the pipeline to Valdez in June 1977. Two species of diver,

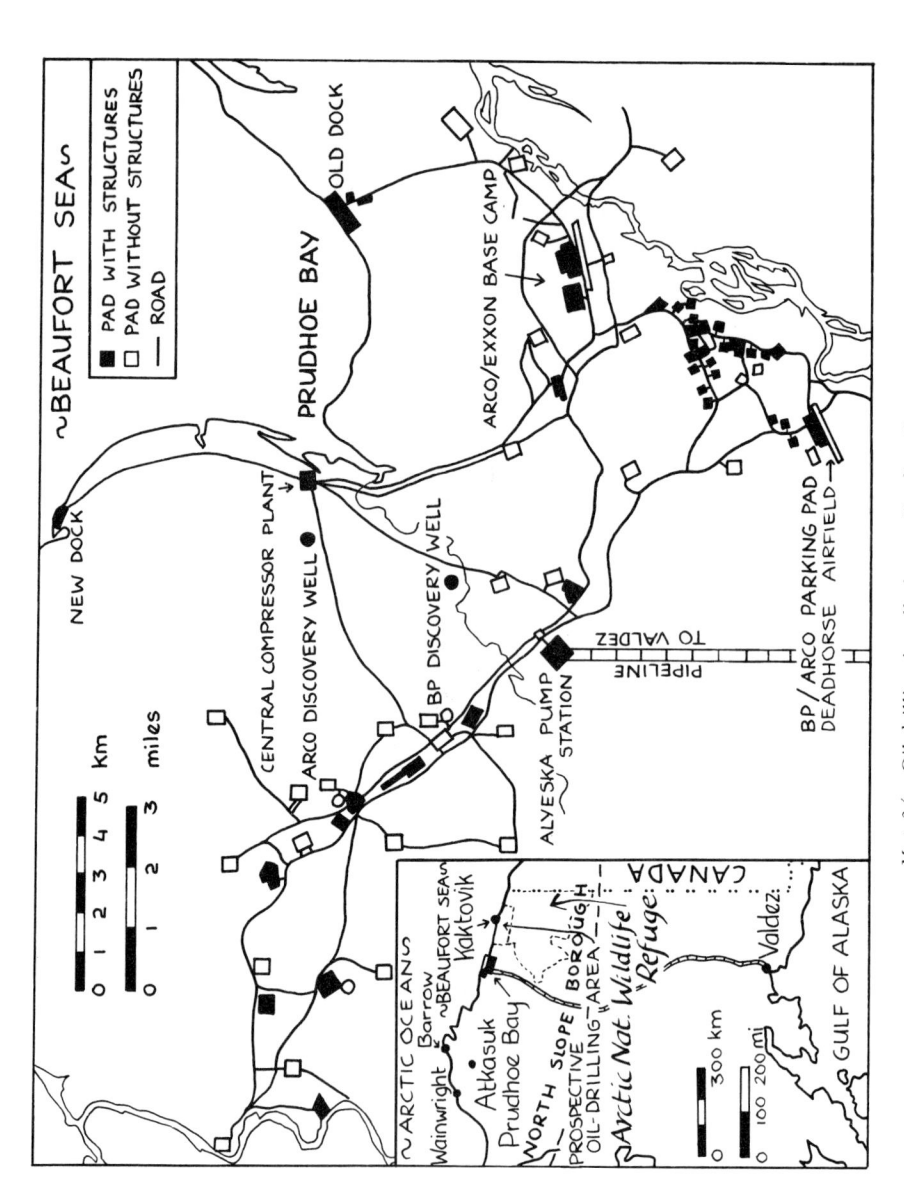

Map 34 Oil drilling installations at Prudhoe Bay.

Sabine's Gulls and Buff-breasted Sandpipers are still to be found in the area, but the scale of landscape destruction can be envisaged from a glance at the map.

In the Spitsbergen Archipelago drilling for oil began in 1965 and was continued in the 1970s on Edge Island. In 1990 the Norwegian state company Statoil was planning to drill in the Ice Fjord area; BP was carrying out extensive seismic surveying in preparation for drilling; and the Russian coalmining concern Arktikugol was drilling in the Van Mijenfjorden area (Map 36) (Hansson *et al.* 1990: 28–30).

In some parts of the Soviet Arctic the quest for oil and gas is currently threatening huge expanses of tundra. What has been happening in the Yamal Peninsula (Map 17) makes Alaska look relatively clean (Eugene Potapov, personal communication; and see Vitebsky 1990). Since the discovery of large deposits of oil and natural gas there in the early 1960s, many hectares of reindeer pasture in the peninsula have been lost through the development of oil and gas fields, industrialisation, and rail links. Overgrazing of the remaining pasture land by reindeer has aggravated the problem. But in 1989 the region's ecological commission turned down an important gas project, and there is hope that the formation of the Association of Northern Ethnic Minorities in the spring of 1990 might permit the Nenets and other peoples in the region to preserve their traditional life-style and natural resources by maintaining most of the Yamal lowlands in their pristine state (Batashev 1990).

Mineral exploitation in the Arctic is not likely to impinge extensively on birds, but there certainly have been important impacts locally and these will doubtless proliferate. In Spitsbergen coal has been mined at several places, in recent years by Russian and Norwegian concerns. In Greenland both the cryolite mine at Ivittuut on the west coast, which was an enormous circular pit with perpendicular sides, and the lead–zinc mine at Mesters Vig have been closed, but mining continues elsewhere. In the Soviet Arctic industrial complexes associated with mineral extraction really have made an impact on the landscape—especially in and around the city of Noril'sk in Siberia, where industry is fed by copper, nickel and coal deposits nearby and where a quarter of a million people live on what was once tundra. The coal mines at Vorkuta at the northern end of the Urals may be extended in such a way as to encroach into the neighbouring tundra. In the Alpine tundra country between the Yana and Indigirka rivers a major tin mining operation began in 1953; and it seems likely that, sooner or later, the copper deposits discovered in 1974 in the Taimyr Peninsula will be exploited (Sage 1981).

TOURISM

It seems absurd to suppose that tourism could modify the Arctic environment adversely enough to affect birds. The most serious danger in the Arctic is presented by cruise ships and other boats passing too close to seabird colonies and even sounding hooters to put the birds off their nests in order to see them in flight. This has been reported from west Greenland (Evans and Nettleship 1985: 455). Some 5000 tourists were thought to have visited the Spitsbergen Archipelago in 1969 on board cruise ships (Norderhaug 1970: 194) but there were no reports of disturbance of birds. In national parks and nature reserves there is no reason why tourism and conservation should not go hand in hand together, but the two have sometimes been in conflict. The Canadian National Parks Act of 1930 paradoxically declared that nature within the parks should be left to herself, but at the same time they were to be used for human enjoyment and recreation. The 400 hikers using the Auyuittuq National Park on Baffin Island in 1983 were concentrated on the trail through Pangnirtung Pass (Map 38). Some of those questioned complained that they had encountered as many as 50 other people during their hike. But little or no harm could have been done to the local bird life. In the early 1980s, when Parks Canada proposed the creation of a new 40 000-km^2 (15 000-square-mile) national park at the northern end of Ellesmere Island, opponents argued that the 50 people per annum who might be attracted to visit it would be more than the environment could tolerate. The threat of tourism to Arctic wildlife is surely real, but future rather than present (Marsh 1987).

RESOURCE MANAGEMENT AND MISMANAGEMENT

Birds and other animals can be an important source of food for humankind but this resource, which is renewable, requires careful management to make it so. Such management has often in the past been absent or inadequate, and even today over-exploitation of particular species occurs and lively debate about what constitutes over-exploitation continues. There are three elements in resource harvesting, two of which are often present together in the use of any one resource. The first is subsistence. Here, the resource is a more or less essential part of the user's diet. For example the 10 000 birds killed and eaten in 1978 by the East Greenlanders at Ammassalik, mentioned in Chapter 2, constituted a so-called "subsistence kill". Secondly, the commercial element can be important: a resource may be harvested mainly for the market. Also mentioned in Chapter 2 were the Nunamiut Eskimos of interior Alaska; they took Gyrfalcons and besides using the feathers for bows and arrows, traded them with the coastal Eskimos. Here the two elements are found together: subsistence and commerce. Thirdly, in the most advanced parts of the world, shooting is recognised as a sport and a resource may be managed with sport hunting in view; here an insignificant subsistence element is usually present too.

Much bitter debate has often accompanied the management of resources for sporting purposes. The North American Tundra Swan, now regarded as a subspecies of Bewick's Swan with an all-black as opposed to part-black beak, is

Plate 71 (*a*) and (*b*) Tundra Swans in their Arctic summer haunts. In 1986, early in July, in the Cambridge Bay area, no evidence of breeding swans was found; the birds were only seen in flocks on the larger lakes. Evidently the breeding lakes had remained frozen so late in the year that the swans, and perhaps divers, simply did not breed that year.

a case in point. Anyone who has seen Tundra Swans on their way to the Arctic flying high overhead, glinting white against the blue sky and honking loudly, as I did in Michigan in spring 1989, is tempted to ask the question, is it really necessary to kill these fine birds? The United States Fish and Wildlife Service says emphatically "yes", and in 1989 authorised the issue of over 12 000 swan-hunting permits distributed between nine states. Among its reasons for doing this, as explained in its own publications, was the desire to control the Tundra Swan population, which had increased to 139 000 birds in 1985, 80 000 of which were in the eastern population. Henceforth this was to be stabilized at 60 000–80 000 birds, permitting a 10% annual kill. Another reason for allowing swan-shooting was "public demand", though less than 20% of Americans had shooting licences. More convincing, perhaps, was damage to agricultural crops, especially in the Chesapeake Bay area where loss of wetland vegetation through industrial and other human activities had caused the swans to resort to farmland. A less convincing reason for allowing swan hunting was damage to aircraft: in 1962 an aircraft flying at 6000 feet over Maryland hit a flock of Tundra Swans and crashed, killing all on board. The U.S. Wildlife Service also justified swan shooting on the grounds that it would "provide a unique recreational opportunity for sport hunting of a trophy species on a limited basis". It would seem hardly necessary to review arguments against shooting swans: one was that large numbers of wildfowl die from the ingestion of spent lead shot which accumulates in wetlands that are annually shot over. In 1973–1974 nearly a thousand Tundra Swans were said to have died from lead poisoning at Mattamuskeet Lake in North Carolina (Heintzelman 1989).

The population of a large, conspicuous bird like the Tundra Swan can be monitored with relative ease and, indeed, granted that just as many people in North America count swans as shoot them, the management of this particular resource harvest should present few problems. The case of the Brünnich's Guillemot, which breeds colonially on the west Greenland coast, is much more complicated. Numbers of this bird underwent a drastic decline in this century before the seriousness of the situation was appreciated. In a 1970 paper on birds useful to man in Greenland, Salomonsen stated that Brünnich's Guillemot was the most economically important birds species in west Greenland. Revising an earlier (1967: 270) estimate of an annual kill of 200 000 birds, Salomonsen now thought that about 750 000 were being shot annually by Greenlanders, mostly for subsistence. But the commercial element was not lacking. He pointed out that a "murre cannery" had been opened at Upernavik which was processing 20 000 birds per annum to supply the southern Greenlandic settlements with meat. Brünnich's Guillemots were also being killed incidentally or accidentally during the harvesting of a quite different resource, for Salomonsen noted that, in 1967, some 15 000 of them died after becoming entangled in fishing nets off southwest Greenland. Within the next few years it became apparent not only that the Brünnich's Guillemot was being grossly over-exploited by the Greenlanders, both in the breeding season and in winter off southwest Greenland, but also that something like a quarter of a million birds were being accidentally killed in some years in monofilament drift-nets used for salmon and cod fishing off the west Greenland coast (Evans and Nettleship 1985: 447–448). The best estimate in 1984 was that there were probably fewer than a quarter of a million pairs of Brünnich's Guillemots left breeding in Greenland of the 2 million estimated by Salomonsen in 1950 (Evans 1984: 69). Efforts, in the shape of both legislation and education, were still being made in 1988 and 1989 to take fully adequate measures to stop the over-exploitation of breeding Brünnich's Guillemots by Greenlanders, and in 1989, it was still uncertain whether the incidental autumn kill of migrating birds would continue and at what level. The latest estimate of the total kill by Greenlandic hunters, for private subsistence and for sale in local markets was of "at least 284 000–388 000 birds" annually (Falk and Durinck 1990). However, neither the figures so far published for the total population of birds nor those for the annual kill carry much conviction.

Brünnich's Guillemot was also a food bird in parts of the Soviet Arctic, but here the exploitation of this natural resource was in large measure commercial, the subsistence take being small or non-existent. Among the most important colonies of Brünnich's Guillemots in this context were those of Novaya Zemlya and in particular Bezymyannaya Bay (Bezymyannaya guba) on the west coast of southern Novaya Zemlya at 72° 52′ N. The birds occupied a total of 23 km (14 miles) of cliff on the south shore of the bay and along the Barents Sea coast

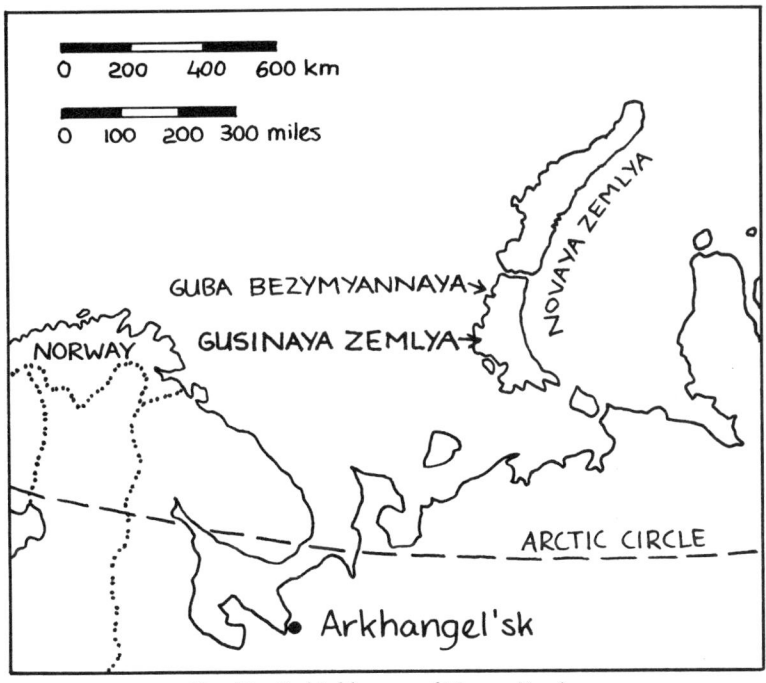

Map 35 Seabird bazaars of Novaya Zemlya.

346 *In search of Arctic birds*

to the north of it and probably formed the largest seabird colony in Russia. Though it entailed a round voyage of some 2400 km (1500 miles) from Archangel (Arkhangel'sk), birds from Novaya Zemlya were being killed, salted down, and sold on the market there in the mid-nineteenth century. There was even talk of using their guano to fertilise the fields around Archangel. Besides merchants from Archangel, Norwegians raided the Novaya Zemlya Brünnich's Guillemot colonies for eggs, but in 1922 all commercial exploitation was banned by the then relatively new Soviet regime. After all, natural resources now belonged to the people and should be exploited rationally for the benefit of all. This meant, first their scientific investigation. A string of biologists were sent to Novaya Zemlya: Gorbunov in 1923; L. A. Portenko in 1930. Finally the government agency in charge of Northern Marine Animal Exploitation (Sevmorzverprom) invited the All-Union Arctic Institute based in Leningrad to mount a combined scientific and commercial expedition to study the bird cliffs, or bazaars as the Russians call them, and the biology of Brünnich's Guillemot, with a view to assessing the possibilities of commercial exploitation. A special ornithological detachment of the institute

Plate 72 Brünnich's Guillemots on the cliffs of Saunders Island, a few miles from Thule Air Base, northwest Greenland, on 19 June 1985. Just a few of the thousands of birds on the cliffs are shown here. The Brünnich's Guillemot breeding ledges are conspicuous because the rocks are stained a pale buff colour not only below, but also above, the birds.

worked in Bezymyannaya Bay in 1933 and 1934 under the leadership of ornithologist S. K. Krasovsky. The results were published by him in 1937 in Russian, in the biological series of the institute's transactions with the title: "Biological foundations for the commercial exploitation of bird bazaars. Studies of the biology of Brünnich's Guillemot". This classic paper is cited over and over again in the literature but may seldom actually have been read (see too Uspenskii 1986: 329–336).

To estimate the total guillemot population of the Bezymyannaya Bay cliffs, Krasovsky counted birds in sample areas in each section of the colony, to estimate the number of birds per square metre, then calculated the area of cliff occupied by guillemots in each section. His calculations resulted in a grand total of 1·6 million birds on the cliffs; to be precise, 1 644 503 birds. The 1933 cull disappointed Sevmorzverprom, which had hoped for 600 000 eggs. It was made up as shown in Table 17.

The 1933 cull had stripped the cliffs bare of eggs, and, after energetic representations by the Arctic Institute, Sevmorzverprom agreed to reduce the next year's by more than half. Krasovsky's results for 1934 looked distinctly better for the guillemots (Table 18). Could the failure to mention predation of eggs

Table 17 Fate of eggs laid in 1933 by Bezymyannaya Bay Brünnich's Guillemots (Krasovsky 1937: 84)

Collected by the expedition	256 000
Taken by local egg gatherers' collective	55 000
Consumed by the expedition's egg gatherers at rate of 10 per person per day during the 45 days of the expedition	31 500
Thought to have been sucked by Glaucous Gulls or the young hatched from them killed and eaten	70 000
Presumed to have resulted in fledged young	10 000
Total production of the bazaar	422 500

Table 18 Fate of eggs laid in 1934 by Bezymyannaya Bay Brünnich's Guillemots (Krasovsky 1937: 85)

Collected by the expedition	111 000
Taken by the local egg-gatherers' collective	81 000
Used on the spot to feed the expedition's egg gatherers	3 850
Reserved for personal necessity	4 400
Minimum presumed to have resulted in fledged young	150 000
Total egg production of the bazaar	350 250

348 *In search of Arctic birds*

by Glaucous Gulls in 1934 be connected with the expedition's hostile action against these birds? In 1933 its members killed 121 of them in the 45 days they were at the bazaar. The much smaller yield of the 1934 operation was diminished still further by the fact that 52 000 eggs were hard-set and therefore unusable. The greater breeding success of the colony in that year was reflected by Krasovsky's ringing activities: in 1934 he and his assistants ringed 795 chicks, against only 17 in 1933.

As a result of the two expeditions, Krasovsky was able to draw up rules for the management of this particular resource. As against Gorbunov's 90% and

Plate 73   Arch-predator of breeding seabirds, the Glaucous Gull breeds commonly throughout the Arctic on sea cliffs and on inland tundra usually on islands in lakes.

Portenko's recommended 25%, Krasovsky thought the proportion of eggs collected should be limited to 10% of the total bird population. He recommended that the duration of egg-collecting should not exceed 16 days, counting from the start of egg-laying, and that thereafter all visits to the colony and all shooting should cease. Since guillemots only begin to lay when 3 years old, Krasovsky advised dividing the colony into three sections and harvesting eggs in one section only each year. But he went further than this. He recommended that off-season building and maintenance work be carried out on the ledges. Many were crumbling and needed reinforcement. And the vegetation should be cleared away in areas where it was overgrowing the ledges. He even recommended the addition of artificial ledges. As for the unpopular and unfortunate Glaucous Gulls, these should be ruthlessly exterminated. In the course of a single night, egg-gatherer Evert reported that all 800 or so eggs in his basket were sucked by the gulls. A Glaucous Gull startled by a shot disgorged two Brünnich's Guillemot chicks around 20 days old. Krasovsky reckoned that each Glaucous Gull, and there were 400 present at the bazaar, could account for 180 guillemot eggs or young in a season. According to Uspensky (1986: 332), something like 150 000 Brünnich's Guillemot eggs were collected annually from the Novaya Zemlya bazaars between 1930 and 1950. Although at the present time the commercial exploitation of seabird colonies is in principle prohibited, it seems that it is allowed in particular cases (compare Uspenskii 1986: 332 with the same p. 322).

ECOSYSTEM CONSERVATION: PARKS AND RESERVES

There certainly are large areas of unspoilt tundra in all parts of the Arctic. But the apparent immense size of the blank spaces in the north of the map in our atlases, a function of many of the projections used, including Mercator's, should not blind us to the relatively small size of the Arctic. Thus Greenland, at 939 870 square miles (2 415 100 km^2), is smaller than Saudi Arabia or the Sudan, and a quarter the size of the United States of America. Even so the complete protection of some very large areas of tundra is a practical possibility, and the steps already taken in most parts of the Arctic towards the creation of such protected areas are welcome. They give hope that the existing Arctic avifauna, along with the flora and fauna, may escape destruction by mankind.

NATIONAL PARKS, NATURE RESERVES AND BIRD SANCTUARIES IN SPITSBERGEN

In 1973 the Norwegian government created three national parks, two nature reserves, and 15 bird sanctuaries in the Spitsbergen Archipelago. Their total area, something like half of which is ice cap, comes to more than 30 000 km^2 (12 000 square miles), which is a little more than half the total area of the archipelago (Norderhaug 1979, 1987). Between them, these national parks

350 *In search of Arctic birds*

and nature reserves cover all the outlying peninsulas and islands, leaving outside their boundaries only the central part of Spitsbergen itself, around Ice Fjord (Isfjorden), including the coal mining and other settlements. A large part of this area, round the heads of the different branches of the fjord, has been made into a plant reserve, where all vegetation is protected. The rationale behind the rather fine distinction between the national parks in the western half of the archipelago and the nature reserves in the eastern is that the nature reserve is somewhat more strictly regulated than the national park: the regulations governing the two are identical except that, in the nature reserves but not in the parks, the governor is empowered to prohibit traffic ashore or offshore, while in the national parks but not in the reserves he may permit the shooting of Ptarmigan in particular cases. (This and the following information is taken from a pamphlet issued by the Norwegian government's Ministry of the Environment entitled *Environmental regulations for Svalbard and Jan Mayen*.)

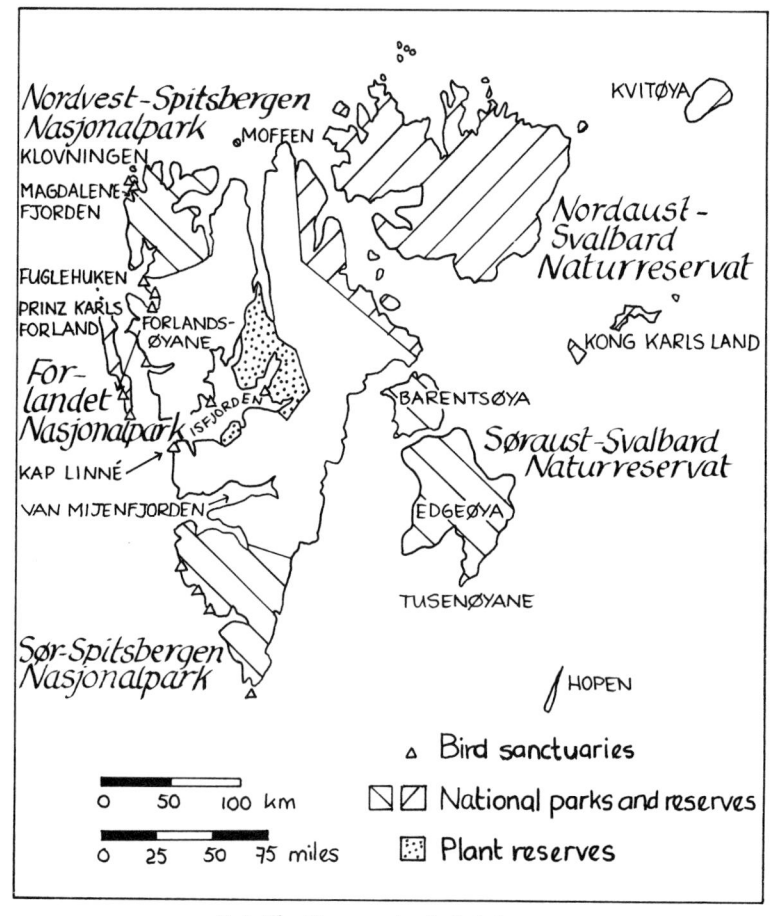

Map 36   Conservation in Spitsbergen.

The rules of these parks and reserves prohibit the erection of buildings, mining and oil drilling, the discarding of waste and rubbish, the use of cross-country vehicles and the landing of aircraft. All mammals, birds and plants are fully protected and may not be disturbed in any way. However, the integrity of both the South Spitsbergen National Park (Sør-Spitsbergen Nasjonalpark) and the Southeast Svalbard Nature Reserve (Søraust-Svalbard Naturreservat) is badly marred if not destroyed by the presence within both of them of numerous enclaves where oil claims have been made or requested. For example, on Edge Island (Edgeøya) there are 16 areas where "claims have been requested" and other "areas with rights relating to 55 find points" all of which are excluded from the nature reserve. Within the South Spitsbergen National Park similar areas forming enclaves within the park and not part of it cover oil claims staked by the Norwegian state, the Soviet coal-mining company on Spitsbergen, Arktikugol, and the multinational oil concern Fina. Naturally these claims are all in the relatively small and few ice-free areas in these two professed conservation areas. The paradox of oil exploitation within conservation areas is easily explained or excused by the 1920 Spitsbergen treaty, which underlined both Norway's responsibility for nature conservation there and the rights of the nine other countries which signed the treaty to drill for oil: "The right of searching for and acquiring and exploiting natural deposits of coal, mineral oils and other minerals and rocks . . . belongs, in addition to the Norwegian state, to: (a) All nationals of those states which have ratified or adhered to the Treaty relating to Spitsbergen. (b) Companies which are domiciled and legally established in any of the said states" (Norderhaug 1987: 583; Østreng 1977: 108).

The Northwest Spitsbergen National Park (Nordvest-Spitsbergen Nasjonalpark) has fine Arctic scenery: mountains, glaciers, fjords and off-shore islands. Within its boundaries is Magdalena Bay (Magdalenefjorden), described in lyrical terms in 1921 by Scottish naturalist Seton Gordon (1922: 93–101):

> The wonders of Magdalena Bay are indeed beyond the power of the pen to describe. We were fortunate to see it under perfect weather conditions, and the picture of that fair country will ever live in the mind's eye. . . .
> The bay, or fjord, was far narrower than any we had hitherto seen, and the conical hills, rising straight from the quiet waters, seemed, on either side the bay, but a stone's throw distant. The waters here were of a curious opaque green colour. On them much ice floated, for no fewer than twelve distinct glaciers could be counted, and some of them, as, for instance, Gully Glacier, were discharging continually a succession of small icebergs into the sea. At the head of the bay was the great Waggonway Glacier, in reality several glaciers almost touching each other. The bay ended in a high ice wall extending out from this great expanse of snow and ice; and what divergent and wonderful colours did this ice wall show—amethystine and ruddy depths, cobalt and pale blue, and (an unlooked-for colour this) russet-brown. . . .
> By now a thin grey mist had settled definitely upon the bay, so we made our way to the higher ground hoping to be able to climb above the fog. Even a few feet above sea level there was still (the date was July 9) much snow, and on a small tarn lingered half-melted snow and ice. It was only here and there that mosses and lichens were appearing through the snow; but a pair of Arctic skuas had already laid their two dark eggs on a little knoll that had been snow-free for a few days. As we passed the skuas called shrilly, and swooped and dived across our path. One can

usually tell when one is near to a skua's nest: the birds settle on the ground and stand on a little knoll, flapping their wings helplessly as though wounded and disabled. When the nest is actually discovered this acting is redoubled, the birds sometimes approaching to within a few yards and flapping their wings distractedly. . . .

As we climbed we gradually, though surely, got above the mist, and soon the sun shone down upon us from an unclouded sky of deepest blue. No wind moved across the ridge; the sun was delightfully warm.

Beneath us the bay was hidden by a grey blanket of fog. As far as the eye could see this mist-layer extended unbrokenly out to sea, appearing, on the horizon, as a low-lying cloud and seeming at that distance of a darker colour.

All around us were the spires of hills, steeply rising from glaciers of dazzling whiteness. There was scarcely any vegetation on these hills; they were barren and devoid of life, and the resting-place of innumerable rocks and boulders.

Where the Gully Glacier entered the bay many miniature icebergs floated on the calm waters. These bergs were being broken off almost continuously from the parent glacier, and drifted with the flowing tide up the bay. The water at the mouth of the glacier was of a pale green colour and curiously milky in appearance.

At the glacier's edge the thin mist gave a wonderfully beautiful effect. In it the hard outline of the ice-wall was softened and rendered dim and indistinct; ice and mist vapours were blended together. Over the glacier a stream hurried, its bed one of ice. There was not a living thing on all this icy expanse save a glaucous gull, which, flying across, rested for a moment on an ice hummock, his white plumage blending with the everlasting ice and snow.

Just east of the glacier was a precipitous hillside. Here were little auks in their thousands.

There is no more restless bird than this small frequenter of the Arctic seas. He is never quiet. Backward and forward, above and across the nesting-cliffs his tribe fly constantly—in companies, in battalions; and all the while their curious laughing cry breaks the stillness of the Arctic.

On the nesting-hill of the little auks the sun shone warmly, so that the air was filled with the scent of Polar plants. Here was the minute willow, *salix polaris*, and the bright yellow buttercup of the Arctic. Here, too, was a charming and delicately coloured lousewort and several species of saxifrages. The air was really warm—it was, one found, only on these steep hillsides which leaned towards the sun that the Spitsbergen air was truly hot—and it was delightful to sit, with no insects to disturb one, and watch guillemots, fulmars, and little auks as they passed by, each intent on his own affairs and unconscious of man's presence.

That evening, in Magdalena Bay, the fog lifted. A great stillness brooded over land and sea.

One looked, and looked again, on the fair scene of almost unearthly beauty—the great hills, the sparkling ice walls, the virgin snows.

And over all, save where the hillsides were so precipitous as to hide his serene rays, the Polar sun shone, as brightly almost as at noontide.

The Northwest Spitsbergen National Park includes Moffen, a low horseshoe shaped island on the edge of the ice pack which has been declared a nature reserve to protect the walruses which haul out there. These animals were nearly exterminated in Spitsbergen by nineteenth-century walrus hunters. When Swedish explorer Adolf Erik Nordenskiöld and his companions landed on Moffen in 1861 they saw no live walruses, but a white mass consisting of hundreds or even thousands of walrus skeletons piled up together was a conspicuous feature on the otherwise bare island (Leslie 1879: 88–89). In 1921, Seton Gordon (1922: 74–82) had a good look around the island and found, besides walrus skulls by the hundred, Eider nests on the bare shingle, Glaucous Gulls with nests of seaweed on the flat ground, a large colony of Arctic Terns, a Red-throated Diver incubating her single egg 20 yards from the water's edge, and three Brent Goose's nests. Still within what is now the park, he paid a midnight visit to Cloven Cliff (Klovningen), a steep offshore islet at the extreme northwestern point of Spitsbergen, and climbed up the steep slopes to photograph the Brünnich's Guillemots that were nesting there, in company with Little Auks, Kittiwakes and Puffins—and the inevitable Snow Buntings (Gordon 1922: 119–123).

The Prince Charles Foreland National Park (Forlandet Nasjonalpark) consisting of the whole of Prince Charles Foreland (Prins Karls Forland) with its offshore islets, contains within its 640 km^2 (247 square miles) snow-covered mountains, level stretches of tundra comprising marshy areas and bare stony flats, coastal lagoons and even some small glaciers. The summer climate is cool and foggy and snow lies late on the ground. But there are birds in plenty. Grey Phalaropes and Purple Sandpipers breed, as well as the Red-throated Divers studied by Huxley. At the northern tip of this park is the great headland of Fuglehuken, now a bird sanctuary where Fulmars, Kittiwakes, Brünnich's Guillemots, Puffins and Black Guillemots breed. Also within the park are the world's most northerly breeding Guillemots (Norderhaug 1987: 585), and all three species of Spitsbergen geese, Brent, Barnacle and Pink-footed, breed on the islets off the west coast which form the Forlandsøyane Bird Sanctuary (Grimmett and Jones 1989: 491).

Most of Spitsbergen's bird sanctuaries are within one or other of the three national parks, and the regulations governing them are virtually identical with those for the parks and nature reserves, except that all traffic, including landing by boat, is prohibited between 15 May and 15 August. All are on the west coast of Spitsbergen and all except Kap Linné are offshore islets or groups of islets. On the whole, the oil threat apart, conservation in Spitsbergen appears adequate. Weak points are the impossibility of ensuring that regulations are respected, and the possibility that increasing numbers of tourists may, perfectly legitimately, land on the eastern islands which make up the Northeast and Southeast Svalbard Nature Reserves. Perhaps some of these, especially the islands making up King Charles Land (Kong Karls Land), where Ivory Gulls breed, and the Thousand Islands (Tusenøyane) south of Edge Island, the most important Brent Goose breeding area in the Spitsbergen Archipelago (Persen 1986), could be made into bird sanctuaries to prevent landings from boats in the breeding season. Relying on their inaccessibility is not entirely satisfactory.

THE WORLD'S GREATEST NATIONAL PARK?

When it was created in 1974 the National Park in North and East Greenland was, at approximately 700 000 km^2 (270 000 square miles), the largest national park in the world. It is likely to remain so. It is uninhabited and almost all of it is Inland Ice. It does however include the largest ice-free area in Greenland, namely Peary Land. In 1989 its western border was moved westwards to include even more Inland Ice, bringing the total size to 972 000 km (375 300 square miles)—a good deal larger than France, England, Scotland, Wales and Ireland put together. On 1 January 1981 the Greenland government took over full responsibility for wildlife and habitat protection from Denmark, and the park has been administered since then by a National Park Board appointed by the Greenland government and parliament. In June 1989 two research teams from five European countries and the United States of America began drilling through Greenland's ice cap or Inland Ice within the boundaries of the park, but, unlike oil drilling, this is scarcely likely to raise objections from environmentalists; nor will the occasional incursions at the edge of the park by Greenlandic hunters, who are permitted free access at any time. They must be from either Avanersuaq (formerly the Thule district) or from Ittoqqortoormiit (Scoresbysund) and be carrying on traditional hunting from dog sledge, kayak or motor boat. The only other persons allowed in the park without a permit obtainable through the Greenland government are "persons fulfilling supervisory or other duties for the public authorities", and the personnel of weather and other stations. Those in the former capacity are the members of the Danish military sledge patrol "Sirius" with headquarters in the park at Daneborg and a beat that carries them up and down the uninhabited east coast of Greenland between there and Mesters Vig. They try to be in contact with expeditions permitted to enter the park (Meyer 1987; Silis 1989; Commission for Scientific Research in Greenland, *Newsletter* 18, July 1989; and leaflet *Guidelines on Greenland Expeditions* issued in 1989 by the Commission and the Danish Polar Centre).

The rules and regulations of the National Park in North and East Greenland protect mammals and birds from any kind of disturbance and camping at or near breeding, foraging, moulting or resting places of mammals and birds is prohibited. If a bear is killed in self-defence the skin and skull, suitably prepared, must be sent to the Greenland government and a report submitted to the chief of police in Greenland. The collection of plants, berries, insects and other invertebrates is permitted so long as they are not for sale. Off-road motor vehicles are prohibited, except on sea ice or snow, and landing by aircraft and the dropping of supplies from the air are likewise prohibited without special permission. Expeditions permitted to visit the national park are expected to do so either via Thule Air Base in Avanersuaq in the north, or via the airstrip at Constable Point (Constable Pynt) on the Hurry Inlet (Hurry Fjord) off Scoresby Sound, in the south. That airstrip is less than 100 miles (160 km) from the national park boundary. It was built in 1985 as part of an oil exploration programme in the Jameson Land area west of Ittoqqortoormiit. In recent years studies of the birds of the park have been continued and extended to little-known areas on its east side. In 1990 the Greenland government's Department of Wildlife Management in Nuuk published bird censuses of 16 different areas within the park. These found Sanderlings and Turnstones to be among the commonest breeding waders. The Knot was less common than expected (Boertmann *et al.* 1990).

The establishment of Greenland's national park in 1974 was the culmination of years of hopes and dreams. The Danish writer-explorer, and ornithologist of the Fifth Thule expedition, Peter Freuchen, had declared on his return from the 1912 First Thule Expedition across the Inland Ice from Avanersuaq to Peary Land (in the northern tip of the park) that north Greenland and the northern part of Greenland's north east coast should become a national park (Pedersen 1934: 147). Danish zoologist Alwin Pedersen spent 5 years in east Greenland and was a member of Lauge Koch's Three-Year Danish East Greenland Expedition from 1931 on. He repeated Freuchen's plea for a national park in northeast Greenland at the end of his book on *Polar animals*, first published in Danish in 1934 and, after going through various editions and translations, eventually finding its way into English in 1962 translated from the French. In Pedersen's vision the park's main function would be to protect walrus, Polar bears and muskoxen. A further plea was made by Knud Oldendow in his book on *Nature conservation in Greenland (Naturfredning i Grønland)* published in 1935.

Leaving aside the National Park in North and East Greenland, the country's largest protected area is the Melville Bay National Wildlife Reserve, which was declared in 1980 on the initiative of the municipal councils or communes of Qaanaaq (Avanersuaq) and Upernavik, whose jurisdictions adjoin in the middle of the reserve. The aim was to conserve the stock of mammals which form the basis of the subsistence hunt of the Greenlandic communities at either end of Melville Bay, namely Qaanaaq and Savissivik to the north and Kullorsuaq and Nuussuaq to the south of the bay. This uninhabitable ice-bound coast is the only polar bear breeding place in west Greenland, and it is also an important breeding area of the ringed seal. The reserve includes the coast, which is mainly an ice cliff, and the many offshore islands. Among these are the Sabine Islands (Sabine Øer), on which Sabine's Gull was first discovered by Edward Sabine in 1818. It still breeds there. The reserve is divided into an inner region along the coast and an outer one comprising the sea and islands offshore. The Greenlanders are allowed free access to travel through the outer zone by dogsled and are allowed to hunt there. But in the inshore zone hunting and egg-collecting, traffic, including motor boats, as well as flying at an altitude of less than 500 m (1600 feet), are prohibited. Although the use of motor boats in the outer zone is not prohibited, landing on the islands is forbidden (Meyer 1987).

Earlier in this chapter the declaration by the Greenland government in 1988 of 11 so-called Ramsar Sites, or Wetlands of International Importance, was mentioned. Two of these are within the national park, but seven are so far unprotected. The remaining two have been declared Breeding Reserves for Birds, where approach of any kind nearer than 500 m (550 yards) between 1 June and 31 August is strictly prohibited. In all there are 12 of these Breeding Reserves for Birds, all on the west coast and all designed to protect colonially breeding seabirds. Except for two, comprising mainly cliffs, they consist of offshore islets (Grimmett and Jones 1989). It is to be hoped that more of these reserves will be declared in due course and that the Greenland parliament, which passed Greenland's first Environmental Protection Act in October

Map 37 Conservation in Greenland.

1988, will not rest content with the present system of nature reserves, good as it is (on conservation in Greenland see Thing 1990b).

CANADA'S MIGRATORY BIRD SANCTUARIES

To understand the present-day status and distribution of conservation areas in Canada one has to go back to the beginning of the present century. The concept of conservation embraced at that time by the Canadian authorities was the protection of wildlife resources for the benefit of the native peoples. It was feared that their subsistence hunting was being jeopardised by commercial hunters and trappers, whalers and explorers, and other whites lured to the north by thoughts of gold and minerals. In accordance with this belief, in 1918 and 1920 Victoria Island and Banks Island respectively were declared native hunting preserves. But when in 1926 these and other reserves were expanded into the Arctic Islands Reserve, the motive of protecting the interests of the native peoples may have been somewhat sullied by the aim of establishing Canadian sovereignty in the Arctic. For it was allegedly the setting up of the Arctic Islands Reserve in 1926 which persuaded the Norwegians not to press their claim for sovereignty over the more northerly islands, later named the Queen Elizabeth Islands, which they had discovered at the end of the nineteenth century. The abolition of the Arctic Islands Reserve in 1966 "brought to an end a remarkable period of conservation aimed at the welfare of the native people" (Kovacs 1987: 533). In this same period a policy was implemented of creating game sanctuaries for particular species—the wood bison, muskox, caribou and beaver—where hunting was prohibited altogether.

Canada's first national park, Banff in the Rocky Mountains, was created in 1885, the second in 1922. Progress remained slow. Only in 1972 was the first national park established in the Arctic, namely the Auyuittuq National Park on Baffin Island's Cumberland Peninsula. But what was to be conserved, in this 21 471-km^2 (9290-square-mile) park, was not wildlife so much as the Penny Ice Cap and its glacial system. Hence the park's Inuit name, which means "the land that does not melt". The park's superb scenery can be enjoyed best from the popular trail over the Pangnirtung Pass. It lies just north of the Arctic Circle almost at the easternmost point of Canada. There are two other national parks in the Canadian Arctic. One is at the other end of the country in the furthest west, bordering on Alaska. This is the 10 000-km^2 (3960-square-mile) Northern Yukon National Park, proposed in 1978 and established in 1984. The other is the Ellesmere Island National Park, which is at the extreme northern tip of Canadian territory, declared in 1985. This comprises 39 500 km^2 (15 250 square miles) of tundra, mountains, ice-fields and glaciers, as well as Lake Hazen where a field research station has been established intermittently in the past. Besides these national parks, an area of 2624 km^2 (1013 square miles) in the centre of Bathurst Island has been declared the Polar Bear Pass National Wildlife Area. Other proposals for Arctic national parks are under consideration. One of the reasons for Canada's

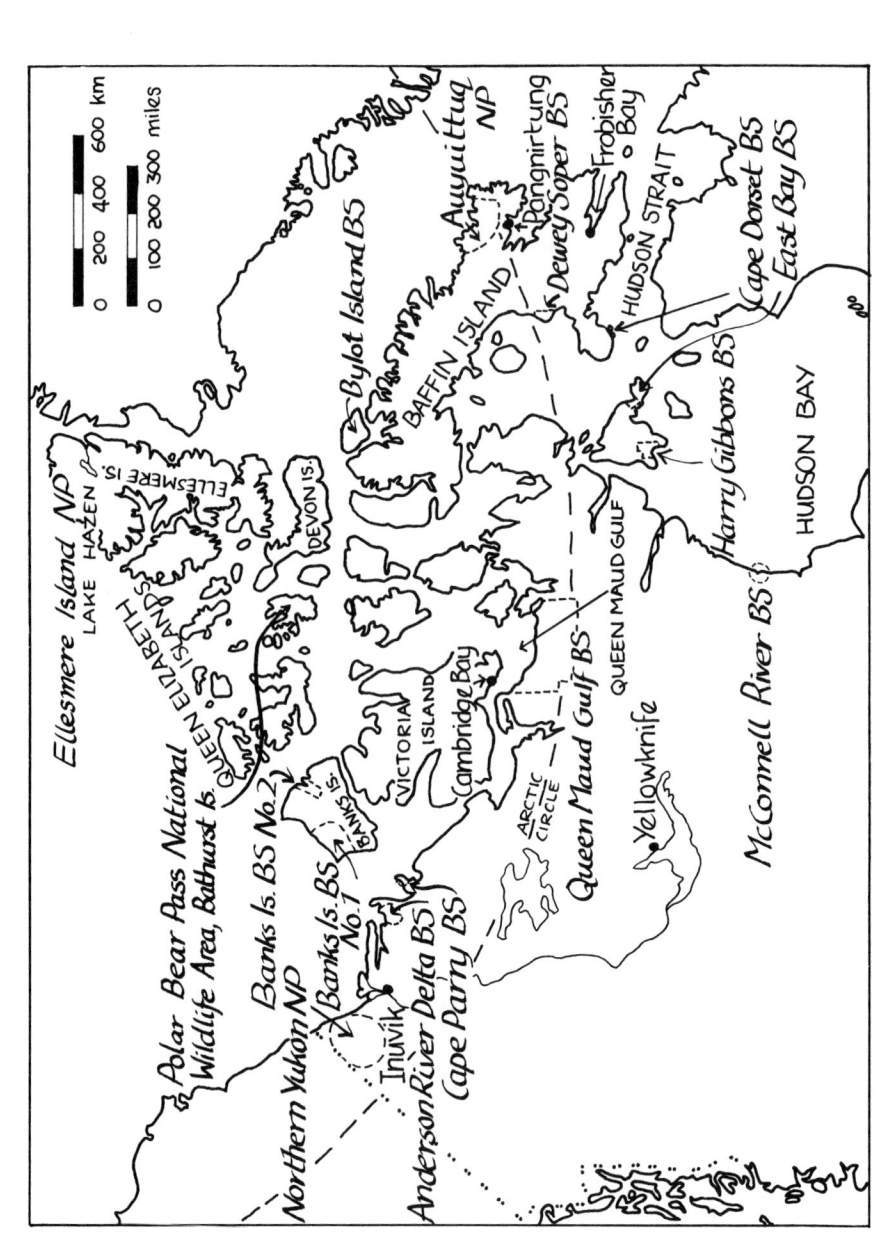

Map 38   National parks (NP) and bird sanctuaries (BS) in Canada.

slow progress in creating national parks in the Arctic is the need to negotiate them with the local Inuit communities through their regional organisations: the Inuvialuit in the west and the Tungavik Federation of Nunavut in the centre and east (Kovacs 1987; Hamre 1987).

More important for the conservation of birds than the far-flung Arctic national parks are the dozen Migratory Bird Sanctuaries established in the Arctic since 1957 and administered by the Canadian Wildlife Service. Their general purpose is to conserve the breeding habitats of birds which migrate to the Arctic in summer, and in particular they have been created to protect certain large and otherwise vulnerable colonies of breeding geese. A network of these sanctuaries was created in southern Canada following the passing of the migratory Birds Convention Act in 1917, which made possible the Convention between the United States and Great Britain for the Protection of Migratory Birds of 1918. Other implications of this convention for conservation have been discussed in Chapter 2, particularly its impact on native peoples and the consequent development of their role in bird conservation in the Canadian Arctic. The first two migratory bird sanctuaries to be established in the Arctic were both created in 1957 in southwest Baffin Island. The larger of the two covers 8160 km^2 (3150 square miles) of low-lying grassy tundra called the Great Plain of the Koukdjuak, where in the 1980s there were thought to be about half a million breeding Snow Geese distributed in three huge colonies. The reserve was named the Dewey Soper Bird Sanctuary after Canadian Wildlife Service biologist J. Dewey Soper. It was he who here first discovered the nesting grounds of the blue phase of the Snow Goose in 1929, as described in Chapter 8. The other was the Cape Dorset Bird Sanctuary on the north shore of Hudson Strait, consisting of an offshore island and two groups of offshore islands. It was created to protect breeding colonies of Eiders. The largest of Canada's Arctic bird sanctuaries is the Queen Maud Gulf Bird Sanctuary, comprising 62 782 km^2 (24 240 square miles) of tundra between the Arctic Circle and the south shore of Queen Maud Gulf. Flowing from the southern border of the sanctuary and debouching into the sea on the northern, and nearly bisecting it, is the Perry River, where the breeding grounds of Ross's Goose were first discovered in 1938 and Peter Scott led an expedition to study them in 1949 (see Chapter 8). The 47 species reported identified within the sanctuary up to 1984 may seem few, but this is quite a respectable number for an Arctic locality. Although most Arctic bird sanctuaries in Canada are there to protect breeding geese and ducks, there is one notable exception: the Cape Parry Bird Sanctuary, only 2 km square and by a long way the smallest, protects the only colony of breeding Brünnich's Guillemots in the western part of the Canadian Arctic (Finlay 1984: 348–351; Hamre 1987: 558–559).

In spite of the huge size of many of them, Canada's Arctic bird sanctuaries are subject to strict regulations, though a permit from the Minister of the Environment via the Canadian Wildlife Service may easily circumvent these. No one is allowed to hunt migratory birds, "disturb, destroy or take" their nests, or have a live or dead migratory bird or its egg in his or her possession "except under authority of a permit therefor". No firearm or dogs at large are permitted in the sanctuaries, but "the Minister may issue a permit authorising any person to have firearms in his possession and to shoot and to have in his possession migratory game birds . . .". It is good to know that "The Minister may refuse to issue a permit to any person" and that one clause virtually makes a permit necessary for any birdwatcher to enter the sanctuary: "No person shall, in a migratory bird sanctuary, carry on any activity that is harmful to migratory birds or the eggs, nest or habitat of migratory birds, except under authority of a permit" (official Migratory Bird Sanctuary Regulations issued by the Canadian Wildlife Service, Environment Canada, Ottawa, July 1990). Their very size, remoteness and inaccessibility gives additional protection to these sanctuaries. If Canada has lagged in the creation of Arctic national parks, her Arctic bird sanctuaries are second to none.

ALASKA: OIL OR WILDLIFE?

On the face of it, Alaska has a complex but admirable system of conservation areas extending through the state and well into the Arctic. But high-sounding titles can be misleading. For example the magnificent Tongass National Forest in Alaska's panhandle is really only a hybrid. While some think of it as a conservation area, the United States Forest Service is currently cutting it down, reputedly selling the timber to loggers at a dollar a tree. Then again, the U.S. Fish and Wildlife Service administers 119 690 square miles (310 000 km^2) of National Wildlife Refuge lands in Alaska, but allows sport hunting as well as subsistence hunting on them. Even within the national parks and preserves, mining, cutting of logs for building cabins, shooting and other exploitation of natural resources are all allowed, though "carefully regulated" (*Alaska almanac* 1989: 126).

In the present context, defining Arctic Alaska as broadly as possible, the most important conservation areas are the two wildlife refuges in the Arctic, the Arctic and Yukon Delta National Wildlife Refuges, two preserves, the Bering Land Bridge and Noatak National Preserves, and two parks, the Gates of the Arctic and Kobuk Valley National Parks. The single area formed by these two parks and the Noatak preserve covers the Brooks Range and many of its valley systems west of the Prudhoe Bay—Valdez oil pipeline. But the Brooks Range is only at the boundary of the Arctic and much of this area is taiga or boreal forest, not tundra, though all of it is north of the Arctic Circle. The two coastal conservation areas include much lowland marshy tundra. The Bering Land Bridge National Preserve covers most of the northern half of the Seward Peninsula beyond the reach of the roads out of Nome. It has been credited with 112 bird species. The Yukon Delta National Wildlife Refuge is a vitally important area for breeding ducks and geese; it was the scene of the Hooper Bay Expedition of 1924 described in Chapter 4, and of the goose management plan worked out in 1984 between local Yupik Eskimo hunters and the U.S. Fish and Wildlife Service, described in Chapter 2.

Alaska's system of conservation areas was set up by the Alaska National Interest Lands Conservation Act, signed by president Jimmy Carter at the end

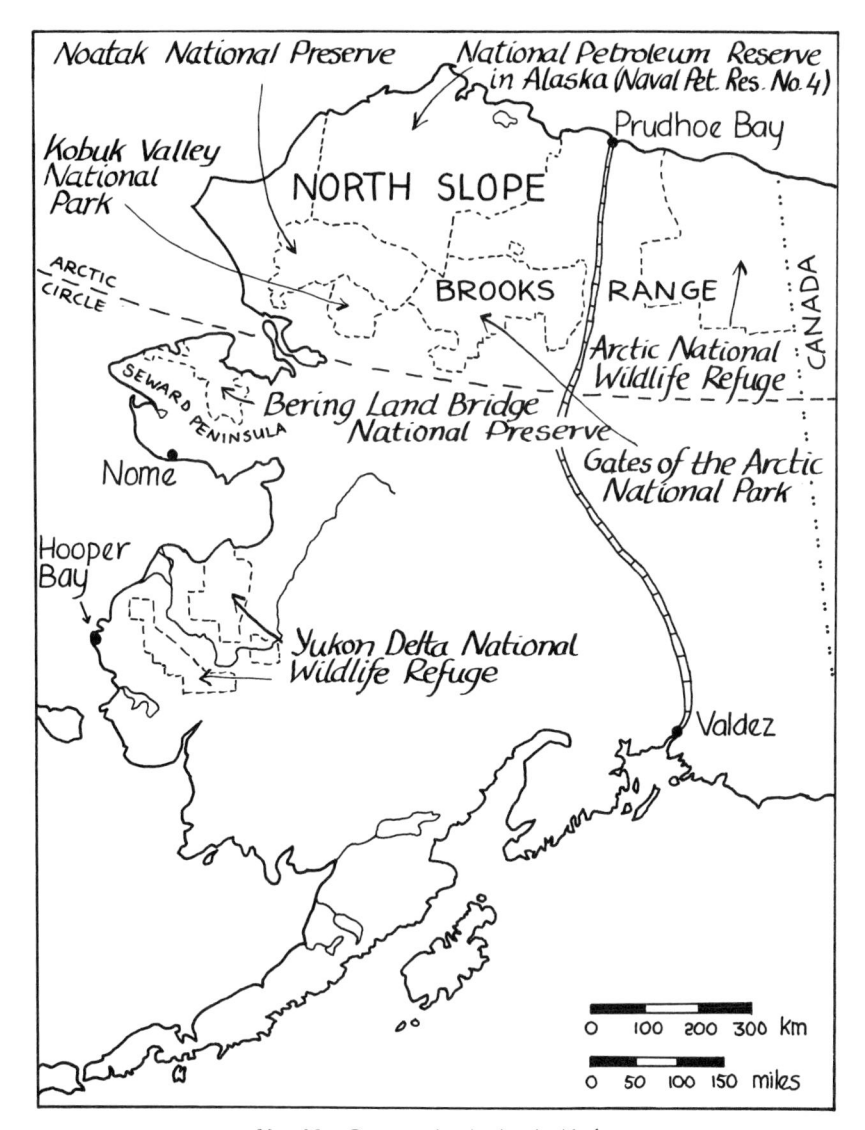

Map 39 Conservation in Arctic Alaska.

of 1980 (Stenmark 1987). It was this act which transformed what had been since its creation in 1960 the Arctic National Wildlife Range into the refuge of the same name, and substantially increased its size. It is on this Arctic National Wildlife Refuge that one must pin one's hopes for the conservation of a significant part of the High Arctic tundra ecosystem of Alaska. For most of the North Slope has already been divided up among the various interest groups. The largest single piece of it, about 250 miles (400 km) long from west to east and reaching from the coast in the north to the Brooks Range in the south, was designated by President Warren G. Harding in 1923 as the Naval Petroleum Reserve No. 4, renamed in 1977 the National Petroleum Reserve in Alaska. Between it and the Arctic Wildlife Refuge, which is about 125 miles (200 km) long from west to east, is a 125-mile (200-km) long stretch of coast and North Slope behind it in the centre of which is Prudhoe Bay, where oil is already being extracted and has been for more than a decade, going south by pipeline over tundra, mountain and forest to Valdez. As if this were not enough, namely making over the central 400 miles (650 km) of the 550-mile (885 km) long North Slope to the oilmen, the Arctic Wildlife Refuge itself is in jeopardy. The Alaska Lands Conservation Act of 1980, while converting the range into a refuge and increasing its size, set aside some two-thirds of the coastal plain within the refuge, that is a coastal strip about 100 miles (160 km) long and 16–34 miles (26–55 km) wide, for future oil exploration (Map 34). After 5 years of field work in the refuge the Fish and Wildlife Service recommended in 1986 to the Secretary of State for the Interior that the United States government should go ahead with oil and gas exploration in this coastal strip. Since then controversy has raged. To obtain a hearing the National Audubon Society had to take the U.S. government to court. The Department of the Interior, the oil companies, and the native corporations allied together in an attempt to bring off a massive land exchange, offering to add almost a million acres to other conservation areas in return for the same amount inside the Arctic National Wildlife Refuge. The battle continues, but few environmentalists are optimistic about the outcome.

ZAPOVEDNIKS AND ZAKAZNIKS

In the Soviet Union, national parks are a comparatively recent development. Since the first was established in Estonia in 1972 others have followed, but none so far in the Arctic. The two types of conservation area that have been established in the Arctic are state nature reserves or *zapovedniki* and sanctuaries or *zakazniki*. The zapovedniks are the more strictly protected: no commercial activity is permitted within them and tourists or non-specialists may only enter if an organised visit or guided tour is arranged for them. Their aim is to conserve an entire ecosystem, or natural complex to use the Russian phrase. In the zakazniks on the other hand, some economic activity may be allowed, carefully regulated shooting or subsistence hunting is possible in some, and full protection may be given only in the breeding season. Regulations governing them differ, according to whether the zakaznik in question has been created to protect a breeding colony of a rare bird or to provide a refuge for wildfowl for the sport shooter's benefit, or for some other limited purpose. Most are relatively small (Syroyechkovskiy and Shtil'mark 1985: 11; Knystautas 1987: 54, 59–60; and see too Pryde 1972). The zakazniks will scarcely concern us further here because there are only a handful in the whole of the Soviet Arctic north of the Arctic Circle, and these are essentially game reserves. Two that come to mind are Vaygach Island, declared in 1963, and the lower Pechora River, which was declared a sanctuary for "water game" in 1967 (Syroyechkovskiy and Shtil'mark 1985: 16).

Although the Arctic zapovedniks have a long history, bureaucratic procrastination has caused delays, and one must admit that the present-day situation leaves much to be desired. Some of the earliest reserves suffered from chopping and changing of the worst kind. The Lappland zapovednik in the Kola Peninsula (Kol'skiy poluostrov), mentioned here because, though in the taiga rather than the tundra, it is north of the Arctic Circle, was created in 1930 with an area of 2000 km^2 (772 square miles). This was reduced to 1310 km^2 (506 square miles) in 1935, and then expanded in 1642 km^2 (63 square miles) in 1950. But in 1951 the reserve was abolished altogether and commercial timber felling began. In 1957 it was restored almost to its former boundaries and logging ceased (Syroyechkovskiy and Shtil'mark 1985: 14), but from the early 1970s onwards attention was repeatedly drawn in the press to the extensive damage being done to its eastern forests by chemical effluents from industrial plant in Monchegorsk.

The next Arctic nature reserve to be established in the Soviet Union was the Kandalaksha Zapovednik, dating from 1934. It has suffered similar bureaucratic vicissitudes, its area having been repeatedly changed. Beginning life as a hunting preserve comprising a group of islands in the Gulf of Kandalaksha (Kandalakshskaya guba), it became a state Zapovednik in 1939 and was enlarged in 1951 when the Seven Islands zapovednik was added to it. The Seven Islands (Sem' ostrovov), off the north coast of the Kola Peninsula (Map 15), had been made into a zapovednik in 1938 to protect the breeding seabirds or bird bazaars on their cliffs. Thousands of Guillemots, Puffins, Kittiwakes, Herring Gulls and Eiders breed here, and it was here that Soviet ornithologist L. O. Belopol'sky carried out most of the field work for his book on Barents Sea seabirds (Belopol'skii 1961; Semenov 1980). The Seven Islands must form the oldest nature reserve of any kind in the Soviet Union. They were protected in the seventeenth century by a special warden, as a Gyrfalcon breeding area, by command of the tsar (Uspensky 1970). The present-day Kandalaksha Zapovednik is a complex of 500 islands and parts of the mainland totalling about 220 square miles (571 km^2) in all along the north and south coasts of the Kola Peninsula (Knystautas 1987: 89; Gaava *et al.* 1987: 593). It is part of the much larger area which has been declared a Wetland of International Importance or Ramsar Site.

Another zapovednik in the Soviet Arctic is made up of Wrangel Island (ostrov Vrangelya) and Herald Island (ostrov Geral'd), out in the Arctic Ocean off the east Siberian coast (Map 16). Beginning life in 1961 as a zakaznik, or game reserve, it became a zapovednik in 1976 and has now been declared a Biosphere Reserve. Its 3072 square miles (7956 km^2) are a mixture of lowland and upland tundra with bird bazaars along the coasts. Mammals include up to 200 female Polar bears, which go ashore on Wrangel Island to make their dens and have cubs, and a flourishing herd of introduced muskoxen, as well as domesticated reindeer. The bird speciality of Wrangel Island is the Lesser Snow Goose, and this is the only place where it nests in any number in the Soviet

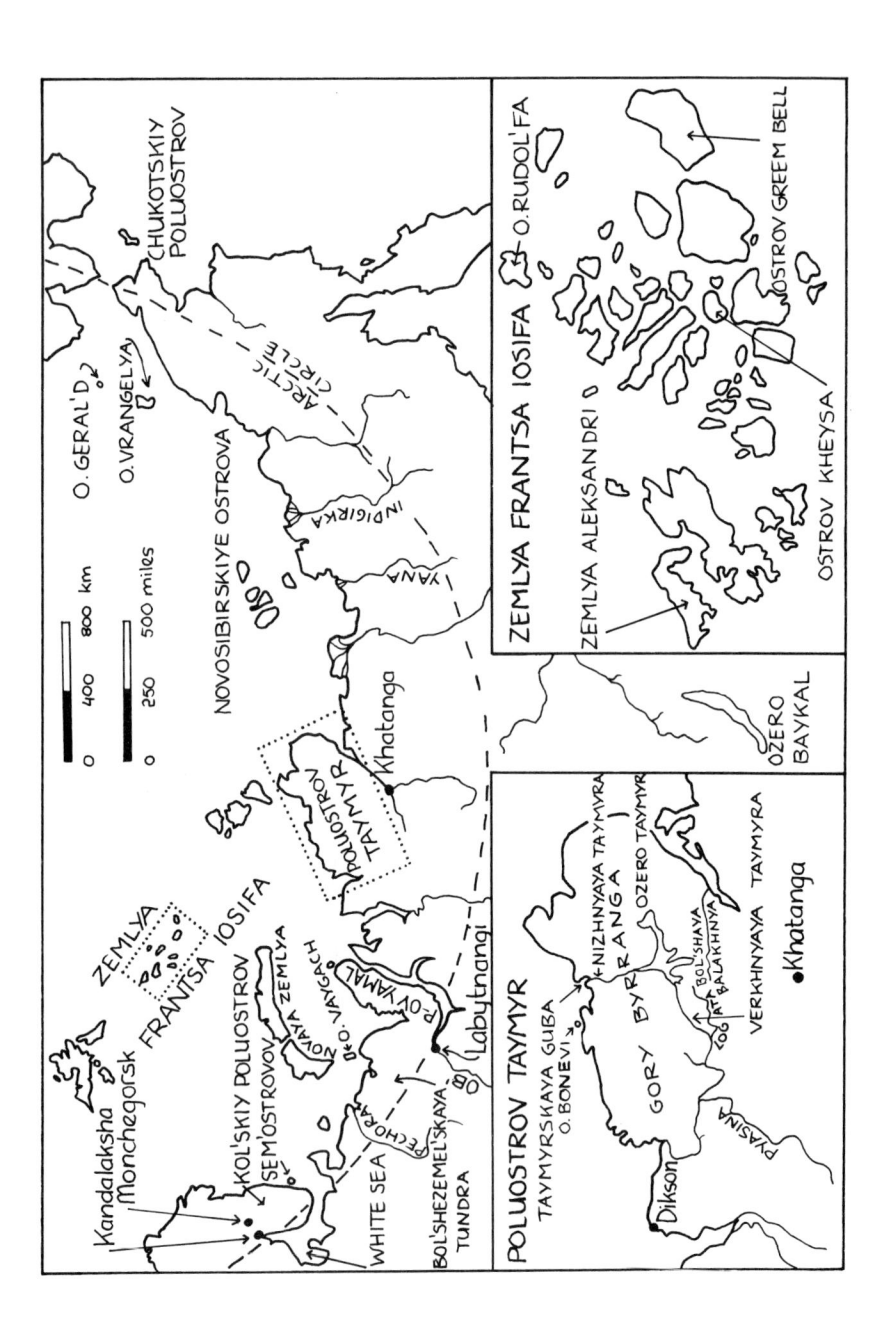

Map 40 Zapovedniks and zakazniks.

Union. The population is separate from the much larger North American one. After a period of decline in the middle years of the twentieth century it now appears to be recovering. A researcher at Moscow's Institute of Evolutionary Animal Morphology and Ecology, Nikita Ovsyanikov, reported that in 1989 there were signs of the geese re-colonising areas long abandoned (Ovsyanikov 1990). Wrangel Island is also an important breeding place for the Snowy Owl and for waders: the Grey Plover, Turnstone, and Knot all nest commonly there.

The latest Arctic reserve to be established, and by far the largest, is the Taimyr Zapovednik, which occupies 5206 square miles (13 483 km^2) in the centre and south of the Taimyr Peninsula (Knystautas 1987: 89; and see Shtil'mark 1977). It is actually the largest reserve in the whole of the Soviet Union. Created in 1979 with headquarters at Khatanga, its heartland comprises the western arms of Lake Taimyr (ozero Taimyr), the valley of the lower Verkhnaya Taimyr River, including spurs of the Byrranga Mountains, and the lower part of the basin of the Logata River. It embraces a remarkable variety of landscapes including hummocky and shrubby moss tundras, as well as dry stony tundras, and it supports abundant animal life. Three species of eider are found here, besides other ducks, and three species of goose. One of the geese is the rare endemic Red-breasted Goose, and the Taimyr Zapovednik has been credited with two or three thousand of them (Knystautas 1987: 89). Sandpipers of several species, Rough-legged Buzzards and Snowy Owls also breed in the reserve. It received considerable publicity when Gerald Durrell visited it in 1985, as described in Chapter 12.

If the Soviet bureaucracy has so far only slowly and haltingly created zapovedniks in the Arctic, there has been no shortage of proposals for them. Conservationist and Arctic expert, V. V. Kryuchkov, who in 1973 had published *The Far North: problems in the rational utilization of natural resources*, was one of those responsible for the establishment of the Taimyr zapovednik. Indeed it was based on a detailed proposal of his. In his *The North: nature and man* published in 1979, he made other proposals that have still not been implemented. He argued that the discovery of reserves of oil and gas in the European Arctic made a Zapovednik in the centre of the Great Land Tundra (Bol'shezemel'skaya tundra) a little north of the Arctic Circle, on the River Moreyu, between the Pechora and Ob' rivers, of the utmost urgency. This would be on the northern fringe of the taiga or in the so-called forest tundra, where Kryuchkov proposed another zapovednik just north of the Arctic Circle in northwestern Siberia on the left bank of the lower Ob' near Labytnangi (Kryuchkov 1979: 114–115). Other proposals of the late 1970s included a giant Yana–Indigirka zapovednik in northeast Siberia to protect the coastal tundra between the two rivers as well as parts of the New Siberian Islands (Novosibirskiye ostrova). Its total proposed area of 20 000 km^2 (7722 square miles) would have made it the Soviet Union's largest nature reserve (Syroyechkovskiy and Shtil'mark 1985: 23). This grandiose plan was shelved in the 1980s, along with a host of lesser ones.

One other proposed zapovednik deserves mention because of its unusual scope and character. In 1967 the conservation authorities of the Soviet Union, represented by the Central Laboratory for the Protection of the Environment of the All-Union Scientific Research Institute for Nature Conservation and Nature Reserves, organised an expedition to Franz Josef Land (Zemlya Frantsa Iosifa) to explore the possibilities for a zapovednik there. Another expedition was mounted in 1980 and 1981 and the participating scientists drew up an elaborate proposal for converting the entire archipelago that makes up Franz Josef Land into a zapovednik. This would include all the islands, forming some 16 090 km^2 (6212 square miles) of terra firma, 85% of it glacier, as well as the permanent or recurring polynyas, or areas of open water, and ice-covered sea between and around them, making the total area of the proposed zapovednik about 42 000 km^2 (16 216 square miles). The archipelago is further north than the Spitsbergen Archipelago and its land area is very much smaller. Still, this zapovednik would include "Polar Bear cubbing dens, walrus haul-out sites, cliff colonies of seabirds, colonies of common Eiders and Ivory Gulls and polynya areas which represent important feeding areas for pinnipeds, whales and bears" (Uspenskiy et al. 1987: 219). The overall aim of the proposed zapovednik was apparently to limit economic and other human activities to the islands where it was already taking place on a significant scale, namely Alexander Land (Zemlya Aleksandri), Rudolph Island (ostrov Rudol'fa), ostrov Kheysa and Graham Bell Island (ostrov Greem Bell), and to protect the so far unaffected parts of those islands from further despoliation.

Though the zapovedniks have been slow in coming and are few and far between, one feature that distinguishes them from many other nature reserves is that each has its own resident staff of research biologists, whose publications have often been of value.

ENDANGERED ARCTIC SPECIES

Besides the need to preserve ecosystems as a whole it is also important to take measures to protect endangered species, both by creating special reserves to cover their breeding, feeding and resting places, and by prohibiting shooting or otherwise disturbing them. Fortunately the list of seriously endangered Arctic species is not a long one. The only two that seem on the verge of extinction or nearly so were mentioned in Chapter 1, namely the Eskimo Curlew in the North American Arctic, and the Siberian White Crane. Soviet ornithologist Savva Uspensky (1986: 342–344; Uspenskiy et al. 1987: 592; see too Solomonov 1987: 117–152 and Chernyavsky and Andreev 1990) has drawn attention to several other Arctic species which, though not on the verge of extinction, merit protection because of their rarity in the Soviet Union. There are three species of goose which have only small and threatened populations in the Soviet Union though they seem to be thriving elsewhere. The Barnacle Geese breeding in the southern island of Novaya Zemlya and on Vaygach Island may number as few as a thousand pairs. The Emperor Geese breeding on the coast of the Chukchi Peninsula are thought to be decreasing in number (Uspenskii 1986: 343); and the Wrangel Island Snow Geese have had to be given special protection. A fourth goose, the Red-breasted, breeds only in mossy and bushy tundras between the Yamal Peninsula and the eastern Taimyr; the total population is small and may be declining. In 1972–1973 an estimate of 7500 nesting pairs was made on the breeding grounds, and a total population of 25 000 birds was suggested (Madge and Burn 1988: 151). A fifth species, the Lesser White-fronted Goose, has become rare throughout its range, which formerly extended throughout the Siberian and Russian Arctic into northern Scandinavia. Nor was Uspensky altogether happy about the status of Bewick's Swan in the Soviet Arctic, where he thought there might only be 10 000–20 000 birds in all. He pointed out that the Gyrfalcon ought to be fully protected as one of his country's rarer breeding birds. The Peregrine, though fairly common, requires protection because various goose and duck species, especially the Red-breasted Goose, tend to breed in close proximity to it, and derive some protection from other predators through its presence near their nests.

DANISH BIRD PROTECTION IN GREENLAND

So far in this chapter emphasis has been on the conservation of Arctic birds at the present time. But a historical postscript seems in order, to draw attention to some of the earliest attempts to conserve the bird life of the Arctic. These initiatives were taken by the Danish administration in Greenland, but the Greenlanders themselves co-operated willingly and for the most part loyally in formulating legislation and seeing to its implementation. One of the first books devoted to nature conservation dealt with Greenland. It was entitled *Naturfredning i Grønland*. Published in 1935, this thoroughly-researched 389-page book was written by Knud Oldendow, a Danish civil servant working in the Greenland administration in Copenhagen who, two years before, had published a book on Greenland's bird life. His ornithological interests did not interfere with his career: from 1938 until 1948 he was director of the Greenland administration (Sørensen 1983).

In Oldendow's day no one ever thought of the abundant Brünnich's Guillemot as in need of protection; for him the birds most threatened by Greenlandic subsistence hunters were the Cormorant, formerly common, now rare, and the White-tailed Eagle. His great success story was the Eider, but already in 1935 Oldendow noted that since the taking of Eider eggs had been prohibited in the 1920s, the Greenlanders had switched to gathering eggs of other species. In the Upernavik district Eider eggs had been an essential part of the local people's diet because they were among the few foodstuffs that could be preserved until the winter. Indeed they were consumed right through till March (Oldendow 1935: 92–93). Now, in the 1930s, although Brünnich's Guillemot's eggs did not keep so well, they and the eggs of other species, including Arctic Tern, were being substituted for Eider eggs, and Oldendow (p. 104) was convinced that the Arctic Terns breeding around Godthåb (Nuuk) were already declining in number as a result.

In the nineteenth century the Eider occurred in enormous numbers on the west coast of Greenland, breeding in huge colonies on offshore skerries as far north as Avanersuaq, where thousands nested on Littleton Island in Smith Sound between 78° and 79° N. Oldendow (pp. 106–107) cites an eyewitness description of the remarkable northward Eider migration in March–April near Nanortalik in south Greenland in about 1885 (Map 37). Offshore in a boat with some Eskimos early one morning in the hope of bagging some Eiders, the writer heard a noise like a storm which he thought was the dreaded fohn, or *Föhn*, wind which sometimes blows from the ice cap down into the fjords. Anxious for the safety of his boat and tent, he was about to insist once more on beating a retreat, when his Eskimo companion pointed out

> a dark streak far up the fjord and said that it was birds; and I too, as soon as I saw it, realised that that was what it was. From the time elapsing between my seeing the birds first and the first bird passing me I reckoned they must have been some two miles away when I first heard them. Now they came pouring past in thousands and hundred thousands, not in many flocks, but in one single large flock, an unceasing broad stream as long as the eye could reach, that certainly was not very dense, yet still in places so dense that at the right distance one could have expected to get four to six birds with a single shot. Unfortunately they were flying so far out that I could not make any use of my shotgun nor of my many cartridges. Still I picked off six birds with my rifle and recovered them all. At the start I was so surprised at the mass of birds that only after part of the flock had passed me did I think of looking at my watch to see how long the birds took to go by. From the moment that I looked at the watch until the last group passed me, one hour and three minutes elapsed.

The over-exploitation of the once so numerous Eider duck became evident during the second half of the nineteenth century. In 1840 Carl Peter Holbøll stated that the most Eider down exported from Greenland in any one year was 8757 pounds. Since twelve nests of down were reckoned to a pound, about 105 000 nests were robbed of down and most of eggs too in that year (Oldendow 1935: 79–80). Henrik Rink (1877) reported in his *Danish Greenland. Its people and its products* (p. 108), that down exports had declined from 5600 pounds to 200 pounds per annum in the twenty years before 1877. Yet there were enough Eiders left for T. N. Krabbe to estimate in 1907 that the annual kill amounted to between 100 000 and 200 000 birds (Oldendow 1935: 85). It was about this time that the hard-pressed Eiders were given protection while breeding. In 1911 the local authorities at Kangeq forbade all egg and down collecting on the neighbouring Kok Islands (Kok Øerne), as well as Eider shooting in the summer. In 1914, the commune or municipality of Fiskenæsset (Qeqertarsuatsiaat) banned all egg collecting. Other places followed suit. The legislative bodies of the two administrative districts into which Danish west Greenland was then divided, namely North and South Greenland, passed laws prohibiting Eider shooting and egging in June and July in 1929 and 1924 respectively. They were given statutory force by the Greenland administration in Copenhagen (Oldendow 1935: 92–98; Salomonsen 1955). These measures were successful and Eider numbers gradually increased, starting with the Kok Islands off Kangeq, where the birds were soon breeding again in annually increasing numbers after the protection given them in 1911.

In 1981 the Greenlanders were thought still to take about 150 000 Eiders annually, providing 250 tons of meat, and the bird was common everywhere on the west coast of Greenland (Salomonsen 1981: 196). The protection given it in the breeding season in the 1920s in Danish west Greenland was supported and paralleled by similar protection in Avanersuaq, northwest Greenland, then administered by Danish explorer—anthropologist Knud Rasmussen and called the Thule district. At that time it was a no-man's-land which belonged to no state. In 1929 a law was drawn up by Knud Rasmussen and approved by a local council of hunters (Vaughan, in press) which gave the Eider full protection from the time of its arrival in the district until 1 September. Once a year only, the collection of eggs and down was allowed under the supervision of the local authorities. The text of the law shows that Rasmussen's ideas about conservation were at one with present-day thinking on the subject (Oldendow 1935: 99):

> . . . every free hunter and trapper may obtain food and skins for himself and his family by hunting and trapping. But game is no longer available in unlimited quantity. Therefore free people throughout the world have agreed that game shall be protected in the breeding season, because otherwise less and less game will remain each year. It is especially important for us here to protect Eiders, Arctic Foxes and Walrus from extinction, and every free hunter should happily accept such protection, for otherwise these animals will have been totally exterminated by the time children born now become grown up.

References

Abbott, C. G. 1929. Ross Gulls for dinner. *Condor* 31: 132.
Abruzzi, L. A. di Savoia, duke of the. 1903. *On the* Polar Star *in the Arctic Sea*. 2 vols. London: Hutchinson.
Alaska Almanac. 1989. Facts about Alaska. Anchorage: GTE Discovery Publications.
Alerstam, T., G. A. Gudmundsson, P. E. Jönsson, J. Karlsson and Å. Lindström. 1990. Orientation, migration routes and flight behaviour of Knots, Turnstones and Brant Geese departing from Iceland in spring. *Arctic* 43: 201–214.
Alerstam, T., C. Hjort, G. Högstedt, P. E. Jönsson, J. Karlsson and B. Larsson. 1986. Spring migration of birds across the Greenland Inland Ice. *Meddelelser on Grønland. Bioscience* 21. 38 pp.
Anderson, J. 1746. *Nachrichten von Island, Grönland und der Strasse Davis*. Hamburg: G. C. Grund.
Anderson, J. 1756. *Beschryving van Ysland, Groenland en de Straat Davis*. Amsterdam: Jan van Dalen.
Andreev, A. V. 1980. *Adaptsii ptits k zimnim usloviyam Subarktiki*. Moscow: Nauka.
Andreyev, A. 1986. Winter and summer in the life of northern birds. *Science in the USSR* 6: 44–55, 104.
Andreev, A. V. and A. Ya. Kondrat'ev. 1981. Novye dannye po biologii rozovoi chaiki (*Rhodostethia rosea*) [New data on the biology of Ross's Gull]. *Zoologicheskiy Zhurnal* 60: 418–425.
Andreev, A. V. 1991. Winter adaptations in the Willow Ptarmigan. *Arctic* 44: 106–114.

Anon. 1852. *Arctic miscellanies*. London: Colburn.
Anon. 1988. Postscripts. Eyes on the Eskimo Curlew. *Natural History*, p. 66.
Armstrong, E. A. 1954. The behaviour of birds in continuous daylight. *Ibis* 96: 1–30.
Armstrong, T. 1990. The Northern Sea Route. *Polar Record* 26: 127–129.
Armstrong, T., G. Rogers and G. Rowley. 1978. *The circumpolar north*. London: Methuen.
Asbirk, S. and N.-E. Franzmann. 1978. Studies of Snow Buntings. In: Green and Greenwood 1978: 132–142.
Asbirk, S. and N-.E. Franzmann. 1979. Observations on the diurnal rhythm of Greenland Wheatears *Oenanthe oe. leucorrhoa* Gm. in continuous daylight. *Dansk Ornithologisk Forenings Tidsskrift* 73: 95–102.
Bailey, A. M. 1948. *Birds of Arctic Alaska*. Denver: Colorado Museum of Natural History.
Bailey, A. M. 1971. *Field work of a museum naturalist. Alaska-southeast; Alaska-far north. 1919–1922*. Denver: Colorado Museum of Natural History.
Baker, J. M. and M. V. Angel. 1987. Marine processes. In Nelson *et al.* 1987: 50–74.
Bannerman, D. A. 1956, 1957, 1963. *The birds of the British Isles, 5, 6, 12*. Edinburgh: Oliver and Boyd.
Bárðarson, H. R. 1986. *Birds of Iceland*. Reykjavik: H. R. Báðotarson.
Barr, W. 1981. Baron Eduard von Toll's last expedition: The Russian Polar Expedition, 1900–1903. *Arctic* 34: 201–224.
Barr, W. 1990. Soviet cruise liner in collision with an ice floe near Svalbard, June 1989. *Polar Record* 26: 238–239.
Barron, W. 1890. *An apprentice's reminiscenses of whaling in Davis's Straits. Narrative of the voyages of the Hull barque* Truelove *from 1848 to 1854*. Hull: M. Waller.
Barry, T. W. 1968. Observations on natural mortality and native use of eider ducks along the Beaufort Sea coast. *Canadian Field Naturalist* 82: 140–144.
Barthelmess, K. 1987. Walfangtechnik vor 375 Jahren. Die Zeichnungen in Robert Fotherbys *Journal* von 1613 und ihr Einfluss auf die Druckgraphik. *Deutsches Schiffahrtsarchiv* 10: 289–324.
Bartonek, J. C. 1986. Waterfowl management and subsistence harvests in Alaska and Canada: an overview. *Transactions of the North American Wildlife and Natural Resources Conference* 51: 459–463.
Batashev, A. 1990. Reindeer people at risk in the land at the edge of the universe. *Soviet Weekly*, 29 November.
Beebe, F. and H. Webster. 1964. *North American falconry and hunting hawks*. Denver.
Beetham, B. 1911. *Photography for bird-lovers. A practical guide*. London: Witherby.
Bellrose, F. C. 1976. *Ducks, geese and swans of North America*. Ed. F. H. Kortright. Harrisburg: Stackpole Books.
Belopol'skii, L. O. 1961. *Ecology of sea colony birds of the Barents Sea*. Jerusalem: Israel Program for Scientific Translations. Translated from *Ekologiya morskikh colonial'nykh ptits Barentsova morya*. Moscow 1957.
Bent, A. C. 1925. *Life histories of North American wildfowl, 2*. Washington, D.C. United States National Museum.
Bent, A. C. 1927. 1929. *Life histories of North American shorebirds, 1, 2*. Washington, D.C.: United States National Museum.
Berge, H. C. ten. 1976. *De raaf in de walvis*. Amsterdam: De Bezige Bij.
Bertelsen, A. 1932. Meddelelser om nogle af de i Vestgrønlands distrikter mellem 60° og 77° N. br. almindeligere forekommende fugle, særlig om deres udbredelsesomraade, deres yngleomraade og deres træk. *Meddelelser om Grønland* 91(4). 75pp.
Bertelsen, A. 1948. Fuglemærkningen i Vestgrønland i årene 1926–1945. *Meddelelser om Grønland* 142(4). 40pp.

Bessels, E. 1875. L'expédition polaire Américaine sous les ordres du Capitaine Hall. *Bulletin de la Société de Geographie* 9: 291–299.
Bessels, E. 1879. *Die amerikanische Nordpol-Expedition.* Leipzig: Wilhelm Engelmann.
Bewick, T. 1804. *History of British birds*, 2. Newcastle: T. Bewick.
Bianki. V. V. 1977. *Gulls, shorebirds and alcids of Kandalaksha Bay*. Jerusalem: Israel Program for Scientific Translations. Translated from *Kuliki, chaiki i chistikovye Kandalakshskogo zaliva.* Murmansk.
Binney, G. 1925. *With seaplane and sledge in the Arctic.* London: Hutchinson.
Bird, C. G. and E. G. Bird. 1940. Some remarks on non-breeding in the Arctic, especially northeast Greenland. *Ibis* (14)4: 671–678.
Birulya, A. 1907. Ocherki iz zhizni ptits polyarnago poberezh'ya Sibiri [Sketches of the bird life of the Polar coast of Siberia]. *Zapiski Imperatorskoi Akademii Nauk, seria* 8, po *Fiziko-Matematicheskomy Otdeleniyu* 18(2). 157 pp.
Blasius, J. H. 1862. *A list of the birds of Europe.* Norwich: Matchett and Stevenson. Translated from the original German privately printed in 1861.
Bliss, L. C., O. W. Heal and J. J. Moore. (eds.) 1981. *Tundra ecosystems: a comparative analysis.* Cambridge: Cambridge University Press.
Blomqvist, Å. (ed.) 1951. Gace de la Buigne, Le Roman des deduis. *Studia romanica holmiensia, 3.* Karlshamn.
Bockstoce, J. (ed.) 1988a and b. *The journal of Rochfort Maguire. 1852–1854.* 2 vols. London: Hakluyt Society.
Boertmann, D., M. Forchhammer and H. Meltofte. 1990. *Biologisk-arkæologisk kortlægning af Grønlands østkyst mellem 75° N og 79°30' N, 2. Optællinger af fugle og pattedyr mellem Besel Fjord og Zachariæ Isstrøm.* Greenland Home Rule, Department of Wildlife Management. Technical report no. 12. Nuuk.
Borup, G. 1911. *A tenderfoot with Peary.* New York: Frederick A. Stokes.
Bourne, W. R. P. 1967. Long-distance vagrancy in the petrels. *Ibis* 109: 141–167.
Bousfield, M. A. and Ye. V. Syroechkovskiy. 1985. A review of Soviet research on the Lesser Snow Goose on Wrangel Island, USSR. *Wildfowl* 36: 13–20.
Boyd, H. 1977. Waterfowl hunting by native peoples in Canada: the case of James Bay and northern Quebec. *International Congress of Game Biologists* 13: 463–473.
Boyd, H. 1987. Do June temperatures affect the breeding success of Dark-bellied Brent Geese *Branta b. bernicla?* *Bird Study* 34: 155–159.
Brandt, H. W. 1943. *Alaska bird trails. Adventures of an expedition by dog sled to the delta of the Yukon river at Hooper Bay.* Cleveland, Ohio: Bird Research Foundation.
Bratrein, H. D. 1986. Falkefangst i Nord-Norge. *Ottar* 162: 34–38.
Bree, C. R. 1864. *A history of the birds of Europe not observed in the British Isles.* 4 vols. First edition. London: George Bell and Sons.
Breuil, M. 1989. *Les oiseaux d'Islande: écologie et biogéographie.* Paris: Chaband.
Brody, H. 1987. *Living Arctic. Hunters of the Canadian north.* London: Faber and Faber.
Brower, C. D. 1950. *Fifty years below zero. A lifetime of adventure in the Far North.* London: Travel Book Club.
Brown, R. N. R. 1920. *Spitsbergen. An account of exploration, hunting, the mineral riches and future potentialities of an Arctic archipelago.* Philadelphia: J. B. Lippincott.
Bruemmer, F. 1987. Life upon the permafrost. *Natural History*, pp. 31–39.
Bruemmer, F. 1989. Island of murres. *Canadian Geographic* 109: 44–51.
Brünnich, M. Th. 1764. *Ornithologia borealis.* Copenhagen.
Bresewitz, G. 1981. *Arktisk sommar. Med Ymer asnom Isharet.* Stockholm: Wahlstrom & Widstrand.
Burnham, W. A. and W. G. Mattox. 1984. Biology of the Peregrine and Gyrfalcon in Greenland. *Meddelelser om Grønland. Bioscience* 14. 25pp.

Butev, V. T. 1959. Zimovka ptits na severe Novoi Zemli [Birds wintering in northern Novaya Zemlya]. *Ornitologiya* 2: 99–101.
Buturlin, S. A. 1906. The breeding-grounds of the Rosy Gull. *Ibis* (8)6: 131–139, 333–337, 661–666.
Cade, T. J. 1982. *The falcons of the world*. Ithaca, New York: Cornell University Press.
Cagni, U. and P. A. Cavalli-Molinelli. (eds.) 1903. *Osservazioni scientifiche eseguite durante la spedizione polare de S. A. R. Luigi Amedeo di Savoia, duca degli Abruzzi. 1899–1900*. Milan.
Campbell, B. and E. Lack. 1985. *A dictionary of birds*. Calton: T. and A. D. Poyser.
Chapman, A. 1897. *Wild Norway: with chapters on Spitsbergen, Denmark, Etc*. London: Edward Arnold.
Chapman, F. M. 1899. Report on birds received through the Peary expeditions to Greenland. *Bulletin of the American Museum of Natural History* 12: 219–244.
Chard, C. S. 1963. The Nganasan: wild Reindeer hunters of the Taimyr Peninsula. *Arctic Anthropology* 1: 105–121.
Chernov, Yu. I. 1985. *The living tundra*. Cambridge: Cambridge University Press. Translated from *Zhizn' tundry*. Moscow, 1980.
Chernov, Yu. I. (ed.) 1989. *Ptitsy v soobshchestvakh tundrovoi zony. Sbornik nauchnykh trudov* [Birds in the natural communities of the tundra zone. Collected papers]. Moscow: Nauka.
Chernyavsky, F. B. and A. V. Andreev. 1990. Studies on the ecology of terrestrial vertebrates in the northeast of the USSR and perspectives for international research cooperation. In: Kotlyakov and Sokolov 1990b: 186–194.
Chernavsky, F. B. and A. V. Tkachev. 1982. *Populyatsionnye tsikli lemmingov v Arktike {Population cycles of lemmings in the Arctic}*. Moscow: Nauka.
Christensen, J. 1979. Den grønlandske Havørns *Haliaeetus albicilla groenlandicus* Brehm ynglebiotop, redeplacering og rede. *Dansk Ornithologisk Forenings Tidsskrift* 73: 131–156.
Clark, R. C. Jr and J. S. Finley. 1982. Occurrence and impact of petroleum on Arctic environments. In: Rey and Stonehouse 1982: 295–341.
Conway, W. M. (ed.) 1904. *Early Dutch and English voyages to Spitsbergen in the seventeenth century*. London: Hakluyt Society.
Connors, P. G. 1983. Taxonomy, distribution and evolution of golden plovers (*Pluvialis dominica* and *Pluvialis fulva*). *Auk* 100: 607–620.
Cott, H. B. 1953. 1954. The exploitation of wild birds for their eggs. *Ibis* 95: 409–449, 643–675; 96; 129–149.
Cournoyea, N. J. and R. G. Bromley. 1986. The role of native people in waterfowl management in Canada. *Transactions of the North American Wildlife and Natural Resources Conference* 51: 507–510.
Court, G. S., C. C. Gates and D. A. Boag. 1988. Natural history of the Peregrine Falcon in the Keewatin District of the Northwest Territories. *Arctic* 41: 17–30.
Cramp, S. and K. E. L. Simmon (eds.) 1980. 1983. *The birds of the western Palearctic*, 2, 3. Oxford: Oxford University Press.
Croxall, J. P., P. G. H. Evans and R. W. Schreiber. 1984. *Status and conservation of the world's seabirds*. Cambridge: International Council for Bird Preservation.
Cullen, J. M. 1954. The diurnal rhythm of birds in the Arctic summer. *Ibis* 96: 31–47.
Cumming, I. G. 1979. Lapland Buntings breeding in Scotland. *British Birds* 72: 53–59.
Custer, T. W. and F. A. Pitelka. 1977. Demographic features of a Lapland Longspur population near Barrow, Alaska. *Auk* 94: 505–525.

Custer, T. W. and F. A. Pitelka. 1987. Nesting by Pomarine Jaegers near Barrow, Alaska, 1971. *Journal of Field Ornithology* 58: 225–230.
Dalgety, C. T. and P. Scott. 1948. A new race of the white-fronted goose. Bulletin of the *Bristol Ornithologists Club* 68: 109–121.
Dalgleish, J. J. 1886. Discovery of the nest of *Larus rossii* in Greenland. *Auk* 3: 273–274.
Damas, D. (ed.) 1984. *Handbook of North American Indians, 5. Arctic.* Washington, D. C: Smithsonian Institution.
Danilov, N. N., V. N. Ryzhanovsky and V. K. Ryabitsev. 1984. *Ptitsy Yamala {Birds of Yamal}.* Moscow: Nanka.
Davis, C. H. (ed.) 1876. *Narrative of the North Polar Expedition. U.S. Ship* Polaris, *Captain Charles Francis Hall commanding.* Washington, D. C.: Government Printing Office.
De Long, E. 1884. 1883. *The voyage of the* Jeannette. *The ship and ice journals of George W. De Long.* 2 vols. Boston: Houghton, Mifflin and Company.
Dement'ev, G. P. and N. A. Gladkov. (eds.) 1966–1970. *Birds of the Soviet Union.* 6 vols. Jerusalem: Israel Program for Scientific Translations. Translated from *Ptitsy Sovetskogo Soyuza.* Moscow 1951–1954.
Dementiew, G. P. 1960. *Der Gerfalke.* Wittenberg: A. Ziemsen Verlag.
Dennis, R. H. 1983. Purple Sandpipers breeding in Scotland. *British Birds* 76: 563–566.
Densley, M. 1979. Ross's Gulls in Alaska. *British Birds* 72: 23–28.
Densley, M. 1988. James Clark Ross and Ross's Gull. A review. *Naturalist* 113: 85–102.
Dhondt, A. A. 1987. Cycles of lemmings and Brent Geese *Branta b. bernicla:* a comment on the hypothesis of Roselaar and Summers. *Bird Study* 34: 151–154.
Dimock, J. F. (ed.) 1867. *Giraldi Cambrensis opera, 5. Topographia Hiberniae.* London: Rolls Series.
Divoky, G. J., G. A. Sanger, S. A. Hatch and J. C. Haney. 1988. *Fall migration of Ross's Gull* (Rhodostethia rosea) *in Alaskan Chukchi and Beaufort Seas.* Anchorage: Alaska Fish and Wildlife Research Centre.
Doughty, R. W. 1979. Eider husbandry in the North Atlantic: trends and prospects. *Polar Record* 19: 447–459.
Dresser, H. E. 1904. On the late Dr. Walter's ornithological researches in the Taimyr Peninsula. *Ibis* (8)4: 228–235.
Driver, P. M. 1974. *In search of the Eider.* London: Saturn Press.
Drolet, J. A. 1986. Land claim settlements and the management of migratory birds, a case history: the James Bay and Northern Quebec Agreement. *Transactions of the North American Wildlife and Natural Resources Conference* 51: 511–515.
Duffey, E., N. Creasey and K. Williamson. 1950. The "rodent-run" distraction-behaviour of certain waders. *Ibis* 92: 27–33.
Dunbar, M. J. 1982. Arctic marine ecosystems. In: Rey and Stonehouse 1982: 233–261.
Dunbar, M. J. 1987. Arctic seas that never freeze. *Natural History,* pp. 50–53.
Durrell, G. and L. 1986. *Durrell in Russia.* London: Macdonald.
Dyck, J. 1979. Winter plumage of the Rock Ptarmigan: structure of the air-filled barbules and function of the white colour. *Dansk Ornithologisk Forenings Tidsskrift* 73: 41–58.
Dymond, J. N., P. A. Fraser and S. J. M. Gantlett. 1989. *Rare birds in Britain and Ireland.* Calton: T. and A. D. Poyser.
Ebbinge, B. S. 1989. A multifactorial explanation for variation in breeding performance of Brent Geese *Branta bernicla. Ibis* 131: 196–204.

Edwards, G. 1743. 1747. 1750. 1751. *A natural history of birds.* 4 vols. London.
Eifert, V. S. 1962. *Men, birds and adventure.* New York: Dodd, Mead.
Ekblaw, W. E. 1918. Finding the nest of the Knot. *Wilson Bulletin* 30: 97–100.
Ekblaw, W. E. 1919. The food birds of the Smith Sound Eskimos. *Wilson Bulletin* 31: 1–5.
Ekblaw, W. E. 1927–1928. *The material response of the Polar Eskimo to their far Arctic environment.* Albany: American Geographic Association.
Elander, M. and S. Blomqvist. 1986. The avifauna of central Northeast Greenland, 73°15′ N–74°05′ N, based on a visit to Myggbukta, May–July 1979. *Meddelelser om Grønland. Bioscience* 19: 44pp.
Evans, P. G. H. 1980. Review of V. V. Bianki's Gulls, shorebirds and alcids of Kandalaksha Bay. *Ibis* 122: 254–255.
Evans, P. G. H. 1981. Ecology and behaviour of the Little Auk *Alle alle* in West Greenland. *Ibis* 123: 1–18.
Evans, P. G. H. 1984. The seabirds of Greenland: their status and conservation. In: Croxall *et al.* 1984: 49–84.
Evans, P. G. H. 1985. Obituary. Professor Dr. Finn Salomonsen. *Ibis* 127: 391–393.
Evans, P. G. H. and D. N. Nettleship. 1985. Conservation of the Atlantic Alcidae. In: Nettleship and Birkhead 1985: 427–488.
Faber, F. 1826. *Das Leben der hochnordischen Vögel.* Leipzig.
Fabricius, O. 1780. See Helms 1929.
Falk, K. and J. Durinck. 1990. *Lomviejagten i Vestgrønland 1988–89.* Greenland Home Rule. Department of Wildlife Management. Technical Report no. 15. Nuuk.
Feilden, H. W. 1877. List of birds observed in Smith Sound and in the Polar Basin during the Arctic Expedition of 1875–76. *Ibis* (4)1: 401–412.
Feilden, H. W. 1878. 1879. Notes from an Arctic journal. *Zoologist* 2: 313–320, 372–384, 407–418, 445–451; 3: 16–24, 50–58, 89–108, 162–170, 200–202.
Feilden, H. W. 1920. Breeding of the Knot in Grinnell Land. *British Birds* 13: 278–282.
Ferdinand, L. 1979a. Finn Salomonsen og fuglene. *Dansk Ornithologisk Forenings Tidsskrift* 73: 5–8.
Ferdinand, L. 1979b. DOF's Havørneekspeditioner til Grønland. *Dansk Ornithologisk Forenings Tidsskrift* 73: 103–105.
Finlay, J. C. 1984. *A bird-finding guide to Canada.* Edmonton: Hurtig.
Finney, G. H. 1990. Native hunting of waterfowl in Canada. In: G. V. T. Matthews (ed.). *Managing waterfowl populations. Proceedings of an International Waterfowl and Wetland Research Bureau Symposium, Astrakhan, USSR, 2–5 October 1989.* IWRB Special Publications 12. Slimbridge, Glos.
[Fisher, A.] [1819] *Journal of a voyage of discovery to the Arctic regions performed between 4 April and 18 November 1818 in H.M.S.* Alexander. London: Richard Phillips.
Fisher, A. 1821. *A journal of a voyage of discovery to the Arctic regions in H.M. ships* Hecla and Griper *in the years 1819 and 1820.* London: Longmann, Hurst, Rees, Orme and Brown.
Fisher, J. 1954. *Birds as animals, 1. A history of birds.* London: Hutchinson.
Fisher, J. 1984. *The fulmar.* London: Collins. [First published in 1952.]
Flint, V. E. (ed.) 1980. *Novoe v izuchenii biologii i rasprostranenii kulikov.* Moscow.
Flint, V. E., R. L. Boehme, Y. V. Kostin and A. A. Kuznetsov. 1984. *A field guide to the birds of the USSR.* Princeton, New Jersey: Princeton University Press.
Flint, V. E., Yu. A. Isakov and V. E. Fokin. 1982. Pamyati Aleksandra Aleksandrovich Kishchinskogo. *Ornithologiya* 17: 192–196.

Flint, V. E. and P. S. Tomkovich. (eds.) 1988. *Kuliki v SSSR: rasprostranenie, biologiya i okhrana.* Moscow: Nauka.

Forster, J. R. 1772. An account of the birds sent from Hudson's Bay. *Philosophical Transactions of the Royal Society from 1665 to 1800* 13: 331–348.

Fox, A. D. and D. A. Stroud. (eds.) 1981. *Report of the 1979 Greenland White-fronted Goose Study Expedition to Eqalungmiut Nunât, west Greenland.* Aberystwyth: Greenland white-fronted Goose Study.

Fox, A. D. and D. A. Stroud. 1988. The breeding biology of the Greenland White-fronted Goose (*Anser albifrons flavirostris*). *Meddelelser om Grønland. Bioscience.* 27: 24pp.

Fox, A. D. and D. A. Stroud. 1989. West Greenland—a bird's eye view! *BTO News. Bulletin of the British Trust for Ornithology* 160: 6–7.

Franklin, J. 1823. *Narrative of a journey to the shores of the Polar Sea in the years 1819–20–21–22.* London: John Murray.

Franklin, J. 1824a and b. *Narrative of a journey to the shores of the Polar Sea in the years 1819–20–21–22.* Second edition. 2 vols. London: John Murray.

Franklin, J. 1828. *Narrative of a second expedition to the shores of the Polar Sea in the years 1825, 1826 and 1827. Including an account of the progress of a detachment to the eastward by John Richardson.* London: John Murray.

Freuchen, P. and F. Salomonsen. 1958. *The Arctic year.* New York: G. P. Putnam's Sons.

Fuller, W. A. and P. J. Kevan. (eds.) 1970. Proceedings of the Conference on Productivity and Conservation in Northern Circumpolar Lands. *International Union for the Conservation of Nature and Natural Resources (IUCN), New Series* 16. Morges.

Gaava, I. A., S. M. Uspenski, and Y. P. Yazan. 1987. National parks (zapovedniks) and other protected areas in the Soviet Arctic and prospects for their development. In: Nelson *et al.* 1987: 591–597.

Gabrielson, I. N. and F. C. Lincoln. 1959. *The birds of Alaska.* Washington, D.C.: The Wildlife Management Institute.

Gaston, A. J. and D. N. Nettleship. 1981. *The Thick-billed Murres of Prince Leopold Island.* Ottawa: Canadian Wildlife Service.

Gavin, A. 1947. Birds of Perry River District, Northwest Territories. *Wilson Bulletin* 59: 195–203.

Geale, J. 1971. Birds of Resolute, Cornwallis Island, N.W.T. *Canadian Field Naturalist* 85: 53–59.

Gelting, P. 1937. Studies on the food of the East Greenland Ptarmigan. *Meddelelser om Grønland* 116. 196pp.

Ginn, H. B. and D. S. Melville. 1983. *Moult in birds.* BTO Guide no. 19. Tring: British Trust for Ornithology.

Glover, R. (ed.) 1958. *A journey from Prince of Wales's Fort in Hudson's Bay to the Northern Ocean. 1769.1770.1771.1772. By Samuel Hearne.* Toronto: Macmillan Company of Canada.

Glutz von Blotzheim, U. N. *et al.* 1971. 1975. *Handbuch der Vögel Mitteleuropas*, 4 and 6(1). Frankfurt am Main: Akademische Verlagsgesellschaft.

Goldman, E. A. 1935. Edward William Nelson—naturalist, 1855–1934. *Auk* 52: 135–148.

Gollop J. B., T. W. Barry and E. H. Iversen. 1986. *Eskimo Curlew. A vanishing species?* Saskatchewan: Saskatchewan Natural History Society.

Goodsell, J. W. 1983. *On Polar trails: the Peary expedition to the North Pole, 1908–09.* Austin, Texas: Eakin Press.

Gorbunov, G. P. 1932. Ptitsy Zemli Frantsa-Iosifa [Birds of Franz Josef Land]. *Trudy Arkticheskogo Instituta* 4: 1–244.

Gordon, S. 1922. *Amid snowy wastes. Wildlife on the Spitsbergen Archipelago.* London.

Graburn, N. H. H. 1969. *Eskimos without igloos. Social and economic development in Sugluk*. Boston, Mass: Little, Brown and Company.
Greely, A. W. 1886a and b. *Three years of Arctic service. An account of the Lady Franklin Bay Expedition of 1881–84*. 2 vols. New York: Charles Scribner's Sons.
Green, G. H. and J. J. D. Greenwood (eds.) 1978. *Joint biological expedition to north east Greenland 1974*. Dundee: Dundee University North East Greenland Expedition.
Greve, T. 1975. *Svalbard. Norway in the Arctic*. Oslo: Grøndahl.
Grieve, S. 1885. *The Great Auk or Garefowl*. London: Thomas C. Jack.
Grimmett, R. F. A. and T. A. Jones. 1989. *Important bird areas in Europe*. Cambridge: International Council for Bird Preservation.
Gubser, N. J. 1965. *The Nunamiut Eskimos, hunters of Caribou*. New Haven: Yale University Press.
Gullestad, N., M. Owen and M. J. Nugent. 1984. Numbers and distribution of Barnacle Geese *Branta leucopsis* on Norwegian staging islands and the importance of the staging area to the Svalbard population. In: Mehlum and Ogilvie 1984: 57–65.
Gulløv, H. C. and H. Kapel. 1979. *Haabetz Colonie, 1721–1728*. Copenhagen: National Museum of Denmark.
Gurney, J. H. 1921. *Early annals of ornithology*. London: H. F. and G. Witherby.
Guttridge, L. F. 1988. *Icebound. The Jeannette expedition's quest for the North Pole*. New York: Paragon House.
Hacquebord, L. 1983. The history of early Dutch whaling: a study from the ecological angle. In: s'Jacob *et al*. 1983: 135–146.
Hacquebord, L. 1984. *Smeerenburg. Het verblijf van Nederlandse walvisvaarders op de westkust van Spitsbergen in de seventiende eeuw*. Groningen: Arctic Centre.
Haftorn, S. 1971. *Norges fugler*. Oslo: Universitetsforlaget.
Haig-Thomas, D. 1939. *Tracks in the snow*. London: Hodder and Stoughton.
Hakluyt, R. 1907. *The principal navigations, voyages, traffiques and discoveries of the English nation, 1*. London: J. M. Dent and Sons.
Hamre, G. M. 1987. Conservation in the Canadian Arctic with special emphasis on territorial parks. In: Nelson *et al*. 1987: 557–566.
Hansen, J. P. H. and H. C. Gulløv. 1989. The mummies from Qilakitsoq—Eskimos in the fifteenth century. *Meddelelser om Grønland. Man and Society* 12. 199pp.
Hansen, J. P. H., J. Meldgaard and J. Nordqvist. 1985. *Qilakitsoq. De grønlandske mumier fra 1400-tallet*. Nuuk and Copenhagen: Grønlands Landsmuseum and Christian Ejlers Forag.
Hansen, K. 1979. Status over bestanden af Havørn *Haliaeetus albicilla groenlandicus* Brehm i Grønland i årene 1972–74. *Dansk Ornithologisk Forenings Tidsskrift* 73: 107–130.
Hansson, R., P. Prestrud and N. A. Øritsland (eds.) 1990. *Assessment system for the environment and industrial activities in Svalbard*. Oslo: Norsk Polarinstit, tutt.
Hare, C. E. 1952. *Bird lore*. London: Country Life.
Harris, J. T. 1979. *The Peregrine Falcon in Greenland*. Columbia, Missouri: University of Missouri Press.
Harrison, P. 1983. *Seabirds. An identification guide*. London: Croom Helm.
Hart, H. C. 1880. Notes on the ornithology of the British Polar Expedition, 1875–6. *Zoologist* (3)4: 121–129, 204–214.
Harting, J. E. 1871. Catalogue of an Arctic collection of birds presented by Mr. John Barrow, F.R.S., to the University Museum of Oxford; with notes on the species. *Proceedings of the Zoological Society of London* 39: 110–123.
Harvie-Brown, J. A. 1905a and b. *Travels of a naturalist in northern Europe*. 2 vols. London: T. Fisher Unwin.
Haviland, M. D. 1914. *The wood-people: and others*. London: Edward Arnold.
Haviland, M. D. 1915a. *A summer on the Yenesei*. London: Edward Arnold.

Haviland, M. D. 1915b. Notes on bird migration at the mouth of the Yenesei River, Siberia, as observed in the autumn of 1914. *Ibis* (10)3: 395–399.

Haviland, M. D. 1915c. Notes on the nestling plumage of the Asiatic Golden Plover. *Ibis* (10)3: 716–717.

Haviland, M. D. 1915d. Notes on the breeding habits of the Curlew-Sandpiper. *British Birds* 8: 178–183.

Haviland, M. D. 1915e. Notes on the breeding habits of the Little Stint. *British Birds* 8: 202–208.

Haviland, M. D. 1915f. Notes on the breeding habits of the Grey Phalarope. *British Birds* 9: 11–16.

Haviland, M. D. 1915g. Notes on the breeding habits of the Asiatic Golden Plover. *British Birds* 9: 82–89.

Haviland, M. D. 1915h. Notes on the Grey Plover on the Yenesei. *British Birds* 9: 161–166.

Haviland, M. D. 1915i. Notes on the Lapland Bunting on the Yenesei River. *British Birds* 9: 230–238.

Haviland, M. D. 1915j. Notes on the breeding habits of the Willow Grouse at the mouth of the Yenesei River, Siberia. *Zoologist* (4)19: 241–244.

Haviland, M. D. 1916. Notes on the breeding habits of Temminck's Stint. *British Birds* 10: 157–165.

Haviland, M. D. 1917. Notes on the breeding habits of the Dotterel on the Yenesei. *British Birds* 11: 6–11.

Hayes, J. G. 1934. *The conquest of the North Pole. Recent Arctic exploration.* London: Thornton Butterworth.

Hearne, S. 1795. See Glover 1958.

Heintzelman, D. S. 1989. *Tundra Swan hunting: a biological, ecological and wildlife crisis.* Allentown, Pennsylvania: Wildlife Information Center Inc.

Helms, O. (ed.) 1929. *Otto Fabricius. Fauna Groenlandica. Pattedyr og fugle.* Copenhagen: Det Grønlandske Selskab.

Helms, P. 1981. Kostundersøgelse i Angmagssalik. *Forskning i Grønland* 1–2: 10–14.

Hensel, W. 1909. Die vögel in der provenzialischen und nordfranzösischen Lyrik des Mittelalters. *Romanische Forschungen* 26: 584–670.

Hewitson, W. C. 1856a and b. *Coloured illustrations of the eggs of British birds.* Third edition. 2 vols. London: John Van Voorst.

Hjort, C. 1985. The early days of Ross's Gull *Rhodostethia rosea* in Greenland. *Dansk Ornithologisk Forenings Tidsskrift* 79: 152–153.

Hobson, W. 1972. The breeding biology of the Knot. *Proceedings of the Western Foundation of Vertebrate Zoology* 2: 5–26.

Hofmann, G. 1957. Falkenjagd und Falkenhandel in den nordischen Ländern während des Mittelalters. *Zeitschrift für deutsches Altertum und deutsche Literatur* 88: 115–149.

Hohn, E. O. 1969. Eskimo bird names at Chesterfield Inlet and Baker Lake, Keewatin, N.W.T. *Arctic* 22: 72–76.

Holk, A. G. F. van, H. K. s'Jacob and A. A. H. J. Temmingh. (eds) 1981. *Early European exploitation of the northern Atlantic. 800–1700.* Groningen: Arctic Centre.

Holloway, C. W. 1970. Threatened vertebrates in northern circumpolar regions. In: Fuller and Kevan 1970: 175–192.

Holtved, E. 1951a and b. The Polar Eskimos. Language and folklore. *Meddelelser om Grønland* 152 (1 and 2). 367 and 153pp.

Holtved, E. 1967. Contributions to Polar Eskimo ethnography. *Meddelelser om Grønland* 182(2). 180pp.

Horrebov, N. 1752. *Tilforladige efterretninger om Island med et nyt landkort.* . . . Copenhagen.

Hørring, R. 1937. *Birds collected on the Fifth Thule Expedition.* Copenhagen: Gyldendalske Boghandel.
Houston, C. S. (ed.) 1974. *To the Arctic by canoe. 1810–1821. The journal and paintings of Robert Hood.* Montreal: McGill–Queen's University Press.
Houston, C. S. (ed.) 1984. *Arctic ordeal. The journal of John Richardson, surgeon-naturalist with Franklin. 1820–1822.* Montreal: McGill–Queen's University Press and Gloucester: Alan Sutton.
Hughes, C. C. 1962. *An Eskimo village in the modern world.* Ithaca, New York: Cornell University Press.
Huish, R. 1835. *The last voyage of Capt. Sir John Ross, Knt. R.N. to the Arctic regions; for the discovery of a North West Passage; performed in the years 1829–30–31–32 and 33.* London: John Saunders.
Hunt, H. J. and R. H. Thompson. 1980. *North to the horizon. Searching for Peary's Crocker Land.* Camden, Maine: Down East Books.
Huxley, J. S. 1923. Courtship activities in the Red-throated Diver. *Journal of the Linnaean Society of London* 35: 253–292.
Il'ičev, V. D. and V. E. Flint. 1985. 1989. 1990. *Handbuch der Vögel der Sowjetunion,* 1, 46(1): Wiesbaden: Aula Verlag and Wittenberg: A. Ziemsen. Translated from *Ptitsy SSSR.* Moscow 1982, Leningrad 1987.
Ingersoll, E. (ed.) 1914. *Alaskan bird-life as depicted by many writers.* New York: National Association of Audubon Societies.
Irving, L. 1953. The naming of birds by Nunamiut Eskimo. *Arctic* 6: 35–43.
Irving, L. 1960. *Birds of Anaktuvuk Pass, Kobuk, and Old Crow. A study in Arctic adaptation.* United States National Museum. Bulletin 217. Washington, D. C.
Irving, L. 1972. *Arctic life of birds and mammals including man.* New York: Springer Verlag.
Iversen, E. 1989. Survival. The uncertain future of the Eskimo Curlew. *Birder's World,* August, pp. 20–23.
Jackson, F. G. 1895. *The Great Frozen Land (Bolshaia Zemelskija tundra). Narrative of a winter journey across the tundras and a sojourn among the Samoyads.* London: Macmillan.
s'Jacob, H. K., K. Snoeijing and R. Vaughan (eds.) 1983. *Arctic whaling.* Groningen: Arctic Centre.
Janovy, J. 1978. *Keith county journal.* New York: St. Martin's Press.
Jenness, D. 1985. *Dawn in Arctic Alaska.* Chicago: University of Chicago Press.
Johansen, H. 1956. 1958. Revision und Entstehung der arktischen Vogelfauna. *Acta Arctica* 8 and 9. Copenhagen: Ejnar Munksgaard.
John, B. 1979. *The world of ice: the natural history of the frozen regions.* London: Orbis.
Johnson, H. 1907. *The life and voyages of Joseph Wiggins, F.R.G.S.* London: John Murray.
Johnson, R. E. 1976. *Sir John Richardson. Arctic explorer, natural historian, naval surgeon.* London: Taylor and Francis.
Johnson, S. R. and D. R. Herter. 1989. *The birds of the Beaufort Sea.* Anchorage: BP Exploration (Alaska).
Jones, G. 1986. *The Norse Atlantic saga.* Oxford: Oxford University Press.
Jones, T. R. (ed.) 1875. *Manual of the natural history, geology, and physics of Greenland . . . together with Instructions for the use of the expedition.* London: Her Majesty's Stationery Office.
Jourdain, F. C. R. 1922. The birds of Spitsbergen and Bear Island. *Ibis* (11)4: 159–179.
Kampp, K. 1988. Grønlands lomvier. *Grønland* 36: 13–27.
Kampp, K. and F. Wille. 1979. Fødevaner hos den grønlandske Havørn *Haliaeetus albicilla groenlandicus* Brehm. *Dansk Ornithologisk Forenings Tidsskrift* 73: 157–164.

Kane, E. K. 1856a and b. *Arctic explorations in the years 1853, '54, '55.* 2 vols. Philadelphia: Childs and Peterson.
Kemp, W. B. 1984. Baffinland Eskimo. In: Damas 1984: 463–475.
Kessel, B. 1989. *The birds of the Seward Peninsula.* Fairbanks: University of Alaska Press.
Kinloch, A. 1898. *History of the Kara Sea trade route to Siberia.* London: A. Kinloch.
Kirby, W. 1837. *Fauna Boreali-Americana or the zoology of the northern parts of British America.* Part 4. The insects. Norwich: J. Fletcher.
Kishchinsky, A. A. 1958. K biologii krecheta (*Falco gyrfalco gyrfalco* L.) na Kol'skom poluostrove [On the biology of the Gyrfalcon on the Kola Peninsula]. *Ornitologiya* 1: 61–75.
Kistchinski, A. A. 1971. Biological notes on the Emperor Goose in north-east Siberia. *Wildfowl* 22: 29–34.
Kistchinski, A. A. 1974. On the biology of the Spectacled Eider. *Wildfowl* 25: 5–15.
Kistchinski, A. A. 1975. Breeding biology and behaviour of the Grey Phalarope *Phalaropus fulicarius* in east Siberia. *Ibis* 117: 285–301.
Kishchinsky, A. A. 1988. *Ornito-fauna severo-vostoka Azii. Istoriya i sovremennoe sostoyanie {Avifauna of northeast Asia. History and present status}.* Moscow: Nauka.
Kishchinsky, A. A. and V. E. Flint. 1983. Taksonomcheskie vzaimootnosheniya v gruppe chernozobuikh gagan [Taxonomic interrelationships in the group of black-throated divers]. *Ornitologiya* 18: 112–123.
Klein, D. R. 1966. Waterfowl in the economy of the Eskimos on the Yukon-Kuskokwim Delta. *Arctic* 19: 319–336.
Knox, A. 1987. Taxonomic status of 'Lesser Golden Plovers'. *British Birds* 80: 482–487.
Knox, A. G. 1988. The taxonomy of redpolls. *Ardea* 76: 1–26.
Knystautas, A. 1987. *The natural history of the USSR.* London: Century Hutchinson.
Koelz, W. 1929. On a collection of Gyrfalcons from Greenland. *Wilson Bulletin* 41: 207–219.
Kondrat'ev, A. Ya. 1982. *Biologiya kulikov v tundrakh severo-vostoka Azii {The biology of waders in the tundra of northeast Asia}.* Moscow: Nauka.
Kondrat'ev, A. Ya. and L. F. Kondrat'eva. 1984. Rost i razvitie ptentsov volokhvostoi chaiki [Growth and development of Sabine's Gull chicks]. *Ornitologiya* 19: 81–88.
Kondrat'ev, A. Ya. and L. F. Kondrat'eva. 1987. Sravnitel'naya kharakteristika ekologii gnezdovaniya rozovoi i vilokhvostoi chaek [Comparative characteristics of breeding ecology of Ross's Gull *Rhodostethia rosea* and Sabine's Gull *Xema sabini*]. *Ornitologiya* 22: 35–50.
Korte, J. de. 1972. Birds observed and collected by "De Nederlandse Spitsbergen Expeditie" in West and East Spitsbergen, 1967 and 1968–69. *Beaufortia* 19: 113–150, 197–232; 20: 23–58.
Korte, J. de. 1977. 1984. 1985. 1986a. 1986b. Ecology of the Long-tailed Skua (*Stercorarius longicaudus* Vieillot, 1819) at Scoresby Sund, East Greenland. *Five parts: Beaufortia* 25: 201–219; *Beaufortia* 34: 1–14; *Beaufortia* 35: 93–127; *Bijdragen tot de Dierkunde* 56: 1–23; and MS. submitted to *Meddelelser om Grønland. Bioscience.*
Korte, J. de. 1986c. Aspects of breeding success in tundra birds. Studies on Long-tailed Skuas and waders at Scoresby Sund, East Greenland. *Seven offprints submitted for a doctorate in lieu of a thesis.* University of Amsterdam.
Korte, J. de and J. Wattel. 1988. Food and breeding success of the Long-tailed Skua at Scoresby Sund, Northeast Greenland. *Ardea* 76: 27–41.
Kotlyakov, V. M. and V. E. Sokolov. 1990a and b. *Arctic research: advances and prospects.* Proceedings of the Conference of Arctic and Nordic Countries on Coordination of Research in the Arctic. 2 vols. Moscow: Nauka.

Kovacs, T. J. 1987. National overview for Canada on national parks and protected areas in the Arctic. In: Nelson *et al.* 1987: 529–556.
Krasovsky, S. K. 1937. Biologicheskie osnovy promyslovogo ispol'zovaniya ptich'ikh bazarov. Etyudy po biologii tolstoklyuvoi kairy (*Uria lomvia* L.) [The biological basis for the rational exploitation of bird bazaars. Studies on the biology of Brünnich's Guillemot]. *Trudy Arkticheskogo Instituta. Biologiya* 77. Leningrad.
Krechmar, A. V., A. V. Andreev and A. Ya. Kondrat'ev. 1978. *Ekologiya i rasprostraneniye ptits na severo-vostoke SSSR* [*Ecology and distribution of birds in the northeastern Soviet Union*]. Moscow: Nauka.
Krechmar, A. V. and A. Ya. Kondrat'ev. 1982. Ekologiya gnezdovaniya gusya-belosheya (*Philacte canadica*) na severe Chukotskogo poluostrova [Breeding ecology of the Emperor Goose in the north of the Chukchi Peninsula]. *Zoologicheskiy Zhurnal* 61: 254–264.
Krechmar, A. V. and V. V. Leonovich. 1967. Rasprostranenie i biologiya krasnozoboi kazarki v gnezdovoi period [Distribution and biology of the Red-breasted Goose in the breeding season]. *Problemy Severa* 11: 229–234.
Kroeber, A. L. 1899. Tales of the Smith Sound Eskimo. *Journal of American Folklore* 12: 166–182.
Kryuchkov, V. V. 1973. *Krainy sever: problemy ratsional'nogo ispol'zovaniya prirodnyikh resursov* [*The Far North: problems in the rational exploitation of natural resources*]. Moscow: Mysl'.
Kryuchkov, V. V. 1979. *Sever: priroda i chelovek* [*The North: nature and man*]. Moscow: Nauka.
Kugler, R. C. 1983. Historical survey of foreign whaling: North America. In: s'Jacob *et al.* 1983: 149–157.
Kumlien, T. L. T. 1879. *Contributions to the natural history of Arctic America in connection with the Howgate Polar Expedition 1877–1878*. Washington, D.C.: Government Printing Office.
Lack, D. 1934. *The birds of Cambridgeshire*. Cambridge: Cambridge Bird Club.
Lapointe, E. 1987. International agreements for conservation: Convention on International Trade in Endangered Species of Wild Fauna and Flora (CITES). In: Nelson *et al.* 1987: 478–485.
Lascelles, G. W. 1892. *Falconry*. Reprinted 1971. London: Neville Spearman.
Le Glay, A., *et al.* 1873. *Inventaire sommaire des Archives Départementales du Nord. Série B*, 4. Lille.
Leonov, N. I. 1967. *Aleksandr Fedorovich Middendorf*. Moscow: Nauka.
Leslie, A. 1879. *The Arctic voyages of Adolf Erik Nordenskiöld. 1858–1879*. London.
Levere, T. H. 1988. Henry Wemyss Feilden, naturalist on HMS *Alert*, 1875–1876. *Polar Record* 24: 307–312.
Lewis, E. 1938. *In search of the Gyrfalcon*. London: Constable and Company.
Ley, W. 1971. *The Poles*. Amsterdam: Time-Life International.
Litvin, K. E. and V. V. Baranyuk. 1989. Razmnozhenie belykh sov (*Nyctea scandiaca*) i chislennost' lemmingov na ostrove Vrangelya [The breeding of Snowy Owls and the number of lemmings on Wrangel Island]. In: Chernov 1989: 112–129.
Lodge, R. B. 1903. *Pictures of bird life*. London: S. H. Bonsfield.
Lok, C. M. and J. A. J. Vink. 1986. Birds at Cambridge Bay, Victoria Island, Northwest Territories, in 1983. *Canadian Field-Naturalist* 100: 315–318.
Long, R. C. and J. C. Barlow. 1986. In memoriam: Lester L. Snyder. *Auk* 103: 809–811.
Loomis, C. C. 1972. *Weird and tragic shores. The story of Charles Francis Hall, explorer*. London: Macmillan.
Løvenskiold, H. L. 1947. *Håndbok over Norges fugler*. Oslo.

Løvenskiold, H. L. 1964. *Avifauna Svalbardensis*. Oslo: Norsk Polarinstitutt.
Lubbock, B. 1937. *The Arctic whalers*. Glasgow: Brown, Son and Ferguson.
MacDonald, S. D. 1981. Scientific progress: terrestrial biology, an overview. In: Zaslow 1981: 171–186.
MacFarlane, R. 1891. Notes on and list of birds and eggs collected in Arctic America, 1861–1866. *Proceedings of the United States National Museum* 14: 413–446.
MacGillivray, W. 1824. Descriptions, characters, and synonyms of the different species of the genus Larus. *Memoirs of the Wernerian Natural History Society* 5(1): 247–276.
MacGillivray, W. 1842. *A manual of British ornithology, 2. The water birds*. London: Scott, Webster and Greary.
MacKay, D. 1937. *The honourable company. A history of the Hudson's Bay Company*. London: Cassell.
MacMillan, D. B. 1918. *Four years in the white north*. New York: Harper and Brothers.
MacMillan, D. B. 1934. *How Peary reached the Pole. The personal story of his assistant*. Boston and New York: Houghton Mifflin Company.
Madge, S. and H. Burn. 1988. *Wildfowl. An identification guide to the ducks, geese and swans of the world*. London: Christopher Helm.
Manniche, A. L. V. 1910. The terrestrial mammals and birds of northeast Greenland. Biological observations. *Meddelelser om Grønland* 45. 200pp.
Manniche, A. L. V. 1911. Nordøstgrønlands fugle. Biologiske undersøgelser. *Dansk Ornithologisk Forenings Tidsskrift* 5: 1–114.
Manning, A. 1990. Nikolaas Tinbergen, F.R.S. (1907–1988). *Ibis* 132: 325–328.
Manning, T. H. 1942. Blue and Lesser Snow Geese on Southampton and Baffin Islands. *Auk* 59: 158–175.
Markham, A. H. 1875. *A whaling cruise to Baffin's Bay*. London: Sampson Low, Marston, Low and Searle.
Markham, A. H. 1878. *The great frozen sea. A personal narrative of the voyage of the* Alert *during the Arctic Expedition of 1875–6*. Third edition. London: Daldy Isbister and Co.
Markham, C. R. (ed.) 1881. *The voyages of William Baffin. 1612–1622*. London: Hakluyt Society.
Marsh, J. S. 1987. Tourism and conservation: case studies in the Canadian north. In: Nelson *et al.* 1987: 298–322.
Marshall, A. J. 1952. Non-breeding among Arctic birds. *Ibis* 94: 310–333.
Martens, F. 1675. *Spitzbergische oder Groenlandische Reise Beschreibung*. Hamburg. *Facsimile*, Berlin 1923.
Martens, F. 1855. Voyage into Spitzbergen and Greenland. In: White 1855: 1–140.
Mattox, W. G. 1970. Bird banding in Greenland. *Arctic* 23: 217–228.
Mattox, W. G. 1990. Report on the Greenland Peregrine Survey. *Newsletter of the Danish Polar Center and the Commission for Scientific Research in Greenland*, 20. December, 26pp.
McEachern, J. 1978. *A survey of resource harvesting, Eskimo Point, N.W.T. 1975–1977*. Prepared for the Polar Gas Project. Delta, British Columbia: Quest Socio-Economic Consultants Inc.
M'Dougall, G. F. 1857. *The eventful voyage of H.M. Discovery Ship* Resolute *to the Arctic regions in search of Sir John Franklin . . . 1852, 1853, 1854*. London: Longman, Brown, Green, Longmans and Roberts.
Mead, C. 1983. *Bird migration*. Feltham: Country Life Books.
Mearns, B. and R. 1988. *Biographies for birdwatchers. The lives of those commemorated in western Palearctic bird names*. London: Academic Press.
Mehlum, F. and M. Ogilvie. (eds.) 1984. Current research on Arctic geese. *Norsk Polarinstitutt Skrifter*, 181. Oslo.

Meldgaard, M. 1988. Isarukitsoq—den stumpvingede. *Grønland* 36: 5–12.
Meltofte, H. 1985. Populations and breeding schedules of waders, Charadrii, in High Arctic Greenland. *Meddelelser om Grønland. Bioscience* 16. 43pp.
Meltofte, H., C. Edelstam, G. Granström, J. Hammar and C. Hjort. 1981. Ross's Gulls in the Arctic pack ice. *British Birds* 74: 316–320.
Mewes, W. and E. F. von Homeyer. 1886. *Ornithologische Beobachtungen grössentheils im Sommer 1869 auf einer Reise im Nordwestlichen Russland gesammelt.* Privately printed. Vienna.
Meyer, H. 1987. Protected areas and national parks in Greenland. In: Nelson *et al.* 1987: 567–575.
Middendorff, A. Th. von. 1843. Bericht über die ornithologischen Ergebnisse der naturhistorischen Reise in Lappland während des Sommers 1840. *Beiträge zur Kentnisse des russischen Reiches* 8: 189–258.
Middendorff, A. Th. von. 1848. *Reise in den äussersten Norden und Osten Sibiriens, 1(1). Einleitung.* St. Petersburg.
Middendorff, A. Th. von. 1853. *Reise in den äussersten Norden und Osten Sibiriens, 2(2). Wirbelthiere.* St. Petersburg.
Mitchell, D. C. 1986. Native subsistence hunting of migratory waterfowl in Alaska: a case study demonstrating why politics and wildlife management don't mix. *Transactions of the North American Wildlife and Natural Resources Conference* 51: 527–534.
Møhl, J. 1979. The bones from Hope Colony. In: Gulløv and Kapel 1979: 215–230.
Møhl, J. 1982. Ressourceudnyttelse fra norrøne og eskimoiske affaldslag belyst gennem knoglematerialet. *Grønland* 30: 286–295.
Muir, J. 1917. *The cruise of the* Corwin. Boston and New York: Houghton Mifflin Company.
Mullens, W. H. 1909. Some early British ornithologists and their works, 6. Thomas Pennant (1726–1798). *British Birds* 2: 259–266.
Mullens, W. H. and H. K. Swann. 1917. *A bibliography of British ornithology.* London: Macmillan.
Murdoch, J. 1885. Birds. In: Ray 1885: 104–128.
Murdoch, J. 1899. A historical notice of Ross's Rosy Gull (*Rhodostethia rosea*). *Auk* 16: 146–155.
Murray, R. 1986. Status of legal issues involving goose populations: Canadian issues. *Transactions of the North American Wildlife and Natural Resources Conference* 51: 535–536.
Muus, B., F. Salomonsen and C. Vibe. 1981. *Grønlands fauna. Fisk. Fugle. Pattedyr.* Copenhagen: Gyldendal.
Myers, J. P. 1989. Making sense of sexual nonsense. *Audubon*, July, pp. 40–45.
Myers, J. P., O. Hilden and P. Tomkovich. 1982. Exotic *Calidris* species of the Siberian tundra. *Ornis Fennica* 59: 175–182.
Mylius-Erichsen, L. and H. Moltke. 1906. *Grønland.* Copenhagen: Gyldendalske Forlag.
Naber, S. P. L'Honoré. (ed.) 1917a and b. *Reizen van Willem Barents, Jacob van Heemskerck, Jan Cornelisz Rijp en anderen naar het Noorden (1594–1597). Verhaald door Gerrit de Veer.* Werken van de Linschoten-Vereeniging, 14, 15. The Hague.
Naber, S. P. L'Honoré. (ed.) 1930. *Walvischvaarten, overwintering en en jachtbedrijven in het Hooge Noorden 1633–1635.* Utrecht: A. Oosthoek.
Nansen, F. (ed.) 1899. *The Norwegian North Polar Expedition. 1893–1896. Scientific results, 4. An account of the birds by R. Collett and F. Nansen.* London: Longmans, Green and Co.
Nanton, P. 1970. *Arctic breakthrough. Franklin's expeditions. 1819–1847.* London.

Nares, G. 1878a and b. *Narrative of a voyage to the Polar Sea during 1875–6.* 2 vols. London: Sampson Low, Marston, Searle and Rivington.

Navid, D. 1987. The Ramsar Convention and Arctic conservation. In: Nelson *et al.* 1987: 486–497.

Nelson, E. W. 1883. Birds of the Bering Sea and the Arctic Ocean. In: *Cruise of the Revenue-steamer* Corwin *in Alaska and the N.W. Arctic Ocean in 1881*, pp. 55–118. Washington. D.C.: Government Printing Office.

Nelson, E. W. 1884. The breeding habits of the Pectoral Sandpiper (*Actodromas maculata*). *Auk* 1: 218–221.

Nelson, E. W. 1887. *Report upon natural history collections made in Alaska between the years 1877 and 1881.* Washington, D.C.: Government Printing Office.

Nelson, E. W. 1899. *The Eskimo about Bering Strait.* Washington. Reprinted 1971. New York and London: Johnson Reprint Corporation.

Nelson, J. G., R. Needham and L. Norton (eds.) 1987. *Arctic heritage.* Proceedings of a symposium August 24–28, 1985. Banff, Alberta, Canada. Ottawa: Association of Canadian Universities for Northern Studies.

Nelson, R. K. 1969. *Hunters of the northern ice.* Chicago: University of Chicago Press.

Nelson-Smith, A. 1982. Biological consequences of oil-spills in Arctic waters. In: Rey and Stonehouse 1982: 275–293.

Nettleship, D. N. 1973. Breeding ecology of Turnstones *Arenaria interpres* at Hazen Camp, Ellesmere Island, N.W.T. *Ibis* 115: 202–211.

Nettleship, D. N. 1974. The breeding of the Knot *Calidris canutus* at Hazen Camp, Ellesmere Island, N.W.T. *Polarforschung* 44: 8–26.

Nettleship, D. N. and T. R. Birkhead (eds.) 1985. *The Atlantic Alcidae.* London: Academic Press.

Nettleship, D. N. and W. J. Maher. 1973. The avifauna of Hazen Camp, Ellesmere Island, N.W.T. *Polarforschung* 43: 66–74.

Newcomb, R. L. 1888. *Our lost explorers: the narrative of the* Jeannette *expedition.* Hartford, Conn.: American Publishing Company.

Newton, A. 1875. Notes on birds which have been found in Greenland. In: Jones 1875: 94–115.

Newton, A. 1896. *A dictionary of birds.* London: Adam and Charles Black.

Newton, A. (ed.) 1864 and 1902. 1905 and 1907. *Ootheca Wolleyana. An illustrated catalogue of the collection of birds' eggs, begun by the late John Wolley Jnr. and continued with additions by Alfred Newton.* Four parts in 2 vols. Parts 1 and 2, paginated as one, published in 1864 and 1902 respectively, forming vol. 1. Parts 3 and 4, forming vol. 2, published 1905 and 1907. London.

Nicholson, E. M. 1930. Field-notes on Greenland birds. *Ibis* (12)6: 280–313, 395–428.

Nicholson, E. M. 1931. *The art of bird watching.* London: H. F. and G. Witherby.

Nielsen, B. P. 1979. Finn Salomonsens arbejde med ringmærkning og fuglefredning i Grønland. *Dansk Ornithologisk Forenings Tidsskrift* 73: 13–24.

Nielsen, C. O. and R. Dietz. 1989. Heavy metals in Greenland seabirds. *Meddelelser om Grønland. Bioscience* 29. 26pp.

Nordenskiöld, A. E. 1880. 1881. *Vegas färd kring Asien och Europa.* 2 vols. Stockholm: F. and G. Beijers Forlag. [English translation: *The voyage of the* Vega *round Asia and Europe.* 2 vols. London 1881a and b.]

Norderhaug, M. 1970. Conservation and wildlife problems in Svalbard. In: Fuller and Kevan 1970: 192–198.

Norderhaug, M. 1979. Problems in Arctic conservation. *Dansk Ornithologisk Forenings Tidsskrift* 73: 59–68.

Norderhaug, M. 1984. The Svalbard geese: an introductory review of research and conservation. In: Mehlum and Ogilvie 1984: 7–10.
Norderhaug, M. 1987. National parks and protected areas in Norway, with particular reference to the Arctic. In: Nelson *et al.* 1987: 576–590.
Norderhaug, M., E. Brun and G. V. Møllen. 1977. *Barentshavets sjøfuglressurser*. Oslo: Norsk Polarinstitutt.
Oggins, R. S. 1980. Albertus Magnus on falcons and hawks. In: Weisheipl 1980: 441–462.
Ogilvie, M. A. 1976. *The winter birds*. London: Michael Joseph.
Ogilvie, M. A. 1990. Photospot. Spectacled Eider. *British Birds* 83: 159–160.
Ogilvie, M. A. and M. Owen. 1984. Some results from the ringing of Barnacle Geese *Branta leucopsis* in Svalbard and Britain. In: Mehlum and Ogilvie 1984: 49–55.
Oldendow, K. 1933. *Fugleliv i Gronland. Det Grønlandske Selskabs Aarskrift 1932–33*. Copenhagen.
Oldendow, K. 1935. *Naturfredning i Grønland. Det Grønlandske Selskabs Skrifter*, 9. Copenhagen.
Oorschot, J. M. P. van. 1974. *Vorstelijke vliegers en valkenswaardse valkeniers sedert de seventiende eeuw*. Tilburg: Stichting Zuidelijk Historisch Contact.
O'Reilly, B. 1818. *Greenland, the adjacent seas, and the Northwest Passage to the Pacific Ocean, illustrated in a voyage to Davis's Strait, during the summer of 1817*. New York: James Eastburn.
Orlov, V. K. 1980. *V prostorakh Taymyra [In the wilds of Taimyr]*. Leningrad: Gidromete oizdat.
Orlov, V. K. 1988. *V krayu bol'shogo medvedya [In the land of the great bear]*. Moscow: Mysl'.
Østreng, W. 1977. *Politics in high latitudes. The Svalbard Archipelago*. London: C. Hurst and Company.
Ovsyanikov, N. 1990. The odd couple. *BBC Wildlife* 8: 748–756.
Owen, M. 1980. *Wild geese of the world. Their life history and ecology*. London: B. T. Batsford.
Owen, M. 1984. Dynamics and age structure of an increasing goose population—the Svalbard Barnacle Goose *Branta leucopsis*. In: Mehlum and Ogilvie 1984: 37–47.
Owen, M. 1986. Hvitkinngjessene på Svalbard—et eksempel på vellykket forvaltning av en gåsebestand. *Vår Fuglefauna* 9: 163–172.
Owen, M. 1987. Brent Geese *Branta b. bernicla* breeding and lemmings—a re-examination. *Bird Study* 34: 147–149.
Owen, M. and N. Gullestad. 1984. Migration routes of Svalbard Barnacle Geese *Branta leucopsis* with a preliminary report on the importance of the Bjørnøya staging area. In: Mehlum and Ogilvie 1984: 67–77.
Owen, M. and M. Norderhaug. 1977. Population dynamics of Barnacle Geese breeding in Svalbard, 1948–1976. *Ornis Scandinavica* 8: 161–174.
Palmén, J. A. 1887. Bidrag till kännedomen om Sibiriska-Ishafskustens fogelfauna enligt *Vega*-Epeditionens jakttagelser och samlingar. In: A. E. Nordenskiöld (ed.) Vega-*Expeditionens vetenskapliga jakttagelser*, 5. Stockholm.
Palmer, T. S. *et al.* 1954. *Biographies of members of the American Ornithologists' Union*. Washington, D.C.
Papanin, I. *et al.* 1977. *The Soviet North*. Moscow: Progress Publishers.
Parkes, K. C. 1970. In memoriam: Walter Edmond Clyde Todd. *Auk* 87: 635–649.
Parmelee, D. F. and R. B. Payne. 1973. On multiple broods and the breeding strategy of Arctic Sanderlings. *Ibis* 115: 218–226.

Parmelee, D. F., H. A. Stephens and R. H. Schmidt. 1967. *The birds of southeastern Victoria Island and adjacent small islands*. Ottawa: National Museum of Canada.
Parry, W. E. 1821. *Journal of a voyage for the discovery of a North-West Passage . . . in 1819–20 in H.M. Ships* Hecla *and* Griper . . . Philadelphia: Abraham Small.
Parry, W. E. 1824. *Journal of a second voyage for the discovery of a North-West Passage . . . in 1821–22–23 in H.M. Ships* Fury *and* Hecla. . . . London: John Murray.
Parry, W. E. 1826. *Journal of a third voyage for the discovery of a North-West Passage . . . in 1824–25 in H.M. Ships* Hecla *and* Fury. . . . London: John Murray.
Pashby, B. S. 1985. *John Cordeaux ornithologist*. Spurn Bird Observatory.
Pattie, D. L. 1990. A sixteen-year record of summer birds on Truelove Lowland, Devon Island, Northwest Territories, Canada. *Arctic* 43: 275–283.
Pearson, H. J. 1899. *"Beyond Petsora eastward". Two summer voyages to Novaya Zemlya and the islands of the Barents Sea*. London: R. H. Porter.
Pearson, H. J. 1904. *Three summers among the birds of the Russian Lapland*. London: R. H. Porter.
Pedersen, A. 1934. *Polardyr*. Copenhagen: Gyldendalske Boghandel. [English translation: *Polar animals*. London 1962].
Pennant, T. 1784. 1785. *Arctic zoology*. 2 vols. London: Henry Hughes.
Persen, E. 1986. Ringgåsa—den norske bestanden fremdales truet. *Vår Fuglefauna* 9: 173–176.
Pettingill, O. S. 1984. In memoriam: George Miksch Sutton. *Auk* 101: 146–152.
Phillips, A. R. 1981. In memoriam: Alfred M. Bailey. *Auk* 98: 173–175.
Pienkowski, M. W. and G. H. Green. 1976. Breeding biology of Sanderlings in north-east Greenland. *British Birds* 69: 165–177.
Phipps, C. J. 1774. *A voyage toward the North Pole undertaken by his majesty's command, 1773*. London. Reprinted 1978. Whitby: Caedmon Press.
Pike, O. 1900. *In bird-land with field-glass and camera*. London: T. Fisher Unwin.
Pimlott, D. H., D. Brown and K. P. Sam. 1976. *Oil under the ice*. Ottawa: Canadian Arctic Resources Committee.
Pitelka, F. A. 1974. An avifaunal review for the Barrow region and North Slope of Arctic Alaska. *Arctic and Alpine Research* 6: 161–184.
Pleske, F. D. 1887. *Kritichesky obsor mlekopitayushchikh i ptits Kolskogo poluostrova*. St. Petersburg.
Pleske, Th. 1928. Birds of the Eurasian tundra. *Memoirs of the Boston Society for Natural History* 6(3). Boston.
Ploeger, P. L. 1968. *Geographical differentiation in Arctic Anatidae as a result of isolation during the last glacial*. Leiden: E. J. Brill.
Poole, K. G. and R. G. Bromley. 1988. Natural history of the Gyrfalcon in the Central Canadian Arctic. *Arctic* 41: 31–38.
Popham, H. L. 1897a. Notes on birds observed on the Yenisei River, Siberia, in 1895. *Ibis* (7)3: 89–108.
Popham, H. L. 1897b. [Exhibition of a clutch of four eggs of the Curlew Sandpiper] *Bulletin of the British Ornithologists' Club* 7: ii.
Popham, H. L. 1898. Further notes on the birds of the Yenisei River, Siberia. *Ibis* (7)4: 489–520.
Popham, H. L. 1901. Supplementary notes on the birds of the Yenisei River. *Ibis* (8)1: 449–458.
Portenko, L. 1981. *Birds of the Chukchi Peninsula, 1*. New Delhi: Amerind Publishing Co. [Translated from: *Ptitsy Chukotskogo poluostrova i ostrova Vrangelya*. 2 vols. Leningrad 1972, 1973.]
Potapov, E. 1990. Birds and brave men in the Arctic north. *Birds International* 2: 73–83.

Powys, L. 1928. *Henry Hudson*. New York: Harper and Brothers.
Prater, A. J. and P. J. Grant. 1982. Waders in Siberia. *British Birds* 75: 272–281.
Prevett, J. P., H. G. Lumsden and F. C. Johnson. 1983. Waterfowl kill by Cree hunters of the Hudson Bay lowland, Ontario. *Arctic* 36: 185–192.
Prop, J., M. R. van Eerden and R. H. Drent. 1984. Reproductive success of the Barnacle Goose *Branta leucopsis* in relation to food exploitation on the breeding grounds, western Spitsbergen. In: Mehlum and Ogilvie 1984: 87–117.
Pryde, P. R. 1972. *Conservation in the Soviet Union*. Cambridge: Cambridge University Press.
Prytherch, R. and M. Everett. 1990. News and comment. Falcon thieves jailed. *British Birds* 83: 517.
Purchas, S. 1906. *Hakluytus Posthumus or Purchas His Pilgrimes*, 13. London: Hakluyt Society.
Rae, J. 1890. Notes on some of the birds and mammals of the Hudson's Bay Company territory and of the Arctic coast of America. *Journal of the Linnaean Society (Zoology)* 20: 136–145.
Rahn, K. A. 1982. On the causes, characteristics and potential environmental effects of aerosol in the Arctic atmosphere. In: Rey and Stonehouse 1982: 163–195.
Rasmussen, K. 1908. *The people of the Polar north*. London: Kegan Paul, Trench, Trübner and Co. [Translated from *Nye mennesker*. Copenhagen 1905.]
Rasmussen, K. 1933. *Across Arctic America. Narrative of the Fifth Thule Expedition*. New York and London: G. P. Putnam's Sons. [Translated from: *Fra Grønland til Stillehavet*. 2 vols. Copenhagen 1925, 1926.]
Ray, D. J. 1975. *The Eskimos of Bering Strait, 1650–1898*. Seattle and London: University of Washington Press.
Ray, D. J. 1984. Bering Strait Eskimo. In: Damas 1984: 285–302.
Ray, P. H. 1885. *Report of the International Polar Expedition to Point Barrow, Alaska*. Washington, D.C.: U.S. Signal Office, Arctic Publications.
Reed, A. (ed.) 1986a. *Eider ducks in Canada*. Ottawa: Canadian Wildlife Service.
Reed, A. 1986b. Eiderdown harvesting and other uses of common Eiders in spring and summer. In: Reed 1986a: 138–146.
Remmert, H. 1980. *Arctic animal ecology*. Berlin: Springer-Verlag.
Rey, L. and B. Stonehouse (eds.) 1982. *The Arctic Ocean. The hydrographic environment and the fate of pollutants*. London: Macmillan Press.
Reynolds, E. E. 1949. *Nansen*. Harmondsworth: Penguin Books.
Rich, E. E. 1958. 1959. *The history of the Hudson's Bay Company. 1670–1870*. 2 vols. London: Hudson's Bay Record Society.
Richards, E. A. 1949. *Arctic mood*. Caldwell Idaho: The Caxton Printers.
Richardson, J. 1825. Zoological appendix to W. E. Parry's *Journal of a second voyage for the discovery of a North-West Passage*. London.
Richardson, J. 1836. *Fauna Boreali-Americana or the zoology of the northern parts of the British America. Part 3. The fish*. London: R. Bentley.
Richardson, J. 1852. *Arctic searching expedition: a journal of a boat-voyage through Rupert's Land and the Arctic Sea in search of the discovery ships under command of Sir John Franklin*. New York: Harper and Brothers.
Richardson and Swainson 1829–1837. *Fauna Boreali-Americana*. 4 parts: Kirby 1837 (Part 4, Insects); Richardson 1836 (Part 3, Fish), Richardson, Swainson and Kirby 1829 (Part 1, Mammals); Swainson and Richardson 1831 (Part 2, Birds).
Richardson, J., W. Swainson and W. Kirby. 1829. *Fauna Boreali-Americana; or the zoology of the northern parts of British America. Mammalia*. London: John Murray.
Rink, H. 1877. *Danish Greenland. Its people and its products*. London: C. Hurst and Co.
Robbins, J. 1985. Anatomy of a sting. *Natural History*, July, pp. 4–10.

Rodahl, K. 1949. *Vitamin sources in Arctic regions*. Oslo: Norsk Polarinstitutt.
Ross, J. 1819. *A voyage of discovery in His Majesty's Ships Isabella and Alexander for the purpose of exploring Baffin's Bay*. London: John Murray.
Ross, J. 1835. *Narrative of a second voyge in search of a North-West Passage during the years 1829, 1830, 1831, 1832, 1833*. London: A. W. Webster.
Ross, J. C. 1826. Appendix on zoology to W. E. Parry's *Journal of a third voyage for the discovery of a North-West Passage*. London: John Murray.
Ross, J. C. 1835. Birds. In: Appendix to J. Ross, *Narrative of a second voyage in search of a North-West Passage*, pp. xxv–xlv. London: A. W. Webster.
Rutilevsky, G. L. and S. M. Uspensky. 1957. Fauna mlekopitayushchikh i ptits tsentral'noi Arktiki (po nablyudeniyam dreifuyushchikh stantsy) [Mammal and bird fauna of the central Arctic according to observations from drifting stations]. *Trudy Arkticheskogo Nauchno-Issledovatelskogo Instituta Glavsevmorputi* 205: 5–18.
Ryabitsev, V. K. 1986. *Ptitsy tundry* [*Birds of the tundra*]. Sverdlovsk: Central Urals Book Publishers.
Ryder, J. P. 1969. Nesting colonies of Ross' Goose. *Auk* 86: 282–292.
Sabine, E. 1818. A memoir on the birds of Greenland, with descriptions and notes on the species observed on the late voyage of discovery in Davis Strait and Baffin's Bay. *Transactions of the Linnaean Society of London* 12: 527–559.
Sabine, E. 1824. Birds. In: *A supplement to the Appendix of Capt. Parry's voyage for the discovery of a North-West Passage in the years 1819–20*, pp. cxciii–ccxi. London.
Sabine, J. 1818. An account of a new species of gull lately discovered on the west coast of Greenland. *Transactions of the Linnaean Society of London* 12: 520–523.
Sabine, J. 1823. Zoological appendix. Birds. In: Franklin 1823: 669–703.
Sage, B. 1981. Conservation of the tundra. In: Bliss *et al*. 1981: 731–746.
Sage, B. 1982. The rape of Alaska. *Wildlife* 24(1): 16–18.
Sage, B. 1986. *The Arctic and its wildlife*. London: Croom Helm.
Saladin d'Anglure, B. 1984. Inuit of Quebec. In: Damas 1984: 476–507.
Salomonsen, F. 1939. *Moults and sequence of plumage in the Rock Ptarmigan*. Copenhagen: P. Haase & Son.
Salomonsen, F. 1948. The distribution of birds and the recent climatic change in the North Atlantic area. *Dansk Ornithologisk Forenings Tidsskrift* 42: 85–99.
Salomonsen, F. 1950. *Grønlands fugle, The birds of Greenland*. Copenhagen: Ejnar Munksgaard.
Salomonsen, F. 1955. Fredning af fuglelivet i Grønland. *Dansk Ornithologisk Forenings Tidsskrift* 49: 3–11.
Salomonsen, F. 1956. The Greenland bird banding system. *Arctic* 9: 258–264.
Salomonsen, F. 1967. *Fuglene på Grønland*. Copenhagen: Rhodos.
Salomonsen, F. 1970. Birds useful to man in Greenland. In: Fuller and Kevan 1970: 169–175.
Salomonsen, F. 1972. Zoo-geographical and ecological problems in Arctic birds. *Proceedings of the International Ornithological Congress* 15: 25–77.
Salomonsen, F. 1979. Ornithological and ecological studies in southwest Greenland. *Meddelelser om Grønland* 204. 214pp.
Salomonsen, F. 1981. Fugle. In: Muus *et al*. 1981: 159–361.
Salvin, F. H. and W. Brodrick. 1855. *Falconry in the British Isles*. London. Reprinted 1971. Maidenhead: Thames Valley Press.
Saunders, H. 1875. On the immature plumage of *Rhodostethia rosea*. *Ibis* (3)5: 484–487.
Schalow, H. 1904. Die Vögel der Arktis. *Fauna Arctica* 4(1). Jena.
Scheel, H. 1927. *Med Lehn Schiøler og Johannes Larsen i Grønland. Fugleekspeditionen 1925*. Copenhagen: Hage & Clausens Forlag.

Schledermann, P. 1990. *Crossroads to Greenland. Three thousand years of prehistory in the eastern High Arctic*. Calgary: Arctic Institute of North America.
Schorger, A. W. 1955. Obituaries. Herbert William Brandt. *Auk* 72: 321–322.
Sclater, P. L. 1875. Instructions for collecting and observing the birds of Greenland. In: Jones 1875: 45–46.
Scott, P. 1951. *Wild geese and Eskimos. A journal of the Perry River Expedition of 1949*. London: Country Life.
Scott, P. and J. Fisher. 1953. *A thousand geese*. London: Collins.
Seebohm, H. 1878. 1879. 1880. Contributions to the ornithology of Siberia. *Ibis* (4)2: 173–184, 322–352; (4)3: 1–18, 147–163; (4)4: 179–195.
Seebohm, H. 1880. *Siberia in Europe*. London: John Murray.
Seebohm, H. 1882. *Siberia in Asia. A visit to the valley of the Yenisey in east Siberia*. London: John Murray.
Seebohm, H. 1887. *The geographical distribution of the family Charadriidae or the plovers, sandpipers, snipes and their allies*. London: Sotheran & Co.
Seebohm, H. 1896. *Coloured figures of the eggs of British birds*. Ed. R. Bowdler Sharpe. Sheffield: Pawson and Brailsford.
Seebohm, H. 1901. *The birds of Siberia. A record of a naturalist's visit to the valleys of the Petchora and Yenesei*. London: John Murray.
Semenov, A. V. 1980. Na ostrove ptich'ikh bazarov. *Chelovek i Stikhiya*, pp. 126–128.
Shackleton, E. 1959. *Nansen the explorer*. London: H. F. and G. Witherby.
Shepherd, C. W. 1867. *The northwest peninsula of Iceland: being the journal of a tour in Iceland in the spring and summer of 1862*. London: Longmans Green and Co.
Sherwood, M. B. 1965. *Exploration of Alaska. 1865–1900*. New Haven: Yale University Press.
Shor, W. 1988. 'Operation Falcon' and the Peregrine. In: T. J. Cade *et al.* (eds.). *Peregrine Falcon populations. Their management and recovery*, pp 831–842. Boise, Idaho.
Shtilmark, F. P. 1977. Is an ecological sanctuary needed on the Taimyr Peninsula? In: Papanin *et al.* 1977: 243–248.
Silis, I. 1989. *The world's greatest national park. North and East Greenland*. Nuuk: The Greenland Home Rule Authorities.
Simms, E. 1978. *British thrushes*. London: Collins.
Slessers, M. 1968. Soviet studies in the northward movement of birds. *Arctic* 21: 201–204.
Smith, D. G. 1984. Mackenzie Delta Eskimo. In: Damas 1984: 347–358.
Snyder, L. L. 1957. *Arctic birds of Canada*. Toronto: University of Toronto Press.
Solomonov, N. G. 1987. *Besedy ob okhrane prirody Severa* [*Discussions about protecting nature in the North*]. Yakutsk: Yakutskoye Knizhnoye Izdatel'stvo.
Soper, J. D. 1942. Life history of the Blue goose. *Proceedings of the Boston Society of Natural History*. 42: 121–225.
Sørensen, A. K. 1983. *Danmark-Grønland i det 20 århundrede. En historisk oversigt*. Copenhagen: Nyt Nordisk Forlag Arnold Busck.
Speck, G. 1963. *Samuel Hearne and the Northwest Passage*. Caldwell, Idaho: The Caxton Printers.
Spencer, K. 1953. *The Lapwing in Britain*. London: A. Brown and Sons.
Stenmark, R. J. 1987. National parks and protected areas in the Arctic: Alaska. In: Nelson *et al.* 1987: 513–528.
Stevens, R. 1953. *Observations on modern falconry*. Privately printed.
Stewart, J. M. 1987. The "lily of birds": the success story of the Siberian White Crane. *Oryx* 21: 6–10.

Stokke, O. S. 1989. Threats to the Arctic environment. *International Challenges* 9: 21–27.
Stone, I. R. 1986a. Profile: Samuel Hearne. *Polar Record* 23: 49–56.
Stone, I. R. 1986b. Profile: Aubyn Trevor-Battye. *Polar Record* 23: 177–181.
Stonehouse, B. 1971. *Animals of the Arctic: the ecology of the Far North.* London: Ward Lock.
Stonehouse, B. (ed.) 1986. *Arctic air pollution.* Cambridge: Cambridge University Press.
Storå, N. 1968. Massfångst av sjöfågel i Nordeurasien. En etnologisk undersökning av fångstmetoderna. *Acta Acadamiae Aboensis, Series A. Humaniora*, 34. Åbo.
Sugden, D. 1982. *Arctic and Antarctic. A modern geographical synthesis.* Oxford: Basil Blackwell.
Summers, R. W. 1986. Breeding production of Dark-bellied Brent Geese *Branta bernicla bernicla* in relation to lemming cycles. *Bird Study* 33: 105–108.
Summers, R. W. and G. H. Green. 1974. Notes on the food of the Gyrfalcon in northeast Greenland. *Dansk Ornithologisk Forenings Tidsskrift* 68: 87–90.
Summers, R. W. and L. G. Underhill. 1987. Factors related to breeding populations of Brent Geese *Branta b. bernicla* and waders (Charadrii) on the Taimyr Peninsula. *Bird Study* 34: 161–171.
Sutton, G. M. 1932. The birds of Southampton Island, Hudson Bay. *Memoirs of the Carnegie Institute.* Pittsburgh.
Sutton, G. M. 1936. *Birds in the wilderness.* New York: Macmillan.
Sutton, G. M. 1961. *Iceland summer. Adventures of a bird painter.* Norman, Oklahoma: University of Oklahoma Press.
Sutton, G. M. 1971. *High Arctic. An expedition to the unspoiled north.* New York: Paul S. Erikssen.
Sutton, G. M. 1980. *Bird student. An autobiography.* Austin, Texas: University of Texas Press.
Swaen, A. E. H. 1937. *De valkerij in de Nederlanden.* Zutphen: W. J. Thieme & Cie.
Swainson, W. and J. Richardson. 1831. *Fauna Boreali-Americana or the zoology of the northern parts of British America. Part 2. The birds.* London: John Murray.
Swann, H. K. 1913. *A dictionary of English and folk-names of British birds.* London: Witherby and Co.
Syroechkovsky, E. V. 1977. Kolonii guseobraznykh okolo gnezd polyarnykh sov na ostrove Vrangelya [Colonies of anserine birds around Snowy Owl's nests on Wrangel Island]. *Ornitologiya* 13: 212–213.
Syroechkovsky, E. V. 1981. Structura kolonii belykh gusei (*Anser caerulescens*) na ostrove Vrangelya i popytka prognoza izmeneniya ikh chislennosti [Structure of the colony of Snow Geese on Wrangel Island and an attempt to forecast changes in their number]. *Zoologicheskiy Zhurnal* 60: 1364–1373.
Syroyechkovskiy, Ye. Ye. and F. P. Shtil'mark. 1985. Nature reserves and sanctuaries in the Soviet Far North. Present status and prospects for development. *Polar Geography and Geology* 9: 9–28.
Teixeira, R. M. 1979. *Atlas van de Nederlandse broedvogels.* The Hague.
Temple, S. A. 1974. Winter food habits of Ravens on the Arctic Slope of Alaska. *Arctic* 27: 41–46.
Thienemann, F. A. L. 1825–1838. *Systematische Darstellung der Fortpflanzung der Vögel Europas mit Abbildung der Eier.* Leipzig.
Thienemann, F. A. L. 1845–1856. *Einhundert Tafeln colorirter Abbildungen van Vogeleiern.* Leipzig: Brockhams.
Thing, H. 1990a. *Encounters with wildlife in Greenland. Animal–human interactions.* Nuuk: Atuakkiorfik.

Thing, H. 1990b. Environmental protection and wildlife management programs in Greenland. In: Kotlyakov and Sokolov 1990b: 116–126.
Thomas, V. G. and S. D. MacDonald. 1987. The breeding distribution and current population status of the Ivory Gull in Canada. *Arctic* 40: 211–218.
Thompson, D. Q. and R. A. Person. 1963. The eider pass at Point Barrow, Alaska. *Journal of Wildlife Management* 27: 348–356.
Tillisch, C. J. 1949. *Falkejagten og Dens historie*. Copenhagen: P. Haase.
Tinbergen, N. 1934. *Eskimoland*. Rotterdam: D. van Sijn & Zonen.
Tinbergen, N. 1935. Field observations of east Greenland birds. 1. The behaviour of the Red-necked Phalarope in spring. *Ardea* 24: 1–42.
Tinbergen, N. 1939. The behaviour of the Snow Bunting in spring. *Transactions of the Linnaean Society of New York*, 5. 92pp.
Tinbergen, N. 1958. *Curious naturalists*. London: Country Life.
Todd, W. E. C. 1963. *Birds of the Labrador Peninsula and adjacent areas. A distributional list*. Toronto: University of Toronto Press.
Toll, E. von. 1909. *Die russische Polarfahrt der Sarja, 1900–1902*. Berlin: Georg Reimer.
Tolmachev, A. I. 1928. K avifaune ostrova Kolgueva [On the avifauna of Kolguyev Island]. *Ezhegodnik Zoologisheskogo Muzeye Akademii Nauk SSSR* 28: 355–365.
Tomkovich, P. S. 1984. Ptitsy ostrova Greem-Bell, Zemlya Frantsa-Iosifa [The birds of Graham Bell Island, Franz Josef Land]. *Ornitologiya* 19: 13–21.
Tomkovich, P. S. 1985. K biologii morskogo pesochnika na Zemle Frantsa-Iosifa [On the biology of the Purple Sandpiper *Calidris maritima* in Franz Josef Land]. *Ornitologiya* 20: 3–17.
Tomkovich, P. S. 1988a. O svoeobrazii biologii belokhvostogo pesochnika na severnom predele areala [Peculiarities of the breeding biology of Temminck's Stint *Calidris temminckii* at the northern limit of its range]. *Ornitologiya* 23: 188–193.
Tomkovich, P. S. 1988b. Waders in the Soviet Union. *BTO News. Bulletin of the British Trust for Ornithology* 159: 8–9.
Tomkovich, P. S. and S. Yu. Fokin. 1983. K ekologii belokhvostogo pesochnika na severo-vostoke Sibiri [The ecology of Temminck's Stint in northeast Siberia]. *Ornitologiya* 18: 40–56.
Trevor-Battye, A. 1895. *Ice-bound on Kolguev*. Westminster: Archibald Constable and Company.
Trevor-Battye, A. (ed.) 1903. *Lord Lilford on birds*. London: Hutchison.
Tugarinow, A. and S. Buturlin. 1925. *Materialen über die Vögel des Jenisseischen Gouvernements*. Halle: Gebauer-Schwetschke A.-G. [Translated from: Tugarinov, A. Ya. and S. Buturlin. *Materialy po ptitsam yeniseyskoy gubernii*. Krasnoyarsk 1911.]
Underhill, L. G. and R. W. Summers. 1990. Multivariate analyses of breeding performance in Dark-bellied Brent Geese *Branta b. bernicla*. *Ibis* 132: 477–482.
Usher, P. J. and G. Wenzel. 1987. Native harvest surveys and statistics: a critique of their construction and use. *Arctic* 40: 145–160.
Uspensky, S. M. 1958a. *Ptitsy Sovetskoi Arktiki* [*Birds of the Soviet Arctic*]. Moscow: Nauka.
Uspensky, S. M. 1958b. *The bird bazaars of Novaya Zemlya*. Ottawa: Canadian Wildlife Service.
Uspensky, S. M. 1964. *Arktika glazami zoologa* [*The Arctic through the eyes of a zoologist*]. Moscow: Nauka.
Uspenski, S. M. 1965. *Die Wildgänse Nordeurasiens*. Wittenberg: A. Ziemsen Verlag.
Uspenski, S. M. 1969. *Die Strandläufer Eurasiens (Gattung* Calidris). Wittenberg: A. Ziemsen Verlag.
Uspensky, S. M. 1970. Problems and forms of fauna conservation in the Soviet Arctic and Subarctic. In: Fuller and Kevan 1970: 199–207.

Uspenski, S. M. 1972. *Die Eiderenten (Gattung* Somateria). Wittenberg: A. Ziemsen Verlag.
Uspensky, S. M. 1977. The firebird of the north. In: Papanin *et al.* 1977: 112–123.
Uspensky, S. M. 1982. Problemy okhrany ptits Arktiki i Subarktiki [Problems of the protection of birds in the Arctic and sub-Arctic]. *Ornitologiya* 17: 18–21.
Uspenskii, S. M. 1986. *Life in high latitudes. A study of bird life.* Rotterdam: A. A. Balkema.
Uspenskiy, S. M., L. S. Govorukha, S. Ye. Belikov and V. I. Bulavintsev. 1987. Proposed protected zones in the Franz Josef Land area. *Polar Geography and Geology* 11: 210–220.
Uspenskiy, S.M. and P. S. Tomkovich. 1987. The birds of Franz-Josef Land and their protection. *Polar Geography and Geology* 11: 221–234.
Vaughan, R. 1952. Accertata nidificazione sul Massicio della Maiella (Abruzzi) del Piviere tortolino (*Charadrius morinellus*, L.). *Rivista Italiana di Ornitologia* 22: 162.
Vaughan, R. 1979. *Arctic summer. Birds in north Norway.* Shrewsbury: Anthony Nelson.
Vaughan, R. 1982. The Arctic in the middle ages. *Journal of Medieval History* 8: 313–342.
Vaughan, R. 1983. Historical survey of the European whaling industry. In: s'Jacob *et al.* 1983: 121–134.
Vaughan, R. 1988. Birds of the Thule district, northwest Greenland. *Arctic* 41: 53–58.
Vaughan, R., in press. *Northwest Greenland: a history.* Orono, Maine: University of Maine Press.
Veer, G. de. 1609. *The True and Perfect Description of Three Voyages. . . . London.* [Translated from: *Waerachtige Beschryvinghe van drie seylagien. . . .* Amsterdam 1598.]
Vink, J. A. J., M. and R. Vaughan and C. M. Lok. 1988. Birds at Cambridge Bay, Victoria Island, in 1986. *Circumpolar Journal* 3: 13–26.
Vitebsky, P. 1990. Gas, environmentalism and native anxieties in the Soviet Arctic: the case of the Yamal peninsula. *Polar Record* 26: 19–26.
Voous, K. H. 1960. *Atlas of European birds.* London: Nelson.
Watson, A. 1957. The behaviour, breeding, and food ecology of the Snowy Owl. *Ibis* 99: 419–462.
Watson, A. 1963. Bird numbers on tundra in Baffin Island. *Arctic* 16: 101–108.
Wayre, P. 1965. *Wind in the reeds.* London: Collins.
Webby, K. 1978. Species diversity and the rate of evolution in the Arctic. In: Green and Greenwood 1978: 232–236.
Weisheipl, J. A. (ed.) 1980. *Albertus Magnus and the sciences.* Toronto: Pontifical Institute of Medieval Studies.
Welty, J. C. 1975. *The life of birds.* Second edition. Philadelphia: W. B. Saunders Company.
White, A. 1855. *A collection of documents on Spitzbergen and Greenland.* London: Hakluyt Society.
Wijngaarden-Bakker, L. H. van. 1984. De dierenresten van Smeerenburg. In: Hacquebord 1984: 279–300.
Wijngaarden-Bakker, L. H. van. 1987. *Zooarchaeological research at Smeerenburg. Smeerenburg Seminar.* Rapportserie. Oslo: Norsk Polarinstitutt.
Wijngaarden-Bakker, L. H. van and J. P. Pals. 1981. Life and work in Smeerenburg. The bio-archaeological aspects. In: van Holk *et al.* 1981: 133–151.
Wilkes, A. H. P. 1922. On the nesting of the Barnacle Goose in Spitsbergen. *Journal of the Museum of Comparative Oology, Santa Barbara, California* 2: 28–30.

Wille, F. 1979. Den grønlandske Havørns *Haliaeetus albicilla groenlandicus* Brehm fødevalg—metode og foreløbige resultater. *Dansk Ornithologisk Forenings Tidsskrift* 73: 165–170.

Winge, H. 1898. Grønlands fugle. *Meddelelser om Grønland* 21. 316pp.

Wood, C. A. and F. M. Fyfe. (eds.) 1943. *The art of falconry, being the* De arte venandi cum avibus *of Frederick II of Hohenstaufen.* Stanford, California: Stanford University Press.

Wright, T. (ed.) 1863. *Alexandri Neckam de naturis rerum.* London: Rolls Series.

Yapp, W. B. 1987. Animals in medieval art: the Bayeux Tapestry as an example. *Journal of medieval history* 13: 15–73.

Yarrell, W. 1882–1884. *A history of British birds, 3.* Fourth Edition. London: John Van Voorst.

Yeates, G. K. 1951. *The land of the loon.* London: Country Life.

Young, S. B. 1989. *To the Arctic. An introduction to the far northern world.* New York: Wiley.

Young, S. C. 1986. *Bibliography on the fate and effects of Arctic marine oil pollution.* Ottawa: Environmental Studies Revolving Funds.

Zaslow, M. (ed.) 1981. *A century of Canada's Arctic islands.* Ottawa: Royal Society of Canada.

Scientific names of plants and animals mentioned in the text

PLANTS

Variegated Horsetail — *Equisetum variegatum*
Cloudberry — *Rubus chamaemorus*
Mountain Avens — *Dryas octopetala* and *integrifolia*
Purple Saxifrage — *Saxifraga oppositifolia*
Alpine Bistort *or* Knotgrass — *Polygonum viviparum*
Arctic Willow — *Salix polaris* and *arctica*
Bearberry — *Arctostaphylos alpina*
Bilberry *or* Blueberry — *Vaccinium uliginosum*
Arctic Bell Heather — *Cassiope tetragona*
Crowberry — *Empetrum nigrum*
Deadly Nightshade — *Atropa belladona*

CRUSTACEANS

Copepod spp. — *Calanus finmarchicus*

FISH

Herring — *Clupea harengus*
Capelin — *Mallotus villosus*
Arctic Cod — *Boreogadus saida*

BIRDS

Gaviidae

Red-throated Diver *or* Loon — *Gavia stellata*
Black-throated Diver — *Gavia arctica*
White-necked *or* Pacific Diver *or* Arctic *or* Pacific Loon — *Gavia pacifica*

Great Northern Diver *or* Common Loon — *Gavia immer*
White-billed Diver *or* Yellow-billed Loon — *Gavia adamsii*

Podicipedidae

Little Grebe *or* Dabchick — *Tachybaptus ruficollis*
Great-crested Grebe — *Podiceps cristatus*

Diomedeidae

Black-browed Albatross *or* Mollymawk — *Diomedea melanophris*

Procellariidae

Fulmar *or* Northern Fulmar — *Fulmarus glacialis*
Great Shearwater — *Puffinus gravis*
Manx Shearwater — *Puffinus puffinus*

Hydrobatidae

Storm Petrel — *Hydrobates pelagicus*

Sulidae

Gannet *or* Northern Gannet — *Sula bassana*

Phalacrocoracidae

Cormorant *or* Great Cormorant — *Phalacrocorax carbo*
Red-faced Cormorant — *Phalacrocorax urile*

Pelecanidae

American White Pelican — *Pelecanus erythrorhynchos*

Anatidae

Mute Swan — *Cygnus olor*
Bewick's *and* Tundra (*formerly* Whistling) Swan — *Cygnus columbianus*
Whooper Swan — *Cygnus cygnus*
Trumpeter Swan — *Cygnus buccinator*
Bean Goose — *Anser fabalis*
Pink-footed Goose — *Anser brachyrhynchus*
White-fronted *and* Greenland White-fronted Goose — *Anser albifrons*
Lesser White-fronted Goose — *Anser erythropus*
Snow Goose *including* Greater *and* Lesser Snow Goose *and* Blue Goose — *Anser caerulescens*
Ross's Goose — *Anser rossii*
Emperor Goose — *Anser canagicus*
Canada Goose — *Branta canadensis*
Barnacle Goose — *Branta leucopsis*
Brent Goose *or* Brant — *Branta bernicla*

Scientific names of plants and animals mentioned in the text

Red-breasted Goose	*Branta ruficollis*
Baikal Teal	*Anas formosa*
Teal	*Anas crecca*
Mallard	*Anas platyrhynchos*
Pintail	*Anas acuta*
Scaup *or* Greater Scaup	*Aythya marila*
Eider	*Somateria mollissima*
King Eider	*Somateria spectabilis*
Spectacled Eider	*Somateria fischeri*
Steller's Eider	*Polysticta stelleri*
Harlequin *or* Harlequin Duck	*Histrionicus histrionicus*
Long-tailed Duck *or* Oldsquaw	*Clangula hyemalis*
Surf Scoter	*Melanitta perspicillata*
Smew	*Mergus albellus*
Red-breasted Merganser	*Mergus serrator*

Accipitridae

White-tailed *or* Sea Eagle	*Haliaeetus albicilla*
Bald Eagle	*Haliaeetus leucocephalus*
Goshawk *or* Northern Goshawk	*Accipiter gentilis*
Rough-legged Buzzard *or* Hawk	*Buteo lagopus*
Golden Eagle	*Aquila chrysaetos*

Pandionidae

Osprey	*Pandion haliaetus*

Falconidae

Merlin *or* Pigeon Hawk	*Falco columbarius*
Gyrfalcon	*Falco rusticolus*
Peregrine	*Falco peregrinus*

Tetraonidae

Spruce Grouse	*Dendragapus canadensis*
Hazel Grouse	*Bonasa bonasia*
Willow Grouse *or* Willow Ptarmigan and Red Grouse	*Lagopus lagopus*
Ptarmigan *or* Rock Ptarmigan	*Lagopus mutus*
Black Grouse	*Tetrao tetrix*
Black-billed Capercaillie	*Tetrao parvirostris*
Greater Prairie Chicken	*Tympanuchus cupido*
Sharp-tailed Grouse	*Tympanuchus phasianellus*

Gruidae

Sandhill Crane	*Grus canadensis*
Whooping Crane	*Grus americana*
Siberian White Crane	*Grus leucogeranus*

Haematopodidae

Oystercatcher — *Haematopus ostralegus*

Charadriidae

Ringed Plover — *Charadrius hiaticula*
Semipalmated Plover — *Charadrius semipalmatus*
Killdeer — *Charadrius vociferus*
Dotterel — *Charadrius morinellus*
American Golden Plover — *Pluvialis dominica*
Pacific Golden Plover — *Pluvialis fulva*
Golden Plover — *Pluvialis apricaria*
Grey *or* Black-bellied Plover — *Pluvialis squatarola*
Lapwing — *Vanellus vanellus*

Scolopacidae

Great Knot — *Calidris tenuirostris*
Knot *or* Red Knot — *Calidris canutus*
Sanderling — *Calidris alba*
Semipalmated Sandpiper — *Calidris pusilla*
Western Sandpiper — *Calidris mauri*
Red- *or* Rufous-necked Stint — *Calidris ruficollis*
Little Stint — *Calidris minuta*
Temminck's Stint — *Calidris temminckii*
Least Sandpiper — *Calidris minutilla*
White-rumped Sandpiper — *Calidris fuscicollis*
Baird's Sandpiper — *Calidris bairdii*
Pectoral Sandpiper — *Calidris melanotos*
Sharp-tailed Sandpiper — *Calidris acuminata*
Curlew Sandpiper — *Calidris ferruginea*
Purple Sandpiper — *Calidris maritima*
Dunlin — *Calidris alpina*
Spoon-billed Sandpiper — *Eurynorhynchus pygmeus*
Broad-billed Sandpiper — *Limicola falcinellus*
Stilt Sandpiper — *Micropalama himantopus*
Buff-breasted Sandpiper — *Tryngites subruficollis*
Ruff — *Philomachus pugnax*
Jack Snipe — *Lymnocryptes minimus*
Great Snipe — *Gallinago media*
Pintail Snipe — *Gallinula stenura*
Long-billed Dowitcher — *Limnodromus scolopaceus*
Hudsonian Godwit — *Limosa haemastica*
Bar-tailed Godwit — *Limosa lapponica*
Eskimo Curlew — *Numenius borealis*
Whimbrel — *Numenius phaeopus*
Bristle-thighed Curlew — *Numenius tahitiensis*
Spotted Redshank — *Tringa erythropus*
Redshank — *Tringa totanus*
Lesser Yellowlegs — *Tringa flavipes*
Wood Sandpiper — *Tringa ochropus*
Terek Sandpiper — *Xenus cinereus*

Scientific names of plants and animals mentioned in the text

Grey-rumped Sandpiper	*Heteroscelus brevipes*
Wandering Tattler	*Heteroscelus incanus*
Turnstone *or* Ruddy Turnstone	*Arenaria interpres*
Black Turnstone	*Arenaria melanocephala*
Red-necked *or* Northern Phalarope	*Phalaropus lobatus*
Grey *or* Red Phalarope	*Phalaropus fulicarius*

Stercorariidae

Pomarine Skua *or* Jaeger	*Stercorarius pomarinus*
Arctic Skua *or* Parasitic Jaeger	*Stercorarius parasiticus*
Long-tailed Skua *or* Jaeger	*Stercorarius longicaudus*
Great Skua	*Stercorarius skua*

Laridae

Sabine's Gull	*Larus sabini*
Bonaparte's Gull	*Larus philadelphia*
Common *or* Mew Gull	*Larus canus*
Herring Gull	*Larus argentatus*
Slaty-backed Gull	*Larus schistisagus*
Glaucous-winged Gull	*Larus glaucescens*
Thayer's Gull	*Larus thayeri*
Iceland Gull	*Larus glaucoides*
Glaucous Gull	*Larus hyperboreus*
Great Black-backed Gull	*Larus marinus*
Ross's Gull	*Rhodostethia rosea*
Kittiwake *or* Black-legged Kittiwake	*Rissa tridactyla*
Ivory Gull	*Pagophila eburnea*

Sternidae

Arctic Tern	*Sterna paradisaea*
Aleutian Tern	*Sterna aleutica*

Alcidae

Guillemot *or* Common Murre	*Uria aalge*
Brünnich's Guillemot *or* Thick-billed Murre	*Uria lomvia*
Razorbill	*Alca torda*
Great Auk	*Pinguinus impennis*
Black Guillemot	*Cepphus grylle*
Pigeon Guillemot	*Cepphus columba*
Little Auk *or* Dovekie	*Alle alle*
Crested Auklet	*Aethia cristatella*
Puffin *or* Atlantic (*formerly* Common) Puffin	*Fratercula arctica*
Horned Puffin	*Fratercula corniculata*

Columbidae

Passenger Pigeon	*Ectopistes migratorius*

Raphidae
Dodo — *Raphus cucullatus*

Strigidae
Great Horned Owl — *Bubo virginianus*
Snowy Owl — *Nyctea scandiaca*
Pygmy Owl — *Glaucidium passerinum*
Great Grey Owl — *Strix nebulosa*

Caprimulgidae
Common Nighthawk — *Chordeiles minor*

Apodidae
Swift — *Apus apus*

Picidae
Black-backed Woodpecker — *Picoides arcticus*

Tyrannidae
Eastern Kingbird — *Tyrannus tyrannus*

Alaudidae
Skylark — *Alauda arvensis*
Shore *or* Horned Lark — *Eremophila alpestris*

Hirundinidae
Sand Martin *or* Bank Swallow — *Riparia riparia*
Northern Rough-winged Swallow — *Stelgidopteryx serripennis*
Tree Swallow — *Tachycineta bicolor*
Swallow *or* Barn Swallow — *Hirundo rustica*

Motacillidae
Olive-backed Pipit — *Anthus trivialis*
Pechora Pipit — *Anthus gustavi*
Meadow Pipit — *Anthus pratensis*
Red-throated Pipit — *Anthus cervinus*
American, Water *or* Buff-breasted Pipit — *Anthus rubescens* (formerly *spinoletta*)
Yellow-headed *or* Citrine Wagtail — *Motacilla citreola*
Yellow *and* Blue-headed Wagtail — *Motacilla flava*
White *and* Pied Wagtail — *Motacilla alba*

Bombycillidae
Waxwing — *Bombycilla garrulus*

Scientific names of plants and animals mentioned in the text 401

Cinclidae

Dipper	*Cinclus cinclus*
American Dipper	*Cinclus mexicanus*

Troglodytidae

Wren	*Troglodytes troglodytes*

Prunellidae

Siberian Accentor	*Prunella montanella*

Turdidae

Siberian Rubythroat	*Luscinia calliope*
Bluethroat *including* Red-spotted Bluethroat	*Luscinia svecica*
Wheatear *or* Northern Wheatear	*Oenanthe oenanthe*
Blackbird	*Turdus merula*
Dusky *and* Naumann's Thrush	*Turdus naumanni*
Fieldfare	*Turdus pilaris*
Redwing	*Turdus iliacus*
American Robin	*Turdus migratorius*

Sylviidae

Sedge Warbler	*Acrocephalus schoenobaenus*
Arctic Warbler	*Phylloscopus borealis*
Chiffchaff	*Phylloscopus collybita*
Willow Warbler	*Phylloscopus trochilus*

Muscicapidae

Red-breasted Flycatcher	*Ficedula parva*

Paridae

Willow Tit	*Parus montanus*
Black-capped Chickadee	*Parus atricapillus*
Siberian Tit	*Parus cinctus*

Laniidae

Brown Shrike	*Lanius cristatus*
Great Grey *or* Northern Shrike	*Lanius excubitor*

Corvidae

Gray Jay	*Perisoreus canadensis*
Siberian Jay	*Perisoreus infaustus*
Carrion *and* Hooded Crow	*Corvus corone*
Raven	*Corvus corax*

Passeridae

House Sparrow — *Passer domesticus*
Tree Sparrow — *Passer montanus*

Fringillidae

Brambling — *Fringilla montifringilla*
Canary — *Serinus canaria*
Pine Siskin — *Carduelis pinus*
Redpoll — *Carduelis flammea*
Arctic *or* Hoary Redpoll — *Carduelis hornemanni*
White-winged *or* Two-barred Crossbill — *Loxia leucoptera*
Rosy Finch — *Leucosticte arctoa*
Bullfinch — *Pyrrhula pyrrhula*

Thraupidae

Western Tanager — *Piranga ludoviciana*
Scarlet Tanager — *Piranga olivacea*

Emberizidae

American Tree Sparrow — *Spizella arborea*
Chipping Sparrow — *Spizella passerina*
Savannah Sparrow — *Ammodramus sandwichensis*
Harris's Sparrow — *Zonotrichia querula*
White-crowned Sparrow — *Zonotrichia leuchophrys*
Smith's Longspur — *Calcarius pictus*
Lapland Bunting *or* Longspur — *Calcarius lapponicus*
Snow Bunting — *Plectrophenax nivalis*
McKay's Bunting — *Plectrophenax hyperboreus*
Little Bunting — *Emberiza pusilla*
Yellow-breasted Bunting — *Emberiza aureola*
Reed Bunting — *Emberiza schoeniclus*
Pallas's Reed Bunting — *Emberiza pallasi*

Icteridae

Brewer's Blackbird — *Euphagus cyanocephalus*
Red-winged Blackbird — *Agelaius phoeniceus*
Yellow-headed Blackbird — *Xanthocephalus xanthocephalus*

MAMMALS

Grizzly Bear — *Ursus arctos*
Polar Bear — *Ursus maritimus*
Weasel — *Mustela nivalis*
Wolf *or* Grey Wolf — *Canis lupus*
Arctic Fox — *Alopex lagopus*
Walrus — *Obodenus rosmarus*
Ringed Seal — *Phoca hispida*
Richardson Ground Squirrel — *Citellus richardsoni* (originally *Arctomys Richardsonii*)

Scientific names of plants and animals mentioned in the text 403

Thirteen-lined Ground Squirrel — *Citellus tridecemlineatus* (originally *Arctomys Hoodii*)

Franklin Ground Squirrel — *Citellus franklini* (originally *Arctomys Franklinii*)

Beaver — *Castor canadensis*
Arctic *or* Collared Lemming — *Dicrostonyx groenlandicus* or *torquatus*
Brown *or* Siberian Lemming — *Lemmus trimucronatus* or *sibiricus*
Arctic Hare — *Lepus arcticus*
Reindeer *or* Caribou — *Rangifer tarandus*
Bison *including* Wood Bison — *Bison bison*
Muskox — *Ovibos moschatus*
Narwhal — *Monodon monoceros*
Bowhead *or* Greenland Right Whale — *Balaena mysticetus*

Index

Abbott, Clinton D. 170
Aberdeen 306
  University 257, 262, 264
Aberystwyth, University College of Wales 284–286
Abruzzi *see* Savoy
Academy Tundra (Akademii tundra), Wrangel Island 232
Abisko, Sweden 291–292
Accentor, Siberian 331, 332
Adams, William 64–65
Aeroflot 308
Akureyri, Iceland 310
Alaska 4, 203, 220, 226, 270,
  breeding birds 10, 12
  conservation 360–362
  Eskimo in 21–29, 33, 34, 35–36
  maps 118, 256, 322, 323, 361
  native hunters in 44
  Nelson in 71–74, 117–120
  research in 242, 256–257, 286–290
  travel in 321–328
Alaska Game Act, 1902 44
Alaska National Interest Lands Conservation Act 360–362
albatross 68–69
Albatross, Black-browed 66
Alberta Province, Canada 240, 273
Albertus Magnus 85
Aleksievka (*former timber port at the mouth of the Pechora*) 137
Alert, HMS 68
Aleutian Islands 2

Alexander Land (Zemlya Aleksandri), Franz Josef Land 366
Alexander, HMS 58–61
Alexei Mikhailovich, tsar of Russia 92
Alezaya River 169
Allen, A. J. 185
Alston, Edward A. 134
Ameralik Fjord 38–40
American Ornithologists' Union 71, 186
Amerind Publishing Co. 222
Amguema River, Chukchi Peninsula 224
Ammassalik 9, 25, 249, 268, 292, 296, 310, 312, 342
Ammassalik Fjord 249
Amsterdam 56, 321
  University Zoological Museum, Institute of Taxonomic Zoology 275
Amsterdam Island (Amsterdamøya), Spitsbergen 15, 52, 53–54
Anadyr' 215, 219, 333
Anaktuvuk Pass, Alaska 24, 34, 256–257, 326–327
Anchorage, Alaska 321, 327
  Alaska Fish and Wildlife Research Centre 170
  Arctic Health Research Centre 257
Anderson, Johann 56–57, 58
Anderson River, Northwest Territories, Canada 178
Andreev, Aleksandr V. 226–227, 229–231
  photographs by 82
Antwerp University 291
Apeldoorn 92

archaeology 26, 38–40, 54
Archangel (Arkhangel'sk) 88, 134, 136, 143, 212, 346
Archer, Colin 77
Arctic 241, 273
Arctic, the, Dundee whaler 64
Arctic Institute of North America (AINA) 180, 239, 240–241, 321, 339
Arctic National Wildlife Refuge, Alaska 360–362
Arctic Ocean 1, 3, 4, 6, 42, 70, 79, 169, 170, 171, 183, 193, 287
 conference 337
 pollution 338–339
 shore 17, 75, 80–81, 101, 105, 107, 113, 205
Arctic Slope, Alaska *see* North Slope
Arctic Slope Regional Corporation 322
Arctic Zone 2, 9, 183, 189
Ardea 251, 275
Arktikugol 341, 351
Armstrong, Edward A. 18, 291–292
Arnhem, Rijksinstituut voor Natuurbeheer 291
Ary-Mas, Taimyr Peninsula 331
Asbirk, Sten 292
Ashwell, Cambs. 85
Assistance, HMS 63–64
Association of Northern Ethnic Minorities (USSR) 341
Atkasuk, Alaska 326
Atlantic, North 10, 12, 147, 261, 280
Audubon, John James 113
Audubon Society, National 120, 362
Auk, the 118, 167, 176, 243
Auk, Great 12, 39, 40
Auk, Little 6, 9
 as food bird 63–64, 67, 68, 74
 and Eskimos 20–21, 25, 26–27, 29, 31, 35
 at sea 61, 76
 bones found 40
 food 216
 in Franz Josef Land 76, 201
 in Greenland 57–59, 261–264, 316
 in Iceland 11, 304
 in Spitsbergen 53, 54, 56, 194, 309, 352, 353
 photograph 263
Auklet, Crested 30, 74
auklets 29, 35, 43
auks 34, 35, 43

Aula Verlag, Wiesbaden 199
Aurora Borealis, newspaper 63–64
Austin, H. W. 63
Austin, Oliver L. 187
Australia 336
Austria 336
Auyuittuq National Park, Baffin Island 342, 357
Avanersuaq (*formerly* the Thule district), northwest Greenland 4–5, 7, 158–159, 315–316, 354, 355
 breeding birds 11, 15, 17, 83, 93, 368, 369
 photographs taken in 7, 13, 16, 32, 65, 72, 124, 128, 160, 177, 190, 201, 234, 246, 247, 250, 254, 263, 265, 267, 283, 294, 300, 313, 314–315, 346, 348
avens, mountain 253–255

Back, George 104, 105, 106
Baer, K. E. von 208
Baffin, William 53
Baffin Bay 58, 59
Baffin Island 4, 6, 187, 257–261, 316, 318, 342, 357, 359
 breeding birds 14, 175–176
Baffinland Eskimo 24, 27, 33
Bailey, Alfred M. 21, 36, 184–186
Baird, P. D. 257
Baird, Spencer Fullerton 68, 113–114, 117, 164, 167
Baker, Mr 133
Banff, Alberta, Canada 337
 National Park 357
Bangsted, Helge 121
Baker Lake, Canada 24
Banks, Joseph 182
Banks Island 10, 316, 321, 357
Bannerman, David 111
Baranyuk, V. V. 232–233
Bárðarson, Hjálmar 303
Barents, Willem 53
Barents Sea 142, 144–152, 214, 215, 223, 304, 329, 338, 346, 363
Barentsburg, Spitsbergen 308
Barentsfjorden (Richard Lagoon), Spitsbergen 244
Barr, James 92
Barr, John 88
Barren Grounds *or* Lands, Canada 8, 12, 99–117

Barrow, John 62
Barrow Jr, John 62–64, 73
Barrow (Ukpeagvik) *and* Point Barrow,
   Alaska 71, 107, 185, 242, 256,
   321–326, 336
  American expedition 1881–1883 27,
    120–121, 166–167
  Arctic haze 339
  breeding birds 10, 46–48, 145, 326
  Eskimos and birds 21, 24, 29, 31, 33,
    34, 42–43, 44
  Lapland Buntings 286–290
  map 323
  photographs taken near 258, 260, 324,
    325
  Ukpeagvik Inupiat Corporation 324
Barrow Strait 64
Bathurst Inlet, Canada 105, 273
  Lodge 320–321
Bathurst Island, Canada 19, 127, 357
Bavo *or* Baaf 85
Bayeux Tapestry 83
bear, grizzly 321
bear, Polar 50, 53, 195, 221, 224, 275,
   302, 321, 354, 355, 363, 366
Bear Island (Bjørnøya) 52, 193, 194, 211,
   281, 306, 308, 338
  map 280
Bear, the USS 185
bearberry 278
Beaufort Sea 14, 43, 49, 339
beaver 103, 357
Beaver County, Pennsylvania 188
Beebe, F. 94
Belaya Zemlya, ostrova, *formerly*
   Hvidtenland 168
Belcher Islands, Hudson Bay 239
Belgium 279
Bell, Alexander Graham 201
Bel'kovskiy Island (ostrov Bel'kovskiy),
   New Siberian Islands 224
Bell Sound (Bellsund), Spitsbergen 51, 52,
   281
Belopol'sky, Leb O. 195, 214–218, 220, 363
Belyaka Spit (kosa Belyaka), Chukchi
   Peninsula 228
  photographs taken at 71, 104, 227
Bennett, James G. 68
Bennett Island (ostrov Bennetta) 70, 78–79
Bent, A. C. 117, 173, 175–176, 178
Bentley, Beetham 155
Bergen, Norway 144

Bering Land Bridge National Preserve,
   Alaska 360
Bering Sea 2, 71, 167, 328, 329
Bering Strait 4, 26–27, 68, 70, 75, 121,
   170, 185, 186, 215, 228, 329
  Eskimo 27, 30, 117
Berkeley, University of California 286, 287
  Museum of Vertebrate Zoology 287
Berlevåg, Norway 304
Berlin 178
Bertelsen, Alfred 191
Bessastaðir, Iceland 90
Bessels, Emil 66–68
Bewick, Thomas 58
Bezymyannaya Bay, Novaya Zemlya 218,
   345–349
Bianki, Vitaly V. 214, 218–219
Bikada River, Taimyr 329–331
bilberry 253, 278
birch 304
Bird Study 291
Bird, C. G. *and* E. G. 290
Birds of Prey International (Canada) 97
Birula *or* Birulya, A. 77–78, 162–163,
   211–212
bison, wood 357
Black Sea 223
Blackbird 14
Blackbird, Brewer's 185, 287
Blackbird, Red-winged 287
Blackbird, Yellow-headed 106
Blasius, J. H. 134
Blencathra, the 143
Bliss, L. C. 336
Blomqvist, Sven 313–314
Bluethroat 257
  Red-spotted 326, 327, 328
Boag, D. A. 273
Boiling, Herr 141
Bolbre, Coloniforsteher 162
Bol'shaya Balakhnya River 169
Bol'shezemel'skaya tundra *see* Great Land
   Tundra
Bonaparte, C. L. 111
Bonevi Island (ostrov Bonevi) 163
Boothia Peninsula 62
Borup, George 159
Boston Society of Natural History 211
Boswall, Jeffery 329
bowhead *or* Greenland right whale 49, 73
Boyd, Hugh 265, 279, 291
Brabant 87

Brambling 144, 292, 304
Brandt, Herbert W. 21, 25, 124–126
Brandt, J. F. 121
Bree, C. R. 134
Breiðamerkursandur, Iceland 304
Brindley, H. H. 156
Britain 9, 14, 57, 97, 104, 140, 239, 243, 262, 281, 306, 336
British admiralty 58, 104, 107, 166
British Arctic Expedition, 1875–1876 66–68, 161
British Birds 156, 171, 202, 222, 227
British Columbia, Canada 113
British Museum (Natural History) 93, 94, 99, 101, 146, 150
British North Greenland Expedition, 1951–1954 312
British Ornithologists' Club 178
British Ornithologists' Union 67, 130–131, 134, 139, 143, 168, 189
British Petroleum (BP) 341
British Trust for Ornithology (BTO) 202, 203, 248
Brooks Range, Alaska 4, 24, 34, 256–257, 287, 321, 326–327, 328, 360, 362
Brower, Charles D. 21, 29, 31, 34, 145, 170, 185
Brower, Tom 256
Brugge, Jacob Segersz van der 53
Brünnich, Morten Thrane 181–182
Brusewitz, Gunnar 171
Brussels 321
Bullfinch 141
Bunge, Aleksandr A. 162
Bunting, Lapland 9, 11, 14, 113, 141, 249, 259, 315, 317, 320
  birds and eggs collected 115, 123, 137, 192, 211
  near Point Barrow 286–290, 326
  photographs 210, 289
  prey of falcons 272, 274
Bunting, Little 135
Bunting, McKay's 328
Bunting, Pallas's Reed 331, 332
Bunting, Reed 144
Bunting, Snow 9, 10, 11, 43, 259, 298, 306
  and Eskimos 20, 23, 24, 35
  arrival 5, 75
  as food bird 64, 137
  at sea 61
  eggs collected 123
  in Alaska 186, 326
  in Canadian Arctic 14, 61–62, 68, 102, 106, 107, 113, 123, 187, 320
  in continuous daylight 292
  in Eurasian Arctic 52, 56, 70, 74, 76, 79, 141, 193, 200, 221, 310, 353
  in Greenland 192, 249–251, 295, 315
  migration 18
  nest sites 45–47
  photographs 124, 250
  prey of falcons 81, 223, 272, 274
  prey of skuas 278
Bunting, Yellow-breasted 135
buntings 62
Bureau (*or* Division) of Biological Survey, *see* U.S. Fish and Wildlife Service
Burgundy, duke of 87
Burnham, William A. 271–273
Butenko, P. T. 219
Butev, V. T. 16, 170
Buturlin, Sergei A. 17, 168–169, 212–213
Buzzard, Rough-legged 12, 15, 83, 274, 365
  breeding 145, 150, 235, 320, 326
  clutches taken 115, 152
  photograph of nest 151
Byrranga Mountains (gory Byrranga), Taimyr 208, 365

Cade, Tom J. 271
Caerlaverock, Scotland 279–281
Cagni, Umberto 76
Caithness 140
Calanus finmarchicus 264
Cambridge 133, 291
  St John's College 153
  Sanctuary Club 156
  Scott Polar Research Institute (SPRI) 337
  Sewage Farm 156
  University 142
  University Press 336
Cambridge Bay (Ikaluktutiak), Victoria Island, Canada 273, 304, 316, 320
  photographs taken near 13, 37, 60, 72, 103, 110, 114, 119, 146, 147, 149, 151, 155, 174, 204, 210, 225, 252, 276, 277, 288, 293, 294, 299, 301–302, 319
Camp, Matthew 65
Canada 4, 8, 27, 188, 241, 272, 287, 321, 336
  National Parks Act, 1930 342

Canadian Arctic 9, 97, 302
  archipelago 4, 6, 10, 127, 316
  Arctic Islands Reserve 357
  breeding birds 11, 12, 41, 98–117,
    121–124, 127–129, 158, 169–170,
    172–179, 296
  conservation 43–45, 357–360
  explorers 60–64
  maps 59, 100, 122, 240, 317, 358
  research in 257–261, 265–268, 273–275
  Resources Committee 339
  travel in 316–321
  whalers in 49
Canadian Government
  Department of Agriculture, Entomology
    Research Institute 266
  Environment Canada 43; see Canadian
    Wildlife Service
  Minister of the Environment 359–360
  Parks Canada 342
Canadian Wildlife Service 42, 117, 175,
    176, 180, 187, 265–266, 291,
    334, 359–360
  Waterfowl Management Plan 45
Canary 102
Cape Dorset Bird Sanctuary, Baffin Island
    359
Cape Farewell (Kap Farvel) 3, 4, 58, 62,
    268
Cape Krusenstern 34
Cape Lisburne 34, 256
Cape Lyamchina, Vaygach Island 11
Cape Parry, Canada 113
  Bird Sanctuary 359
Cape Schmidt (mys Schmidta *or* Shmidta),
    Chukchi Peninsula 221, 224, 332
Cape Serdtse-Kamen' 221
Cape Sparbo, Devon Island 321
Cape York, northwest Greenland 67
Cape Zhelaniya, Novaya Zemlya 3, 16, 170
capelins 216
capercaillies 81
Capercaillie, Black-billed 229–230
caribou *or* reindeer 8, 33, 38, 53, 64, 78,
    143, 232, 284, 308, 327, 341,
    357, 363
Caribou Eskimo 22, 24, 30
Carlsberg Foundation 191, 270
Carlton House, Saskatchewan, Canada 109
Carter, Jimmy 360
Cartwright, Newfoundland 116
Cary Islands (Carey Øer) 64

Caspian Sea 18, 87
Cassin, John 114, 178
Catalogue of birds in the British Museum 188
Catherine the Great, tsaritsa of Russia 207
Cavalli-Molinelli, P. A. 76
Chapman, Abel 15
Chaun Bay (Chaunskaya guba) 219, 226,
    228, 332
Chaun River 169
Chelyuskin, the 215, 220–221
Chernyavsky, F. B. 233
Cherskiy 332
Chesapeake Bay 344
Chesterfield Inlet, Canada 24
Chicago 117
  Academy of Sciences 186
  Field Museum of Natural History 21,
    125, 186
Chickadee, Black-capped 106, 257
Chicken, Prairie 120
Chiffchaff, 9
China 12, 257
Chipewyan Indians 103
Chokurdakh 332
Christianshåb *see* Qasigiannguit
Chukchi Autonomous Okrug *or* Region
    197, 332
Chukchi Mountains (Chukotskiy khrebet)
    219
Chukchi Peninsula (Chukotskiy poluostrov)
    4, 29, 63, 75, 197, 211, 212, 236,
    238, 328, 332, 366
  breeding birds 10, 71–73, 121, 169,
    202–203, 227, 306
  expeditions to 215, 219–222, 224
  map 220
Chukchi Sea 71, 215, 220, 326
Chukchi-Anadyr' Expedition 219
Chukchis 22, 31
Churchill, Manitoba 102, 103, 114, 123,
    127, 172, 302, 316, 317
Churchill River 101, 172
Churchill, the 100–101
Ciesielksi, Konrad, Marcus *and* Lothar
    96–97
Circumpolar Journal, the 241
Clavering Island 253
Cleveland, Ohio 125
Cloven Cliff (Klovningen), Spitsbergen 353
Coburg Island, Canada 64
cod, Arctic 6, 268
Collett, Robert 75–76, 135

Colorado 186, 271
Colville River, Alaska 21, 242, 271
Commander Islands (Komandorskiye ostrove) 86, 229
Condor, the 170
Conover, Henry Boardman 125–126
Constable Point, east Greenland 310, 354
Constantinople, sultan of 91
Convention Concerning the Conservation of Migratory Birds and their Environment, 1978 44
Convention for the Protection of Migratory Birds, 1916 *and* 1918 43–45, 359
Convention on International Trade in Endangered Species of Wild Fauna and Flora (CITES) 335–336
Convention on Wetlands of International Importance Especially as Waterfowl Habitat (Ramsar Convention) 336
Conway, Martin 144
Copenhagen 90, 162, 251, 310, 313, 321
  Commission for Scientific Research in Greenland 241
  Danish Government, Ministry for Greenland 191, 312, 367, 368
  Danish Polar Centre 241, 312
  Greenland Travel Bureau 312
Copenhagen University 181, 189, 251
  Zoological Museum 39, 121, 167–168, 189, 191, 270, 312
Copper Indians 109
Coppermine, Northwest Territories, Canada 12
Coppermine River 101, 104, 106, 107, 111
Coppinger, R. W. 67
Cordeaux, John 140, 142
Cormorant 29, 40, 42, 58, 367
Cormorant, Red-faced 74
cormorants 29, 221
Cornell University, Ithaca, New York 126, 127
Cornwallis Island 318
Coronation Gulf, Canada 21, 105
Corwin, the USS 70–74
Cott, H. B. 40
cotton-grass 285
Coues, Elliott 114, 167
Council, Alaska 322
Court, G. S. 273
Crane, Sandhill 21, 102, 105, 112, 221, 332
  and Eskimos 23, 24, 35–36, 41
  eggs taken 123
  in Greenland 313
Crane, Siberian White 12, 226, 332, 366
Crane, Whooping 12, 102, 105, 112
Cree Indians 22, 43
Creswell Bay, Somerset Island 318
Crimea, khan of the 91
Crocker Land Expedition 158–159
Cromvoirt, Holland 89
Cross Bay (zaliv Kresta) 333
Crossbill, Two-barred 106
crossbills 105
Crow, Hooded 152
crowberry 109, 278
Cullen, J. M. 18, 291–292
Cumberland House, Saskatchewan, Canada 106
Cumberland Peninsula, Baffin Island 259, 357
Cumberland Sound, Baffin Island 27, 45, 187
Curlew, Bristle-thighed 328
Curlew, Eskimo 12, 105, 106, 109–110, 116–117, 366
Curtis, Dora 152
Custer, Thomas W. 287–290
Czaplicka, M. A. 152

Dalrymple Rock, northwest Greenland 17, 64–66
Dalton Highway 328
Daneborg, east Greenland 354
Danilov, Nikolay 236
Danish East Greenland Expedition, 1931–1934 251, 355
Danish Ethnographical Expedition to Arctic North America 121–124, 355
Danish Ornithologists' Union (Dansk Ornithologisk Forening) 189, 270
Danmark, the 192
Dartmouth College, Hanover, New Hampshire 271
Davis Strait 57, 59
Decius, emperor 85
De Long, George W. 68–69, 166
Demarcation Point 185
Dement'ev, Georgy Petrovich 86, 199, 200, 215, 223
Dempster Highway 316
Denali National Park and Preserve 81
Denmark 15, 25, 133, 138, 189, 271, 284, 306, 354
  import of Gyrfalcons 87–91

Denmark, king of 90, 91
Denmark Strait 87, 304
Densley, Michael 168, 170–171, 242
Denver (*formerly* Colorado) Museum of Natural History 184–185, 186
Devon Island, Canada 14, 240–241, 318, 321
Dewey Soper Bird Sanctuary, Baffin Island 359
Dezhnev 202, 219
Dhondt, André A. 291
Dhuleep Singh, maharajah 92
Diabasøya, Spitsbergen 282–283
Diana, the, Dundee whaler 65–66
Digges Islands, Canada 339
Dikson 78, 205, 329, 332
Dipper 11
Dipper, American 328
Discovery, HMS 68
Disko, west Greenland 39, 83, 87, 162, 166, 336
Disko Bay *or* Bugt 167, 268, 271, 296
Distant Early Warning Line (DEW-Line) 292–296
Diver, Black-throated 12–14, 144, 199, 200, 201, 332
Diver, Great Northern 11, 182, 199, 303, 314–315, 320
 as food bird 39
 Eskimo name 24
 in Canadian Arctic 63
Diver, Red-throated 23, 25, 113, 199, 200, 314–315
 as food bird 39, 42, 63
 clutches taken 115
 Gyrfalcon prey 223
 in Spitsbergen 242–248, 309, 353
 skins used for clothing 29
 photographs 246, 247
Diver, White-billed 21, 105, 107, 186, 199, 219, 320, 326, 332
 Eskimo names 23, 24
 photograph 104
Diver, White-necked *or* Pacific 14, 23, 25, 30, 106, 199, 320, 326, 328
 birds and eggs collected 115, 125
divers (unspecified) 11, 20, 23, 29, 41, 68, 101, 107, 118, 290, 296, 328
Dodo 164
Dolgaya Bay, Vaygach Island 150
Dolgiy Island (ostrov Dolgiy) 150
dolphins 338

Donovan, Ernest 179
Dorogoy, I. V. 233
Dotterel 18, 145, 156, 326, 331
 photograph of nest 145
Doronin, Ivan 221
Dovrefjell 152
Dowitcher, Long-billed 10, 227, 326, 328, 332
Drent, Rudi H. 279, 282–283
Dresser, Henry Eeles 134, 135, 164
Driver, Peter M. 239
Duck, Harlequin 315
Duck, Long-tailed 9, 298, 306, 309, 313, 314, 339
 and Eskimos 23–24, 26, 31, 35–36, 42
 as food bird 63, 68, 78
 birds and eggs collected 115, 123, 125, 138
 bones found 40
 described by Sutton 128–129
 in Alaska 328
 in Canadian Arctic 62, 105, 106, 318
 in Chukchi Peninsula 220
 in Iceland 11
 in winter 16
 north of Asia 70
 off Spitsbergen 51
 on Wrangel Island 221
 photographs 128
 prey of falcons 223, 272
ducks (unspecified) 34, 35–36, 38, 43, 62, 64, 107, 193, 292, 296, 339, 365
 eggs 138
Ducks Unlimited 44
Dudinka 141, 332
Dumfries 279
Dundee 64–65
 University 292
Dunipace, Scotland 138
Dunlin 10, 18, 120, 125, 137, 141, 194, 213
Dunøyane, Spitsbergen 279
Durrell, Gerald *and* Lee 329, 331, 365
Dutch Greenland Expedition to Scoresby Sound, 1973–1975 275
Dutch Ornithologists' Union (Nederlandes Ornithologische Unie) 251, 275
Duti of Lincolnshire 85
Dvina River *and* Delta 134–135, 136, 137
DYE 2, DYE 3 *and* DYE 4 292–296
Dyfed 85
Dyfi estuary 284

Eagle, Bald 112, 270
Eagle, Golden 21, 83, 274, 328
  clutches taken 115
  feathers 27
Eagle, White-tailed 39, 42, 43, 79, 268–271, 315, 367
East Siberian Sea 70
Ebbinge, Barwolt S. 291
Eclipse, the, Peterhead and Dundee whaler 65–66
Edge Island (Edgeøya), Spitsbergen Archipelago 193, 275, 306, 351
Edinburgh University 108
  Museum 113, 166
Edmonton, Alberta, Canada 316, 318
  Conference on Productivity and Conservation in Northern Circumpolar Lands, 1969 335
  University 273, 321
Edwards, George 99
Eerden, M. R. van 281–283
Egede, Hans 38–39
Egvekinot 333
Eider 9, 25, 37, 148, 181, 186, 187, 219, 326, 338, 363
  and Eskimos 23–24, 26, 29–30, 35–36, 41, 42
  as food bird 63, 68, 148
  clutches taken 115
  eggs eaten 41, 64–66, 94
  food of 216–217
  Gyrfalcon prey 94
  in Canadian Arctic 107, 123, 239, 318, 320, 321, 359
  in Chukchi Peninsula 221
  in Franz Josef Land 201, 366
  in Greenland 58, 367–369
  in Spitsbergen 15, 16, 53, 353
  migration 17, 368
  on Wrangel Island 221, 234–235
  photographs 28, 37, 217
  seen from the *Fram* 76
Eider, King 8, 74, 221, 306, 313, 316
  and Eskimos 23–24, 26, 29–30, 36
  as food bird 63, 68, 121, 179–180
  birds and eggs collected 115, 123, 150
  in Alaska 186, 322, 326
  in Canadian Arctic 62, 107, 187, 317, 318, 320, 321
  in Soviet Arctic 18, 148, 331, 332
  migration 14, 121, 125
  photographs 149, 313

Eider, Spectacled 25, 29, 121, 125, 186, 220, 221, 222, 326, 332
  Kishchinsky on 224, 226
Eider, Steller's 10, 23, 125, 131, 186, 208, 306, 326, 332
  photograph 209
eiderdown 27–28, 368
eiders (unspecified) 9, 20–21, 23–24, 29, 31, 70, 195, 328, 339, 365
  and Eskimos 20–21, 23–24, 29, 31, 41, 42–43, 44
  bones found 40
Eindhoven 89
Ekblaw, W. E. 25, 34–35, 158–159, 163
Elander, Magnus 313–314
Ellesmere Island 4, 6, 8, 26, 68, 176, 296, 316, 318, 321, 342
  breeding Knots 68, 159, 161, 266
  National Park 357
Elton, Charles 243, 244
Elveden, Suffolk, England 92
England 10, 51, 55, 83, 85, 101, 131, 133, 140, 141, 199, 279
Envall, A. 74
Equaluit, west Greenland 39
Eqalummiut Nunaat, west Greenland 284–286, 336
Erik the Red 38, 313
Erik, the, Dundee whaler 65
Erskine, Angus B. 312–313
Eskimo Point, Canada 41
Eskimonæs, Clavering Island, east Greenland 253, 255
Esquimaux, the, Dundee whaler 64–65
Estonia 11, 77, 85, 208, 362
Eton College 94, 131
Evans, Peter G. H. 262
Expéditions Polaires Françaises 292

Fabricius, Otto 58
Faeroe *or* Faroe Islands 31, 181, 306
  map 307
Fafard, E. 24
Fairbanks, Alaska 125, 326, 328
  University of Alaska Museum 171
Fair Haven, Spitsbergen 50, 52, 56
Falconers' Association of North America 97
Falconry Club 88
falcons 20–21
Falk, Knud 312
Fauna Boreali-Americana, 108–113, 173

Feilden, Henry W. 30, 67–68, 135, 142–143, 144–150, 161
Fen Ditton, Cambs. 153
Fifth Thule Expedition, 121–124, 355
Figgins, J. D. 184
Fina 351
Finch, Rosy 332
Finland 131, 134, 336
Finlay, J. Cam 316, 318, 321
Finney, George 43
Finnmark 87
Fischer, J. G. 121
Fisher, Alexander 58, 61–62
Fisher, James 303
Fiskenæsset, west Greenland 368
Flagstaff Island 179
Flakkerhuk, Disko 162
Flanders 85
Flint, Vladimir E. 199, 200, 203, 224
Florida 297
Flycatcher, Red-breasted 328
Førlandsøyane Bird Sanctuary, Spitsbergen 353
Forster, J. R. 99
Fort Albany, Ontario, Canada 99, 173, 178
Fort Anderson, Northwest Territories, Canada 115–117, 178
Fort Enterprise, Northwest Territories 105
Fort Good Hope, Northwest Territories 115
Fort Norman, Northwest Territories 113
Fort Prince of Wales, Manitoba, Canada 101, 103, 172
Fort Resolution, Northwest Territories 113
Fort Severn, Ontario, Canada 99
Fotherby, Robert 53
fox, Arctic 8, 53, 81, 234–235, 264, 285, 290, 291, 296, 369
    photographs 234, 265
Fox, A. D. 284–286
Fram, the 75–76
France 57, 306
Frankfurt 321
Franklin, John 104–111
Franklin Bay, Northwest Territories, Canada 116
Franz Josef Land (Zemlya Frantsa Iosifa) 4, 17, 66, 75, 164, 168, 170, 171, 309, 329, 366
    early records 76
    map 364
    Russian expedition 200–202, 203–205, 366

Franzmann, Niels-Erik 292
Frederick II of Hohenstaufen, emperor 83, 85
Frederikshhåb *see* Paamiut
French Revolution 90, 92, 104
Freuchen, Peter 5, 42, 121–124, 355
Frobisher Bay *see* Iqaluit
Fuertes, Louis Agassiz 126
Fuglehuken, Prince Charles Foreland, Spitsbergen 353
Fulmar 5, 12, 16, 54, 61, 292, 309, 316, 352, 353
    and Eskimos 24, 35
    early accounts 55–58
    in Arctic Ocean 76
    in Barents Sea 215, 216
    in Canadian Arctic 265–266, 318
    in Franz Josef Land 76, 201
    photograph 55

Galapagos Islands 297
Gambell (Sivokak), St Lawrence Island, Alaska 321, 328
Gannet 55–56, 62, 304
Gaston, A. J. 266
Gates, C. C. 273
Gates of the Arctic National Park, Alaska 326, 360
    Iniakuk Lake Lodge 328
Gätke, Heinrich 142
Gavin, Angus 179–180
Geale, John 318
geese (unspecified) 20, 74–75, 85, 101, 105, 193, 195, 292, 328, 339, 365
    crossing Arctic Ocean 79
    described by Hearne 102–103
    hunted by native peoples 33–34, 35–36, 38
    shot 64, 78
Gelting, Paul 251–256
Gensbøl, Benny 270–271
George I, king of England 57
Gerald of Wales 84–85
German Ornithological Association (Deutsche Ornithologische Gesellschaft) 182
Germania Land, Greenland 83
Giesecke, Karl Ludwig 165
Gilham, Charles A. 179
Gipsdalen, Spitsbergen 279
Gitz-Johansen, Mr 192
Gladkov, N. A. 199

Glasgow, 303
Glinski, Mr 141
Godhavn *see* Qeqertarsuaq
Godthåb *see* Nuuk
Godwit, Black-tailed 156
Godwit, Bar-tailed 131, 153, 257, 304, 326, 328, 331
Godwit, Hudsonian 317
Gol'chikha 141, 152–153, 156, 164, 332
Goodsell, John W. 161
Gooilust, Holland 178
Goose, Barnacle 8, 9, 54, 243–244, 279–283, 309, 313, 314, 353, 366
Goose, Bean 30, 33–34, 137
Goose, Brent 9, 200, 244, 290, 332
  as food bird 26, 35, 36, 44, 54, 63, 67, 68
  autumn counts 14–15
  clutches taken 115
  hunting 33–34, 35–36
  in Alaska 326, 339
  in Canadian Arctic 62, 68, 173–175
  in Chukchi Peninsula 220
  in Greenland 292–296
  in Spitsbergen 50–51, 53, 56, 279, 307, 351, 353
  in Taimyr 290–291
  on Wrangel Island 234–235
  photograph 51
Goose, Canada 25, 107, 123, 172–173, 175, 179–180
  and Eskimos 35–36, 41, 44
Goose, Emperor 29, 118, 172, 224, 328, 333, 367
  and Eskimos 35–36, 41, 44
  birds and eggs collected 125
  breeding biology 227–228
  photograph 227
Goose, Hutchins's *or* Richardson's 173–175, 320
  photographs 174
Goose, Lesser White-fronted 131, 367
Goose, Pink-footed 279, 303, 304, 309, 313, 314, 353
Goose, Red-breasted 8, 200, 207, 213, 329, 331, 332, 365, 367
  Krechmar's studies 227, 235
Goose, Ross's 102–103, 114, 172–173, 178–180, 336, 359
Goose, Snow 10, 12, 207
  and Cree Indians 43
  and Eskimos 24, 26, 35–36, 40–41
  in Canadian Arctic 43, 102, 105, 106, 123, 172–173, 175–176, 178, 318, 320, 321, 336, 359
  in Chukchi Peninsula 220
  in Greenland 32, 312, 313, 316
  in Mackenzie Delta 339
  on Wrangel Island 195, 219, 221, 231–235, 363–365, 366–367
  photographs 32, 177
Goose, White-fronted 23, 44, 172–173, 179, 320
  and Eskimos 29, 30, 42
  as food bird 26, 35–36, 39, 78, 179–180
  caught by Samoyeds 33–34
Goose, Greenland White-fronted 283–286, 292–296
Gorbunov, G. P. 200, 346, 348
Gordon, Seton 243, 351 352
Göteberg, Sweden 304
Goshawk 83, 85
Graham, Andrew 99
Graham Bell Island (ostrov Greem Bell), Franz Josef Land, 201–202, 203–205, 366
Gray, David 66
Great Bear Lake 105
Great Falls, Montana 97
Great Lakes 18
Great Land Tundra (Bol'shezemel'skaya tundra) 135–138, 150, 195, 236, 365
Great Slave Lake 105, 112, 113, 114
Grebe, Great Crested 244, 247
Grebe, Little 244
Greely, Adolphus W. 161
Greenland 4, 6, 181, 184, 189–190, 192–193, 302, 304, 339, 342, 349
  albatrosses off coast 66
  birds and Greenlanders 20–21, 25, 26–27, 29–30
  breeding birds 11, 12, 19, 165, 166, 169–170, 176, 203, 306
  conservation 335, 336, 344–345, 353–357, 367–369
  egging 41
  eiderdown from 203, 306
  European settlers 38–40
  explorers off coast 49, 57–58
  Gyrfalcons 81–83, 86, 87, 93
  Knots 158–161
  maps 269, 280, 311, 356
  migration 17, 18

mineral exploitation 341
Polar bears 302
research 241–242, 248–256, 261–264, 268–273, 275–278, 283–286
ringing 41–42, 43, 191–192, 279
travel in 310–316
whalers off coast 49, 57–58
Greenlanders 249, 274, 278, 342, 345, 354, 355, 367–369; *see too* Greenland, birds and
Grønlandsfly (Greenlandair) 310, 316
Grimsey, Iceland 304
Grise Fjord, Ellesmere Island 321
Groningen University
  Arctic Centre 241–242, 275, 307
  Zoological Laboratory 279
grosbeaks 105
grouse 62
Grouse, Black 94
Grouse, Hazel 81, 230
Grouse, Red 94, 259, 334
Grouse, Sharp-tailed 105, 106
Grouse, Spruce 126
Grouse, Willow 11, 102, 186, 191
  as food bird 63
  birds and eggs collected 41, 115, 125, 138
  Eskimo names 24–25
  Gyrfalcon prey 81, 223
  snared by Eskimos 34
  winter life 229–231
Guðmundsson, Finnur 303
Guillemot 182, 304, 353
Guillemot, Black *or* Tystie 36, 39, 40, 81
  as food bird 63, 67, 68
  at sea 61
  in Canadian Arctic 62, 318
  in European Arctic 76, 145, 201, 216, 218
  in Greenland 57, 58, 338
  in Spitsbergen 15–16, 53, 54, 56, 353
  north of Asia 70, 74, 79
  photographs 16, 216
  prey of Gyrfalcon 94
Guillemot, Brünnich's 9, 145, 182, 194, 304, 359
  as food bird 26, 31, 35, 39, 41, 54, 63, 64, 70, 74
  conservation 339, 344–349
  eggshell thickness 215
  in Franz Josef Land 201
  in Greenland 41–42, 58, 312, 316, 344–345, 367

in Novaya Zemlya 148, 218, 345–349
in Spitsbergen 15–16, 51, 52, 53, 56, 309, 353
migration 17
north of Asia 74, 79
on Jan Mayen 292
on Murman Coast 363
photographs 267, 346
Prince Leopold Island 265–268, 318
skin used for clothing 29
Guillemot, Pigeon 74
guillemots (unspecified) 29, 43, 57, 68, 216, 264
  egg shape 217–218
  Gyrfalcon prey 223
Gulf of Taimyr (Taymyrskaya guba) 208
Gulf of Yenisey (Yeniseyskiy zaliv) 213, 332
Gulf Stream 214, 304, 306
Gull, Bonaparte's 115, 317
Gull, Common *or* Mew 218, 317
Gull, Glaucous 9, 74, 315, 352, 353
  as food bird 34, 36, 40, 54, 74
  early mentions 58
  Eskimo name 23
  Gyrfalcon prey 223
  in Canadian Arctic 62, 107, 113, 317, 318
  in Eurasian Arctic 53, 56, 76, 148, 150, 201, 221, 310
  its prey 59, 283, 347–349
  photograph 348
Gull, Glaucous-winged 74
Gull, Great Black-backed 40, 57, 94, 216
Gull, Herring 14, 63, 81, 201, 216, 218, 223, 235, 317, 363
Gull, Iceland 39, 40, 315
Gull, Ivory 8, 9, 18, 57, 183, 185, 322
  as food bird 26, 54, 64
  at sea 61, 76, 329
  eggs 164
  in Canadian Arctic 62, 107, 113, 318, 321
  in Franz Josef Land 201–202, 366
  in Greenland 58, 67
  in Spitsbergen 53, 56, 194, 307, 309, 353
  in winter 16
  Newcomb's records 70
  on Wrangel Island 221
  photograph 202
Gull, Ross's 9, 67, 76, 157, 164–171, 185, 212–213, 228, 313, 329, 332

Gull, Ross's (*continued*)
  at Barrow 121, 167, 170, 242, 322
  at Churchill 317
  Bessel's record 67
  in Arctic Ocean 76, 329
  in Canadian Arctic 318, 321
  in Chukchi Peninsula 220, 226
  in Franz Josef Land 201
  map 165
  migration 17, 170
  on Bennett Island 79
  on the Yana 195
  photograph 169
  Soviet research 229
Gull, Sabine's 10, 67, 125, 157, 168, 326, 341
  discovery 59
  eggs eaten 41
  Eskimo name 23
  in Canadian Arctic 113, 317, 320
  in Eurasian Arctic 76, 219, 221, 228, 331–332
  in Greenland 68, 355
  photograph 60
Gull, Slaty-backed 123
Gull, Thayer's 123, 316, 318
Gullestad, Nils 280–281
gulls (unspecified) 20–21, 23, 40, 62, 67, 70, 76, 101, 107
  Hearne on 102
  on Bennett Island 79
Gusinaya Zemlya (Goose Land) 74–75
Guyana 156
Gyda Peninsula (Gydanskiy poluostrov) 236, 238
Gyrfalcon 8, 12, 80–97, 127, 145, 208, 220, 304, 332, 335–336, 367
  bones found 40
  clutches taken 115
  Eskimo names 24
  feathers used 27, 342
  in Canadian Arctic 26, 68, 97, 123, 273–275, 320
  in Greenland 81, 83, 87, 93, 249, 268, 271, 273, 285, 312, 314, 316
  in Kola Peninsula 88, 223, 363
  photographs 82
  shot 64

Hacquebord, Louwrens 54, 241
Haftorn, Svein 88, 230
Hague, The, Holland 189, 249

Haig-Thomas, David 45
Hakluyt Island, northwest Greenland 266
Hall, C. F. 66–67
Hall, H. V. 152
Hall Land, northwest Greenland 66–68
Hamburg 54–57, 307
Hansen, Mr 150
Hanson, Harold 179–180
Harding, Warren G. 362
hare, Arctic 8, 64, 309
Harold, king of England 83
Harris, James T. 271–273
Hart, H. C. 67, 68, 161
Harting, James E. 73
Harvie-Brown, John A. 66, 133–140, 143, 157, 164, 208
Haviland, John 153
Haviland, Maud D. 152–156, 157, 213, 332
Hearne, Samuel 100–104, 105, 114, 172–174
heather, Arctic bell 160–161, 253–255, 298
Heck, Dr 178
Hedtoft, Hans 192
Heidelberg University 66
Helgeland, Norway 280
Heligoland 142
Helms, Peder 25
Helsingør, Denmark 138
Hendee, Russell W. 184–186
Henry II, king of England 85
Herald Island (ostrov Geral'd) 70–74, 363
herrings 216
's-Hertogenbosch 89
Hesse, landgrave of 90
Hewitson, William 133
Hill, C. B. 164
Hobson, William 161
Hohn, E. O. 24–25
Holbøll, Carl P. 368
Holland *or* The Netherlands 50, 53, 54, 57, 199, 279, 303, 306
Holloway, C. W. 335
Hood, Robert 104–106
Hooper Bay (Napakiakamut), Alaska 25, 29, 124–125
Hooper Bay Agreement 44
Hooper Bay Eskimos 30, 34, 45
Hooper Bay Expedition 25, 125–126, 360
Hope Colony (Håbets Koloni), west Greenland 39–40

Hopewell, the 51
Horn Sound, Spitsbergen 279
Hornøya, Varanger Peninsula, Norway 266, 304
Horrebov, Niels 87
Horse Head Island (Appalersalik), west Greenland 262
horsetail, variegated 282–283
Hotellneset Airport *near* Longyearbyen, Spitsbergen 308
Houston, C. Stuart 105–106
Howard, Eliot 249
Hoy, Mr 133
Hudson, Henry 49, 51
Hudson Bay 4, 10, 29, 30, 41, 98, 99–104, 109, 128, 173, 178, 187, 239, 273, 302, 316, 339
Hudson Strait 27, 33, 359
Hudson's Bay Company, the 98–104, 108, 113–116, 173, 176, 178–180, 182
Huish, Robert 62
Hull, England 57, 58, 64, 138
Hungary, king of 58
Hunt, Harrison J. 159
Hunton, H. E. 295–296
Hurry Inlet *or* Fjord, east Greenland 354
Husum, Germany 312
Hutchins, Thomas 99–100, 173
Huxley, Julian S. 242–248, 353
Hyland, Thomas 143–144
Hørring, Richard 121–123

Ibis, the 130–131
  papers appearing in 142, 153, 156, 163, 168, 178, 189, 213, 222, 243, 248, 259, 262, 290, 291
Ibis, the schooner 141
Ice Fjord (Isfjorden), Spitsbergen 281, 308, 341, 350
Iceland 2, 4, 8, 57, 127, 144, 286, 296, 302–304, 306, 307
  breeding birds 9, 11, 12, 164, 269
  eiderdown from 27–28, 56
  Gyrfalcon in 81, 83–94
  map 303
Icy Cape, Alaska 36
Iditarod River, Alaska 126
Igarka 141
Igloolik, Northwest Territories, Canada 165
Iita, Avanersuaq 24, 27, 93, 158–159
Ikaluktutiak *see* Cambridge Bay
Ikamiut, Disko Bay, west Greenland 167

Illinois 120
  Natural History Survey 179–180
Ilychev, V. D. 199, 200
India 73, 222
Indians in North America 43–45, 101, 109, 112, 116, 257; *see* Chipewyan, Copper, Cree
Indigirka River 169, 202, 224, 236, 238, 332, 341, 365
Inland Ice, Greenland 284, 292–296, 310, 353–354, 355
International Biological Programme 335, 336
International Ornithological Congress 189, 199
International Polar Year 1882–1883
  American Expedition to Point Barrow 27, 120–121, 166–167
International Polar Year 1932–1933
  Dutch Expedition to East Greenland 249–251
International Union for the Conservation of Nature and Natural Resources (IUCN) 335
Inuhuit Eskimos, Avanersuaq, northwest Greenland 5, 21, 22, 66, 159, 161
  bird names 24
  birdskin coat 29
  food birds 25, 27, 34–35, 41
Inuit *or* Eskimos 20–48, 22 (map), 109, 179; *see* Baffinland Eskimo, Bering Strait Eskimo, Caribou Eskimo, Greenlanders, Hooper Bay Eskimo, Inuhuit, Inupiat, Mackenzie Delta Eskimo, Nunamiut, Takamiut, Yupik
Inupiat Eskimos 21, 22, 29, 185, 186, 322, 324
  bird names 23–24
  bird catching and shooting 31, 33, 34, 42, 44
Inuvialuit 359
Inuvik, Northwest Territories, Canada 3, 12, 316–317
Iowa City 184
  University of Iowa 184, 186
Iqaluit, *formerly* Frobisher Bay 310, 316
Ireland 87, 153, 283, 336
Irkutsk 227
Irving, Laurence 24, 242, 256–257, 287, 326
Isabella, HMS 58–59

Isafjörður, Iceland 94
Isakov, Yu. A. 236–238
Isham, James 99
Islay 94
Israel 90, 297
Italy 198, 306
Ittoqqortoormitt, *formerly* Scoresbysund 275, 310, 354
Ivan III, tsar of Russia 88
Ivan IV the Terrible, tsar of Russia 85
Ivittuut, west Greenland 341

Jackson, F. G. 135–136, 164
Jacobson, Billy 117
James VI, king of Scotland 244
James Bay, Canada 43, 99, 178
James Bay and Northern Quebec Agreement, 1975 45
Jameson, Robert 108
Jameson Land, east Greenland 312, 336, 354
Jan Mayen 290, 292, 306
Japan 228, 306
Jay, Gray 105, 126
Jay, Siberian 137
Jeannette, the 68–71, 166
Jena University 66
Jenness, Diamond 21–22, 27
Johansen, Hans 9, 183
Johansen, Hjalmar 75
John, king of England 85
Jonas im Walfisch, the 54
Jones Sound 64
Jourdain, F. C. R. 243
Juneau, Alaska 44

Kaftanovsky, Yu. M. 217–218
Kalmbach, E. R. 184
Kamchatka Peninsula 2, 81, 86, 182, 223
Kampp, Kaj 312
Kandalaksha Bay (Kandalakshskaya guba) 218–219, 336, 363
Kandalaksha Coast (Kandalakshskiy bereg) 214
Kandalaksha Zapovednik 219, 336, 363
Kane, E. K. 27
Kane Basin 26
Kangeq, west Greenland 368
Kangerlussuaq *see* Søndre Strømfjord
Kanin Peninsula (poluostrov Kanin) 86, 306
Kansas 116
Kara Sea 78, 141, 142, 150, 329

Karlsøy, Norway 88
Kaujuitoq *see* Resolute
Kautokeino, Norway 93
Kay, Joe 257
Kazach'ye 78
Kearton, Cherry *and* Richard 155
Keenan Island 168
Kemi River, Finland 132
Kennedy, J. F. 44
Kennedy International Airport, New York 97
Kennicott, Robert 113–115
Khabarovo 150
Khalerchinskaya tundra 229
Kharlov Island (*one of the* Seven Islands) 81, 83, 216–217
Khatanga 329–331, 365
Khaypudyrskaya Bay 150
Kheysa Island, Franz Josef Land 366
Kholmogory 88
Kiev 219
Killdeer 185, 287
King Charles Land (Kong Karis Land), Spitsbergen Archipelago 353
King Island 21–22
Kingbird, Eastern 287
King's Bay *see* Kongsfjorden
Kirby, William 108
Kirkenes, Norway 304
Kishchinsky, Aleksandr A. 169, 222–226
Kittiwake 10, 61, 70, 74, 81, 290
  as food bird 39, 40, 63
  Gyrfalcon prey 94, 223
  in Barents Sea 215, 216, 217
  in Canadian Arctic 266, 318
  in Eurasian Arctic 76, 150, 201, 363
  in Spitsbergen 54, 56, 353
  on Jan Mayen 292
  photograph 201
Klavdiya Elenskaya, the 329
Klein, D. R. 35–36
Knot 18, 67, 68, 78, 194, 211, 221, 238, 298, 318, 365
  breeding in Iceland 304
  in Greenland 312, 313, 314, 316, 354
  migration 292–296
  nest and eggs 158–163, 192, 208
  photographs 160, 294
Knot, Great 186, 202, 203, 226, 328, 329, 332
knotgrass, alpine 255
Knutsen, Willie 292

Kobuk, Alaska 257
Kobuk Valley National Park, Alaska 360
Koch, Lauge 251, 355
Kok Islands *or* Øerne, west Greenland 368
Kola Peninsula (Kol'skiy poluostrov) 144–145, 197, 207, 208, 214–219, 223, 236, 238, 331, 363
  map 214
Kolguyev Island 45, 142–144, 212, 214
  goose hunt on 33–34
Kolmya River *and* Delta 17, 162, 168–169, 212–213, 227, 236, 238, 332
  field station 226, 229–230
Kolyuchin Bay (Kolyuchinskaya guba) 75, 226, 227–228
Kolyuchin Island 221, 227
Komandorskiye Ostrova *see* Commander Islands
Komsomol (Young Communist League) 229
Kondrat'ev, A. V. 228–229
Kondrat'ev, A. Ya. 207, 226–229
Kongsfjorden, Spitsbergen 51, 52, 309
Kongsøya, Spitsbergen Archipelago 309
Koomyuk, David 43
Koren, J. 152
Korobko, S. I. 221
Korotaikha River 3, 11
Korte, J. de 241–242, 275–278, 307
Koryak Mountains (Koryakskiy khrebet) 202, 223
Kostin, Yu. V. 199
Kostin Shar Strait (proliv Kostin Shar), southern Novaya Zemlya 148
Kotel'nyy Island (ostrov Kotel'nyy), New Siberian Islands 70, 77–78, 162
Koukdjuak, Great Plain of the, Baffin Island 359
Krabbe, T. N. 368
Krasnoyarsk 141, 213
Krasovsky, S. K. 347–349
Krechmar, Arseny V. 226–228, 229, 231
Kronshtadt 138
Kryuchkov, V. U. 199, 365
Kublai Khan 85
Kullorsuaq, west Greenland 310, 355
Kulusuk Island, east Greenland 296, 310
Kumlien, Ludwig 45
Kureyka 140–141
Kuummiut, east Greenland 249–250
Kuvaev, Oleg 169

Labrador 86, 116, 125, 187–189, 203, 316
Labytnangi 365
Lack, David 156
Lake Baikal 2
Lake Hazen, Ellesmere Island 266, 357
Lake Mývatn, Iceland 304
Lake Taimyr (ozero Taymyr) 208, 331, 365
Lamb, Alexander 65
Lancaster Sound 61, 65
Lappland 131–133, 155, 291–292, 363
  map 132
Lapps 131–133
Lapwing 14, 187
Larch 331
Larionov, V. F. 200
Lark, Shore 10, 11, 14, 123, 274, 306, 320
Latham, John 100
Látrabjarg, Iceland 304
Latvia 11
Laura, the 150
Laysan Island, Hawaii 184
Lea, Michael *and* Katherine 312
Leeds, duke of 92
Leiden University 248, 249
Leith 138
Leitzoff, Nathalie 135
lemmings 8, 67, 253, 287, 290, 291
  prey of falcons 81, 123, 223, 232–233, 274
  prey of owls 208, 259, 261
  prey of Ravens 16
  prey of skuas 278
Lena River *and* Delta 4, 70, 86, 139, 162, 166, 169, 205, 236, 332
Leningrad *or* St Petersburg 77, 134, 140, 168, 213, 229
  All-Union Arctic Institute 219, 346–347
  Soviet Academy of Sciences, Ornithological Laboratory of the Zoological Institute 196, 200, 208, 219, 222
  Soviet Academy of Sciences, Zoological Institute 223
Lenivaya River, Taimyr 205
Leonovich, V. V. 235
Leopold, duke of Saxe-Coburg-Saalfeld 265
Lepekhin, I. I. 207
Liefdefjorden, Spitsbergen 309
Light, Alexander 99
Light, William 62
Lilford, Lord, *alias* Thomas Littleton Powys 143, 161–162

Lilford Hall 162
Lincolnshire 85, 140
Linnaean Society of London, the 59, 108
Linnaeus, Carl 57, 58
Linné, Kap, Spitsbergen 353
Lithuania 11
Littleton Island, northwest Greenland 368
Litvin, K. E. 232–233
Lodge, R. B. 155
Lofoten Islands 66
London, England 2, 49, 57, 59, 131, 133, 140, 155, 303, 321
  specimens sent to 99–100, 108, 109
Longspur, Smith's 115, 317
Longyearbyen, Spitsbergen 243, 308, 310
Longyeardalen, Spitsbergen 243
Loo Hawking Club 92
Lorenz, Konrad 248
Louisiana 184
Luckman, Glen 97
Lund University, Department of Ecology 296
Løvenskiold, Herman 56, 184, 193–194, 241

MacDonald, Stuart D. 321
MacFarlane, Roderick 113–117, 164, 173–174, 178
MacGillivray, William 165, 166
Mackenzie Delta 107, 113, 256, 339
  Eskimos 40, 45
Mackenzie River 107, 111, 113, 115
MacMillan, Donald B. 24, 159, 163
Magadan 199, 229, 332
  Soviet Academy of Sciences, Far Eastern Scientific Centre, Institute of Biological Problems of the North 226–231, 332
Magdalena Bay (Magdalenefjorden), Spitsbergen 351–352
Maher, W. J. 266
Maidstone, Kent 97
Mainz 166
Malaya Logata River 331
Mallard 29, 35–36, 81, 94, 148, 295
Malmgren, Anders J. 164
Malygin, the 329
mammoth 221
Manitoba Province, Canada 317
Manniche, A. L. V. 83, 192, 241
Manning, Tom H. 176
Marco Polo 85

Maria Theresa, empress 90
MarkAir 326
Markham, Albert H. 64
Marsden, Mr 164
Marshall, A. J. 290
Marshall, Alaska 126
Martens, Friedrich 54–56
Martin, Humphrey 99
Martin, Sand 62, 107, 113
Maryland 344
Matochkin Shar Strait (proliv Matochkin Shar) 150, 152
Mattamuskeet Lake, North Carolina 344
Mattox, William G. 271–273
Maxim Gorky, the 307
McKay, Captain 65
McPartlin, J. J. 96–97
Meade River, Alaska 326
Meddelelser om Grønland 241, 251, 271
Mediterranean Fleet 108
Melville Bay, Greenland 3, 6, 11, 29, 59
  National Wildlife Reserve 355
Melville Island, Canada 61–62, 64
Melville Peninsula, Canada 123, 165
Merganser, Red-breasted 40, 45, 94, 105, 223, 313
Merlin 83
Messager Ornithologique, the 213
Mesters Vig, east Greenland 292, 310, 313, 314, 354
Meves, W. 135, 164
Mexico 127
Mezen, Russia 136
Michigan 344
  University of, at Ann Arbor 93
Middleburg tryworks, Spitsbergen 53
Middendorf, Aleksandr F. 18, 131, 158, 164, 207–208
Middle East 96–97
Milan, duke of 88
Miller, T. F. 65
Milne, Captain 65
Moffen, Spitsbergen Archipelago 352–353
Monaco, Comité Arctique International 337
Monchegorsk 363
Montana 96–97
Monte Amaro, Italy 145
Montreal 116, 316, 318
  McGill University 239, 257
Montreux, Switzerland 212
Moore, Thomas E. L. 63
Moose Factory, Ontario, Canada 178

Moreyu River 365
Morges, Switzerland 335
Mortensen, Hans Christian 191
Moscow (Moskva) 85, 88, 92, 199, 203, 213, 222, 223, 229, 308, 332
  All-Union Scientific Research Institute for Nature Conservation and Nature Reserves 200, 224, 366
  Municipal Pedagogical Institute 194
  Soviet Academy of Sciences, Institute of Evolutionary Animal Morphology and Ecology 226, 231, 265
Moscow State University 218–219, 224
  Faculty of Biological Sciences 222–223
  Zoological Museum 196, 200, 213, 223, 224, 331
Moscow-Archangel railway 134
mosquitoes 8, 107, 115, 301–302, 318
Moss, E. L. 67
Mount McKinley National Park see Denali
Mount Pelly, Victoria Island, Canada 151, 320
Muonio, Finland 131
Muonio River (Muonio Älv) 131, 133
Muonioniska 133
Muoniovaara 93, 131, 133
Murchison Fjord, Northeast Land, Spitsbergen Archipelago 164
Murdoch, John 120–121, 167, 168, 170
Murie, Olaus J. 125–126
Murman (or Murmansk) Coast (Murmanskiy bereg) 3, 17, 81, 88, 92, 150, 194, 214–218
Murmansk 214, 215, 219, 329, 331
murre see guillemot
muskox 8, 26, 64, 103, 309, 320, 329, 355, 357, 363
Myggbukta, east Greenland 312, 313–314
Mykines or Myggenæs, Faeroes 31, 216, 307
Møhl, Jeppe 39

Nagozruk, Arthur 185
Nanortalik, south Greenland 310, 368
Nansen, Fridtjof 75–76, 168
Napoleonic Wars 92, 104, 108, 200
Naprudny, Moscow 85
Nares Strait 68, 106
Narsarsuaq 310, 313, 314–315
Narvik 308
narwhal 5
National Geographic Society 201

National Wildlife Legislative Fund of America 44
Nature Conservancy (Britain) 248, 257, 284
Nebraska 116
Neckham, Alexander 85
Needham, R. 337
Nelson, Edward W. 21, 29, 71–74, 117–120, 124, 184, 248
Nelson, J. G. 337
Nelson, R. K. 23–24, 30–31, 36
Nenets 341
Nettleship, David N. 266–268, 318
Neue Brehm-Bücherei, Die 195
New Orleans, Louisiana State Museum 184
New Siberia Island (ostrov Novaya Sibir) 78, 163
New Siberian Islands (Novosibirskiye ostrova) 4, 70, 75, 77–78, 139, 158, 162, 163, 211, 224, 236, 290, 365
New York, American Museum of Natural History 158, 159, 184
New York Herald, the 68
Newcomb, Raymond L. 68–69
Newcome, E. C. 88
Newfoundland 12, 338
Newton, Alfred 61, 131–133, 134
  egg collection 161–162, 163–164
  on Greenland birds 67–68
  on Seebohm 142
Newton, Edward 133, 164
Nganasan 22, 36–38
Niaqussat, west Greenland 39
Nicaea 85
Nicholson, E. Max 248
Niedrach, R. J. 186
Nielsen, Bent 192
Nielsen, Hans 17
Nighthawk, Common 287
Nikol'skoye 3, 11
Nipaitsoq, west Greenland 39
Noatak National Preserve, Alaska 360
Nobel Prize 248
Nome, Alaska 21–22, 321, 327–328, 333
Noord-Brabant 89–90, 133
Nordenskiöld, A. E. 73–76, 352–353
Nordenskiöld Coast, Spitsbergen 281–283
Norderhaug, Magnar 261, 281, 335
Noril'sk 332, 341
North American Waterfowl Management Plan 176

North American Wildlife and Natural
    Resources Conference 42
North Atlantic Treaty Organisation
    (NATO) 279, 284
North Cape 55, 150
North Pole 2, 10, 49–50, 66–68, 75, 76,
    304, 329
North Rona 61, 262, 307
North Sea 54
North (*or* Arctic) Slope, Alaska 4, 184,
    185, 256, 286–290, 360–362
    map 361
North Slope Borough, Department of
    Wildlife 47, 324, 326
North Water 6
Northamptonshire 162
Northeast Land (Nordaustlandet),
    Spitsbergen Archipelago 164, 193
Northeast Passage 49–50, 74–76
Northern Sea Route 215
Northwest Passage 49–50, 59, 61–64, 104
Northwest Territories 273, 316
Norton, L. 337
Norton, L. 337
Norton, Moses 103, 114
Norton Sound, Alaska 123
Norway 4, 66, 75, 131–134, 144, 152,
    181, 193, 304–306, 336, 338, 351
    geese in 280–281
    Gyrfalcons in 81, 83, 85, 87–88, 89, 92–93
    king of 85
Norwegian Government, Department of the
    Environment 279
Nova Zembla, the, Dundee whaler 65
Novaya Zemlya 4, 6, 8, 9, 16, 30, 50, 74,
    81, 143, 150, 164, 194, 212
    Brünnich's Guillemots in 17, 26, 345–349
    map 345
    Pearson in 144, 145, 148
    Russian expeditions 215, 219
Novgorod 88
Novosibirsk 229
Novosibirskiye ostrova *see* New Siberian
    Islands
Nunamiut Eskimo 22, 24, 27, 34,
    256–257, 342
Nunavut, Tungavik Federation of 359
Nutauge Lagoon, Chukchi Peninsula 224
Nuuk (*formerly* Godthåb) 38–40, 189, 302,
    310, 313, 354, 367
Nuussuaq, west Greenland 355
Ny Ålesund, Spitsbergen 308

Ob' River 4, 12, 142, 365
Ogilvie, Malcolm A. 280
Okhotsk, Sea of 170, 332
Oklahoma 127
    University of, at Norman 126
Old Crow, Alaska 257
Oldendow, Knud 191, 192, 355, 367–368
Olenek, Upper 236
Omaha, Nebraska 116
Omolon River 226, 227
Ontario Province, Canada 43, 187
Oodaaq 161
Oologist, the 188
Oorschot, J. M. P. van 90–91
Ootheca Wolleyana 131, 133, 163
Operation Falcon 96–97
O'Reilly, Bernard 58
Orlov, Valery 170, 199
Ornis Fennica 202
Ornitologiya 200, 228
Oslo 193, 241, 304, 308
    Norwegian Polar Institute 193, 241,
        261–262, 278, 279, 281
    University 193
Osprey 105–106, 112
Ottawa, Museum of Natural Sciences 321
Ovsyanikov, Nikita 365
Owen, Myrfyn 280–281, 291
Owl, Pygmy 108
Owl, Great Grey 109
Owl, Great Horned Owl 109, 126
Owl, Snowy 8, 9, 11, 18, 43, 125, 183,
    187, 292, 302, 304, 306
    as food bird 27, 64
    Eskimo names 23, 24
    Freuchen on 123–124
    Gyrfalcon prey 81
    in Alaska 186, 290, 324, 326
    in Canadian Arctic 68, 101–102, 109,
        257–261, 318, 320, 321
    in Eurasian Arctic 15–16, 144, 148,
        208, 220, 365
    north of Asia 70, 74
    on Wrangel Island 221, 365
    photographs 258, 260
    skin used by Eskimos 30
Oxford 14, 291
Oxford University 152, 248
    Bureau of Animal Population 243
    Edward Grey Institute of Field
        Ornithology 231, 262, 266
    Greenland Expedition, 1928 248

Jan Mayen Expedition 296
  Museum 62–63
  Press 243
  Spitsbergen Expedition, 1921 243, 244
Oystercatcher 218

Paamiut (*formerly* Frederikshåb) 302
Pallas, Peter Simon 207
Palmén, J. A. 75
Paneak, Simon 24, 43
Pangnirtung Fjord, Baffin Island 187
Pangnirtung Pass, Baffin Island 259, 357
Panora, Iowa 186
Paris 156, 306, 321, 351
  Natural History Museum 73
Parmelee, David F. 43, 127, 318, 321
Parry, William Edward 61, 165–166, 265
Patrikeyev, Trifon 85
Pearson, Charles 144–152
Pearson, Henry J. 30, 144–152, 164, 195
Peary, Robert 159, 161, 184
Peary Land 10, 15, 268, 355
Pechora River 4, 86, 135, 136–139, 164, 363, 365
Pedersen, Alwin 355
Pelican, American White 105, 112–113
Pells, John 92
Pembrokeshire 85
Pennant, Thomas 99, 100, 103, 182
Pennsylvania 127, 188
Penny Ice Cap, Baffin Island 357
Peregrine 10, 14, 79, 89, 144, 220, 235, 332, 334
  eggs taken 123
  in Canadian Arctic 97, 273–275, 320, 321
  in Greenland 11, 268, 271–273, 312, 315
Perry River, Northwest Territories, Canada 179–180, 359
Perry River Post 179
Persia, shah of 91
Person, R. A. 42
pesticides 271–273, 274, 334, 338
Peter the Great, tsar of Russia 200
Peterhead, Scotland 66
Petrel, Storm 61
Pevek 332
Phalarope, Grey 53, 56, 67, 156, 227, 304
  breeding biology 224, 226
  in Alaska 120, 125
  in Canadian Arctic 68, 105, 187, 318, 320

in Soviet Arctic 153, 155, 156, 194, 213, 332, 353
  north of Asia 70, 73, 74
  photographs 72, 225
  seen from the *Fram* 76
Phalarope, Red-necked 141, 213, 310
  clutches taken 115, 153
  in Alaska 120, 326
  in Canadian Arctic 105
  in Greenland 251, 272, 295, 300, 315
  north of Asia 73
  photographs 72, 300
phalaropes (unspecified) 70, 301, 320
Philadelphia University Museum 152
Philanthus 249
Philip *and* Mary, king *and* queen of England 92
Philippines 257
Pigeon, Passenger 116, 187
Pigniq, Point Barrow, Alaska 31, 42, 323
Pike, Arnold 15–16, 165
Pike, Oliver 155
Pintail 23, 35–36, 313
Piottuch, Ignati N. 135, 137, 141
Pipit, Meadow 11, 223
Pipit, Olive-backed 331
Pipit, Pechora 333
Pipit, Red-throated 131, 138
Pipit, Water *or* Buff-breasted 14, 274
pipits 221
Pitelka, Frank Alois 286–290
Pittsburgh, Pennsylvania, Carnegie Museum 126, 188
Pituffik *see* Thule Air Base
Plancius, the 306–307
Pleske, Fedor *or* Theodore 78, 163, 208–212
Pleyce, John 51
Plover, HMS 31, 63
Plover, American Golden 12, 105, 317, 326, 328
  clutches taken 41, 115
  Eskimo names 23, 24
  explorers' records 62
  in Greenland 67
  photographs 319
Plover, Golden 18, 213, 223, 296, 331
Plover, Grey 10, 18, 67, 109, 213, 320, 321, 326, 332, 365
  and Eskimos 21, 23, 41
  birds and eggs collected 125, 135, 137, 144, 145, 146–147, 153, 154–155, 158

Plover, Grey (*continued*)
  nest and eggs discovered 100, 208
  photographs 146, 147, 153, 293
Plover, Lesser Golden 12, 200, 328
Plover, Pacific Golden 12, 74, 141, 153, 155, 292, 328, 331, 332
Plover, Ringed 9, 12–13, 14, 194, 218, 296
  breeding 15
  in Canadian Arctic 61–62, 66–68
  in Eskimo legend 21
  photograph 13
  seen from the *Fram* 76
Plover, Semipalmated 12–13, 100, 109, 115, 123, 317, 320, 326
  photograph 13
plovers 62, 64
Point Lake, Northwest Territories, Canada 111
Pokhodsk 3, 17, 168
Pokhodskaya Edoma 82
Poland, king of 91
Polar Bear Pass, Bathurst Island 321
  National Wildlife Area 336, 357
Polar Gas Project, Toronto 41
Polaris, the USS 66
Polyarnyy, Murman Biological Station 215
Polynia, the, Dundee whaler 64
Ponsonby, Mrs 143
Poole, Jonas 52–53
Poole, K. G. 273
Popham, Hugh Leyborne 153
Popham, Mrs Leyborne 143
Popov, A. A. 36–38
Portenko, Leonid A. 29, 31, 75, 219–222, 223, 224, 227, 231, 346, 349
Portugal, king of 90
Potapov, Eugene R. 231
Povungnituk, Quebec, Canada 22, 33
Powys, Mervyn 143, 144
Prince Charles Foreland (Prins Karls Forland), Spitsbergen 244
  National Park (Forlandet Nasjonalpark) 353
Prince Leopold Island, Canada 265, 318
Princeton University Press 199
Problemy Severa (Problems of the North) 235
Prokosch, Peter 312
Prop, J. 281–283
Providence Bay (bukhta Provideniya) 63, 71–73
Provideniya 71, 333

Prudhoe Bay, Alaska 326, 328, 339, 360, 362
  map 340
Ptarmigan 10, 18, 37, 89, 102, 183, 187, 259, 292, 300, 301, 304, 350
  and native peoples 21, 22–23, 24–25, 26, 27, 34, 38
  as food bird 35, 39, 39–40, 41, 63, 68, 74
  falcon prey 81, 94, 223, 272
  food 251–256
  in Canadian Arctic 127, 318, 321
  in Chukchi Peninsula 221
  in Spitsbergen 52, 53, 56, 309
  in winter 15–16, 249
  nest 57
  on Commander Islands 229
  plumage 189–191, 208
  photographs 37, 190, 252
Puffin 31, 43, 51, 53, 56, 81, 304, 353, 363
  Gyrfalcon prey 94
  seen from the *Fram* 76
Puffin, Horned 74
Pyasina River, West Taimyr 235

Qaanaaq, Avanersuaq, northwest Greenland 247, 310, 313, 316, 355
Qasigiannguit (*formerly* Christianshåb) 167–168
Qeqertarsuaq (*formerly* Godhavn) 162, 165, 189, 302
Qeqertarsuatsiaat (*formerly* Fiskenæsset), west Greenland 368
Quebec Province, Canada 25, 29, 30, 33, 316
Queen Elizabeth Islands 357
Queen Maud Gulf 180, 273
  Migratory Bird Sanctuary 336, 359
Queneau, Paul 179–180
Qilakitsoq, west Greenland 29

Rae, John 111–113, 176–178
Ramsar Convention 336
  Sites 336, 355, 363
Rankin Inlet, Hudson Bay 273–275
Rasmussen, Knud 121–124, 369
Raven 11, 18, 285, 326
  and Eskimos 20–21, 24, 27, 35, 39
  bones found 40
  in Canadian Arctic 61–62, 102, 123–124, 274

in winter 16, 53, 249
nests used by Gyrfalcons 83, 274
Ray, P. Henry 120
Razorbill 39, 81, 148, 216, 218, 264, 304
Redpoll 12, 123, 137, 144, 192, 272, 317, 327
Redpoll, Arctic 11, 12, 317, 326, 327
  in winter 16, 106
  nest sites 46, 48
  photographs 46
redpolls (unspecified) 24, 141, 186, 223, 315
Redshank 145
Redshank, Spotted 131, 133, 152, 331, 332
Redwing 137, 223, 304, 315
Reindalen, Spitsbergen 279
Resolute (Kaujuitoq), Cornwallis Island, Northwest Territories, Canada 316, 318
Resolute, HMS 64
Rey, Louis 337
Reykjavik 90, 94, 303, 310
  Natural History Museum 303
Richards, Eva 23
Richardson, John 61, 100, 104–113, 166, 173
ringing 41–42, 43, 215, 274–275, 279–280, 282, 285–286, 288, 292–294, 303, 313
Rink, Henrik 368
Riyadh 97
Robin, American 105, 123
Rocky Mountains, the 178, 257, 357
Rodgers Bay (bukhta Rodzhersa), Wrangel Island 221–222, 232
Romania 156
Ross, Bernhard R. 113–114, 178
Ross, James Clark 61, 165, 166
Ross, John 58–61, 62, 63, 187
Rossia, the 329
Rovaniemi 132
Royal Air Force 284
Royal Artillery 59
Royal Geographical Society 62, 64, 143
Royal Greenland Trading Company 27, 29
Royal Navy 58–64, 67–68, 100, 104
Royal Society 59, 99, 279
Rubythroat, Siberian 333
Rudolf (*or* Prince Rudolf) Island (ostrov Rudol'fa), Franz Josef Land 76, 366
Ruff 141, 213

Rupert River, Quebec, Canada 178
Russia 91–92, 134–140
  All-Russian Society for the Protection of Nature 222
  October Revolution, 1917 213
  Russian Polar Expedition, 1900–1903 76–79, 211–212
  Russian Soviet Federated Socialist Republic 4, 197
Rutten, Elizabeth 249
Ryabitsev, V. K. 199, 236

Sabine, Edward 59, 61, 182, 355
Sabine, Joseph 59, 105, 106, 108
Sabine Islands (*or* Øer) 59, 355
Sachs Harbour, Banks Island 321
Sage, Bryan 9, 11, 183, 339
St Agnes, Scilly Isles 292–294
St Kilda 307
St Lawrence Island 71, 328
St Lawrence River 176
St Michael, Alaska 71, 117–120, 121
St Petersburg 138, 140, 207
  Natural History Society 208
St Petersburg, Imperial Academy of Sciences 77, 121, 158, 207
  Zoological Museum 77, 78, 162, 163, 200, 208
Salix see willow, Arctic
Salomonsen, Finn 5, 9, 10, 184, 189–193, 270
  and ringing in Greenland 42, 43
  Birds of Greenland cited 93, 162
  on Brünnich's Guillemot 41, 42, 344–345
  on conservation 335
  on eiderdown 27
  on Gyrfalcon nests 83
  on migration in Greenland 292–296
Salvadori, Tommaso 76
Samoyeds 22, 27, 30, 143–144, 146
  goose hunt 33–34
San Diego Society of Natural History 170
San Francisco 68, 117
San Joaquin Valley 180
sand-eels 216
Sanderling 15, 18–19, 67, 238, 295, 296, 312
  in Canadian Arctic 62, 68, 109, 187, 318, 321
  in Franz Josef Land 201
  in Greenland 313, 314, 354

Sanderling (*continued*)
  in West Taimyr 163
  nest discovered 116, 164
  north of Asia 70
Sanders, Armytage 155
Sandpiper, Baird's 10, 187, 202, 318, 320, 321, 326
  clutches taken 115
  in Chukchi Peninsula 219, 333
  in Greenland 15, 316
  named 114
  photographs 114, 294, 314–315
Sandpiper, Broad-billed 332
Sandpiper, Buff-breasted 115, 120–121, 326, 341
Sandpiper, Curlew 10, 19, 78, 153, 156, 163, 212, 213, 238, 329, 332
  at Barrow 121, 326
  photograph 153
Sandpiper, Grey-rumped 333
Sandpiper, Least 115
Sandpiper, Pectoral 118–120, 123, 287, 313, 317, 320, 332
  in Soviet Arctic 331
  photographs 119
Sandpiper, Purple 11, 14, 42, 78, 296, 304
  in Canadian Arctic 127, 187, 318
  in Franz Josef Land 200, 202, 203–205
  in Greenland 68, 249, 295, 315
  in Spitsbergen 53, 56, 194, 310, 353
  photograph 204
Sandpiper, Semipalmated 100, 109, 203, 317, 320, 327
  photographs 206
Sandpiper, Sharp-tailed 202, 203, 329, 332
Sandpiper, Spoon-billed 63, 71–73, 202, 219, 226, 227, 238, 329, 333
  as food bird 75
  photograph 71
Sandpiper, Stilt 10, 111, 123, 298, 320
  photographs 110
Sandpiper, Terek 135, 328
Sandpiper, Western 120, 125, 202, 203, 227, 326
Sandpiper, White-rumped 318, 321, 326
Sandpiper, Wood 152, 313
sandpipers (unspecified) 70, 183, 195, 365
Sannikov, Yakov 76–77
Sannikov Land 77, 78
Saskatchewan River 106, 109, 111
Saskatoon, University of Saskatchewan 105
Saudi Arabia 97

Saunders, Howard 133, 143, 166, 168
Saunders Island, Avanersuaq, northwest Greenland 346
Savissivik, Avanersuaq 355
Savoy, Prince Luigi Amedeo of, duke of the Abruzzi 76
saxifrage, purple 253, 255, 282–283
Saxon, the yacht 143, 144, 145, 148
Scandinavia 86, 131–134, 296
Scandinavian Airlines System (SAS) 308
Scaup 144
scaups 29, 317, 339
Schaanning, H. T. L. 152
Schalow, Herman 2, 182–183
Science in the U.S.S.R. 230–231
Schiermonnikoog 51
Schiøler, E. Lehn 189
Schreiber, Carl 165
Scilly Isles 292–294
Sclater, P. L. 67
Scoresby Sound 14, 87, 275–278, 304, 354
Scoresbysund *see* Ittoqqortoormiit
Scoter, Surf 328
scoters 43, 317, 339
Scotland 11, 12, 51, 94, 140, 243, 283, 286
  photograph taken in 55
Scott, Mr 161
Scott, Peter 179–180, 200, 265, 283–284, 303, 329, 359
Scottish Office 286
seal, ringed 355
sealskin 21
Seebohm, Henry 135–142, 143, 153, 155, 157, 158, 208
  his Knot's egg 162, 167
Segnit, R. W. 244
Seven Islands (Sem' ostrovov), Murman Coast 81, 88
  Zapovednik 194, 215–217, 363
Severnaya Zemlya 4, 158, 306, 309
Sevmorzverprom (Northern Marine Animal Exploitation) 346–348
Seward Peninsula, Alaska 321, 327–328, 360
Sharpe, R, Bowdler 143
Shearwater, Great 39–40, 58, 62
Shearwater, Manx 182
shearwaters 61
Sheffield 138, 142
Shelduck 54
Sherborne School, Dorset 279

Shergold, Mr 135
Shortt, Terence M. 187
shrews 223
Shrike, Brown 73
Shrike, Great Grey 89
Sibiryakov, the 215
Simpson, John 31
Sirius sledge patrol 354
Siskin, Pine 326
Sivokak *see* Gambell
Skeiðarársandur, Iceland 304
Skua, Arctic 10, 58, 200
  at sea 61
  in Canadian Arctic 62, 105, 111, 187, 318
  in European Arctic 53, 56, 76, 145, 310
  in Greenland 68, 313
  in Spitsbergen 351–352
  photograph 112
Skua, Great 304
Skua, Long-tailed 111, 115, 138, 145, 187, 306, 313, 318, 320
  breeding 242, 275–278
  photographs 276, 277
Skua, Pomarine 10, 74, 111, 123, 125, 187, 332
  at Barrow 288, 290, 326
  in European Arctic 76, 145
  photographs 324, 325
skuas (unspecified) 9, 12, 20–21, 76, 111, 123, 223, 290, 320, 321
  Eskimo names 23–24
Skylark 83
Sladen, W. J. L. 226
Sledge Island, Alaska 123
Slimbridge, Gloucestershire 180
Smeerenburg, Spitsbergen 53–54
Smew 138
Smith Sound 6, 66–68, 368
Snipe, Great 137, 145
Snipe, Jack 133, 182
Snipe, Pintail 331, 332
Snyder, Lester L. 9, 10, 11, 184, 186–187
Societas Artica Scandinavica 183
Soissons 156
Solovetskiye ostrova *or* Islands 207
Solway Firth 279, 281
Somerset Island, Canada 318
Søndre Strømfjord (*the fjord*) 271–273
Søndre Strømfjord Air Base (*also called* Kangerlussuaq) 284, 286, 295, 310
Soper, J. Dewey 175–176, 359

Southampton Island, Canada 126, 127–129, 175–176
Soviet Academy of Sciences 196, 199, 221
  and see Leningrad, Magadan, Moscow
  Kamchatka Expedition 223
Soviet Arctic 168–169, 194–196, 197, 200–238, 328–333
  conservation 335, 336, 362–366
  Gyrfalcons in 81, 87, 88–89, 92
  maps 136, 165, 198, 214, 220, 237, 330, 364
  oil exploration 341
Soviet Union 2–3, 8, 31, 306
  ornithology 197–200
Soviet Weekly, the 329
Sparrow, American Tree 115, 328
Sparrow, Chipping 185
Sparrow, Harris's 127, 317
Sparrow, House 11, 137, 188
Sparrow, Savannah 326, 328
Sparrow, White-crowned 111–112, 318, 328
Sperr, Rudolf 97
Spitsbergen Archipelago (Svalbard) 4, 6, 66, 68, 74, 170, 171, 211, 241, 275, 302, 304, 306–310, 329
  albatross off 66
  avifauna 193–194, 241
  breeding birds 158, 170, 224
  Brünnich's Guillemot in 17
  conservation 335, 349–353
  explorers and whalers in 49–57, 144
  Fram near 75, 76
  goose research 242–248, 275–283
  maps 52, 245, 307, 308, 309, 350
  mining and oil drilling 341, 351
  tourism 342
  wintering birds 15–16
squirrels, ground 108
Stanley, Henry M. 68
Station Centrale 292
Statoil 341
Stella Polare, the 76
Stevens, J. C. 131, 133
Stint, Little 14, 135, 141, 155, 213, 238, 306, 332
  birds and eggs collected 144, 145, 146, 150, 152, 153, 158, 164
  breeding 138–139, 156, 208
  on migration 203
  photographs 138, 139

Stint, Red-necked 202, 203, 326, 328, 332
Stint, Temminck's 18–19, 133, 137, 145, 153, 155, 205–207, 213, 304
  photograph 206
Stirlingshire 138
Stockholm Natural History Museum 164
Stonehouse, Bernard 337
Store Ekkerøya, Varanger Peninsula, Norway 217
Strizhev, Petr 163
Stroud, David A. 284–286
Stuttgart University 66
Sula Sgeir 262, 307
Summerhayes, V. S. 244
Summers, R. W. 291
Sundevall, C. J. 164
Sutton, George Miksch 98, 126–129, 175–176, 186, 303, 321
Sværholtklubben 150
Sverdlovsk 199, 236
Svyatoy Nos, mys (*or* Holy Cape), Murman Coast 144–145
Swainson, William 100, 108–111, 173
Swallow 200
Swallow, Northern Rough-winged 113
Swallow, Tree 328
Swan, Bewick's 10, 14, 144, 228, 332, 367
  birds and eggs collected 138, 146, 158
  nests in Novaya Zemlya 74–75
Swan, Mute 10
Swan, Trumpeter 10, 12, 102
Swan, Tundra 10, 14, 102, 105, 187, 228, 317, 320, 326
  and Eskimos 23, 35, 41
  eggs collected 115
  photographs 343
  shooting 344
Swan, Whooper 10, 39, 304
swans 20, 23, 29, 81, 101, 107, 290
  hunted by Eskimos 35
Swarthmore College, Swarthmore, Pennsylvania, Edward Martin Biological Laboratory 242
Sweden 85, 131, 133, 199, 306, 336
Swedish Northeast Greenland Expedition, 1979 313
Swedish Spitsbergen Expeditions
  1861 164
  1872–1873 74
Swift 200
Switzerland 57
Syroechkovsky, Eugene E. 238

Syroechkovsky, Eugene V. 231–238
Syltefjordstauran, Varanger Peninsula 304
Sytygan-Tala Bay (bukhta Sytygan-Tala) 205, 207

Taimyr Peninsula (poluostrov Taymyr) 4, 36, 199, 236, 238, 309, 336, 341, 367
  breeding birds 10, 121, 145, 153, 158, 169, 290, 306, 329, 331, 332
  coast 78, 205
  Krechmar in 227, 235
  map 364
  Middendorf in 18, 158, 164, 208
  Toll in 162–163, 211–212
Taimyr River, Lower (Nizhnyaya Taymyra) *and* Upper (Verkhnaya Taymyra) 18, 163, 208, 211, 238, 365
Taimyr Zapovednik 331, 365
Takamiut Eskimos 22, 30
Tallinn 77
Tana River 131
Tanager, Scarlet 287
Tanager, Western 287
Tartu (*formerly* Dorpat), Estonia, University of 207–208
Tattler, Wandering 327, 328
Taylor, Alaska 322
teal (unspecified) 105
Teal 208, 223
Teal, Baikal 185
Teller, Alaska 322
Temple, S. A. 16
Teriberka River, Kola Peninsula 223
Tern, Aleutian 328
Tern, Arctic 10, 26, 168, 201
  as food bird 63–64, 251
  eggs collected 123
  Eskimo name 23
  in Canadian Arctic 68, 102, 105, 318
  in Greenland 59, 167, 367
  in Spitsbergen 54, 56, 310, 353
  in White Sea 218
  migration 17
  photograph 103
  prey of falcon 223
  seen from the *Fram* 76
Tevuk, Dwight 185
Texas 12
Thames, the 140–141
Thayer, John E. 159
Thienemann, F. A. L. 131, 208

Thing, Henning 302
Thomas, the, Hull whaler 58
Thompson, D. Q. 42
Thousand Islands (Tusenøyane), Spitsbergen Archipelago 353
Thrush, Dusky 141, 328, 331, 332
thrushes 221
Thule Air Base (*also called* Pituffik), northwest Greenland 15, 234, 262, 283, 310, 313, 315, 320, 346, 354
　North Mountain 159, 161, 301
Thule district 5; *see* Avanersuaq
Thule Trading Station, Avanersuaq, northwest Greenland 17, 121
Tiksi 332
Timanskaya tundra 207
Times, the 306
Tinbergen, Jan 248
Tinbergen, Nikolaas 248–251
Tit, Siberian 230, 304
Tit, Willow 257, 292
Tkachev, A. V. 233
Tobol'sk 88, 89
Todd, W. E. Clyde 126, 184, 187–189
Toll, Eduard Vasil'evich *alias* Eduard von Toll 76–79, 162–163, 211
Toll, Emmy von 163
Tolmachev, A. I. 144, 214
Tolstov, Sergei 163
Tomkovich, Pavel 200–207, 331–333
　photographs by 71, 104, 153, 169, 202, 227, 238
Tongass National Forest, Alaska 360
Tooruk 23
Toronto 41
　Royal Ontario Museum 186–187
Toronto Daily Star, the 187
Transactions of the Linnaean Society of New York, the 249
Trevor-Battye, A. 33–34, 45, 142–144, 145, 146, 150
Triad, the 138
Trifon, St 85
Tristan da Cunha 39, 58
Troms County, Norway 87
Tromsø 74, 144, 150, 306, 308
　Museum 164
Truelove Lowland, Devon Island 240–241, 321, 336
Truelove, the, Hull whaler 64
Tuktoyaktuk, Northwest Territories, Canada 316, 317

Tundra River (reka Tundrovaya), Wrangel Island 232
Turner, Lucien McShan 117
Turnstone 9, 10, 15, 24, 74, 162, 301, 318, 365
　at Lake Hazen 266
　in Alaska 326
　in Greenland 354
　in Taimyr 212
　in Varanger Peninsula 304
　in White Sea 218
　migration 292–296
　nest and eggs 100, 144
　photograph 299
Turnstone, Black 125
Turukhansk 3, 11
Tyumen' 88, 89

Uboynaya *and* Uboynaya River, Taimyr 153, 205
Uelen 31, 202, 291–221
Ukpeagvik *see* Barrow
Underhill, L. G. 291
Ungava Peninsula *see* Labrador
Union of Soviet (*or Sovereign*) Socialist Republics (*USSR*) *see* Soviet Union
United Nations Educational Scientific and Cultural Organisation (UNESCO) 243
United States of America (USA) 12, 67–68, 71, 90, 166, 188, 243, 272, 306, 321, 336, 354
　Air Force 96, 256, 295
　Army Signal Service 71, 117, 120, 161
　Fish and Wildlife Service 36, 42, 44, 96, 120, 124–126, 179, 184, 188, 274, 288, 344, 360, 362
　Forest Service 360
　Naval (*or* National) Arctic Research Laboratory (NARL), Point Barrow, Alaska 46–48, 242, 256, 286–288, 324, 326
　Naval Petroleum Reserve No. 4 (*renamed the* National Petroleum Reserve in Alaska) 242, 362
　Office of Naval Research 242
Upernavik, west Greenland 64, 262, 268, 344, 355, 367
Ural Mountains (Ural'skiy khrebet) 4, 89, 92, 135, 195, 197, 236, 238
Ushakovskiy, Wrangel Island 221, 232
Uspensky, Savva M. 9, 40, 169, 184, 200, 207, 215, 218, 224

Uspensky, Savva M. (*continued*)
  Birds of the Soviet Arctic 194–196
  in Novaya Zemlya 218
  on conservation 335, 349, 366–367
Ust' Chaun 332
Ust' Tsil'ma 136–137
Utte-Veem River, Chukchi Peninsula 221
Uummannaq, Avanersuaq 124, 159
Uummannaq, west Greenland 29, 83, 191

Vadsø, Finnmark, Norway 131, 304, 306
Valdez, Alaska 339, 360
Valkenswaard, Holland 89–90, 133
Vankarem, Chukchi Peninsula 221, 224
Van Mijenfjorden, Spitsbergen 341
Vanna, Norway 88
Vansittart Island, Canada 123–124
Varanger Peninsula, Finnmark, Norway 88, 131, 135, 144, 302, 304–306, 331–332
  map 305
  photographs taken in 28, 37, 112, 138, 139, 145, 206, 209, 267, 299, 327
Vardø, Finnmark, Norway 143, 144–145, 304
Vaygach Island (ostrov Vaygach) 3, 11, 14, 150, 164, 195, 366
  zakaznik 363
Veer, Gerrit de 50–51
Vega, the 73, 74–76
Venus, transit 99
Verslev, Mr 162
Vesey *alias* Lewis, Ernest 94
Viborg, Denmark 191
Victoria Island (ostrov Viktoriya) 170
Victoria Island, Northwest Territories, Canada 316, 318, 320, 357
Victory, the 187
Vienna (Wien) 90, 135
  Hofmuseum 165
Vigur, Iceland 28
Vikings 22, 28, 38–39, 310, 313
Vladivostok 168
voles 16, 223
Volosovich, Constantin 77
Vorkuta 341
Voron'ya River, Kola Peninsula 88, 223
Vosnesensky, I. G. 121

Wadden Sea 291
Wader Study Groups
  England 292

Soviet Union 203
Wager Bay, Canada 123
Wagtail, Citrine 137, 138
Wagtail, White 9, 10
Wagtail, Yellow 11, 70, 257
wagtails 221
Wainwright, Alaska 21, 23, 30–31, 36, 185–186, 326
Wales 283
Wales, Alaska 185–186
Wales, William 99
walrus 5, 31, 220, 352–353, 355, 366, 369
Walter, Hermann 77–78, 162–163
Warbler, Arctic 75, 257, 331
Warbler, Sedge 9
Warbler, Willow 9, 292
warblers 221, 317
Warkworth, the 142
Waterfowl Habitat Owners' Alliance 44
Washington, D.C. 97, 125
  Smithsonian Institution 113–116, 117–118, 164, 166, 173, 178, 257
  United States National Museum 257
Watson, Adam 14, 257–261, 304
Watson, George E. 222
Wattel, Jan 275
Waxwing 132–133
weasel 290
Webster, H. 94–96
Wellman, Walter 201
Welty, J. C. 17
Wernerian Society 166
Wheatear 10, 11, 123, 144, 192, 249, 257, 272, 292, 295, 296
Whimbrel 12, 94, 317, 326
White, Gilbert 182
White Sea (Beloye more) 92, 134, 143, 214, 218–219, 336
Wieringen, Holland 50
Wiesbaden 199
Wiggins, Joseph 140–142
Wigeon 137
Wildfowl 224
Wildfowl and Wetlands Trust (*formerly the* Severn Wildfowl Trust) 180, 265, 279–281, 284, 291, 303, 334
Wildfowlers' Association of Great Britain and Ireland 284
Wilkes, A. H. P. 243
willow, Arctic (*Salix* sp.) 253, 255, 282–283, 298, 352
  photograph 283

Wilson, Alexander 100
Wilson Bulletin, the 158
Windward, the 75
Winge, Herluf 192
Winnipeg 316
Witherby, H. F. and G. 155
Wittenberg 195
Wolley, John 81, 92–93, 131–134, 155, 163, 164
Wolstenholme Fjord 7
Wolstenholme Island, northwest Greenland 64
wolves 8, 83
Woodpecker, Black-backed 106
woodpeckers 105
World Wildlife Fund 270
Wrangel Island (ostrov Vrangelya) 4, 70–74, 114, 158, 166, 167, 176, 195, 236, 238
  field station 226
  Kishchinsky on 224
  Portenko on 221–222
  Snow Geese 229, 231–235, 366–367
  Snowy Owls 232–235
  Zapovednik 231, 263–265
Wren 287
Wryneck 185

Yakutia (*or* the Yakutskaya Autonomous Republic) 12, 197
Yakuts 77
Yakutsk 77, 168, 199, 332
Yamal Peninsula (poluostrov Yamal) 226, 236, 341, 367
Yana River 78, 169, 224, 236, 238, 341, 365

Yarrell, William 133
Yeates, G. K. 303
Yellowknife 316, 318
Yellowlegs, Lesser 327, 328
Yenisey River 4, 11, 86, 140–142, 152–153, 164, 213, 332
Yeniseysk 141
Yevdokimov, Nesterka 88
Ymer, the 171
Yokohama 74
York Factory, Manitoba, Canada 99, 104, 109, 111
Yorkshire 138, 148
Yugorskiy Peninsula *or* poluostrov 11, 195
Yukon Delta National Wildlife Refuge, Alaska 360
Yukon River 125, 126
Yukon Territory, Canada 316
  Northern Yukon National Park 357
Yukon-Kuskokwim Delta 33, 35–36, 41, 117–120, 124–126, 321
  Emperor Goose in 228
  Goose Management Plan 44, 360
Yupik Eskimo 22, 35–36, 41, 44, 328, 360

Zaliv Kresta *see* Cross Bay
Zarya, the 76–79, 162–163
Zemlya Frantsa Iosifa *see* Franz Josef Land
Ziemsen, A. 195, 199
Zoological Society of London 99, 113, 132, 243
Zoologicheskiy Zhurnal 229
Zoologist, the 156